CALCULATING BRILLIANCE

CALCULATING BRILLIANCE

An Intellectual History of Mayan Astronomy at Chich'en Itza

GERARDO ALDANA Y VILLALOBOS

THE UNIVERSITY OF
ARIZONA PRESS

TUCSON

The University of Arizona Press
www.uapress.arizona.edu

We respectfully acknowledge the University of Arizona is on the land and territories of Indigenous peoples. Today, Arizona is home to twenty-two federally recognized tribes, with Tucson being home to the O'odham and the Yaqui. The University strives to build sustainable relationships with sovereign Native Nations and Indigenous communities through education offerings, partnerships, and community service.

ISBN-13: 978-0-8165-4220-8 (hardcover)
ISBN-13: 978-0-8165-5703-5 (paperback)
ISBN-13: 978-0-8165-4421-9 (ebook)

Cover design by Leigh McDonald
Cover illustration by Gerardo Aldana
Typeset and designed by Sara Thaxton in 10/14 Warnock Pro with Landa and Blakely

Unless otherwise noted, all maps, tables, diagrams, photographs, and drawings are by the author.

Publication of this book is made possible in part by the proceeds of a permanent endowment created with the assistance of a Challenge Grant from the National Endowment for the Humanities, a federal agency.

Library of Congress Cataloging-in-Publication Data
Names: Aldana y Villalobos, Gerardo, author.
Title: Calculating brilliance : an intellectual history of Mayan astronomy at Chich'en Itza / Gerardo Aldana y
    Villalobos.
Description: Tucson : University of Arizona Press, 2021. | Includes bibliographical references and index.
Identifiers: LCCN 2021021352 | ISBN 9780816542208 (hardcover)
Subjects: LCSH: Codex Dresdensis Maya. | Maya astronomy—Mexico—Chichén Itzá Site. | Venus
    (Planet)—Observations
Classification: LCC F1435.3.C14 A4285 2021 | DDC 972.81/016—dc23
LC record available at https://lccn.loc.gov/2021021352

Printed in the United States of America
♾ This paper meets the requirements of ANSI/NISO Z39.48-1992 (Permanence of Paper).

CONTENTS

PREFACE

Tobacco filled the cab, enveloping us as we chain-smoked on our trip across town. The chassis of the old Toyota pickup danced as we bumped along the cobblestone streets to a small, relatively new-looking building in the western part of the village. In 2004, I was in the Mazatec highlands of Oaxaca working on a small reforestation project led by a local indigenous herbalist, Adriana and her Catholic priest collaborator. In her own work, Adriana was exploring the "backyard" cultivation of medicinal plants that traditionally were harvested from the forest. The priest was himself of an indigenous background, and I learned in a later discussion that he maintained inclinations to indigenous ritual ideologies alongside his Catholicism. Our countless conversations were rich, broad in scope, and lively, generally occurring at breakfast in the church rectory or in the evening over dinners of tacos in the street. After one long day of constructing seedbed planters for an elementary school, the priest invited me to accompany him to a mass he was presiding over at a local church. That invitation put us in his pickup for the short but corporeally demanding trip.

A local artist had been commissioned to paint the new-looking church in bright colors, with a mural depicting community life and an idealized reliance on maize agriculture. I smiled at the imagery as we entered and walked up the main aisle. The priest directed me to the pews with a toss of his chin, and he went around back to get dressed and prepare for the ceremony. Sitting in a pew, I was comfortable, having been raised by strict Catholic parents who brought their Mexican heritage to engage the majority Euroamerican demographic of the Southern California community we lived in. I felt a strong familiarity with the church and the practices it housed, but there was also a visceral difference. In this highland indigenous community, Catholicism had been appropriated into local lifeways over centuries to take on distinctly Mesoamerican forms (cf. Carlsen and Prechtel 1991; Clendinnen 1989; Farris 1984). The architecture was somewhat modern, but the celebration was very traditional, ossified against the revisions of the Second Vatican Council of the 1960s and newer Mormon and Evangelical Christian pressures. Most notably local to the Mesoamerican highlands was the heavy reliance on copal—an incense derived from a tree (*Protium copal*) found throughout the Mayan region. This incense had been used ritually long before contact with the Catholic Church throughout Mesoamerica, and it is recorded in Mayan hieroglyphic inscriptions, as we will see in later chapters of this book.

The priest used a censer swinging from a long chain to fill the center aisle of the church with a cloud of fine white smoke, which dissipated into the pews around me and the other parishioners. The scent of copal was thick in the air as the ceremony in this church ended, at which point several elders moved in concert to the altar and removed a statue from its resting place. Dressed with fresh flowers, the statue's throne was lifted up and carried down the main aisle and out the front door with great reverence. Parishioners filed out of the pews to initiate a procession, meeting members of another community in the street in front of the church. Combining to form one large crowd, community members accompanied the statue as it was transported through the streets on its way to a much larger and older church—this one more traditionally colonial in architecture.

As I walked along near the tail end of the procession, taking care not to trip on the cobblestones, I was reminded vividly of a documentary that I have used repeatedly in my classes: *The Tree of Life*, directed by Pacho Lane. This 1976 film records the felling of a ten-meter tree by a rural community in Puebla for use as the main post in a ritual, which itself appears to have gone back to Precontact times. Community members raised this post in front of their highland Catholic Church and watched as four voladores, hanging from their feet, spiraled from its top to the ground. Lane's film does a provocative job of visually tracking the means by which ritual was woven into daily life in that community. He allows the viewer to follow the construction of elaborate candelabras, which relies on both the expertise of an individual and the helping hands of apprentices. Children play with their parents to imitate the roles they would take on as voladores themselves when reaching the appropriate age, laughing as they learn. I recalled that the voladores ceremony, too, was accompanied by ritual processions, and as documented in Lane's film, these processions served their town's patron saint. In both cases, in Puebla and in the procession I was a part of in the Mazatec highlands—and as is familiar to tourists and anthropologists alike who travel through Mexico and Central America—the movement of a sacred statue from one location to another symbolized far more than mere transport.

These processions also bring to mind the records left by colonial Europeans, by sixteenth-century antecedents of my Mazatec highland Catholic priest host, capturing native religiosity in written words—albeit with a clear intent to subvert them and convert their practitioners to Christianity. Such documents describe the Precontact ritual processions of ceramic deity effigies—religious statues—around circuits through cities to and from symbolic structures. These records take me to the scenes painted on ceramic chocolate-drinking cups excavated from Classic Mayan royal burials. Polychrome pottery pieces painted by expert scribes depict processions to and from sacred bundles, which wear masks and sit on altars. As I walked along the cobblestone streets in the Mazatec procession, I suspected some underlying continuity across these ritual expressions, although I could not readily reduce that continuity to a formula or anthropological conceptual tool.

What did snag my heavily biased attention was that these colonial and modern rituals were all timed according to the Gregorian calendar. They represented ritual weavings of space and time, but during an era in which time itself was prescribed by a calendar imposed on indigenous communities by the Spanish priests of the sixteenth century. Ancient Mesoamericans, of course, had different ways of accounting for time before contact with Europe. As we will review in more detail in the chapters of this book, Mayan communities across regions and eras used a ritual calendar of 260 days in conjunction with other cycles, which meant that not all ceremonies had to be tied to the cycle of the solar year. In fact it is this responsibility of constructing ceremonial time, combined with the people charged with that responsibility and the processions that facilitated ritual events, that serves as the primary content of this book.

Over the now several years I've worked on it, I've tried often to imagine one member of that intellectual genealogy of ritual timekeepers—the author of the ancient Mayan hieroglyphic codex at this book's core. Now known as the "Dresden Codex," at its time it was one of many codices dedicated to preserving temporally driven ritual knowledge. And this author was far from alone as a ritual specialist with vast calendric knowledge. She or he was part of a community of intellectuals, charged with guiding the moral bearing of the community in which they lived. Accordingly, all manner of questions have crossed my mind. What kind of family was this woman from—if indeed she were a woman? How many brothers or sisters did she have? Was she young when she penned the pages of my particular interest—adding to the list of genius youth in human history? Did she walk through cornfields to the edges of the rainforest in her free time, or did she prefer the crowded walkways of the marketplace, smelling foaming chocolate as it was poured by vendors, or chatting with relatives trading chili peppers? How similar were her markets to the ones I had strolled through in Highland Guatemala or in Oaxaca? Was she actually a he? A bookworm, happiest in a library of stacked screen-fold, bark-paper books? Or a recluse during the day, only to find joy in staying up all night to observe the residents of the celestial realm?

We do know something about our author's intellectual genealogy even if we don't have access to his personal experiences. Because he lived in the Mesoamerican Postclassic period (AD 1000 to 1500) and studied the movements of the planet Venus, we can be sure that some of the scholars from whose work he drew would have come from Chich'en Itza. At that great Terminal Classic Yucatec city (AD 800 to 1000), as covered deeply in the later chapters of this book, we will encounter a specific member of our author's intellectual genealogy whose image was preserved in stone (figure P.1).

The Caracol Disk is a large carved circular stone that adorned the upper façade of the structure at Chich'en Itza known as the Caracol, or sometimes as "the Observatory." On

FIGURE P.1 The Caracol Disk from Terminal Classic Chich'en Itza. Courtesy Getty Images

the Disk, a Terminal Classic artist carved a scene including the image of a person dressed in a long skirt, standing near the head of a procession, along with a few other individuals. We should recognize at the outset that this is not the only figure so dressed in Chich'en Itza art, nor in Mayan artwork more broadly across the region. In fact, Waxaklajun Ub'aah K'awiil, the thirteenth ruler of Copán—whom we will see in chapter 4—was mistaken for a woman by early archaeologists because he wore a similar, very long skirt. So the attire does not by itself identify this figure as a woman; on the other hand, there is corroborating evidence at Chich'en Itza that the figure on the Caracol Disk uncomplicatedly is a woman wearing a long skirt as is traditional for the region. And if so, this was a woman of particular note.

As we will see in the coming chapters of this book, it appears that she was not serving in a role or office always held by a woman, but instead she was a woman who made a name for herself such that her image merited historical preservation on the Caracol structure itself. By the final chapters, I aim to show that this woman's renown resulted from her tremendous astronomical and calendric discovery and innovation and that her scholarly work converted a Venus Almanac into the now famous Dresden Codex Venus Table.

We will return to the hieroglyphically recorded history that this woman of Chich'en Itza took part in with particular focus in chapters 6, 7, and 8. Here it is worth noting that some aspects of this woman's work—as well as that of her copyist—do not require significant speculation. What we do not have to wonder as much about, for example—what seems much more securely known—is the part where our Postclassic scribe, the author of the final copy of that woman's manuscript, dips his brush into half-shell bowls filled with black and red ink and deftly presses it on the blank pages of a long sheet of bark paper. In his hieroglyphic script, he was following the same tradition as the one we find gracing the stone monuments of the Classic period, centuries before his and even her time. That Mayan hieroglyphic writing system had been well developed by the end of the Roman Empire across the Atlantic, and it had been adapted to various media: painted on paper, walls, clay vessels; carved into stone; modeled out of stucco. By our scribe's time, some of his local Yucatecan language had crept into the writing system, though, so while he certainly maintained a respect for tradition in his work, his own book reflected a changing world.

That specific changing world occurred during what we refer to as the Postclassic period in ancient Mesoamerica, when a type of globalization had taken hold, bringing the elite from the Mayan region into close contact with their counterparts in Oaxaca and those to their north in the Basin of Mexico. Through increasing contact, what has been referred to as an "international" artistic style developed throughout Mesoamerica (M. Smith and Berdan 2010)—an imagery reflecting borrowings and lendings between Mayans, Zapotecs, Mixtecs, and Toltecs. Such stylistic intervention made its way even into the illustrations painted by our scribe. He very likely first encountered this international style in his youth in murals painted on architectural walls, though he wouldn't have recognized its internationalism until he came across polychrome ceramic serving ware at great feasts. Bowls, platters for tamales, jars for serving beverages—all the images on pottery would have been familiar to him and other elite regardless of place of origin within Mesoamerica, even when they carried regional stylistic variation.

By the time he sat down to write his own manuscript, his illustration of an Aztec deity, Xiutecuhtli, reflected much of this interplay within artistic standardizations (figure P.2). On page 49 of his codex, for instance, the body proportions of his drawing adhere to local Mayan tradition, with Mayan forms coming closer to being representational, while

FIGURE P.2 *(left)* Xiutecuhtli on page 49 of the Dresden Codex; *(right)* Xiutecuhtli on the frontispiece of the Féjérvary-Mayer Codex. Courtesy Sächsische Landesbibliothek-Staats-und Universitätsbibliothek Dresden (SLUB)

the "Mixteca-Puebla" international style exaggerates the head, hands, and feet. But the face paint, the weapons carried, and the clothing worn make clear that it represents the same figure found in the Aztec "Féjérvary-Mayer Codex," painted by our scribe's contemporary in the Basin of Mexico (V. Bricker 2010, 315–16).

By the time our scribe painted his version of Xiutecuhtli, or whom he named in glyphs "Chak Xiwite," he was definitely a master of the artistic style and the hieroglyphic script. Because of what is popularly known about the Dresden Codex, though, it may be easier for the reader to focus on the scribe's carefully populating the pages of his book with numbers and calendric symbols, copied from an earlier manuscript. The concentration in reproducing subtle lines, his slight smile of recognition that the mathematics he copied were elegant and subtle—this is much easier for me to visualize than his figural illustrations, and it is the material for which his manuscript is best known. And it was a long tradition that our author took part in as scribe and astronomer—one that has been extensively explored by modern scholars (Aldana 2007; Aveni 2001; Bricker and Bricker 2011; Lounsbury 1983; Thompson 1972; Vail and Hernandez 2010) and one that takes central concern in the pages of this book.

◇◇◇◇◇◇◇◇◇◇◇◇◇◇◇◇◇◇

The final painting of the Dresden Codex relied on a community of scholars reaching across time, as well as a contemporary network of family, friends, and associates. In that sense, my own book is no different. For the exploration that follows, I am indebted to numerous colleagues, communities, and institutions. To those whose names I have inadvertently left out of these acknowledgments, I apologize in advance, but I hope that they recognize their influence within the work itself and that they find it worthwhile.

Of inestimable impact on this work have been my interactions and conversations with indigenous intellectuals in Mexico, the United States, and Central America. Pedro Pablo Chuc Pech has been a true and generous colleague over decades, and I will never be able to repay my debts to him, his family, and the community of Popolá. Likewise, Salome Gutierrez Morales and Adriana Gris have welcomed me into their local community-service efforts, thus sharing the riches of rural life in Veracruz and Oaxaca. The growing scholarly community of Mayan Mayistas—those whom I've met or listened to at international conferences over the years and most recently have been able to work with through the organization MAM (Mayas for Ancient Mayan)—has already impacted the field and provides tremendous promise for the future of Mesoamerican history and archaeology. Unfortunately, our dear colleague and pillar of the study of Native American religions here at UC Santa Barbara, Profesora Ines Talamantez, passed away before I was able to share this completed work with her. I have no doubt that she would find much from our conversations within it—and as much to challenge and question me on.

My colleagues at UC Santa Barbara have provided boundless fuel for my intellectual fire—some directly in line with my research efforts and others indirectly or tangentially. In the Chicana/o Studies Department, I continue to be fortunate in having junior and senior intellectual mentors: Dolores Inés Casillas, Aída Hurtado, Chela Sandoval, Daina Sanchez, and San Juanita Garcia. They have been both generous and fierce in intellect and in spirit in equal measures, and they provide me daily inspiration.

On methodological and theoretical levels, I've enjoyed lively debate with graduate students working in several fields. Felicia Lopez and Natalie Avalos expanded the scope of my interests at this work's original formulation, and our discussions directly impacted the inclusion—both explicit and implicit—of material from the Borgia Group of Central Mexican codices as well as modern Native North American scholarship. I am specially indebted to Andrea Medina, who shares my deep interest in ancient, historical, and modern indigenous communities in Mesoamerica. She has pushed me to engage the multiple audiences impacted by the work that we do with both purpose and passion.

The work in this manuscript has also benefitted from my participation in the Programme for Belize (PfB) research community, on the Rio Bravo Conservation Area, in the northwestern region of the country. Fred Valdez's generosity and acumen as the director has been inspiring as well as comforting in finding that he encountered much of what I did at Harvard, though just a little before me. My initial visits to PfB were instigated by Toni Gonzalez, whose work has transformed my conception of undergraduate field research in archaeology—in particular with respect to creating new opportunities for first-generation and Latinx students. She and Samantha Lorenz have enriched my thinking on the field and the possibilities of archaeological research. PfB has also provided the opportunity to start new conversations and continue old ones, with Spencer Mitchell, Risa Trachmann, Thomas Hart, Anastasia Kotsoglu, and my graduate-school compatriot Sarah Jackson.

My first vetting of the core arguments in chapters 5 and 6 of this book benefitted tremendously from colleagues in the International Society of Archaeoastronomy and Astronomy in Culture: Jarita Holbrook, Annette Lee, Barbara and Dennis Tedlock, Duane Hammacher, and Stan Iwanisewski. That astronomical focus has been complemented by conversations and correspondence with Michael Grofe, Marc Zender, David Stuart, Stan Guenter, Marcelo Canuto, Hal Green, and Takeshi Inomata. I learn something every time I encounter them at conferences, workshops, or randomly in hotel lobbies as we pass through the same portals traveling in and out of Mesoamerica. Tony Aveni and I have corresponded and conversed numerous times since Owen Gingerich first introduced us in the late 1990s. Although we've had our share of disagreements, I am profoundly indebted to his considerable contributions to the field, of course, but also his thoughtful counsel. I genuinely appreciate our last conversation, which ended with his generous smile and his advice to "keep doin' what you're doin.'" Other supportive colleagues here at UC Santa Barbara who have inspired me to continue the work even though they may not have done so intentionally include Paul Spickard, Rudy Busto, Harry Nelson, and Laury Oaks.

Although I didn't intend it as such, I have found that my own administrative work within the university has informed my interpretation of historical administrative labor. In my work with Bruce Tiffney and the staff of the College of Creative Studies and with Joann Erving, Shariq Hashmi, and Mayra Villanueva in the Department of Chicana/o Studies, I have been confronted with questions of messaging, of negotiating conflicting interests, and of managing (very) limited resources that seem to resonate with the circumstances faced by the ajk'ins, ajk'uhuuns, and even k'uhulajaws of the Classic period. The analogy only goes so far, of course, but I do believe this administrative labor has added nuance to my representation of ancient Mesoamerican governance.

The editorial staff at the University of Arizona Press deserve special recognition. Kristen Buckles and Allyson Carter believed in the manuscript from the start and have patiently nurtured it through completion. I owe a great debt to Shannon Lee, who meticulously edited a manuscript that was long and unwieldy. Also critical to this book has been the generosity of Sarah Applegate and the Ancient Americas archive at the Los Angeles County Museum of Art. In particular, they have graciously stewarded Linda Schele's illustration and photograph collections upon which I know myself and so many of my colleagues have relied.

While it could not have been completed without travel to and through Mesoamerica, the vast majority of the labor on this book has been conducted on the ancestral land of California Central Coast Chumash communities. As I run, cycle, drive, kayak, and surf across these land- and water-scapes, I am nourished by glimpses of what they have enjoyed and am inspired to write back against some of the injustices they have experienced in kind and in chronology with Mayan and so many other indigenous communities of the Americas. Finally, I am grateful that my dear family—Santiago, Seri, Iliana Luise, and Cita—has been able to grow and learn with me from this place in profound ways.

NOTE ON LANGUAGE

It is now conventional to include a note on Mayan languages within English-language publications intended to reach beyond the academic specialist audience. Here I do so to review the basic aspects that may be new to the reader, as well as to address an additional, long-standing convention in the field.

In this book, I follow the orthography standardized by the Academia de Lenguas Mayas de Guatemala (ALMG). For the English-language reader, this orthography has the effect of treating vowels and some consonants as though they were in Spanish:

a – as *a* in *father*

e – as *e* in *set*

h – soft aspiration

i – as *ee* in *seen*

j – as *h* in *house*

o – as *o* in *omega*

u – as *oo* in *oops*

A second orthographic convention is to include a single apostrophe to indicate a glottal stop. In English, this sound is familiar as the vocal break within in the utterance "uh oh." That second vowel would be rendered *'o* accordingly. Additionally, in Mayan languages, consonants may be plain or glottal such that *ki* and *k'i* carry different sounds and meanings.

As for the names of places, this book uses the historical and more common versions as opposed to those that would follow ALMG orthography in order to facilitate the reader's ability to engage related literatures. Thus, for example, I use Calakmul rather than Kalakmul and Uaxactun rather than Waxaktun.

Regarding a further convention, I do not contest that the tradition in the field of Maya Archaeology is that one uses "Mayan" in reference to the language only and "Maya" for everything else. I recognize that as the accepted academic convention. However, this practice has been both explicitly and indirectly challenged in recent years.

In 2007, linguist John Justeson and historian David Tavárez explicitly commented that this was poor practice, since the only other cases in English in which we use the singular to refer to the plural is for animals (e.g., deer, fish). They write:

> We . . . avoid an affectation that developed in early academic anthropology: systematically using morphologically singular forms of count nouns to refer in the plural to members of certain ethnic groups, as in "the Olmec," "the Zapotec," or "the Maya." This usage effectively marks those to whom it is applied as having less humanity than Europeans, to whom it is *never* applied: on the one hand, it is ungrammatical to say, for example, "The Pict were immigrants to Britain"; and, outside of this pattern, in English nouns the use of a

morphological singular form as the plural of a count noun is systematic only in references to animals, especially as game or food. (Justeson and Tavárez 2007, original emphasis)

I concur that to have this singular practice for "the Maya" is to perpetuate the "Othering" of them, as well as to represent a false homogeneity to the diversity of communities behind the term. Their recommendation is that we simply go with what is "natural" in English: Mayan, Mayans, far beyond just in reference to the language.

Furthermore, it is common outside of archaeology, in other academic fields (i.e., in ethnography and ethnic studies), to use "Mayan" much more freely. In my dissertation (straddling archaeology and the history of science), I argued that Mayan communities used astronomy as a language, and for that reason I referred to it as "Mayan Astronomy"—at that point sticking to the principle, but expanding the application.

For this book, I recognize that there are both political and academic reasons to make use of "Maya," "Mayan," and "Mayans," and so I adopt all three, which I hope causes neither strife nor confusion.

CALCULATING BRILLIANCE

Introduction

The woman on the Caracol Disk mentioned in the preface of this book also shows up in another visually preserved procession at Chich'en Itza. Within the structure known as the Lower Temple of the Jaguar, this dignitary, who we will refer to as K'uk'ul Ek' Tuyilaj, stands some few paces behind the main event. From this vantage point, she would have been in a great position to take in the details of the ceremony. The scent of copal would have been rich; drums, whistles, and trumpets all sounding nearby; feathers in regalia rustling; jade celts clinking against one another. As for the motivation behind this specific ceremony, however—that question has vexed modern scholars for over a century.

To the modern eye, the architects at Chich'en Itza produced some of the most enigmatic structures in ancient Mesoamerica. Hypotheses about the purposes of and the cultural influences behind this architecture abound within popular and scholarly literatures from the late nineteenth century to the present day. Within this book, we will spend some time investigating the lives of K'uk'ul Ek' Tuyilaj, her neighbors, and colleagues of this great Yucatecan city along with the architecture they moved through, but here, to initiate this study, we focus on one figure and one structure in particular. The figure in question was carved by the artists of Chich'en Itza and given an affiliation with a theme of great renown. Here, he is at the center of the ritual activity that K'uk'ul Ek' Tuyilaj witnessed, within one of the more elaborate and complex murals of the city. Adorning the interior of the Lower Temple of the Jaguar, the mural resided within a structure built into the southern end of what is known today as the Great Ball Court (figure I.1).

Scribes created this mural to cover the walls and ceiling of the structure. To do so, they divided the walls vertically into five stacked bands, each populated with the men

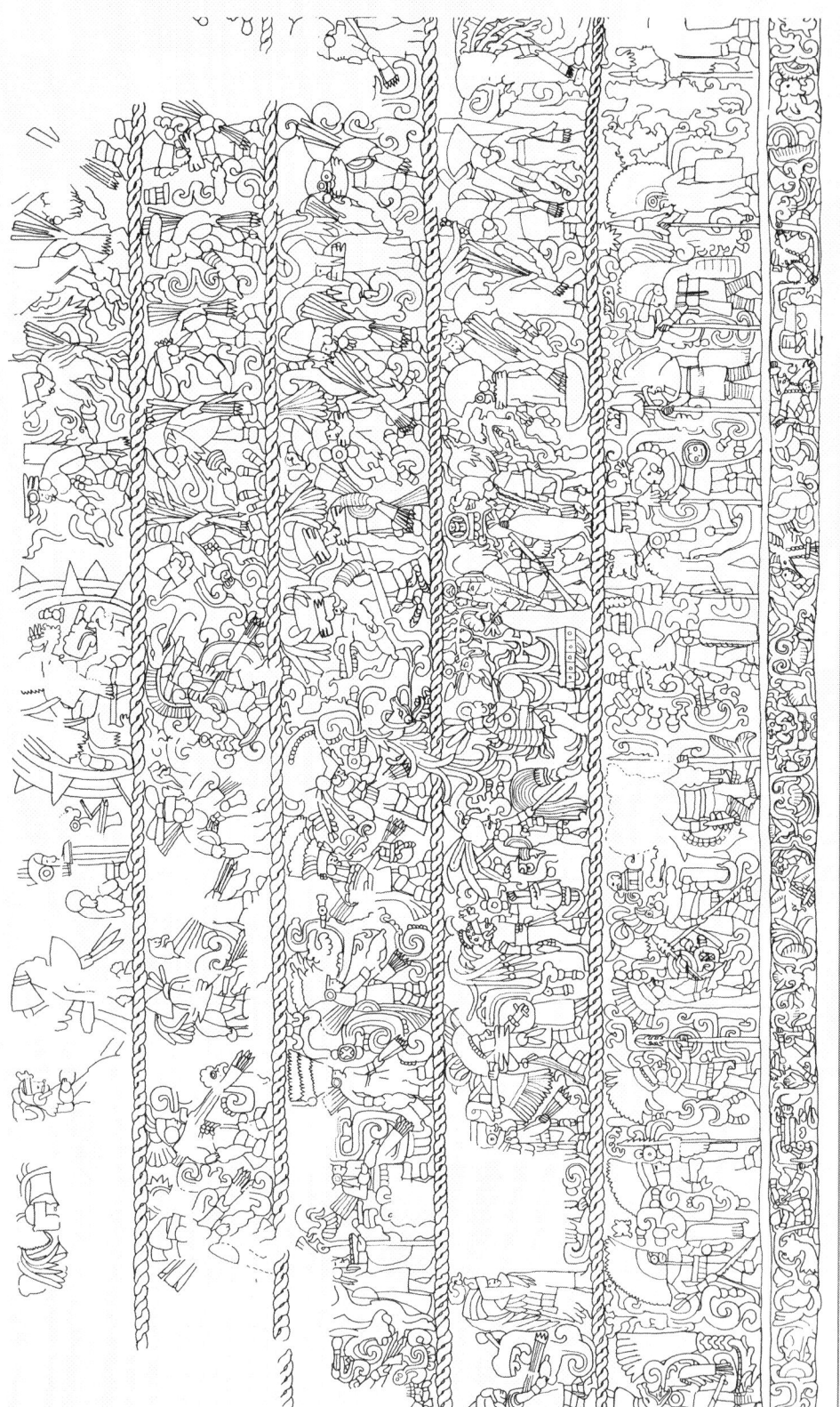

FIGURE 1.1 Central portion of a mural in the Lower Temple of the Jaguar in the Great Ball Court at Chich'en Itza. Drawing by Linda Schele © David Schele; photo courtesy Ancient Americas at LACMA (ancientamericas.org)

and women accompanying K'uk'ul Ek' Tuyilaj lined up in what looks like a series of processions. The figure who appears to enjoy the focus of attention stands left of the central event within the second band from the bottom. His costume is appropriately extravagant, but perhaps most arresting are his facial ornaments. He wears large rings encircling his eyes and an oval-shaped ring around his mouth. These ornaments bear great resemblance to artifacts extracted by archaeologists in the early twentieth century from the Great Cenote of Chich'en Itza, located a few hundred meters from the Great Ball Court (Schele and Mathews 1999, 223) (figure I.2). The dredging of the cenote produced two flat gold rings that could be worn around the eyes, each carrying along the top a serpent with feathers on its body and a flat oval ring—also of gold—with whiskers attached at each pointed end.

The "goggle eyes" extracted from the Great Cenote have long been understood—during ancient times and to modern scholars—as a visual referent to Central Mexico and the great Early Classic period metropolis of Teotihuacan (Schele and Mathews 1999, 223). In the mural at Chich'en Itza, there is further connection to Central Mexico: a creature rises behind the goggle-wearing figure identifiable as the same "feathered serpent" as that on the gold rings. While the referent is visual here in these archaeological contexts, the concept of a "feathered serpent" entered the modern scholarly literature through a Central Mexican deity name: Quetzalcoatl (Lopez Austin 2015).

The name is made up of two Nahuatl nouns: *quetzal*, referring to a bird of brilliant green plumage and extraordinarily long tail, is combined with *coatl*—"serpent." In Yucatan, K'uk'ulkan was an equivalent name and was used to reference a historical leader (Tozzer 1941, 20–26). *K'uk'* is Yucatec Mayan for "quetzal" and *kan* the equivalent of "serpent." One of the key questions we are confronted with in this mural—and our touchstone for this book—is how this imagery within this Ball Court structure might bring together the regionally distinct figures of K'uk'ulkan and Quetzalcoatl. This is a question with a very long and developed scholarly history (Carrasco 2001; Lopez Austin 2015; Ringle 1990, 2009), which we will not attempt to answer here at the beginning of the first substantive chapter of this book. For now it is enough to recognize the central place of the feathered serpent within the elaborate processional event depicted in the Lower Temple of the Jaguar—a procession that included men and women, young and old, carrying various accoutrements, some wearing masks, all in formations resonating with the processions referred to in the preface to this book.

The Chich'en Itza artists also included a date in the ritual calendar of prominent size within the mural scene, adjacent to the central figure. And here again, they did so using the symbols of Central Mexico, not their own Yucatec or inherited Peten Mayan symbols. This might be interpreted as a purely symbolic appeal to the "international style" of Postclassic Mesoamerica. But it also may signify that the event occurred on a specific date within the ritual calendar, which in turn inspires speculation about

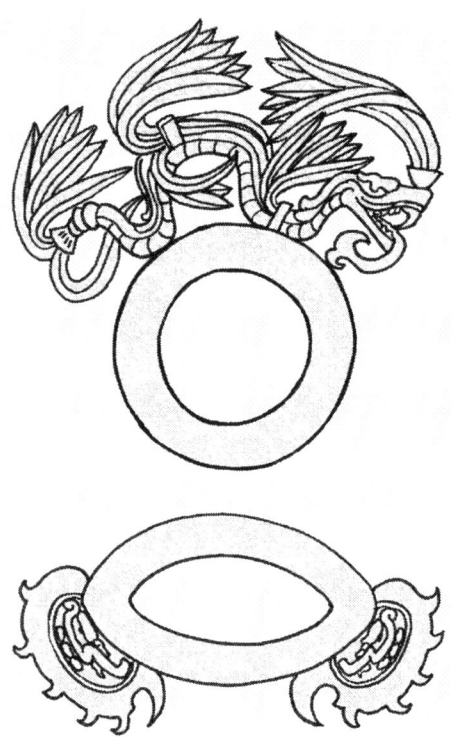

FIGURE I.2 Feathered Serpent imagery incorporated into gold facial ornaments dredged from the Great Cenote of Chich'en Itza.

whether the event was deliberately timed. Feathered serpents, ritual processions, and political alliances are all implied in this imagery, and they will concern us throughout the following pages, but so too will the question of timing and the astronomer-priests charged with such duties. In order to access the intent of the event in this mural along with the relevance of its timing, this book takes a long journey through history, archaeology, and multiple forms of scientific endeavor. In this chapter, we turn to a contextualization of astronomy in ancient Mesoamerica by way of a detour through another mural representation of a feathered serpent.

MESOAMERICAN NATIONALIST PROJECTS

In the late 1920s, the great Mexican muralist Diego Rivera confronted a considerable challenge—one akin to that taken up by the muralists of Chich'en Itza's Ball Court structures. Rivera had been given a commission in the National Palace—not only in a manner worthy of its august location, but also in a way that would speak to the nationalist agenda of his patron, the president of Mexico. To be sure, politics were complicated and unstable in the aftermath of the Mexican Revolution, but the interest in using publicly accessible murals carrying themes that would enable the diverse population to support the nation-state persisted after its conception by the secretary of education, Jose Vasconcelos. Rivera tackled his charge by including images throughout the palace representing the ancient past, the colonial past, and the new republic context. He sought to forge a national identity by creating a visual narrative tying the mythological and historical pasts to the modern context. Public art was a tool of the state, building narratives across time in ways that resonate, as we will see, with earlier Mesoamerican uses of public astronomy. On the north wall of the main stairwell of the Mexican National Palace, Rivera included the visual representation of an indigenous legend—the story of Quetzalcoatl.

A sixteenth-century manuscript provided one version of the inspiration for this part of Rivera's painting. Written in Nahuatl—the language spoken within the Aztec Triple Alliance—but using the Latin alphabet, the *Anales de Cuauhtitlan* recorded a history of Quetzalcoatl within the more provincial origin story of the Cuauhtitlan nobility (figure I.3). Like Rivera's project, and as we will find to be common within indigenous

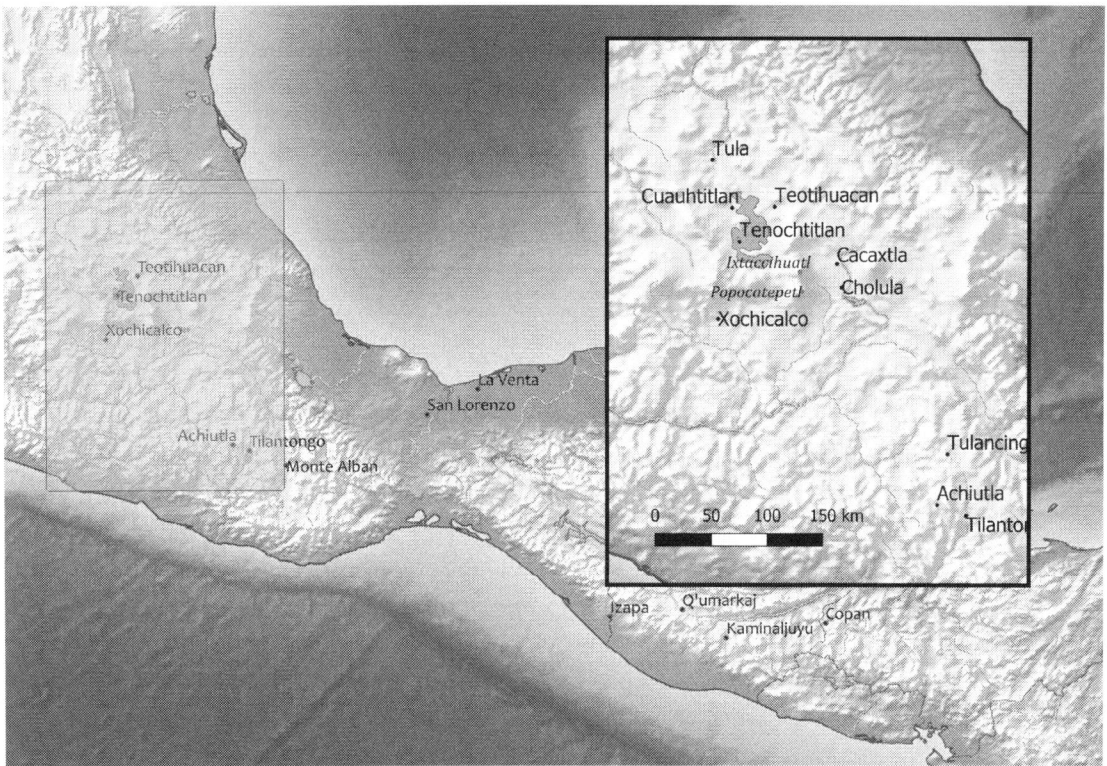

FIGURE 1.3 Sixteenth-century cities of the Basin of Mexico

Mesoamerican literature, this text includes mythological content on its way to establishing local genealogical history.

The *Anales* passage tells us that although he ruled Tollan as a peaceful king, Quetzalcoatl was tricked and shamed, and he eventually banished himself from his home.

> Then Quetzalcoatl departed. He got up, called together his pages, and wept over them. Then they set out, heading for Tlillan, Tlapallan, Tlatlayan. . . .
>
> Now, this year, 1 Reed, is when he got to the ocean, the seashore, so it is told and related. Then he halted and wept and gathered up his attire, putting on his head fan, his turquoise mask, and so forth. And as soon as he was dressed, he set himself on fire and cremated himself. And so the place where Quetzalcoatl was cremated is named Tlatlayan [land of burning].
>
> And they say as he burned, his ashes arose. And what appeared and what they saw were all the precious birds, rising into the sky. They saw roseate spoonbills, cotingas, trogons, herons, green parrots, scarlet macaws, white-fronted parrots, and all the other precious birds.
>
> And as soon as his ashes had been consumed, they saw the heart of a quetzal rising upward. And so they knew he had gone to the sky, had entered the sky.

The old people said he was changed into the star that appears at dawn. Therefore they say it came forth when Quetzalcoatl died, and they called him Lord of the Dawn.

What they said is that when he died he disappeared for four days. They said he went to the dead land then. And he spent four more days making darts for himself. So it was after eight days that the morning star came out, which they said was Quetzalcoatl. It was then that he became lord, they said. (Bierhorst 1998, 36)

For his visual rendition of this passage, Rivera focused on Quetzalcoatl's apotheosis as "Lord of the Dawn," in Nahuatl Tlahuizcalpanteuctli, or what has become understood as the planet Venus as Morningstar. In his mural, Rivera depicts Quetzalcoatl as a bearded man riding a flying rattlesnake, quetzal feathers streaming from his headdress. He carries a shield marked with a half-conch shell and travels east toward Tlillan Tlapallan, the Land of Red and Black, as the Mayan region was known.

We will return to Quetzalcoatl, Venus, and a somewhat parallel Yucatec "nationalist agenda" in detail in chapter 7. For now it is curious to note that as a politically motivated artist, Rivera chose to include Quetzalcoatl as a symbol toward Mexican national unity. Moreover, Rivera's inclusion of the narrative in the National Palace itself represented only one of a series of efforts at negotiating political strife at that specific location—art and architecture with political ends. And this structure takes us on our next step toward unpacking the context of the Mesoamerican astronomer.

After the war between Motecuhzoma's and Hernan Cortes's armies, the Spanish administration chose the site of Motecuhzoma's New Palace for the residence of the new viceroyalty of New Spain. For the National Palace's construction in the sixteenth century, indigenous masons—as subjects of the Spanish Crown—reused the stone and construction material taken from the architecture of Tenochtitlan. These earlier structures made up the residence of the last emperor of the Aztecs, Motecuhzoma Xocoyotzin. Time and again, art and architecture symbolized Mesoamerican efforts toward cultural nationalism.

Antonio de Mendoza y Pacheco was the first viceroy of New Spain; as he contemplated the National Palace's new form, he also found himself in a politically tenuous position. Mendoza confronted the challenges of keeping the conquistador Hernan Cortes and his allies in check while maintaining good relations with the Tlaxcalan nobility and other indigenous allies to the Spanish Crown, all while developing the infrastructure for the colonial seat of the Spanish Empire. In this effort, too, a work of art was commissioned to address the problem of nation-building.

In part, the work of Mendoza's colonial administration required the collection of data in various forms. These constitute the primary archival records of the time. One document produced toward this end carries the viceroy's name: the Codex Mendoza. Indigenous artists were recruited to provide illustrations for it using imagery that drew upon but was modified from previous traditions. Further breaking from indigenous tradition, each painted image was accompanied by its description using Spanish text. While these glosses were intended for the Spanish Crown's administrative purposes, they have since served as an aid for modern scholars' interpretations of Precontact art.

The work in the Codex Mendoza for which these artists are probably best known is the frontispiece depicting the founding of Tenochtitlan itself. In the rest of the codex, though, these scribes went far beyond the city's founding to document topics varying from history through imperial tribute to everyday life; these artists even included an illustration of Motecuhzoma's New Palace as it had been before destruction (Berdan and Anawalt 1997). Most importantly for our purposes, though, the artists of the Codex Mendoza contributed to the primary task of this chapter. On page 63r, they included a depiction of one of the basic duties assigned to

FIGURE I.4 Detail after page 63r of the Codex Mendoza: an astronomer tells the time at night

their astronomer colleagues at the Aztec capital of Tenochtitlan: tracking time at night (Aveni 2001, 19–21) (figure I.4).

MESOAMERICAN ASTRONOMERS

As an astronomer of the Mesoamerican Postclassic period at Chich'en Itza, K'uk'ul Ek' Tuyilaj made use of a standardized practice for telling time by virtue of an astrological construct with origins thousands of years earlier in the Mesoamerican Preclassic period (Aldana 2008; Aveni 2001, 131–48; Rice 2007, 46). Like the figure in the Codex Mendoza, she and other official timekeepers over the years observed nine celestial "seats" spread out evenly across the night sky, in line with the ecliptic—that is, the path across the sky, traveled by the Sun, Moon, and planets (see figure I.5). These seats provided a structure for measuring the procession of celestial bodies across the night sky.

On any given evening, K'uk'ul Ek' Tuyilaj would take up a position at the top of an appropriately dedicated ritual structure with a clear view of the horizon. From such a perch, as the Sun set in the west, she would identify the brightest celestial body visible at the opposite side of the sky, along the eastern horizon; this brightest celestial body is what modern K'iche' timekeepers call the *retal ak'ab'* or the "sign of the night" (B. Tedlock 1982, 180–83). Our timekeeper would track this retal ak'ab' as it climbed toward its highest point in the sky, along the way passing by the deities who occupied the first four celestial seats. These deities were known as the Yohualteuctin in Nahuatl, which has been translated as the "Lords of the Night"—even though not all of them were male. In Mayan hieroglyphic writing, parallel deities held the title *ti'huun*, for "speaker of the paramount," in what is commonly referred to as Glyph F of the Supplementary Series (Aldana 2008). These ti'huuns occupied celestial seats along the ecliptic such that the retal ak'ab passed by all nine each night, marking off nine hours. At the fifth hour, around midnight, the retal ak'ab' would "meet" the ti'huun residing in the highest celestial seat. By daybreak, the sign of the night met the ti'huun in the ninth seat at the western horizon.

FIGURE 1.5 The Aztec period Yohualteuctin positioned in the nine celestial seats for telling time at night, from Tlazolteotl in the east to Piltzinteuctli at zenith and Tepeyollotl in the west

FIGURE 1.6 Deities with celestial affiliation on the back and in the open mouth of the Celestial Dragon (K'awiil, Ik', Ix Chel, Itzamnaaj, Chaak, and K'inich Ajaw) meet lunar deities (the Jaguar God of the Underworld, the Moon Goddess, and the Death Moon) being transported in a canoe by the Stingray and Jaguar Paddlers

In ancient Mesoamerica, telling time at night was not a strictly practical duty. The meeting of the "star of the night" with the occupant of the fifth seat generated an omen, which would maintain its astrological effect through the rest of the night and morning until midday (Aldana 2008; Caso 1967; Léon y Gama 1832). A timekeeper would have made note of this omen and then followed the celestial body's movement over the rest of the night through the remaining four celestial seats. She would have communicated the omen to her peers and the nobility of her community for deliberation in the morning.

Yet with this simple duty alone, initiated in the first centuries BC, timekeepers ancestral to K'uk'ul Ek' Tuyilaj would have witnessed countless other patterns in the night sky. Over the course of the tropical year, they would have noticed that different stars took over the duty of serving as the retal ak'ab'. That is, each evening, the "star of the night" from the previous evening would appear slightly higher above the eastern horizon at sunset. This height increased until another prominent celestial body would appear at sunset closer to the horizon. During the Mesomerican Formative Period, timekeepers living in the Isthmus of Tehuantepec systematized these observations, allowing each retal ak'ab' to play the role for twenty days. In turn, this generated a partitioning of the year into eighteen months, with five "extra" days, providing the observational basis for the Mesoamerican 365-Day Count (Aldana 2008, 66–73).

Timekeepers staying up all night (in light-pollution-free environments) also witnessed numerous other celestial phenomena worthy of note. Meteors interpreted as "arrows of the celestial deities" would certainly have been observed—sometimes in isolation, other times in "showers"—and communicated to the appropriate authorities for interpretation (Aldana 2005; Stuart 1995, 310–11). They noted and accounted the wanderings of the planets, following their own paths past the "stars of the night" and other celestial bodies that remained "fixed" relative to each other (Aveni 2001, 80–94). With the ti'huuns in their seats, with personified planets and with the Moon's and the Sun's arrival or departure over the horizons, the celestial realm would have been viewed as a rich skyscape populated by a dynamic community, laden with interactions—predictable and not—to be interpreted by astronomers and priests (figure I.6).

All of these patterns were certainly among the training and duties of the timekeepers working over the millennia from the Middle Preclassic through K'uk'ul Ek' Tuyilaj's time and up to Aztec times when scribes depicted a skywatcher in the Codex Mendoza. It is worth emphasizing, though, that, like Diego Rivera's murals in the National Palace, the work of Mesoamerican astronomers was not disinterested. Dedicated skywatching relied on a stratified society, fostering various specializations of labor. Astronomers held their jobs because they demonstrated the requisite skills, but also because the system of governance sufficiently valued the products of their labor to allocate them the resources they needed to live comfortably, without having to provide their own caloric sustenance. By the Late Postclassic period, astronomical observation and astrological interpretation had become part of societal infrastructure as described in the sixteenth century by the quixotic priest and bishop of Yucatan, Diego de Landa. Through Landa, that is, we gain access to a view of lived Mesoamerican communities and the roles that astronomers played within them.

Referring to indigenous priests in the Yucatec communities of the region around Izamal, Landa wrote:

> They taught the sons of the other priests and the second sons of the lords who brought them for this purpose from their infancy, if they saw that they had an inclination for this profession. The sciences which they taught were the computation of the years, months and days, the festivals and ceremonies, the administration of the sacraments, the fateful days and seasons, their methods of divination and their prophecies, events and the cures for diseases and their antiquities and how to read and write with the letters and characters, with which they wrote and drawings which illustrate the meaning of the writings. (Tozzer 1941, 27–28)

Tozzer's English translation preserves the Spanish gendering of male children as the targets of priestly training, but it doesn't reflect the ambiguity in the reference. Since Spanish uses the masculine for the "default," Landa could very well have intended to be describing both boy and girl acolytes entering the priesthood with the term *hijos*. On the other hand, given the patriarchy of sixteenth-century European culture, he may also have been only referring to sons out of cultural bias. Either way, we do now have records of women priests and scribes throughout Precontact times, so K'uk'ul Ek' Tuyilaj would not have been an isolated anomaly in her position as astronomer (Coe and Kerr 1998, 99; Bruhns and Stothert 1999, 162; cf. Sharer 2006, 140).

Landa further noted that priests kept records in specific types of books, like the one that will occupy our primary concern in the following chapters.

> Their books were written on a large sheet doubled in folds, which was enclosed entirely between two boards, which they decorated, and they wrote on both sides in columns following the order of the folds. And they made this paper of the roots of a tree and gave it a white gloss upon which it was easy to write. And some of the principal lords learned

about these sciences from curiosity and were very highly thought of on this account, although they never made use of them publicly. (Tozzer 1941, 28–29)

Within his extensive description, Landa additionally provided us a window into the societal differentiation of Precontact Mayan society.

> The natives of Yucatan were as attentive to the matters of religion as to those of government and they had a high priest whom they called *Ah Kin* Mai and by another name *Ahau Can* Mai, which means the Priest Mai, or the High Priest Mai. He was very much respected by the lords and had no *repartimiento* of Indians, but besides the offerings, the lords made him presents and all the priests of the towns brought contributions to him, and his sons or his nearest relatives succeeded him in his office. In him was the key of their learning and it was to these matters that they dedicated themselves mostly; and they gave advice to the lords and replies to their questions. . . . They provided priests for the towns when they were needed, examining them in the sciences and ceremonies, and committed to them the duties of their office, and the good example to people and provided them with books and sent them forth. And they employed themselves in the duties of the temples and in teaching their sciences as well as in writing books about them. (Tozzer 1941, 27)

From Landa's sixteenth-century document, then, we find that in Precontact traditions kinship provided some basis for participation among the ceremonial and political leaders of Yucatec communities, defining classes of nobility and commoners within genealogical structures.

The consensus among modern scholars is that the underlying structure of Mayan society by the Postclassic is represented reasonably well by the concept of "houses" (Sharer 2006, 695). In an abstracted sense, intended to capture similar relationships found in societies from Western Europe to Southeast Asia, the term "house" here refers to organization by genealogy, while also including differential access to resources—political or economic. In the Mayan highlands of Chiapas, ethnographer Evon Vogt witnessed a similar structure in the 1960s, which he referred to with the Tzotzil Mayan term *sna* (1969, 140). These were extended families living in residential compounds, further organized by access to a common water source. While majority patrilineal, some snas also followed matrilines. Archaeologist Robert Sharer notes that common practice in Yucatan at the time of Contact was to take surnames from both the patriline, in Yucatec *ch'ibal*, and the matriline, or *ts'akab* (2006, 693). Those born of prominent lineages on both sides and so of "noble house" were referred to in Yucatec as *almehen*, which brings together *al*, the term for "child of mother," and *mehen*, the term for "child of father" (Barrera Vásquez 1980, 14). To be almehen was to be of publicly recognized father and mother.

As we consider, then, to whom astronomers would have been giving counsel, we find that in the responsibilities expected of the leaders of houses—men and women—there is also resemblance to Native North American clans (cf. Chambers 2010). House leaders throughout the Mayan region sat on councils, which met in structures referred to as *popol nah* (mat house), *nicteil nah* (flower house), or *yotoch cah* (house of the community) (Jackson 2011, 696). Archaeologists William and Barbara Fash have identified one such *Popol Nah* as far back as Late Classic Copán, suggesting that the societal structure Landa encountered in the sixteenth century had deep roots in Mayan cultures (B. Fash et al. 1992). Hieroglyphic texts from those earlier times attest that a "ruler" was selected from one of the almehen houses. Within the language of the hieroglyphic writing system, this ruler was the *k'uhulajaw*, while during the Late Postclassic in Yucatan on a smaller scale it was the *halach uinic* (Jackson 2011). We will see in the coming chapters that political differences and alliances among houses—in selecting rulers, but also in visions of leadership—impacted the fortunes of entire cities, but also shaped astronomical investigation and application.

We also saw in Landa's description above that some priestly positions were inherited, certainly complicating the relationships between politics and religion. But given the structural similarity to clans, it may have at the same time allowed for some social mobility. In other words, those families within houses, but without sufficient status, fit into either a middle class (of substantial size during the Late Classic) or a commoner class—*macehuales* in the language of the sixteenth century—but they were still members of ch'ibals or ts'akabs (Sharer 2006, 710). While they may not have been in frequent social contact at either end of the class spectrum, Postcontact evidence suggests that they were dependent on each other for ceremonial activity. House-based religious activity for common revered ancestors, or citywide religious activity for deity celebrations, would have been sponsored by the noble-class families within a house. While the almehen members would have feasted together, acted within priestly roles, and served within theatrical performances, at the same time the sweeping and decorating of streets, the preparation of food, and participation in processions would have drawn from the labor of middle- and commoner-class families within those houses. In other words "houses" were not simply political structures; they also brought families together for ritual activity across class lines.

There is also strong evidence that these "house" structures allowed society to replicate ritual activity at different scales. For example, each extended family unit living within the same residential compound would have its own revered ancestor(s), while the larger house that they belonged to would have maintained a hierarchically more prestigious revered ancestor (McAnany 1995). The royal family, accordingly, would have had the access to the most renowned ancestors, which may even have extended to the Creation deities of mythological times (Lopez Austin 2015, 43). The principles of ancestor veneration were the same across classes; it was mainly the scale of ceremony that differentiated them.

Other "deities," such as Chaak, received community-wide attention across houses. As rain deities, the Chaaks have been associated with caves, perhaps as far back as the Middle Preclassic. Cave archaeologist Holley Moyes has argued that this may have been related to the observable phenomenon of clouds emerging from caves as mist under the right conditions of humidity and temperature (Moyes and Brady 2012). Corroborating this association is the fact that scribes and sculptors of the Classic and Postclassic periods represented Chaak with caves frequently in inscriptions and artwork; they show up

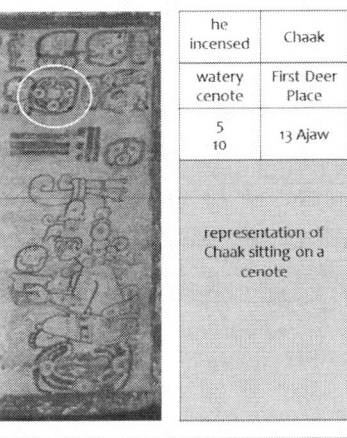

he incensed	Chaak
watery cenote	First Deer Place
5 10	13 Ajaw

glyph for cenote

representation of Chaak sitting on a cenote

FIGURE I.7 Chaak sits on a kab-ch'en (or cenote) on Dresden Codex page 39. Courtesy SLUB

regularly in the pages of the Dresden Codex—as we will see in later chapters of this book—where caves are notable as physical locations or animated mouths (figure I.7). The upshot is that ancestors and caves were key to political house structures, but they were also core to annual ritual activity. Politics and religion—the middle world and the Underworld—were intimately related and familiar within daily life. Of course revering ancestors did not entail simply remembering them, but ritually interacting with them according to small-scale and large-scale seasonally timed ceremonial activity. And this meant access to the domain of the ancestors: the Underworld.

In the language of the inscriptions, caves or any such subterranean spaces were referred to as *ch'en* (as in *chi-ch'en* or "mouth of the cave" in the name Chich'en Itza). Epigrapher Alexandre Tokovinine has recently found various uses of both *chan-ch'en* ("sky" or "high" caves) and *kab-ch'en* ("earth" caves) in hieroglyphic texts as far back as the Early Classic period (Tokovinine 2013). He finds that the importance of such caves is highlighted by textual references to polities as "their land and their cenotes" (2013, 25). Certainly in Yucatan this would have related to the water-source function of cenotes, but it also most likely was related conceptually to the fact that the Underworld served as the residence of the spirits of the ancestors. As archaeologist Patricia McAnany has developed extensively, ancestors were not buried and forgotten in ancient Mesoamerica; rather, the souls of the dead maintained lives in the Underworld, allowing them to interact on special occasions with the living (1995). In the Mayan region, such interactions were guided by the landscape.

In the highlands as far back as the early Preclassic, caves were abundant. Many of these are used today, exhibiting a continuity in ritual activity—including processions—that goes deep into the past. Caves in the highlands, though, are notably often in locations above residential centers. As mountain caves, they look down on villages and out over vast landscapes, aptly fitting the denomination of chan-ch'en, or "high" caves. In the lowlands, however, caves are frequently at ground level, either open as cenotes or deep as tunnels traveling miles underground. These fit the term kab-ch'en or ground-level caves. If we follow the iconography, then the large pyramid-based temples constructed during the Middle Preclassic through the Classic period may well have been conceptualized as human-created mountains (Schele and Mathews 1999, 43; Moyes 2012, 9), with the temples at their tops serving as man-made chan-ch'en (Nick Hopkins, Barbara MacLeod, pers. comm., 2018) (figure I.8). The point is that subterranean space—atop pyramids, on hills, or adjacent to homes—was part of the inhabited landscape, preserving genealogical ties to the construction and maintenance of place.

The need for contributions of communal labor—at the scale of great pyramid mountains in earlier times or locally used public plazas in later times—also served the purpose

FIGURE 1.8 (*top*) Temple V at Tikal as a chan-ch'en; (*bottom*) an active twenty-first-century ceremonial cave at Q'umarkaj

of facilitating the integration of "foreigners" into existing communities. In Yucatan during the sixteenth and seventeenth centuries, newcomers were indebted to the community with prescribed forms of labor, which conferred membership when accomplished (Redfield and Villa Rojas 1962, 6). Bioarchaeological studies in southeastern Peten suggest that families who moved into the area from another city may have constituted between 10 and 25 percent of town populations during the Classic period (cf. Prufer et al. 2017). Eventually, these indigenous foreigners would have married into ch'ibals or ts'akabs affiliated with specific houses and so been integrated further into the community. The point is that city populations of the Classic and Postclassic periods were thus dynamic and certainly cosmopolitan at the larger scales, but also diverse and yielding opportunities for interactions across social classes.

Of course, external conflict also fostered collective action and interaction. During the Classic period, k'uhulajaws themselves acquired status by capturing enemies in battle (Martin and Grube 2000, 130; cf. Baudez and Latsanopolous 2010; Ringle 2009). We will see indirect evidence in chapter 2 that the military forces accompanying rulers into battle may have been organized by house as well. War against another polity would have required recruiting young men from across houses to go into battle. In such case, each house would have a military head to supervise training in the crafting of weapons and the techniques of hand-to-hand combat but also to represent their interests when war was proposed (cf. Sharer 2006, 709; chapter 2). It would be fascinating to learn more about the ruling dynasty's military force and whether it was made up specifically of those within a lineage, if it recruited from all houses, or even if the dynasty had no army of its own, but required the consent of the nobles to send their armies off to war. It does appear, though, that prowess in war led to promotions for commoners within house armies as well and so provided another means of acquiring status in society (Sharer 2006, 709).

Making our way back to ceremonial concerns, we find that similarly, priesthoods appear to have been several, each guided by commitment to a specific deity or community of deities. Landa referred to the title *ajk'in* (or in an earlier orthography *ah kin*, translated as "daykeeper" or "diviner") to describe a single high priest, Ah Kin Mai, but he also noted that the ajk'ins of other orders would contribute according to the ceremony at hand. In Aztec society, for example, priesthoods were ranked, with the highest order being that of Quetzalcoatl—the figure of Rivera's mural in the National Palace of Mexico. Membership in a given ajk'in collective, again as attested by Landa, most often came from a demonstration of aptitude early in life and so entrance into a dedicated school. It is also quite possible that membership was granted for political reasons, or alliances between members of an almehen house and members of a specific ajk'in school. Collectively, their charge would be to conduct the proper rituals at the appropriate time to ensure the health of the city, which would have required contributions from several (if not all) houses, depending on the size of the ceremony. Again, we will return to this aspect of ajk'in work and its attendant political complications in detail in chapters 5 and 6.

It is within this broader context of social, political, and religious relationships that we consider in this book the case of K'uk'ul Ek' Tuyilaj, to whom we give the title "*ajk'uhuun*," which is related to "ajk'in," as we will see in chapter 5. The Ajk'uhuun K'uk'ul Ek' Tuyilaj

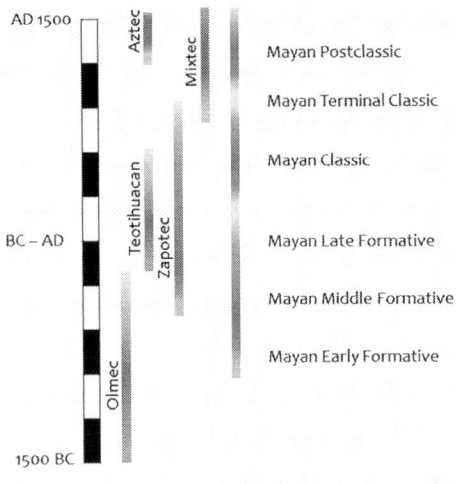

AD 1500

Aztec
Mixtec
Teotihuacan
Zapotec
Olmec

Mayan Postclassic

Mayan Terminal Classic

Mayan Classic

BC – AD

Mayan Late Formative

Mayan Middle Formative

Mayan Early Formative

1500 BC

FIGURE I.9 Timeline of Ancient Mesoamerica

was the author of a hieroglyphic manuscript, who viewed the night sky from the top of a structure in the northern Yucatan Peninsula, gazing out over the green sea of floral canopy within a very particular time period (figure I.9). In this book we focus even further on our ajk'uhuun's accounting of one planet in particular—a celestial body privileged by centuries of her predecessors' work.

We only have a copy of K'uk'ul Ek' Tuyilaj's astronomical work, and it is framed in a hieroglyphic manuscript by content dedicated to different "almanacs" organized by the 260-day calendric device known today as the *chol qiij* or *tzolkin*. Within this content, on six specific pages of text, a copyist of her work hieroglyphically recorded the phrase "Chak Ek'" repeatedly (figure I.10). Often translated as "Great Star," modern scholars understand the referent as the planet we refer to today as Venus (Aveni 2001, 184–96; Förstemann 1906; Thompson 1972). In this book we take up a deep exploration of a copy of K'uk'ul Ek' Tuyilaj's manuscript now commonly known not by an affluent individual, like the Codex Mendoza, but after the city in which it resides: the Dresden Codex.

In the following chapters, this book considers the cultural and documentary contexts from which the Dresden Codex emerged, the challenges scholars have faced over the last two hundred years in attempting to interpret it, and the newly deciphered scientific innovation that I have found buried within it. The chapters herein go back and forth between the words written by ancient Mayan scribes and the interpretations developed by modern scholars, critically examining and contextualizing the work on both sides of the temporal divide. By expanding the scope of the community of discussion around an indigenous science, this book finds gaps that provide new spaces for our ability to represent ancient Mayan scribes, their communities, and their collective endeavors.

ASTRONOMY AND THE DRESDEN CODEX

The personal name of the copyist of K'uk'ul Ek' Tuyilaj's work was lost when the manuscript traveled by colonial ship across the Atlantic in the sixteenth century. Here, we will refer to him as *Tawiskal Uwoojil*—a descriptive name in the tradition of Classic Mayan royal personal names—and he too carries the ajk'uhuun title. How Tawiskal Uwoojil's screen-fold book found its way out of Yucatan and across the Atlantic Ocean is unknown, though Pedro Sanchez de Aguilar's description of the books he confiscated as a Catholic priest in the sixteenth century matches its dimensions well (Thompson 1972, 6) (figure I.11). Two other Precontact manuscripts, the Madrid and Paris Codices, were of similar dimensions, and the former was originally thought to have come from Hernan

FIGURE I.10 The Venus pages of the Dresden Codex, pages 24 and 46–50. Courtesy SLUB

FIGURE I.11 Mayan hieroglyphic codices depicted within an instructional context; Kerr Vase 1196. Courtesy mayavase.com

Cortes. So it is possible that all three were looted from a Mayan library by Francisco de Montejo and his mercenaries during the early sixteenth century and shipped back to Spain as booty (cf. Thompson 1972, 17).

Regardless of the specific route it took, Tawiskal Uwoojil's manuscript was held in private European collections for some two hundred years, through the Colonial period, until Johann Christian Götze purchased it in Vienna for the Dresden Electoral Library in Germany. Cyrus Thomas translated Götze's words from his *Curiosities of the Royal Library at Dresden, First Collection*, as follows: "A Mexican book with unknown characters and hieroglyphic figures, written on both sides and painted in all sorts of colors, in long octavo, laid orderly in folds of 39 leaves, which, when spread out lengthwise, make more than 6 yards" (Thomas 1888, 261). He then referred to the bargain he struck, commenting: "Our royal library has this superiority over all others, that it possesses this rare treasure. It was obtained a few years ago at Vienna from a private person, for nothing, as being an unknown thing. It is doubtless from the personal effects of a Spaniard, who had either been in Mexico himself or whose ancestors had been there" (Thomas 1888, 262). This codex, which had been prized for the ritual and astronomical records it contained while still in Yucatan, was by the eighteenth century now an enigma and an idle curiosity in Europe.

With Germany itself broken up into several kingdoms, and the city of Dresden held by the House of Wettin, Tawiskal Uwoojil's now anonymous book was kept in the repository of the Royal Library when the whole collection was moved to the new Japan Palace down the street; it became known thereafter as the "Dresden Codex." It was in this library, some fifty years after Götze purchased it, that the book took on a new life, under the attention of Ernst Wilhelm Förstemann—a librarian who had just moved over from a post as archivist for the Count of Stolberg-Wernigerode, in central Germany.

Whereas the Codex Mendoza recorded Aztec knowledge and was written explicitly for the purposes of cultural translation and political use, the Dresden Codex was never intended for European eyes. In fact Tawiskal Uwoojil probably did not mean for it to be seen by the eyes of all members of the Yucatec Mayan community he lived in. As we saw in Landa's comments, this was a text for the priesthood, containing privileged knowledge. From Förstemann's perspective, then, it must have been quite the enigma. By his time, the ability to read and write the Mayan hieroglyphic script had been lost (or persecuted out of existence) for over a century. Unlike the Codex Mendoza, no alphabetic glosses were available to make sense of the characters and images he perused.

The latest hieroglyphic manuscripts in use by Mayan communities that we know of were witnessed by non-Mayans during Andrés de Avendaño's seventeenth-century trips from Yucatan to Nohpeten in central Guatemala (figure I.12). Avendaño described his

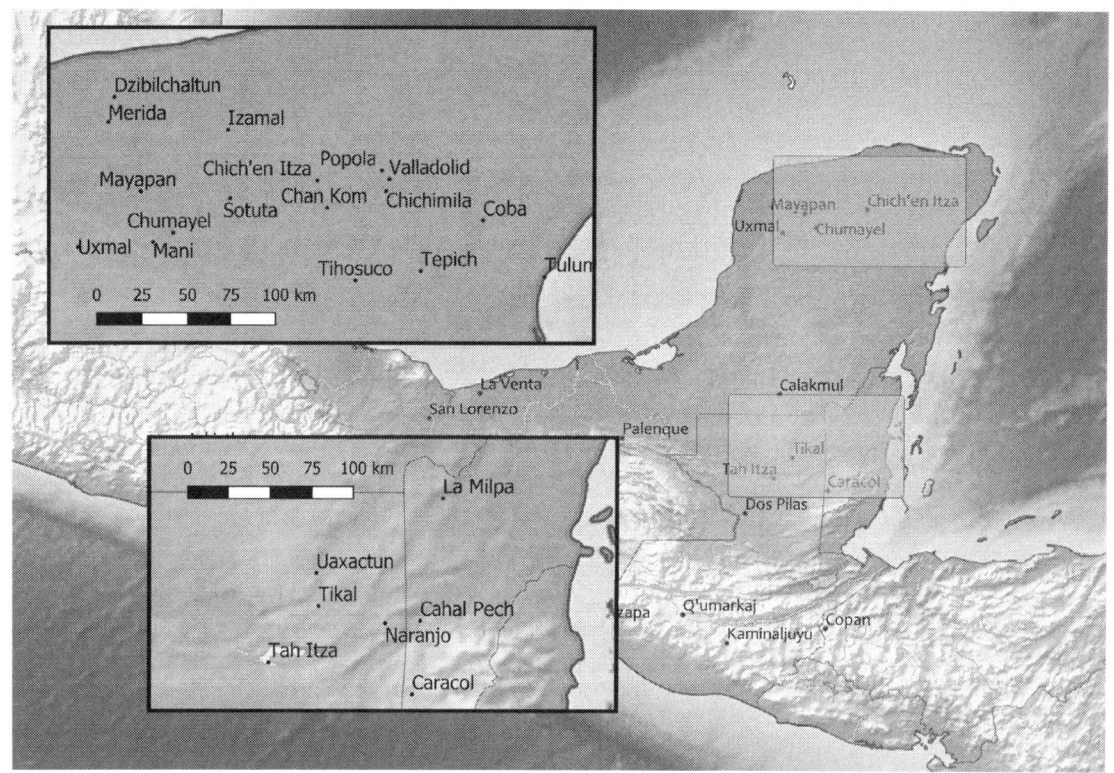

FIGURE I.12 Izamal region in the Yucatan and Tah Itza in the Peten

interest in the books kept by the priests of Tah Itza (Tayasal, or Flores as it is now known) in his accounts of his travels.

> I told them that I wished to speak to them of the old manner of reckoning which they use, both of days, months and years and of the ages, and to find out what age the present one might be (since for them one age consists only of twenty years) and what prophecy there was about the said year and age; for it is all recorded in certain books of a quarter of a yard high and about five fingers broad, made of the bark of trees, folded from one side to the other like screens; each leaf of the thickness of a Mexican *Real* of eight. These are painted on both sides with a variety of figures and characters (of the same kind as the Mexican Indians also used in their old times), which show not only the count of the said days, months and years, but also the ages and prophecies which their idols and images announced to them. (Means 1917, 141)

Avendaño referred not only to the maintenance of these books by Mayan priests in 1695, but also of his own ability to read them:

> Because from the time that I had convinced them by their own ancient computations,—a thing that they considered impossible for any other man except their priests to learn,— they began to love and fear me at the same time; saying that I was undoubtedly a great personage in the service of my Gods, since I had succeeded in learning the language of their ancestors and their own, for from no one else of all these neighboring natives had they heard it, nor did they have any information that the Spaniards who subjugated their lands knew it. (Means 1917, 146)

The ability of Mayans or Spaniards to read books like the Dresden Codex, though, lasted only into the early eighteenth century. With the move of Mayan elites to subaltern positions after the fall of Tah Itza during the Colonial period, leaders of Mayan communities lost the resources necessary to pursue the arts and sciences of socioeconomic luxury, leaving relatively little opportunity for the preservation of hieroglyphic literacy (Farriss 1984). More to the point, the abovementioned Avendaño, Sahagún, Landa, and their compatriot Christian missionaries pressed the project of converting all indigenous people to their religion, which they understood to require the destruction of—among other things—indigenous books. So when Förstemann puzzled through the manuscript in the nineteenth century, there was much for European "Mayanists" to rediscover.

To Förstemann the problem of decipherment was not merely one of cracking a code. During his time, proposals floated about in European and American popular imaginations of Mesoamerican cultures associated with "lost civilizations" of Atlantis, or of Mu (Coe 1999, 106–9). The latter came from the "adventurer" Augustus Le Plongeon, who, unlike Förstemann, managed to travel to Yucatan. Le Plongeon performed his own rudimentary "excavations" in the 1870s with his wife, Alice, and even claimed to be able to read the Mayan hieroglyphic script. Based on his translations, Le Plongeon claimed that the name of a reclining stone human figure was an ancient king named Chaacmol (Desmond 2001, 168). It was also this proposed decipherment that led him to claim that a second Mayan codex, the Troano Codex (one of only three known at the time), contained in it the name of the lost continent "Mu."

Neither Le Plongeon's proposed decipherment nor his proposal of the continent of Mu stood the test of time, but his writings were in circulation when Förstemann ran his fingers over the pages of Tawiskal Uwoojil's manuscript. And they have left a lasting impression—to this day the reclining stone figures at Chich'en Itza are referred to as "chacmools" (Desmond 2001). Similarly, in the United States, Joseph Smith and Brigham Young were incorporating indigenous Americans and specifically Mayan civilization into their own Mormon origin stories throughout the nineteenth century at the time of Förstemann's work. While uncorroborated by secular academic scholarship, Mormon scholars have continued to link ancient Mayans with their "Nephites" up to the modern day (Clark 2005).

Working in the Dresden Library in Germany, without visiting the Western Hemisphere, Förstemann found a way to set much of this popular context aside and, working with this one indigenous manuscript, along with a few secondary sources, he created an interpretive portal deep into the now-unknown Mayan past. His success was tremendous, as we will explore in detail in the chapters that follow. It was certainly recognized contemporaneously as well; by the time of his death his colleagues held the fruits of his labor in the highest regard. In 1907, professor of anthropology at Harvard University Alfred Tozzer wrote:

> We, as American anthropologists, owe a great debt to Dr Ernst Forstemann, the foremost worker in the field of the hieroglyphic writing of Central America, and it is with very great regret that we learn of his death on November 4, 1906, in Charlottenburg, Germany. (Tozzer 1907)

Certainly merited, even Tozzer's estimation would pale in comparison to the impact that Förstemann's work would have on the field of Maya archaeology over the course of much of the twentieth century. Through his study of the Dresden Codex, the German librarian provided convincing evidence to the Western world that whatever else people might want to believe, a mathematically subtle astronomy was important to ancient Mayan religion and cultural production.

It is worth recognizing, then, that if the hieroglyphic writing system had already been deciphered when Förstemann made his calendric discoveries, had it not been lost in the first place, or had he known as much about our scribe's civilization as we do now, the subsequent impact of his work on interpretations of Mayan astronomy would have been much different. The vast majority of the hieroglyphic inscriptions that have surfaced through the archaeological excavations of the nineteenth and early twentieth centuries, for example, would have been seen then (as they are now) as historical in content, describing the births of rulers, their reigns, and their deaths (cf. Martin and Grube 2000). Records of royal visits, meetings of courtiers, across and within political lines, rituals performed to complete time periods, and wars that strained alliances and created others were hieroglyphically carved into stone and placed in both public and private places. Rulers such as Yuhknoom Ch'en, Chak Toh Ich'aak, and Janaab' Pakal—famous throughout Mayan lands in their own times—would have formed the backdrop for Förstemann's work. This in turn would have allowed for the astronomy Förstemann found in the Dresden Codex to be viewed by modern archaeologists as a specialized tool for one segment of a religious (and political) community.

Instead, for Förstemann and his colleagues, the decipherment of the nonnumerical and noncalendric material written by our scribe was still over half a century away, and professional Mesoamerican archaeology was still in its infancy; there was little else to contextualize his finds. And so the opposite extreme took hold. Proposals developed taking the accounting of time and calendric manipulations as the entire relevance of hieroglyphic inscriptions. Ancient Mayan civilization was "obsessed with time" and represented an idealized society in which priests controlled ceremonial centers, and all citizens were humble farmers, coming together at temples only to participate in astronomically and calendrically determined rituals (Aveni 2001, 129; Coe 1999; W. Fash 1991). Under the prevailing view that took hold during the nascent twentieth century, and alongside the fantastic appropriations of Mayan Otherness, the inscriptions—*by inference*—were understood by Europeans and their descendants to record the events of the gods, timed by the movements of the planets, uniquely centered on astronomy.

Aside from generating an inviting esoteric Otherness to explore, this astronomy-centered interpretation produced the convenient implication for many researchers that, even without the ability to read the texts, scholars could find much to interpret in the hieroglyphic record. Looking through inscriptions for astronomical periodicities, for example, would shed light on as much of the written content as many scholars cared to understand. The double-edged sword, then, was that Förstemann's work revealed an impressive astronomical knowledge at the same time it heavily biased future investigation.

Archaeologist and U.S. intelligence officer Sylvanus Morley, for one, took up this interpretation and managed his efforts accordingly. Describing the results of his first journey to Copán (Honduras) in 1912, Morley stated explicitly:

> Unlike the inscriptions of every other people of antiquity, the Maya records on stone do not appear to have been concerned—at least primarily—with the exploits of man, such as the achievements of rulers, priests, or warriors—in short, with the purely personal phenomena of life; on the contrary, *time in its many manifestations was their chief content.* (Morley 1920, 33, emphasis added)

By the first decades of the twentieth century, the work that upheld Förstemann's esteemed reputation had been appropriated into a growing model of Mayan civilization as being centered on astronomy, influencing even the methods researchers built into their further investigations.

In his insider's history of the decipherment of Mayan hieroglyphic writing, *Breaking the Maya Code*, Michael Coe (1999) attests the impact of Morley's astronomical focus on his final reports on Copán and his subsequent forays into the region of the Classic period florescence, the Guatemalan Peten—that area around Tah Itza, which Avendano had visited two hundred years earlier. Coe notes that Morley photographed some hieroglyphic texts (albeit with lower quality than his peers), but in other cases he provided only substandard line drawings (Coe 1999, 129). And rather than include complete records of the inscriptions he recovered—in the style of his predecessors Alfred Maudslay or Teobert Maler—Morley preserved only the portions of the inscriptions that contained dates (Coe 1999, 129). Instead of providing a robust set of archaeological records, Morley prioritized the recovery of just enough data to further his and his colleagues' astronomical hypotheses.

For some time thereafter, well into the first half of the twentieth century, astronomers, engineers, and enthusiasts of various backgrounds joined forces to comb through the dates that Morley and his colleagues accumulated, finding within them what looked like even more celestial information. As we will see in the chapters of this book, for instance, the chemical engineer John Teeple tinkered with patterns in the inscriptions at Morley's urging during the mid-1920s (Coe 1999, 130). Teeple quickly made a name for himself as a Mayanist, alongside his professional success as a chemist, by deciphering lunar records (1925a) without reading a single noncalendric hieroglyph; the numbers and calendric symbols gave him all he needed.

Sylvanus Morley's personal impact was even greater than that of the methods he took into the field. Morley extended his perspective of Mayan civilization further into the twentieth century as he led the Carnegie Institution's archaeology division and excavations at Chich'en Itza for the next two decades. The archaeological projects he authorized and the archaeologists he hired perpetuated his approaches. Morley eventually passed the torch of placing astronomy at the center of Mayan culture to Eric Thompson, whom he had hired to work on architectural reconstruction at the site of Chich'en Itza in northern Yucatan. Thompson gained and then fiercely held on to a central position of authority within Maya Studies through much of the middle and late twentieth century (Coe 1999, 123). From this position, we hear echoes of Morley's perspective in Thompson's major oevre, *Maya Hieroglyphic Writing: An Introduction*:

> [B]ecause of the deprecatory attitude toward individual assertiveness which characterizes Maya culture, glyphic inscriptions on the monuments, unlike those of almost every other civilization in the history of mankind, almost certainly do not record the deeds of individuals; instead, they are utterly impersonal records of calendarial and astronomical data and of religious matters. (1978, 15)

Thompson had spent much more effort than his former boss on the epigraphy of the writing system, but this did not impact his emphasis on astronomy.

> It is my conviction that we shall interpret the glyphs only by relying heavily on the beliefs, the religious symbolism, the mythology, and, to a lesser extent, the everyday activities

of the Maya, because such concepts surely are imbedded in the structure of each glyph. (1978, 35).

Since from this perspective there was no way to read the text other than to interpret the esoteric ideological meaning hidden within it, the assumption that the esotera was anchored to astronomy could not be readily challenged. This helps to explain, for instance, how Thompson initially discredited the Russian linguist Yuri Knorosov's hypothesis concerning the hieroglyphic writing system. By finding even small internal discrepancies in Knorosov's work, Thompson could argue that its logic was flawed (Coe 1999, 152). And since there was already an impressive popular and internally consistent interpretation that the material Knorosov was trying to read was "demonstrated" to be astronomical, it would be difficult for anyone to sustain any challenge.

And so it wasn't until Tatiana Proskouriakoff "beat them at their own game" that the "ideographic," nonlinguistic interpretation of the hieroglyphic writing system was breached legitimately and the release of Mayan civilization from astronomically centered interpretations was made possible (cf. Coe 1999, 167–84).

In the late 1950s, Proskouriakoff followed the practices of her predecessors—of Teeple, Morley, Thompson, and others—to initiate her study by collecting a set of dates and looking for patterns within them. But Proskouriakoff was far subtler in approach. With her artistic eye, she brought together those dates from public stone monuments (*stelae* pl.; *stela* sing.) at Piedras Negras bearing the same visual theme: "astronomical gods" sitting in celestial thrones (1960, 455). She writes:

> The niche is framed by a "sky band" of astronomical signs, with a grotesque bird above, and a two-headed monster below, a combination of symbolic figures that I will refer to as the "cosmic motif." (Proskouriakoff 1960, 455)

When she cross-referenced the time intervals against specific hieroglyphs on monuments constrained by iconographic theme, Proskouriakoff found history—not astronomy. Specifically, Proskouriakoff found that the time intervals recorded on these stelae were interpreted better as representing periods between royal births and accessions, between one accession and the next, between one death and the accession of a successor (1960, 460). While she did not challenge the fact that these and other hieroglyphic inscriptions certainly included records of the Moon—Teeple's Lunar Series, for example, did not go away—that was not the monuments' reason for being. Under Proskouriakoff's analysis, astronomical records in the inscriptions provided data for more "complete" dates to accompany the *historical* events of importance in rulers' lives.

Having personally witnessed Proskouriakoff working on her argument, Michael Coe found it extremely compelling, as did many of his colleagues in the 1960s (Coe 1999, 171–76). Even Thompson conceded at the end of his career that Proskouriakoff was right. But the point for our work here is that it wasn't until the latter third of the twentieth century that a different framework could emerge to set a new context for "Mayan

astronomy." For the scholarship of the second half of the twentieth century, Proskouriakoff's work countered the assumption that the dates only corresponded to religious
records; this in turn allowed the "ideographic" hypothesis to give way to a phonetic
hypothesis in line with the beleaguered Knorosov's intervention. The latter permitted a
new generation of epigraphers to begin reading the names of rulers, their genealogies,
and the various events they participated in (Coe 1999). When the decipherment resulting from the insights of Knorosov, Heinrich Berlin, and Proskouriakoff were adopted
by a new generation of epigraphers and matured in the 1960s, 1970s, and 1980s, the
prevailing astronomical hypothesis was finally unseated.

Throughout this period of change, however, Förstemann's work on the Venus Table
was not challenged. In fact his basic interpretations of what would become known as
the Venus and Eclipse Tables in the Dresden Codex were consolidated and extended
several times well beyond the initial decipherment era, even into the early 2000s. To
some degree, this has allowed palimpsests of early twentieth-century astronomically
centered interpretations to remain in some scholarly literature even today, and they
certainly feed representations in popular media. In addition, scholars have continued to
find patterns that they argue were astronomically motivated underlying political events,
even when the deciphered texts themselves make no such mention (Bricker and Bricker
1986a; Closs 1978; Dütting et al. 1982; Milbrath 1995; Schele and Freidel 1990). Behind
these studies—sometimes deeply implicit, other times explicit—the Venus Table and its
observational basis in the Dresden Codex serve as the legitimation. It is both the explicit
and deeply implicit biases that I work toward recalibrating with this book.

TRANSLATING SCIENCE

Given its place in history and the attention it has received, one might be tempted to
believe that the early twentieth-century interpretation of the Dresden Codex Venus
Table is ironclad. But there are still unsolved puzzles within it (Thompson 1972; Bricker
and Bricker 2011). And there are incongruities that really do not stand up to close scrutiny, which have yet to be addressed. This book shows that some of these questions can
now be answered through a more robust use of the decipherment of the hieroglyphic
writing system.

Chapter 1 initiates the project by providing the basic interpretation of the Dresden
Codex Venus Table generated by Förstemann at the end of the nineteenth century.
Rather than approach it as an astronomical device that can be explained as a sort of
technical manual, this chapter nests each major interpretive insight within the historical
context of the investigator generating it. It begins, therefore, with the earliest numerical decipherment of Dresden Codex pages 46–50 as a "Venus Table" by Förstemann,
walking through the Mayan calendric system and the naked-eye astronomy necessary
to understanding its operation. The chapter moves through the early twentieth century to consider the significant contributions of Mesoamericanist Eduard Seler and the

chemical engineer John Teeple in order to get to the basic interpretation of the Venus Table that is current in the literature today.

Chapter 2 shifts to the hieroglyphic history recorded by the ancient Mayan rulers of the Classic period sites known today as Tikal and Dos Pilas, providing an appreciation of their culture that Förstemann did not have access to. Here we consider the political history well known to modern epigraphers that treats the betrayal of stepbrother by stepbrother with the assistance of the most powerful k'uhulajaw in the Mayan realm. For our purposes, B'ahlaj Chan K'awiil's hieroglyphic record of military strife provides insight into one role of astronomy within royal self-representation and so within Late Classic period intellectual history. It therefore provides a context for the scientific material we should expect to see leading up to the material recorded by Tawiskal Uwoojil and the Mayan astronomers of the Late Classic and the Postclassic periods. In the process of examining B'ahlaj Chan K'awiil's use of astronomy for political reasons, we encounter astronomical records at other Classic Mayan sites, which together demonstrate a trajectory of activity, speaking to the sense that, during the Late Classic, references to the celestial realm had become fashionable among the Mayan elite. In other words, we develop evidence that astronomical activity most likely was motivated not by some (pre-) modern sensibility of scientific accuracy but by the esoteric extravagances of a significantly expanding almehen class.

The representation of astronomy coming out of chapter 2 appears to counter the interpretation of Venus astronomy consolidated by John Teeple in the 1920s, which suggested a more mechanical, ephemeris-based Mayan astronomy. Chapter 3 considers the history of the Venus Table's academic treatment after Teeple's work to investigate this apparent contradiction, demonstrating that an academic complication occurred during midcentury. Here we find that through Eric Thompson's intervention, the Venus Table was no longer considered of its own accord; it had been appropriated and implemented as a tool within the larger (modern) problem of correlating the ancient Mayan calendar with Christian (Julian and Gregorian) chronologies. With a brief digression into nuclear physics and the invention of radiocarbon dating, we find in chapter 4 that Thompson's move has since prevented substantive revision of the interpretation of the Dresden Codex Venus Table because one element of Teeple's interpretation became necessary for the proposed solution to the Maya Calendar Correlation Problem. With subsequent scholars taking Thompson's solution as proven, new interpretations were built on it, further sedimenting his interpretation of calendrics, astronomy, and the Dresden Codex Venus Table. Chapters 3 and 4, therefore, suggest that the modern interpretation of the Venus Table has been unnecessarily encumbered by the Calendar Correlation Problem and might be better served by consideration free of it.

The investigation of hieroglyphic references to astronomy along with the possibility of considering Venus apart from the Calendar Correlation Problem yields the opportunity in the next chapter to reconsider a specific Late Classic period Venus record. In order to explore the possibilities of interpretation without the anchor to modern methodological interests, chapter 5 looks to other independent hieroglyphic records that provide explicit historical context for the Venus Table. A critical one comes from the Late Classic site of Copán and the reign of the sixteenth member of the local dynasty, Yax Pahsaj Chan Yopaat. His architectural patronage made use of the same hieroglyphic phrase to record a Venus event as the one Tawiskal Uwoojil used hundreds of years later, demonstrating that it held a broader cosmological scope than has been previously recognized.

Yax Pahsaj Chan Yopaat's text sets up an exploration of the operative verb behind the Dresden Codex Venus Table in other linguistic applications to find that it represents a ritual concept resonating throughout Mesoamerica. In doing so, it reintroduces the practice of indigenous ceremonial processions. And so here we recognize that the decipherment of the writing system does shed important new light on Tawiskal Uwoojil's Venus Table. By reading the text, we find that the description of Venus in the Dresden Codex fits within the description of various other Classic Mayan ceremonial activities—it is not a radical outlier. By the end of this chapter, we find that at the end of the Late Classic, there appears to have been a close connection between Yax Pahsaj Chan Yopat's Venus astronomy at Copán and the astronomical practices documented in political activities of other rulers, such as those exemplified in chapter 2. The Dresden Codex Venus Table at this point appears aligned much better with the types of politics in which B'ahlaj Chan K'awiil found himself embroiled, and it seems similar to a ceremonial practice widespread throughout Mesoamerica.

Having pulled the interpretation of the Dresden Codex Venus Table away from the needs of modern scholarship and having presented a consistent use of astronomy through the Late Classic, next we more closely examine the noncalendric content of the Venus Table. Chapter 6 goes back to the hieroglyphic text in the manuscript to place its content and accompanying illustrations into dialogue with the material culture of Tawiskal Uwoojil's world during the Postclassic period. We find there that the manuscript references the types of incense burners that have been archaeologically excavated from the Terminal Classic and Postclassic sites of Chich'en Itza, Uxmal, and Mayapán. This leads us to other pages in the Dresden Codex written by Tawiskal Uwoojil and his peers, which refer to similar rituals and similar material culture, now bringing diachronic cultural continuity in line with the synchronic continuity found throughout the text. We also return to the connection to the ritual practices encountered by the sixteenth-century bishop of Yucatan, Diego de Landa, and the processions still enacted within rural indigenous communities in Mexico and Central America. All of this material leads to the conclusion that what we have in Tawiskal Uwoojil's Venus Table is not an ephemeris of Venus per se, but an application of Venus periods to the construction of a calendar of ceremonial events performed at Chich'en Itza in accord with Postclassic international ritual activity. And such ritual activity would perforce have been patronized by the almehen class and the ceremonial leaders of the day—the Dresden Codex Venus Table would not have been immune to political influence and effect.

So it is within a ceremonial context that chapter 7 considers the different forms of activity that the Venus events described in the Dresden Codex would have supported.

Going back to the (rest of the now deciphered) text, we see that the outcome of all of the Venus events is represented by the same glyph: *muuk*, or "omen." Like the nightly practice of timekeeping, the Venus events in the Dresden Codex all generate oracular knowledge, which then requires specific ritual response. Here, then, we turn to consider Mayan indigenous divination practices documented from the time of Contact through the twentieth century (in the ethnography, for example, of Barbara Tedlock among the communities of Highland Guatemala). Resonating with the interpretation of Mayan astronomy that I investigated at Palenque in my first book (Aldana 2007), we again find that there are both public aspects of these ritual Venus events at Chich'en Itza and more private elite uses. Both are explored in this chapter.

Chapter 8 brings the revised interpretation of Tawiskal Uwoojil's Venus Table to a new purpose. Here we find that page 24 of the Venus Table preserves the record of an actual astronomical discovery at Chich'en Itza. Even though it appears that we have moved away from Mayan astronomy as "science" proper, we find that the recovered ritual practice required careful observation and record keeping—consonant with centuries of timekeeping and other basic astronomical duties reviewed at the beginning of this introduction. These in turn would have necessitated some level of mathematical computation and correction over long periods of time. In fact it is precisely these characteristics that produced the basis for Mayan astronomical discovery—not unlike how Renaissance European needs to reform the Julian calendar led to Nicolas Copernicus's "mathematical hypothesis" of heliocentrism. This chapter walks through the modern discovery of K'uk'ul Ek' Tuyilaj's ancient Mayan discovery of Venusian movement, as well as an ingenious method for taking into account the variability of Venus observability over the short run, while maintaining considerable accuracy on average over the long run.

In chapter 9, we recognize that K'uk'ul Ek' Tuyilaj's depiction on the Caracol Disk ties her astronomical observations to a pan-Mesoamerican cultural phenomenon sweeping through the region. The Quetzalcoatl affair has drawn attention from scholars and enthusiasts for centuries, and with a treatment of Venus astronomy at Chich'en Itza, we find ourselves drawn into the debate. By contextualizing K'uk'ul Ek' Tuyilaj's efforts within the political, religious, and scientific developments of the Terminal Classic, this chapter sheds new light on the cultural innovation of the Feathered Serpent. Here we follow recent scholarship arguing for a Quetzalcoatl priestly sodality emerging in Central Mexico and spreading east into the Mayan region. Now, though, we see that it was likely drawing on K'uk'ul Ek' Tuyilaj's astronomical discovery that Chich'en Itza's nobility chose to participate in a Feathered Serpent–inspired hegemony, thus converting a small city into a regional civic-ceremonial center.

Chapter 10 follows up on the fact that the Venus Table contains both a copy of an earlier table of dates and an updated version. That updated version brings us to the Postclassic city of Mayapán and a political context that would have supported the type of ceremonial activity recorded in the final version of the Dresden Codex. That is to say that in this chapter we encounter a historical context that supported the construction and use of Tawiskal Uwoojil's manuscript. In this chapter, the codex comes closer to looking like a Postclassic Mayan document, not a curiosity found on a shelf within a European library.

This book closes with chapter 11 as a more reflective piece, considering the implications of taking up cultural translations of science. Here I suggest that there is an opportunity in the investigation of cultural translations of science—and not simply to pat our (collective post/modern) selves on the back and suggest that other cultures were attempting to create proto-science versions of our own. Instead, we can appreciate and learn from sciences within their own contexts and so find the values and cultural characteristics embedded within them. This in turn (as with all substantive ethnic studies and anthropological scholarship) aims to shed new and critical light on our own contemporary situation and the use of science and technology within modern cultures and societies.

I refer to this project as an attempted recalibration, then, because it entails much more than simply filling in a scaffold of dates with newly deciphered events. As has been recognized since the adoption of Knorosov's phonetic hypothesis in the late twentieth century, it is one thing to decipher rigorously a hieroglyphic phrase, sentence, or paragraph on a linguistic basis, but it is another project to access the meaning of that text. To access meaning, we have to be able to translate across cultures, across time, and across authorial interests.

Every scholar of ancient Mayan astronomy has struggled with a resonant problem at some level. We can see mathematical consistency generating patterns, but we have trouble accessing that pattern's meaning. Eric Thompson attempted to explain away complications by appealing to astrology. In his final publication on Mayan hieroglyphic writing, Thompson writes:

> I further believe that the calculations they made to find precedents for the various combinations which would influence the end of a katun occupy the bulk of the inscriptions. That is to say they were using astronomy to develop laws of astrology. . . . It was a scientific approach, but the original premises were false. (1978, 64)

But is "astrology" the right term either? And should we expect that they were looking for its "laws"? To make this distinction is already to impose our own (post-)modern categories of knowledge on ancient Mayan scribes. There was, as one parallel, no distinction in ancient Greece between the terms *astrologia* (αστρολογία) and *astronomia* (αστρονομία), nor does Tawiskal Uwoojil include any implication that he was working with such a distinction.

On the other hand, as we work within a postdecipherment consideration of Mayan intellectual activity, we should not beleaguer Thompson as though he were completely unaware of the challenges he accepted in constructing his interpretations. At some level, he understood the problem very well:

I am not unmindful of the pitfalls in the path of one who would stray in the tangled woods of Maya mythology. There is too much danger of finding what one seeks, for many opposed ideas exist in the religious concepts of the peoples of Middle America, and one is free to pick and choose. (1978, 35)

Thompson is here referring to a problem of representation. As Subaltern Studies scholar Gayatri Spivak has pointed out, there is a natural dual purpose to any attempt at representation: on one side, it illustrates (image as a representation); on the other, it advocates (replacement as representative). This duality is present in the earliest European records of Mesoamerica (cf. Restall 1999) through today.

Scholars who are interested now in constructing representations of any indigenous peoples of the Americas are often pulled by one side to counter negative representations, which have become imbued within Western common sense. Neither indigenous cultures nor communities are served by popular culture representations, for example, that focus solely on human sacrifice, esoteric rituals, and/or uncontextualized astronomy. But there is another tension arising from the "other side." There is an entire industry of representations capitalizing on the uncritical celebration of indigenous people as well. "The Maya" are perhaps the greatest victims/beneficiaries of this modern cultural phenomenon. In popular representations, they are celebrated as so advanced that they were not actually human, but alien—a higher form of consciousness arising within an alternate species. In other cases, their scientific knowledge is said to have outpaced even our own modern science, and they were able to develop astronomical models that predicted events thousands of years into their future with accuracies to within minutes.

These romanticized interpretations of ancient indigenous people to either extreme have left modern scholars with a relatively narrow band of banality within which to explore ancient cultural expressions. Some scholars certainly have tested the boundaries of this band—such as Linda Schele and David Freidel, calling themselves "edgewalkers"— but most scholarly representations remain safely within interpretations that don't stray far from our long-term "commonsense" notions of these cultures.

In my efforts to take the problem of representation seriously, I have relied in this book on a three-pronged approach. The first is certainly a reliance on mathematical and astronomical coherence in the treatment of indigenous scientific activity. The second is a robust contextualization anchored firmly to hieroglyphic texts, but reaching out into other forms of cultural production. The third prong is a connection to lived indigenous experiences. While the first two are found in most every archaeoastronomical study, the third presents a greater challenge, especially when the work aims to incorporate modern Mayan or indigenous Mesoamerican experiences. And while I am not alone in suggesting that some of this access should come from a connection to lived indigeneity today, by no means am I suggesting here the type of straightforward cultural continuity proposed, for example, by David Freidel and Linda Schele in their attempt to reconstruct a pan-Mayan religion:

> Modern Maya live in a metaphysical, philosophical, and religious tradition bridging the
> ages, from the time of the Classic kings to the shamans and ritual leaders of today; from
> sovereign states of old with their huge royal capitals embracing tens of thousands to the
> villages and small towns that now encompass the lives of most Maya. . . . The Preco-
> lumbian world of Middle American gave birth to a great universalizing tradition just as
> successful as that of the European Old World, and the Maya cultivated one of the most
> eloquent of its variants. (Freidel et al. 1993, 40)

My own work breaks from such an essentializing approach to seek a form of continuity in subtler form—one of cultural palimpsests.

We will see in the chapters that follow a foundational concept underlying the ritual maintenance of space and time that speaks directly to the kind of continuity "gluing" together the vignettes of the preface with the rituals supervised by Tawiskal Uwoojil and K'uk'ul Ek' Tuyilaj. The goal is to bring them into conversation with each other, without essentializing them through their dependence on one calendric or astronomical concept or practice. In some ways, then, a primary goal of this book is to use this opening through the 260-Day Count and its relationship to ceremonial activity to work toward a more robust cultural translation, which on the one hand requires a reconsideration of even the assumptions behind current work, and on the other hand opens up a new perspective on ancient Mayan cultures.

Ernst Förstemann and the Dresden Codex Venus Table

To close his landmark 1894 article on the Dresden Codex Venus Table, Ernst Förstemann imagined its author at work. Förstemann's herculean intellectual labor had turned a barely known manuscript into a powerful tool for accessing ancient Mayan culture. As he summarized this work, he attempted a first-person summary of what he thought the Mayan "writer desire[d] to say":

> I am here treating especially the periods consisting of five successive Venus years, bringing them into harmony with the solar year and the [260-Day Count]. I am at the same time considering a second important period, that in which the two heavenly bodies of the second class, the moon and Mercury, come together in their orbits, a period made up of four unequal parts. Just in the same way is each individual Venus year divided into four unequal parts, which appertain to the east, north, west, and south and are ruled by certain deities, which I can mention only in part, owing to lack of space. Lastly, I would add that each of the five Venus years of a period is dominated as a whole by a deity, and the signs of these I give here. (1894, 443)

Förstemann's imagination was powerful enough to avoid the exoticism of Mayan cultures common at the time to capture much of what constitutes the basic interpretation of the Venus Table today. Although the connections to Mercury and the Moon have mostly faded away, the (1) division of Venus's period into observation-based subperiods, the (2) assignment of those subperiods to the cardinal directions (north, west, south, and east), and (3) an emphasis on one deity for each of five Venus Rounds—these are at the core of modern interpretations (Bricker and Bricker 2011; Lounsbury 1978; Thompson 1972). In this chapter, we must forget any connection to Tawiskal Uwoojil or K'uk'ul Ek' Tuyilaj and the fruits of late twentieth-century decipherments. We now turn to the close of the nineteenth century to see how Förstemann came to formulate his interpretation and how others then extended it.

INDIGENOUS TEXTS

The seventy-four-page "Mexican book" in the Dresden Library that Förstemann began working on in the 1880s had already been partially published in 1810 by Alexander von Humboldt. In hindsight, we might wonder why Humboldt wasn't the European scholar to decipher the Venus Table instead of Förstemann. Having returned from a tour through Latin America where he met members of various indigenous communities in South America and Mexico and made a name for himself as a naturalist and geographer, Humboldt wrote up his experiences in *Vue des Cordilleres* (1810) to great acclaim. His reputation as an explorer and scientist made him a celebrity in Europe and the Americas, and his work set new standards in the natural sciences (Humboldt et al. 2012).

As he moved over land and sea, Humboldt didn't keep his eyes at the terrestrial level; he also took careful measure of celestial events in order to improve European understandings of the geography of the Americas.

> From the whole of the observations which I made in 1799 and 1800 it follows, that the latitude of the great square at Cumana [Venezuela] is 10 deg 27' 52", and its longitude 66 deg 30' 2". This longitude is founded on the difference of time, on lunar distances, on the eclipse of the Sun on the 28th of October, 1799, and on ten immersions of Jupiter's satellites, compared with observations made in Europe. (Humboldt and Bonpland 1815, 394)

Venus too captured Humboldt's careful gaze, leading him to note: "I sometimes even observed, between the fifteenth degree of latitude and the equator, small haloes around the planet Venus; the purple, orange and violet were distinctly perceived" (Humboldt and Bonpland 1815, 395).

Humboldt, therefore, would have had an excellent set of experiences and expertise from which to draw when he encountered the Dresden Codex; from it he included facsimiles of pages 47, 48, and 50 through 52 in *Vue des Cordilleres*. In fact he was even in an amenable frame of mind, challenging European scholars who denigrated the cultures of the "New World":

> Since the turn of the century, a happy revolution has taken place in our conception of the civilizations of different peoples. . . . The monuments of Egypt, which are nowadays described with admirable exactitude, have been compared to monuments in the most distant lands, and my study of the indigenous peoples of the Americas appears at a time when we no longer consider as unworthy of our attention anything that diverges from the style that the Greeks bequeathed to us through their inimitable models. (Humboldt et al. 2012, xv [and 2]).

So Humboldt had been exposed to a multitude of indigenous languages, he was sufficiently versed in astronomy to participate in accurate recordings, and he was invested in finding worthy scholarly material in the cultures of indigenous peoples. But when it came to his description of the very pages in the Dresden Codex that Förstemann would identify as the Venus pages, he had very little to say.

Humboldt's manuscript described the purchase of the codex by Götze and the book's overall physical dimensions. He noted that the characters looked more like Egyptian hieroglyphs or Chinese characters than anything in the other Mexican codices he had seen. He did venture so far as to propose that the bars and dots might record numbers (Humboldt and Williams 1814, 146–47), but his reluctance to speculate in the manner of the scholars he criticized kept him from commenting any further. "Some scholars have devoted their time to generating theories that, although brilliant, are founded on shaky ground, and they have therefore drawn general conclusions from a small number of isolated facts" (2012, 4). Humboldt's genius and reputation would be built on his observations of the natural world and not in the careful analysis the manuscript would have required.

It was left to Förstemann the librarian, then, drawing on his father's specialization (a teacher of math far to the northeast of Dresden at Gdansk), to actually decipher Mayan numeration, which would lead to the Venus patterns contained on the pages Humboldt reproduced (figure 1.1). Fortunately, in coming to terms with math and calendrics,

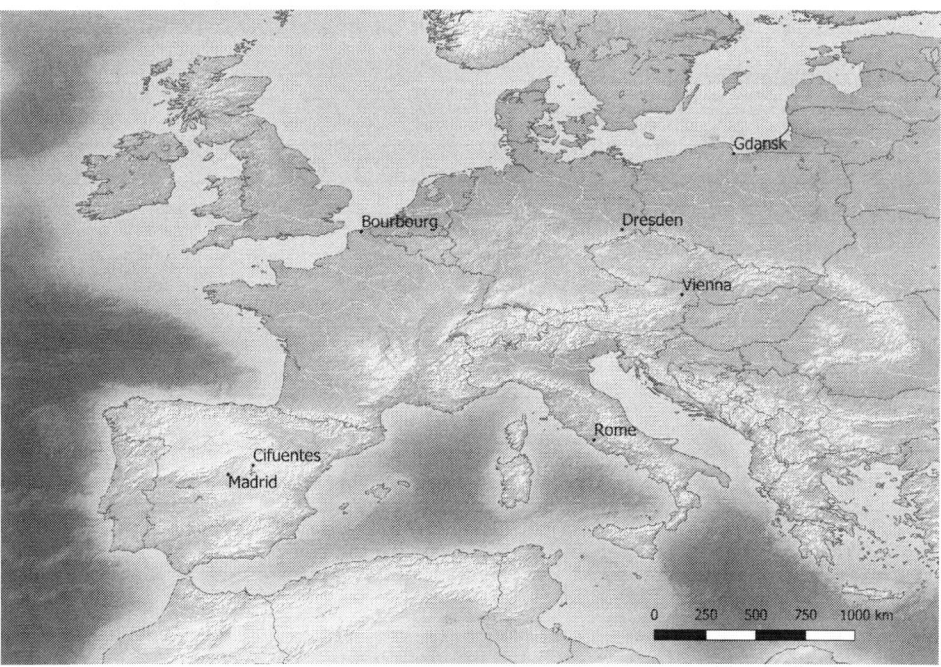

FIGURE 1.1 Förstemann was raised in Gdansk, then moved to Dresden. Also noted are Brasseur de Bourbourg's home in France, Landa's hometown of Cifuentes, and Madrid where Landa's *Relación de las cosas de Yucatán* was found.

Förstemann did not have to start from scratch. In fact, not long before he took over the duties of the Royal Librarian, a French missionary, Charles Etienne Brasseur de Bourbourg, had published three critical documents also brought back to Europe, with origins in the Mayan region. Two of these manuscripts had been written in the sixteenth century—well over a century after Tawiskal Uwoojil wrote up his Venus pages—and the third was a transcription of an oral narrative.

While Brasseur de Bourbourg's contributions have been important to serious scholarship since the nineteenth century, his own personal motivations again breached into what we now consider the nonacademic. Brasseur de Bourbourg left his home in France as an adult, and as an already accomplished writer, to become ordained as a Catholic priest in Rome. He then traveled to Canada before pursuing his interest in moving to Central America as a missionary (Coe 1999, 99). In Mexico and Central America, he combined the tactics of the travel adventurers, such as Le Plongeon and Stephens and Catherwood, with the methods of his priesthood predecessors, studying and recording indigenous languages. These pursuits came together for Brasseur de Bourbourg in the unique form of seeing Mayan civilization as the link between the Old World and the New via the lost continent of Atlantis (Coe 1999, 106). His descriptions of the site of Palenque in Chiapas, Mexico, illustrated with the drawings of Frederick Waldeck, demonstrated what Humboldt had noted decades earlier and what still plagues popular media today: it was nearly impossible for these men to appreciate the indigenous cultures they encountered without tracing them back to Egypt or Greece.

Nevertheless, Brasseur de Bourbourg's travels and linguistic efforts did provide the documents that would become essential to scholarly treatments of Mayan cultures since he encountered them in the 1860s. Perhaps best known is a copy of the *Popol Vuh*— the Creation mythology of the K'iche' Mayan elders written shortly after the conquest of their kingdom at Q'umarkaj, adjacent to Santa Cruz K'iche', in what is now highland Guatemala (figure 1.2). Another, earlier priest, Francisco Ximenez, had encountered a version of this manuscript not far south of Q'umarkaj in Chichicastenango during the first decade of the eighteenth century (cf. Coe 1999, 99). Ximenez transcribed the text in the original K'iche' and translated it into Spanish, but it remained relatively unnoticed for another century. While staying in Guatemala City, less than one hundred kilometers southeast of Chichicastenango, Brasseur de Bourbourg encountered Ximenez's manuscript at the library of the University of San Carlos (D. Tedlock 1985, 30). The *Popol Vuh* has since become an anchor for scholarly consideration, although it is often mischaracterized as "the Mayan Bible" in part because it contains a long description of the Creation of the World (cf. Coe 1999, 99). We will return to the *Popol Vuh* in chapter 8 of this book, with special attention to its reference to a Feathered Serpent leader in its historical section.

A second text came from another K'iche' source in Guatemala: Brasseur de Bourbourg transcribed a play—now known as the *Rabinal Achi*—from a native speaker (Coe 1999, 99; Graham 2002, 143). Much more important for Förstemann back in Dresden was that, upon his return to Europe, Brasseur de Bourbourg found an abridged and

FIGURE 1.2 Santa Cruz K'iche', Guatemala City, Chichicastenango, and Rabinal in the highlands of Guatemala

edited copy of Diego de Landa's *Relación de las cosas de Yucatán* in an archive in Madrid (Coe 1999, 100).

Even though Brasseur de Bourbourg is a complicated figure in Maya Studies, the bishop Diego de Landa is perhaps even more so. Landa was only twenty-five years old in 1549 when he left Cifuentes (some eighty miles east of Madrid, Spain) and arrived in Yucatán. Once there, he took up residence at a monastery under construction atop the abandoned buildings of the Precontact Mayan religious center of Izamal (see figure I.12). The Catholic Church's practice of building religious architecture on locally sacred ground was already common evangelical practice by Landa's time. The Spanish governance of Yucatan was no exception, with churches built in Maní, Mérida, and Izamal by the end of the sixteenth century.

In 1563, however, Landa was forcibly removed from New Spain and condemned by the Council of the Indies in Seville for the violent abuse of indigenous people under "his jurisdiction" (Coe 1999, 101; Tozzer 1941, x). Curiously, only nine years later, he was absolved of his crimes and sent back to serve as bishop of Yucatan. So while his peers themselves had trouble characterizing Landa's "contributions," so have modern scholars. What no one contests across the temporal divide is that Landa's persecution of indigenous religious leaders was extreme and excessive; his book-burning auto-da-fé

FIGURE 1.3 Landa's "alphabet" of Mayan writing in his *Relación de las cosas de Yucatán*, along with two sentences: *ma inkati*, meaning "I do not want to," and *e le e, le*, for the spelling of the word *le* for *lasso*.

is legendary. Of course, neither can any contest that the records he produced made possible the recovery of much of the very material he aimed to destroy.

For his acquisition of information, Landa relied on local experts. Two Mayan nobles in particular were critical to his efforts. From a lineage originating in the Postclassic city of Uxmal and settling in the west of Yucatan around the town of Maní, the local indigenous priest and teacher of Latin, Gaspar Antonio Xiu, was summoned by Landa to work with him at Izamal (Tozzer 1941, vii) (see figure I.12). Landa also explicitly referred in *Relación* to his conversations with Juan Nachi Cocom, who traced his lineage back to the eastern city of Chich'en Itza, but who, in the sixteenth century, resided only twenty kilometers south of Izamal in the town of Sotuta (Tozzer 1941, vii). From discussions with these two men, Landa went beyond his own moralizing and descriptions of local phenomena; he included what he could understand of Mayan hieroglyphic writing, including the "alphabet" that Brasseur de Bourbourg and Le Plongeon used for their interpretive work (figure 1.3). Alongside this material—and critical for Förstemann—Landa included a description of the local calendric system.

Förstemann acknowledged his debt to Landa explicitly:

> In [Landa's] manuscript were found the signs of the numerals from 1 to 19, the twenty day signs of the 20-day period, and the eighteen signs of the periods of this kind which make up the year. All these signs, apart from numerous variants, were actually met with again on the inscriptions and in the manuscripts, so that by the discovery of [Landa's] manuscript the corner stone was laid, and building could proceed. (Förstemann 1904, 501)

In Landa's text, Förstemann found what he would need to investigate the numerical and calendric patterns in Tawiskal Uwoojil's manuscript. It certainly took some doing—

going back and forth, visually comparing the drawings in Landa's text against the hiero-glyphs of the bark-paper book—but the calendar, including part of the numerical system, is probably the material that first became visible. Here we work through this back-and-forth on our way to reconstructing Förstemann's work on the Venus Table.

THE 260-DAY COUNT

For the "signs of the numerals," Förstemann was referring to the Mayan practice of using dots to represent "1" each, and a bar to replace five dots—that is, to hold the value of "5." Förstemann found them throughout the codex, and they are what Humboldt referred to in his brief description in *Vue des Cordilleres* as noted above. With this notation, Tawiskal Uwoojil and his colleague scribes painted the number 13, for example, with two bars lying horizontally, one above the other, and three dots in a row above and parallel to the two bars (figure 1.4). Depending on the hieroglyphic context, a scribe might rotate this representation ninety degrees counterclockwise to stand the bars on end with vertical paint strokes and the dots aligned vertically on their left.

the Mayan Number 13

Förstemann's indebtedness to Landa also included his familiarity with the calendric "260-Day Count" made up of a combination of thirteen numerals with a set of twenty "Day Signs." Still used by *ajqiij* ("daykeep-ers"; cognate of *ajk'in*) in the highlands of Guatemala, the 260-Day Count is called the *chol qiij* ("ordering of the days") in modern

FIGURE 1.4 Example of the Mayan number 13 on page 32 of the Dresden Codex. Courtesy SLUB

K'iche' Mayan, and referred to as the *tzolkin* in much of the modern scholarly literature (Aveni 2001; Lounsbury 1978; D. Tedlock 1985; Thompson 1972). In his treatise, Landa included a daily sequence of 260-Day Count dates correlated to a full Julian year from the first of January through the thirty-first of December, which therefore shows the count in operation (figure 1.5).

This list shows that the 260-Day Count worked by having the numerical coefficient and the Day Sign each advance daily. Starting with the date made up of the coefficient "1" and the Day Sign "Imix," for example (in Landa's manuscript falling on January 29), the next day would be 2 Ik' (on January 30), followed by 3 Ak'bal, 4 K'an, 5 Chikchan, and so on until reaching 13 Ben (on February 10). With only thirteen numbers available as coefficients, the date after 13 Ben would be 1 Ix—that is, the coefficient resets to 1 while the Day Sign advances to Ix. The latter is only the fourteenth Day Sign, so the progression continues on February 12 with 2 Men, then 3 Kib, 4 Kaban, and so on until 7 Ajaw (on February 17). Now Ajaw is the twentieth Day Sign, so the next date would

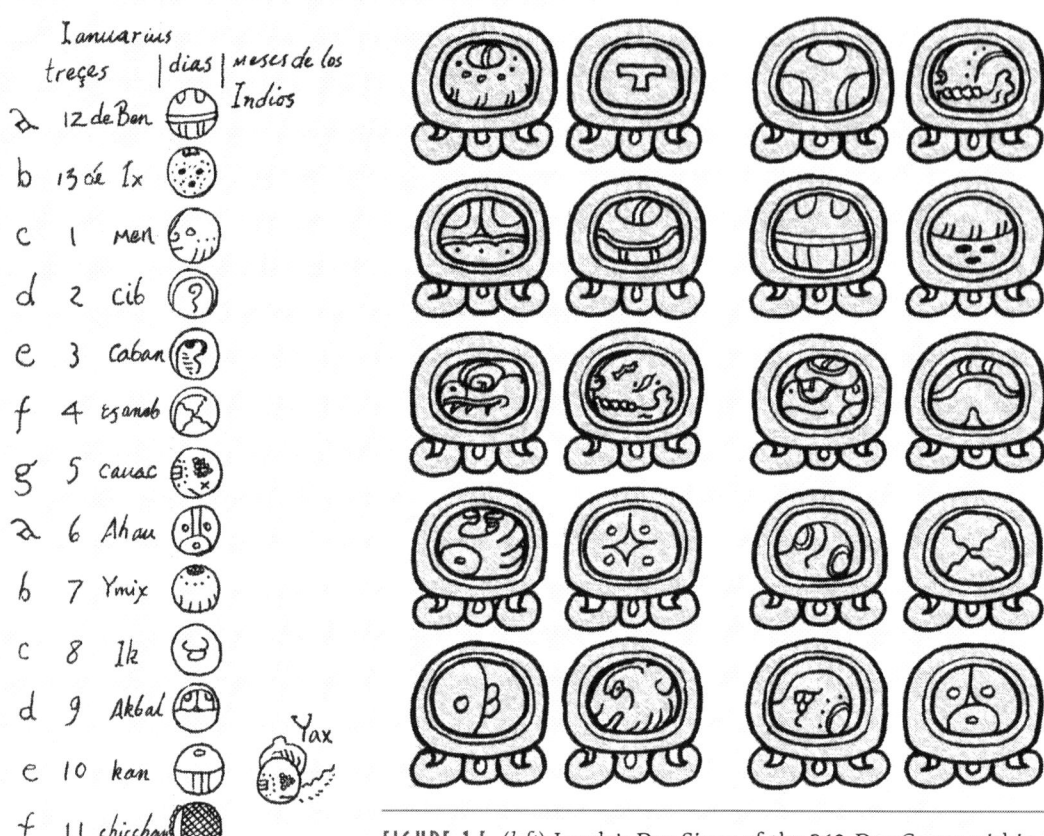

Ianuarius
treçes | dias | meses de los Indios
a 12 de Ben
b 13 de Ix
c 1 Men
d 2 cib
e 3 Caban
f 4 tzanab
g 5 cavac
a 6 Ahau
b 7 Ymix
c 8 Ik
d 9 Akbal
e 10 kan
f 11 chicchan
g 12 cimij
a 13 Manik
b 1 lamat

Yax

FIGURE 1.5 (*left*) Landa's Day Signs of the 260-Day Count within his list of dates in *Relación de las cosas de Yucatán*; (*above*) examples of the Day Signs of the 260-Day Count from Classic period carved inscriptions in paired columns. The first two columns read in zig-zag fashion: Imix, Ik', Ak'bal, K'an, Chikchan, Kimi, Manik, Lamat, Muluk, and Ok. The third and fourth columns also in zig-zag fashion read: Chuwen, Eb, Ben, Ix, Men, Kib, Kaban, Etz'nab, Kawak, and Ajaw.

start the Day Signs over to move to 8 Imix, which is then followed by 9 Ik', 10 Ak'bal, 11 K'an. Two hundred and fifty-nine days after 1 Imix, the count would come to 13 Ajaw, to be followed by a return to 1 Imix after two hundred and sixty days. In Landa's sequence, this would have occurred on October 16—260 days after January 29.

In the middle section of pages 29 and 30 of the Dresden Codex, referred to in the modern literature as the material on pages 29b–30b, Förstemann encountered dates in the same count that Landa had learned from Gaspar Antonio Xiu and Juan Nachi Cocom (figure 1.6). Although the Day Signs in the manuscript were not identical to those appearing in the Dresden Codex, several were similar enough to make the case clear. Comparing Landa's drawings to the middle Day Sign on page 29b, the third from the top, would demonstrate the similarity between the renditions of the Day Sign Etz'nab in both cases (figure 1.7). Förstemann could then use the rest of the text and imagery of

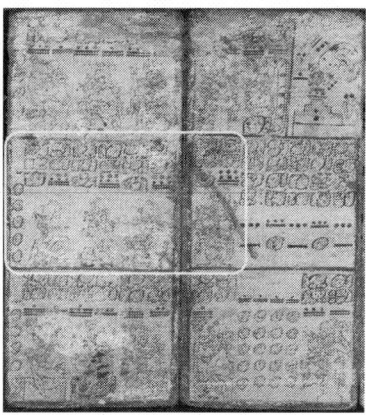

almanac for Chaak,
Pages 29b – 30b

FIGURE 1.6 Dresden Codex pages 29b–30b. The Chaak almanac runs across the full length of the middle of page 29, but then takes up only the first register of page 30b. The second part of 30b—with hieroglyphic text and numbers only—initiates a distinct almanac. Courtesy SLUB

the Day Sign
Etz'nab

FIGURE 1.7 Identification of Day Sign Etz'nab within a column of Day Signs on page 29b of the Dresden Codex. Only two dots are shown, but the coefficient was originally three, with the leftmost dot eroded. It is therefore denoted 3* in the description below. Courtesy SLUB

pages 29b–30b to verify that the functioning of the 260-Day Count in the Dresden Codex was the same as that in Landa's manuscript.

Specifically, in the middle register of page 29, Förstemann could see that the scribe painted a column of Day Signs positioned below a row of three red dots. If he took these red dots to be the coefficients of the Day Signs, then the first date in the sequence would be 3 Ix. On this date, he would look to the right to encounter the first pairing of hieroglyphic text with corresponding image below it. The image that Tawiskal Uwoojil painted here is of the storm deity hieroglyphically named Chaak (rendered: CHAK-ki). Chaak is recognizable by his long nose, the dotted curl below his eye, his sloping forehead with a single dot in his brow, and the catfish-looking barbell curving out of the side of his mouth (figure 1.8). Here Chaak sits with crossed arms on a bench that is marked with two symbols: *ak'bal* for darkness and a crossed bands icon representing the sky. For the text above Chaak, Tawiskal Uwoojil wrote *walaj lak'in Bakab Chaak* (wa-la-ja la-K'IN-ni ba-ba-ka CHAAK-ki), which we can translate now as "Chaak the Bakab was stood up in the East," though of course Förstemann would not have known that. The text was written in conventional zig-zag reading order in paired columns. For the third row of text, there is a hieroglyph for

		was stood up	in the East
		Bakab	Chaak
	3*	bread	3 13
	Ix		
	Kimi	representation of Chaak sitting on a skyband seat	
	Etz'nab		
	Ok		
	Ik'		

	is stood up	in the East	is stood up	in the North	is stood up	in the West	is stood up	in the South
	Bakab	Chaak	Bakab	Chaak	Bakab	Chaak	Bakab	Chaak
3*	bread	3 [Manik] after 13 days	fish	3 [Ajaw] after 13 days	iguana bread	3 [Ben] after 13 days	turkey bread	3 [Kimi] after 13 days
Ix								

FIGURE 1.8 We can now read Dresden Codex Pages 29b–30b: (*top*) First register of the almanac along with the column of Day Signs, anchoring the progression. Chaak is depicted seated on a throne; the text tells us that he is "stood up" in the east with the title Bakab. The food offered to him is bread, and the event occurs on 3 Ix. (*bottom*) The Day Signs through the sequence are suppressed through each run, but confirmed by each anchor of 52 days (or 4 × 13). A narrative read through the text would be "On 3 Ix, Chaak as Bakab is enthroned in the east. His offering is bread. After 13 days, on 3 [Manik], Chaak as Bakab is enthroned in the north. His offering is fish. After 13 days, on 3 [Ajaw], Chaak as Bakab is enthroned in the west. His offering is iguana bread. After 13 days, on 3 [Ben], Chaak as Bakab is enthroned in the south. After 13 days, on 3 [Kimi]." By ending on the suppressed Day Sign Kimi, we are directed to the next row of the Day Sign list in the first column to find that Kimi is just below Ix. The red coefficient of 3 thus corresponds to Kimi for this row of 52 days, and all remains the same with the exception of the four suppressed Day Signs, which now are Kawak, Eb, Chikchan, finishing with Etz'nab, which initiates the next row. It takes a total of 260 days (5 rows of 52 days each) to return to the initial date of 3 Ix. Images courtesy SLUB

bread (as in tamal or tortilla) with something stuffed into it in the left column. To the right, there are two numbers: a red "three" above a black "thirteen." The numbers and the calendric sequence are what Förstemann would have been able to crack.

Taking the black 13 as a time interval, Förstemann added thirteen days to the first date of the column "3 Ix" to move past the 260-Day Count dates 4 Men, 5 Kib, 6 Kaban, 7 Etz'nab, 8 Kawak, 9 Ajaw, 10 Imix, 11 Ik', 12 Ak'bal, 13 K'an, 1 Chikchan, and 2 Kimi, arriving at the date 3 Manik. Here he would have noticed some useful patterns. For one, adding thirteen to the coefficient is the same as adding zero: the 3 as coefficient of Ix and Manik doesn't change. Also, moving 13 Day Signs after Ix, we come to Manik, having passed through the last Day Sign, Ajaw. In shorthand, then, Förstemann found that the text could be "read" as: 3 Ix + 13 = 3 Manik.

The next text-image register of Tawiskal Uwoojil's manuscript gives us another red 3 and black 13. Following the same practice, we advance 13 days after 3 Manik to the date 3 Ajaw. The sequence continues through the next two registers, yielding 3 Ben for the third and then 3 Kimi for the fourth. This fourth 260-Day Count date is the final one in the sequence. Förstemann may have hypothesized that he should go back to the beginning—to the first column—and start again. If we do, we find that we now have moved from 3 Ix to 3 Kimi by going through four time intervals of 13 days each for (4 × 13 =) 52 days. This fourth register, therefore, would have revealed to Förstemann the operation of this small section of the codex.

Coming all the way back to the first column of dates on page 29b, underneath the red number 3 and the Day Sign Ix, Tawiskal Uwoojil painted the Day Sign Kimi. If we take the red number 3 at the top of this column as assigned to each Day Sign, then this means that after 3 Ix, we arrive at 3 Kimi, which is precisely what we get when we add four intervals of thirteen to the date 3 Ix.

For this sequence of dates, then, each Day Sign in the column serves as the anchor for a progression through the registers of hieroglyphic text and images in a type of almanac. Since there are four registers with 13-day intervals in each, that gives us a 52-day time interval between rows in the first column, and sure enough, Förstemann found that as dates, the column could be reconstructed as 3 Ix + 52 days gives 3 Kimi + 52 days gives 3 Etznab + 52 days gives 3 Ok + 52 days gives 3 Ik'. This would have confirmed his reliance on Landa's text for the continued interpretation of his Dresden Codex and given him confidence that there was a coherent mathematical structure underlying much of the text in the book. The author that Förstemann had imagined was adept in the same calendric system as Landa's consultants. Overall, Förstemann could make significant headway without having to rely on a decipherment of the language itself.

We will see in later chapters that the hieroglyphic text of these passages is as (or more) important as the mathematical coherence in peering into Tawiskal Uwoojil's world. For now, it is worth recognizing that for Landa's advisors, Xiu and Cocom, the 260-Day Count was not an abstract, inert computational tool. Dates in the 260-Day Count carried oracular qualities beyond their timekeeping capabilities (cf. Aveni 2001, 140). For comparison, we might analogize the Day Signs to Gregorian days of the week, which themselves carry "omens" of sorts: "TGIF," Monday as the first day of the workweek, Sunday as a day of rest. There is also something of a numerical analogue in that, in the United States, Friday the thirteenth carries a particular cultural meaning distinct from, say, Friday the fourteenth or Tuesday the thirteenth (La Farge 1947, 9).

This analogy only goes so far, however, as notable in the divination practices recorded by European witnesses during the Colonial period, and by ethnographers of the Mayan highlands from the early twentieth century through today. In these cases, the Day Signs are seen as being animate—they are spoken to directly through ritual practice with expectations of response. Barbara Tedlock recorded one type of divination practiced in Guatemala during the early 1980s. She writes:

> When all the seeds of the first arrangement have been set out, the diviner addresses the first group of seeds by the name of the day on which he is divining, for example, . . . "Come here Lord 1 Quej [1 Manik], you are being spoken to." . . . After the first day has been addressed and summoned, the diviner will repeat or allude to the question being asked. In a marriage divination, one might say, . . . "1 Quej, you are being spoken to about the seven goodnesses, seven fatnesses" (B. Tedlock 1992, 162).

Through divination rituals, the Day Signs are given agency within the lives of community members. Codices from Central Mexico attest to similar divination practices with the 260-Day Count deep into Precontact times (Jansen and Perez Jimenez 2005, 17–19). We will return to consideration of the divinatory nature of the 260-Day Count in chapter 6.

These symbolic and oracular values likely are what have made the 260-Day Count the most robust of calendric records, existing in the earliest of records from Preclassic Oaxaca (around the third century BC) and still being used in the highlands of Guatemala today, often without any other vestiges of the "full" Mayan calendar. In the case of the Dresden Codex, the 260-Day Count dominates, speaking to its purpose of governing the timing of ceremonial events and hence its "sacred" character. Most of Förstemann's exploration of Tawiskal Uwoojil's codex was utterly dependent on patterned progressions through the 260-Day Count.

THE 365-DAY COUNT

There was a second calendric tool described by Landa and critical to Förstemann's work on Venus: the 365-Day Count. This count is analogous to the Gregorian tropical year and is often considered to be the "secular" counterpart to the 260-Day "Sacred" Round (Aveni 2001, 145) (figure 1.9). We encountered it briefly in the last chapter relative to timekeeping practices at night, which speaks to its development. Landa's full run through a Julian calendar year, noted above, demonstrates the persistence of the same structure over time. It is clear that in it eighteen "months" contain twenty days each to produce a period of 360 days. A final period of five days completed the cycle.

FIGURE 1.9 Classic period inscriptional representations of the months of the 365-Day Count. Read in paired columns, the sequence depicts the 20-day months: Pohp, Wo, Sip, Sots', Tsek, Xul, Yaxk'in, Mol, Ch'en, Yax, Sak, Keh, Mak, K'ank'in, Muwaan, Pax, K'ayab and Kumk'u. The 5-day period Wayeb is the glyph of the final row of the third column.

It is worth noting explicitly that the operation of the 365-Day Count differs from that of the 260-Day Count. Whereas both parts of the 260-Day Count advance each day, only the coefficients advance daily within a given month. The following sequence illustrates the difference:

4 Ajaw 8 Kumk'u

+ 1 day

5 Imix 9 Kumk'u

+ 1 day

6 Ik' 10 Kumk'u

+ 9 days

2 Chuwen 19 Kumk'u

+ 1 day

3 Eb Seating of Wayeb

+ 1 day

4 Ben 1 Wayeb

+ 3 days

7 Kib 4 Wayeb

+ 1 day

8 Kaban Seating of Pohp

+ 1 day

9 Etz'nab 1 Pohp

Notice that the "Seating" of a month corresponds to its inception and in the scholarly literature is often given the shorthand notation of a "0" coefficient. In the latter run, either the Seating of Pohp or 1 Pohp may have been considered the first day of the New Year.

Unfortunately, Landa's narrative of the 365-Day Count is rather convoluted, which may reflect the editing of later copyist-compilers who produced the version that Brasseur de Bourbourg retrieved from the Madrid archive (Restall and Chuchiak 2002; Tozzer 1941, viii). The text, for example, states that leap years were built into the Mayan year—that is, that an extra day was inserted into every fourth year in order to preserve a specific date relative to the solstices and equinoxes. While it is clear that tropical years were recognized and probably held ritual significance as far back as the Preclassic—as exemplified architecturally (see, e.g., Aveni 2001, 288–93)—the 365-Day Count records we have in hieroglyphic texts demonstrate clearly that it was allowed to wander through the tropical year during Precontact times (Aveni 2001, 145). Precontact progressions through the Mayan calendar, that is, did not include leap-year accommodations. A second complication in the Landa manuscript concerns the "beginning" of the year, but that had no impact on Förstemann's work, so we will return to that issue in a later chapter as appropriate.

In addition to the list of dates, Landa's manuscript included Xiu's and Cocom's descriptions of the rituals that were to occur during each twenty-day month of the 365-Day Count during the Late Postclassic. The first month of the year, Pohp, for example, was a time of renewal in which "they swept out their houses, and the sweepings and the old utensils they threw out on the waste heap outside of town, and no one, even were he in need of it, touched it" (Tozzer 1941, 151–52). Continuing through the activities of all eighteen months, Landa notes that the last five-day "month" was a time of rest (Tozzer 1941, 166).

With all of this calendric material, thanks to Brasseur de Bourbourg's discovery of the Landa manuscript, Förstemann was able to make relatively quick headway into a number of calendric patterns within Tawiskal Uwoojil's codex without ever leaving his library desk in the city of Dresden.

CRUNCHING NUMBERS WITH FÖRSTEMANN

The key to Förstemann's lasting contribution—the discovery that raised his reputation to that lauded by Alfred Tozzer—was his find that in addition to the combination of the 260-Day Count and the 365-Day Count into a "Calendar Round," Xiu and Cocom described to Landa their use of large numbers (i.e., greater than 19). While Landa recorded no graphic examples, he did note that numbers were written according to powers of twenty.

	two rows of hieroglyphic text	
four columns of hieroglyphic text		
	illustration of a (wounded) reclining figure	

11	4	12	0
16	10	10	8

FIGURE 1.10 Examples of "stacked numbers" in the Dresden Codex at the bottom of page 49. The same numbers appear in the same order and the same positions on each of pages 46–50. Images courtesy SLUB

Their count is by fives up to twenty, and by twenties up to one hundred and by hundreds up to four hundred, and by four hundreds up to eight thousand; and they used this method of counting very often in the cacao trading (Tozzer 1941, 98).

This alone would have been useful to Förstemann in his recovery of what we now refer to as the Mayan Long Count.

Throughout the Dresden Codex, Förstemann encountered Mayan bars and dots, not serving as calendric coefficients, but arranged in stacks (figure 1.10). Within these stacks, he was able to work out that their arrangement reflected a positional base-20 notation as Landa described. Unlike the red and black numbers we saw in the earlier example, these were stacked numbers all in the same color. Across the bottom of page 49, for example, Förstemann could decipher four stacks (in red paint) as an 11 above a 16 (i.e., two bars and one dot above three bars and one dot), a 4 above a 10, a 12 above a 10, and a zero above an 8. Although there are many possible interpretations, one readily recoverable pattern starts with the lower number residing within a "ones" place (20^0) and the upper number within a "twenties" place (20^1). The first number composed of $(11 \times 20) + (16 \times 1)$ gave him $220 + 16$ or a sum of 236. A little more arithmetic would reveal to him that the bottom section of numbers was equivalent to the sequence 236, 90, 250, and 8. This interpretation made sense of Landa as reference and of the numbers given. It would not, however, have given him any proof that his reconstruction was accurate.

Above these four stacks, though, in the far-left column, Förstemann found another 11 above a 16, which seemed to repeat the number below it. But the next column did not repeat the 4 above a 10; instead he encountered a 16 above a 6. Following the base-20 positional notation, this translated to 326. In Arabic numbers the pattern was intriguing: the second column in the top row reflected the sum of the bottom two intervals: $326 = 236 + 90$ (figure 1.11).

The next interval, however, introduced an important complication. His first look probably had Förstemann see the third time interval in the upper section as an 11 above a 16. That interpretation, though, was betrayed by an apparent gap between the two bars, representing 10, and the single dot above it. The representational suggestion was

		1	1
11	16	10	11
16	6	16	4
6 Sak	18 K'ank'in	4 Yax	12 Yax

four rows of hieroglyphic text			

16 K'ayab	4 Sots'	14 Pax	2 K'ayab
11	4	12	0
16	10	10	8

236	326	576	584
6 Sak	18 K'ank'in	4 Yax	12 Yax

four rows of hieroglyphic text			

16 K'ayab	4 Sots'	14 Pax	2 K'ayab
236	90	250	8

FIGURE 1.11 The bottom left-hand side of page 46 of the Dresden Codex table. Images courtesy SLUB

that the single dot represented a third register—the Mayan equivalent of a "hundreds place," but here the four-hundreds place (20^2). Part of that hunch was corroborated by a glance to the right. The next number was clearly a 1 above an 11 above a 4. To this point, he probably thought he could simply extend the pattern he reconstructed. But the math would not have worked out straightforwardly in either case. If he took the gap as meaningful and the dot within the register for the next power of 20 (reflecting Landa's text), then he would have $1 \times 400 + 10 \times 20 + 16 \times 1 = 616$. This was not the sum of 326 and 250, though, which would be 576. Fortunately, Förstemann needed only a little algebra for his critical insight.

Assuming the first two columns provided the right pattern, Förstemann had a coefficient of 1 for an unknown interval to be summed with the result of $10 \times 20 + 16 \times 1$. This would follow the pattern if it were equivalent to the sum of 326 and the bottom interval of 250—that is, 576. Algebraically, he confronted: $1 \times T + 216 = 576$, where T is the unknown base of the third register. Solving this equation would have given Förstemann $T = 360$. He could then check this hypothesis against the interval of the next column: 1.11.4. Using T as 360, the time interval in the far right column is $1 \times 360 + 11 \times 20 + 4 = 584$, which is the sum resulting from the pattern, $576 + 8$. This discovery ended up working in every representation of these large numbers throughout the manuscript. Förstemann thus worked through the anomaly of the third register in this positional notation to rediscover the Mayan "Long Count" (table 1.1).

Förstemann therefore realized that the stacks of bars and dots throughout the Dresden Codex could be translated easily into tables of numbers. Once he identified their structure, Förstemann pushed further to find three different uses of Long Count numbers (cf. Aveni 2001, 136–39): (1) Calendar Round associated (CRa) dates, (2) time inter-

TABLE 1.1 The Mayan Long Count

Traditional Mayanist Terms	Relative Value	Hieroglyphic Terms	Modern Equivalents
1 k'in		= 1 k'in	= 1 day
1 winal	= 20 k'in	= 1 winik	= 20 days
1 tun	= 18 winal	= 1 haab	= 360 days
1 katun	= 20 tun	= 1 winikhaab	= 7,200 days
1 baktun	= 20 katun	= 1 pih	= 144,000 days

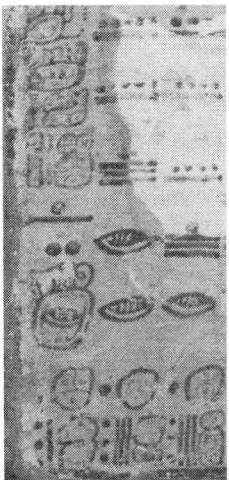

		9	9
partial column of text		9	9
		16	9
	6 2	0	16
	0	0	0
	4 Ajaw	1 Ajaw	1 Ajaw
	8 Kumk'u	18 K'ayab	18 Wo

FIGURE 1.12 Three forms of Long Count representations: a Ring Number, translated as an underlined "0," or 6.2.0, counted from the anchor of 4 Ajaw 8 Kumk'u; a time interval of 9.9.16.0.0; and a Long Count with Calendar Round of 9.9.9.16.0 1 Ajaw 18 K'ayab. Courtesy SLUB

vals, and (3) negative numbers. Each of these types of numbers played an important role in Förstemann's work on the Venus Table.

The CRa dates were characterized by having five periods (i.e., baktun through k'in), by having an 8, 9, or a 10 in the baktun register, and by being accompanied by a Calendar Round date (figure 1.12). These CRa dates were all anchored to a "zero date" falling on 4 Ajaw 8 Kumk'u. To see a progression with this count, we note that one baktun after that "zero date" would have been *jun pih, mih winikhaab, mih haab, mih winik, mih k'in*, or one baktun, zero katun, zero tun, zero winal, and zero k'in on the Calendar Round 3 Ajaw 13 Ch'en. Scholars quickly developed a "shorthand" notation for Long Count dates in alphanumeric characters as pih.winikhaab.haab.winik.k'in. Hence, one pih after the "zero date" would be found in modern scholarly literature as the date 1.0.0.0.0 3 Ajaw 13 Ch'en. This date would be followed in another 144,000 days by 2.0.0.0.0 2 Ajaw 3 Wayeb and eventually 8.0.0.0.0 9 Ajaw 3 Wo. The historical period of the Classic Maya started around 8.15.0.0.0 after their mythological zero date, or 1,260,000 days after. The last days of the Classic period ended around 10.0.0.0.0, meaning that Classic Mayan history lasted for approximately 180,720 days, or just shy of five hundred years.

The second type of Long Count record in the Codex was the "time interval" (also known in the literature as "distance numbers"). These could be of any length, from only days (k'in) mediating between 260-Day Count dates (see figure 1.10), to the huge "Serpent Numbers" on pages 61 and 62, reaching 12,488,821 days or over 34,000 years. As on page 49, Tawiskal Uwoojil had written four of these time intervals at the bottom left of page 46, aligned within the four columns of the entire page. Above these, Förstemann noticed that three rows of 365-Day Count dates were aligned with them (figure 1.13). Moreover, Förstemann found that this association extended from page 46 into page 47 and on to page 50. He could now work through more complicated sequences of the calendric material in Tawiskal Uwoojil's manuscript.

On page 46, for example, in the upper row of 365-Day Count dates, Tawiskal Uwoojil painted 4 Yaxk'in and 14 Sak in the first two columns. Underneath the 14 Sak date, in the same column, he painted a time interval of 90 days (four dots in the upper register and two bars in the lower register). By adding 90 days to 4 Yaxk'in, we arrive at 14 Sak. This pattern extends across all five pages, until we reach 13 Mak in the last column of the row on page 50. But this simply starts the cycle all over again since returning to the first column of the row on page 46, we find the time interval of 236 below the 4 Yaxk'in date and 13 Mak + 236 days yields 4 Yaxk'in. This too Förstemann should have expected

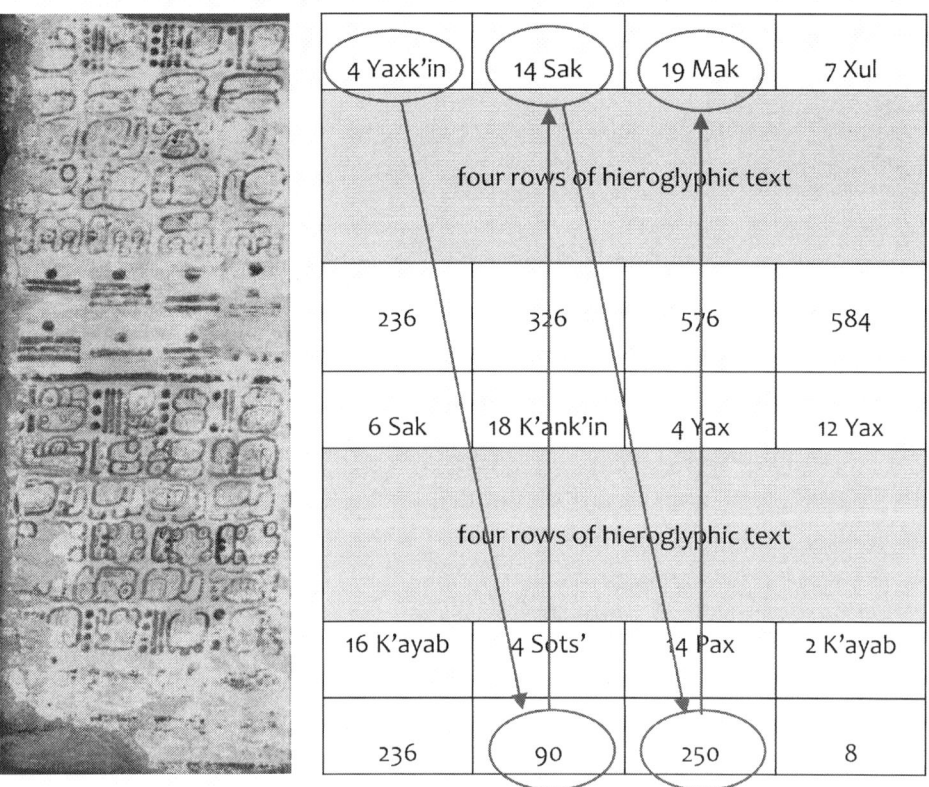

FIGURE 1.13 Working with 365-Day Count dates and time intervals on pages 46–50. Adding 90 days to 4 Yaxk'in leads to 14 Sak; then adding 250 days to 14 Sak leads to 19 Mak. Courtesy SLUB

since the full sequence of intervals across the five pages covered 2,920 or 5 × 584 days. Since 2,920 is also a multiple of 365 (by 8), the full row over five pages could be recycled for any 365-Day Count date—again, like the almanac on pages 29 and 30 treated above.

Förstemann actually found that he was able to extend this pattern further to include every Calendar Round date in these four columns on the left-hand side of the page. The upper block of text—the first thirteen rows—constituted a table of 260-Day Count dates. These too were separated by the time intervals he had recovered, so that this block of dates does for the 260-Day Count what the row below it does for the 365-Day Count. The date in the first row and first column on page 46, for example, is 3 Kib; the date in the column to its right is 2 Kimi (figure 1.14). As above, 3 Kib summed with the time interval of 90 days yields 2 Kimi. Again, the pattern extends until the last date in the first row on page 50, which is 9 Ajaw.

The main difference between the block of 260-Day Count dates and the row of 365-Day Count dates is that 73 is a common factor of 584 and 365, but 584 and 260 do not

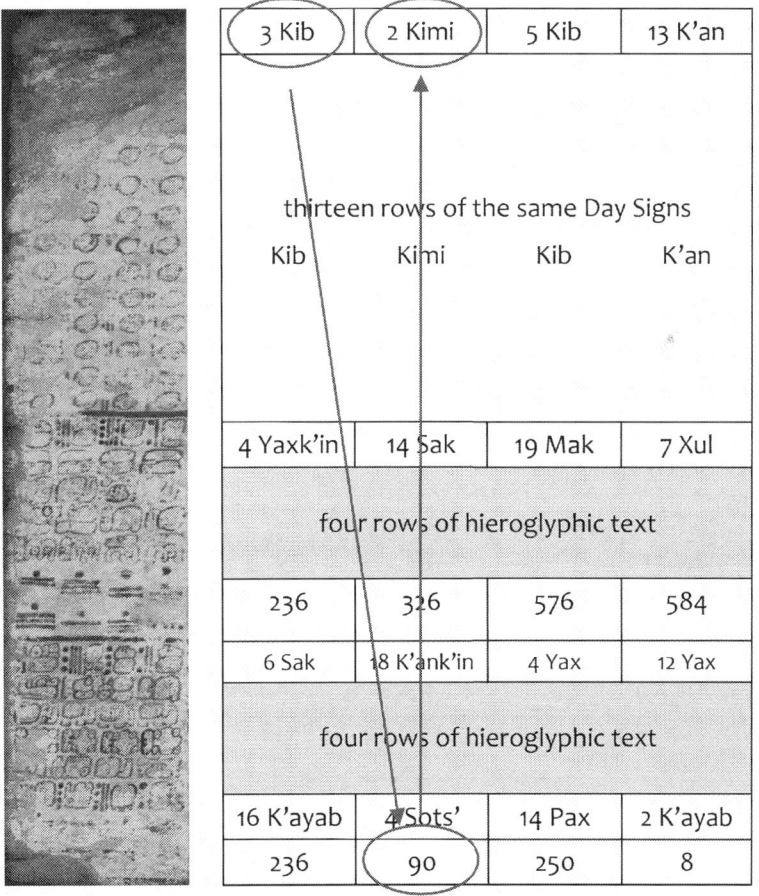

FIGURE 1.14 Working with 260-Day Count dates and time intervals on pages 46–50. Adding 90 days to 3 Kib leads to 2 Kimi. Courtesy SLUB

recycle for 37,960 days (or about 104 years); the latter amounts to 146 rounds through the 260-Day Count. This is why so much of each page is dedicated to 260-Day Count dates. Only one row of 365-Day Count dates is necessary because one row captures 5 × 584 = 8 × 365 = 2,920. But every 2,920 days, the 260-Day Count shifts in coefficient while keeping the Day Sign consistent. All thirteen coefficients must be cycled through, hence the thirteen rows of 260-Day Count dates and only one row of 365-Day Count dates.

Looking at the big picture of these pages, each 260-Day Count date in a column would have been paired with the 365-Day Count date in the same column to create a valid Calendar Round date. The time interval at the bottom of the page could be added to that Calendar Round date to generate the next Calendar Round date to its immediate right (Förstemann 1891, 118) (figure 1.15). To aggregate our above work, then, the first

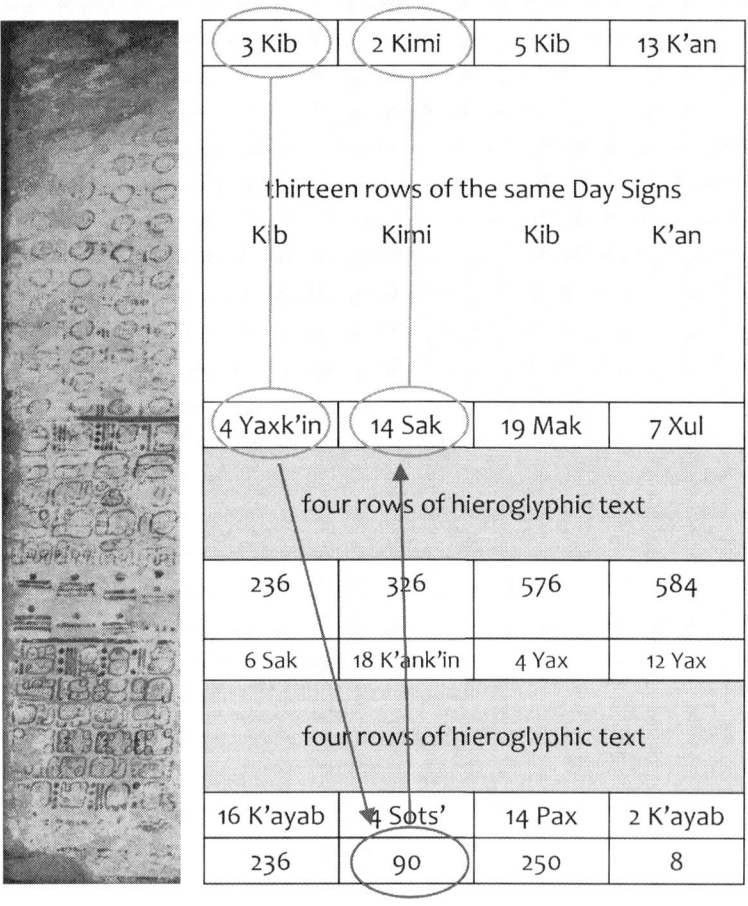

FIGURE 1.15 Assembling Calendar Round date progressions through pages 46–50. Adding 90 days to the Calendar Round 3 Kib 4 Yaxk'in leads to 2 Kimi 14 Sak. Similarly, 2 Kimi 14 Sak + 250 days = 5 Kib 19 Mak, and so on. At the end of the first row of 260-Day Count dates on page 50, the sequence continues with the second row on page 46. Courtesy SLUB

| 6 |
| 2 |
| 0 |
| 4 Ajaw |
| 8 Kumk'u |

FIGURE 1.16 "Ring Number" on Dresden Codex page 24. The ring in red ink signifies that the Long Count number counts back from the Calendar Round date below it. It represents the date 6 haab, 2 winik, and 0 k'in before 4 Ajaw 8 Kumk'u, or the date at −6.2.0. For the computation using the Ring Number base on page 24: to the Long Count date 6 haab 2 winik 0 k'in before 4 Ajaw 8 Kumk'u, add 9 pih 9 winikhaab 16 haab 0 winik 0 k'in to reach 9 pih 9 winikhaab 9 haab 16 winik 0 k'in, which is the Calendar Round 1 Ajaw 18 K'ayab. Or −6.2.0 + 9.9.16.0.0 = 9.9.9.16.0 1 Ajaw 18 K'ayab. (The 1 Ajaw 18 Wo date corresponds to another computation on the page, which we will see in a later chapter.) Courtesy SLUB

Calendar Round date in the table on page 46 is 3 Kib 4 Yaxk'in. Adding the time interval of 90 days to this date results in 2 Kimi 14 Sak. With some patience, anyone could use this method to work through the entire table of dates. For his part, Förstemann now was able to make sense of almost half of the total content on each of these five pages without having read a single hieroglyphic word.

Finally, in dealing with another page, Förstemann came across a third type of Long Count number in the manuscript. Specifically, he found that some time intervals were marked with a red loop encircling the k'in (or "day") register (figure 1.16). Their repeated use demonstrated to him that these time intervals were intended to go backward from the "zero" Calendar Round base date of 4 Ajaw 8 Kumk'u. In other words, he recognized that Tawiskal Uwoojil's scribal community was working with a system of negative numbers. On page 24, Förstemann found a provocative example illustrating this entire complex of Long Count dates with Calendar Rounds, related to the negative time intervals, which he called "Ring Numbers."

At the bottom left of page 24, below a column of hieroglyphic text, he found the number '6' above a '2' above a '0,' with the zero circled by a red "ring." Below this Ring Number was the base Calendar Round 4 Ajaw 8 Kumk'u. In the next column to the right, Förstemann identified a time interval of 9.9.16.0.0, followed by a Calendar Round of 1 Ajaw 18 Kayab (figure 1.12). The next column gave a Long Count date of 9.9.9.16.0. These come together since 0.0.0.0.0 4 Ajaw 8 Kumk'u − 6.2.0 + 9.9.16.0.0 = 9.9.9.16.0 1 Ajaw 18 K'ayab.

Sitting at his desk in the Japan Palace in Dresden, Germany, Förstemann now had the mathematical logic underneath most of six pages of the Codex. And even though he was not able to read the hieroglyphic text itself, he found that an association with the celestial realm put him in a good position to hypothesize what it all meant.

THE VENUS TABLE—DRESDEN CODEX PAGES 24 AND 46–50

With these calendric and mathematical tools, Förstemann rigorously recovered the quantitative patterns that would make possible his proposed connection to Venus. The latter came from the sequence of time intervals at the bottom of page 46, which was the same sequence on each of the following pages through page 50. The sum of the four intervals on each page was 584, which Förstemann noticed is very close to the synodic period of Venus, computed by his contemporary astronomers as "583 days and 22 hours" (or 583.917 days) (1886, 120).

As a planet with its orbit closer to the Sun than that of the Earth, Venus is described as an "interior" planet (cf. Aveni 2001, 80–94). This gives it (along with Mercury) a visibility pattern distinct from the other planets readily observable by the unaided eye: Mars, Jupiter, and Saturn (figure 1.17). Whereas the exterior planets can be seen anywhere along the zodiacal band over the course of their revolutions, Venus is restricted in its visibility from Earth to regions near the horizon in the morning and evening. These periods of visibility are not identical from one to the next due to the fact that the orbits of Earth and Venus are neither perfectly circular, nor do they lie in the same plane. But on average, the behavior is sufficiently consistent for it to be captured well by simple arithmetic. The

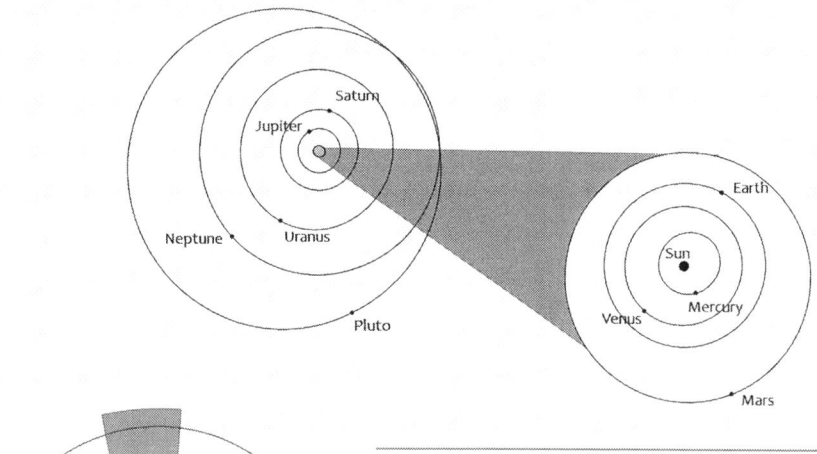

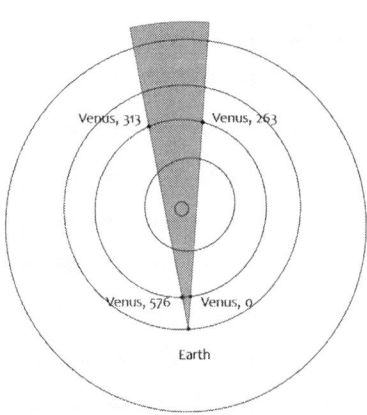

FIGURE 1.17 (*above*) Schematic representation of Venus as an interior planet relative to Earth, impacting its visibility relative to the exterior planets such as Mars; (*left*) fixing Earth in its orbit, four distinguishable visibilities of Venus result from its position relative to the Sun: 0 corresponds to first morning visibility, where it can be seen just above the eastern horizon as the sun rises; 0 to 263 corresponds to Venus's Morningstar visibility in the east; 263 to 313 represents Venus's invisibility at superior conjunction, where it is lost in the brightness of the Sun; 313 to 576 corresponds to Venus's visibility as Eveningstar over the western horizon; 576 to 0 represents invisibility at inferior conjunction.

584-day period is an average, but it is one that works well as an approximation in the short term and the long term.

Förstemann followed up on this hypothesis by recording his own observations between 1882 and 1884. From his balcony, he witnessed the initiation of a given Venus period with its "first morning visibility" over the eastern horizon just ahead of sunrise. Over the course of the next 242 days, it rose first slightly earlier in the morning and just a bit north of its previous rising position along the horizon. Some 198 days after its first morning appearance, Venus reached its maximum separation from the Sun, rising 1 hour and 15 mintues before sunrise (figure 1.18). After reaching this peak, Förstemann watched Venus rise later and later—or closer and closer to sunrise—as its angular separation from the Sun dropped. It was then so close to the Sun (in an angular sense) that it was lost in its glare for a period of 90 days. Thereafter, or about 333 days after its first morning visibility, Förstemann shifted his gaze to witness Venus appear over the western horizon just as the Sun set. Now he tracked its wandering in the west for 243 days until again it dropped into the Sun's glare. The latter transition transpired much more

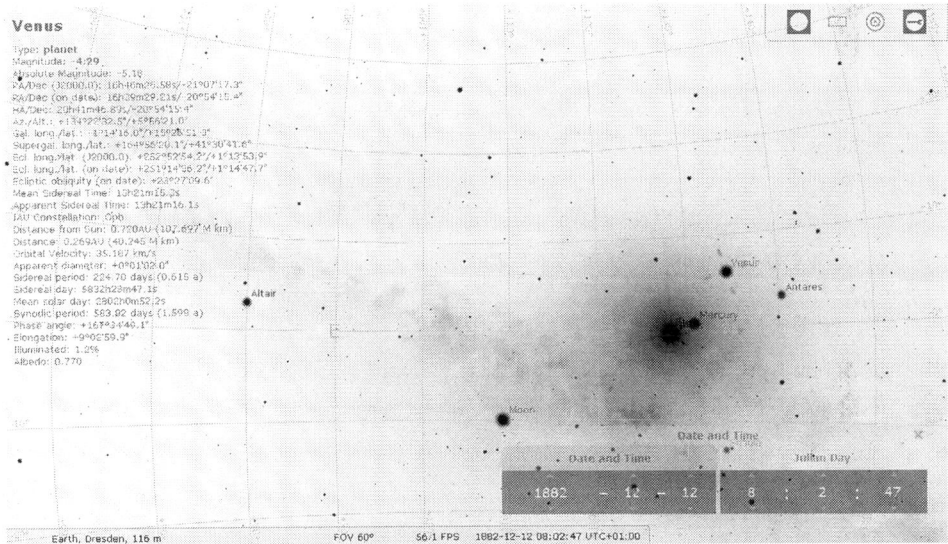

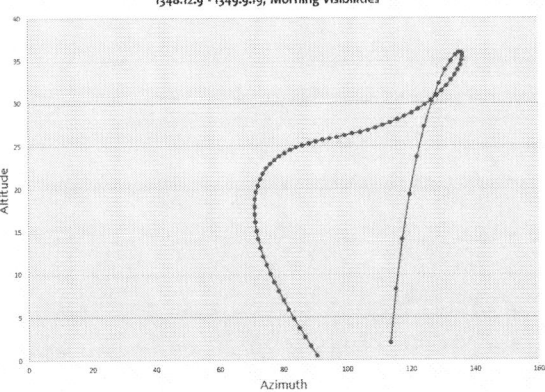

1348.12.9 - 1349.9.19, Morning Visibilities

FIGURE 1.18 (*above*) Reconstruction of Förstemann's view of Venus from Dresden in the late nineteenth century, exemplifying one of Venus's observable trajectories over the eastern horizon, courtesy Stellarium; (*left*) plot of a sequence of observations over the full period of morning visibility, with each dot separated by a time interval of three days. Increased angular separation between Venus and the Sun in 1.17b corresponds to the altitude of Venus's position above the horizon in 1.18b.

quickly than the parallel event in the east; now Venus was only lost in the Sun's glare for 8 days before completing its cycle and again becoming visible over the eastern horizon ahead of the Sun.

Anchored to his own observations, Förstemann considered the time intervals of 8 and 90 days recorded in the Dresden Codex to be reasonable accounts for Venus's invisibility (inferior and superior conjunction respectively). But he had to acknowledge a lack of comprehension regarding the asymmetry between the other two intervals:

> As to the evening star period of 250, and the morning star period of 236 days, I confess that my astronomical knowledge is too small for me to be able to explain this inequality; in reality it is usual to give these two periods, which are not exactly equal, 243 days each. (Förstemann 1891, 121)

In his later work—as we saw at the beginning of this chapter—Förstemann would speculate that the difference might be attributed to an attempt to integrate either Mercurian or lunar periods into the pattern (1894, 443). Of more lasting importance, however, was the overall clear logic within this set of pages. That is, even though the correspondence between specific intervals and Venus's observable motion was not extremely precise, the 8-day period of invisibility and the overall 584-day periods provided strong numerical evidence that Venus was the intended referent. Better quantitative data eventually shifted Förstemann's observational data to put morning and evening visibilities at 263 days on average, with periods of invisibility at 8 and 50 days (Aveni 1980, 187). But these did not change his primary conclusions.

Förstemann noticed that the 8-day time interval corresponded to the final column of dates on each of pages 46–50 and so also the final column in the whole table. In turn, the final date in the table—the bottom right 260-Day Count date on page 50, just above the 365-Day Count of 13 Mak—was 1 Ajaw. Förstemann realized, therefore, that all five pages were built to return the table to the Calendar Round date 1 Ajaw 13 Mak on a first morning visibility of Venus every 104 years. With the Long Count, the 260-Day Count, and the 365-Day Count, alone, therefore, Förstemann was able to account for most of the information on pages 46–50 and to hypothesize that it all related to the tracking of the planet Venus. He had deciphered the quantitative core of the Venus Table.

BEYOND NUMERICAL PATTERNS

The rest of the interpretation that Förstemann provided in the quotation at the beginning of this chapter came from what he was able to interpret of the hieroglyphic text and the imagery of the left-hand side of these pages. He admitted, though, that it wasn't entirely secure. Regarding "the glyphs," he wrote: "Here we enter a mysterious realm, where conjectures occupy a greater space than actual facts" (Förstemann 1894, 438). Nevertheless, some of it seemed accessible.

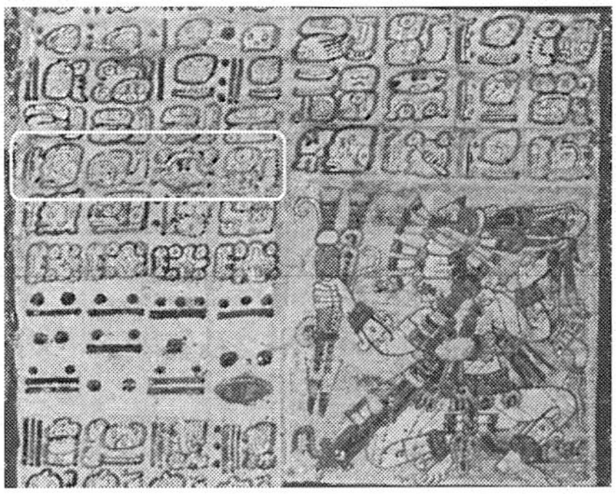

2 Eb	3 Ik'	6 Eb	1 Ajaw	
10 K'ank'in	0 Wayeb	5 Mak	13 Mak	three rows of hieroglyphic text
glyph	glyph	glyph	glyph	
North	West	South	East	
glyph	glyph	glyph	glyph	
Venus	Venus	Venus	Venus	illustration of warrior figure holding an atlatl and darts
7	7	8	8	
2	7	1	2	
12	2	12	0	
15 Kumk'u	0 Tzek	10 K'ayab	18 K'ayab	

FIGURE 1.19 Cardinal directions in the Venus pages. The translation on the right reflects the impressive level of calendric decipherment Förstemann achieved by the end of the nineteenth century. The numbers and dates were deciphered as were the directions and the glyph for Venus. Courtesy SLUB

Among the dates and numbers were two sections of text, also arranged vertically in columns. Within the upper section, Förstemann was quick to draw from the 1876 published work of his colleague Cyrus Thomas, a professor of ethnology at Southern Illinois University in the United States, to identify hieroglyphic representations of the four cosmic regions (or "cardinal directions"), east, north, west, and south (Förstemann 1894, 438) (figure 1.19). Thomas noticed that these four glyphs showed up together variously throughout another set of pages in the Dresden Codex, and in other places these glyphs were arranged spatially to suggest the four cosmic regions (Thompson 1978).

Bringing Thomas's interpretation back to the Dresden Codex, and reading the text in columns set by the time intervals and Calendar Round dates, Förstemann could see that each of Venus's subperiods was linked to one of the cosmic regions. In the first column, the north was associated with the end of morning visibility and superior conjunction; in the second, the west was associated with evening visibility; in the third, the south was associated with the invisibility of inferior conjunction; and in the fourth, the east was associated with first morning visibility. It is worth pointing out here that the association had to be symbolic. It is not that the planet's motion took it physically through these regions in this order—it does not. This would have to be a symbolic affiliation based on the ritual purpose of the text.

Förstemann also recognized that another glyph repeated in the same position in every column. He inferred (and it has been borne out since) that this is the name of Venus, which we now read as Chak Ek'—Great Star (Förstemann 1881, 118; cf. Seler's reading of *citlalpol* or *uei citlalin* as Nahuatl for "the great star" [1904, 358]). Finally, Förstemann drew on the work of his countryman Paul Schellhas, who also found an interest in the Dresden Codex. Schellhas noticed that the same hieroglyphs showed up in different parts of the codex and may even have been connected to glyphs in the Classic period inscriptions. His contribution is still in use today: he went through as many examples as possible and identified images of deities by their consistent iconographic composition, giving them the generic "names" God A, God B, God C . . . through God P (Förstemann 1894, 438). These provided Förstemann with the "names" for the deities associated with each cosmic region and each Venus subperiod, from page 46 to page 50.

Putting all of this together, Förstemann attempted to reconstruct Tawiskal Uwoojil's intent as we saw at the beginning of this chapter: Venus Rounds captured in Long Count time intervals and calendric cycles linked to deities, associated with cosmic regions. This constituted his core interpretation of the Venus Table, and it can be found still in publications today (Aveni 2001; Bricker and Bricker 2011; Coe 1999; Thompson 1972). More importantly, this demonstrated to Förstemann and generations of Mayanists that Mayan scribes paid close attention to Venus and astronomy, setting a key trajectory of scholarship for the next century.

AFTER FÖRSTEMANN

Another of Förstemann's German colleagues, Eduard Seler, is probably best known today for his erudite descriptions and analyses of Central Mexican codices and not his work on Mayan cultures. His profile portrait, frequently reprinted in black and white, emphasizing his age and his austere white beard, graces countless popular texts on Mesoamerica, carrying along the implied authority of "Old World scholarship." Seler's facsimile of and commentary on the Borgia Codex—painted in the "Mixteca-Puebla" or "International" style—is deeply thoughtful and holds much even for modern researchers reading it one hundred years after its composition.

Seler's first job out of university, though—where he had studied broadly in math, the "hard sciences," and botany—was as a high school teacher in Berlin. After only four years, he left for reasons of poor health and dedicated himself to an academic life, which included the study of comparative linguistics and archaeology. In 1887, just after Förstemann published his first essay on "Maya Scripts and Maya Gods," Seler completed his doctorate on *The Conjugation System of the Mayan Languages* at thirty-eight years of age at the University of Leipzig, one hundred kilometers west of Dresden (Rivet 1923, 281). At the beginning of the twentieth century, a new hot spot for studying Mayan ceremonial knowledge had emerged across the Atlantic, in Germany.

Seler soon moved beyond his doctoral studies to produce a monograph titled *The Mexican Picture Writings of Alexander von Humboldt in the Royal Library in Berlin*— addressing fragments of "a Central Mexican codex." The point here is that even though his regard is tied most often to his work on Central Mexico, it was only relatively late, after taking his doctorate, that Seler turned to the study of Central Mexican manuscripts, including the Codex Borgia and others in the similar style, from the Cospi to the Aubin and the Féjérvary-Mayer. And even while he tackled this Central Mexican material, Seler didn't completely leave Mayan cultures behind—in his scholarship or in his travels.

In 1899, now fifty years of age, Seler was honored with an endowed professorship at the University of Berlin by the wealthy American Joseph Florimond. Florimond had been given the title "Duke of Loubat" by Pope Leo XIII six years earlier (Rivet 1923, 281; Coe 1999, 120), and he used it to become a strong patron of Mesoamerican archaeology broadly. (He also funded Marshall Saville's professorship at Columbia University.) In particular, he provided Seler with travel funds and support for the acquisition of artifacts from collections in Mexico and Central America (McVicker 1989, 120). Thus supported, Seler's travel took him throughout the Western Hemisphere, including multiple stops in Yucatan between 1887 and 1911. His visits to one site in particular, Chich'en Itza, actually turned out to be quite opportune in timing relative to his interest in the Dresden Codex, for the development of Maya archaeology and for the advancement of Förstemann's astronomical hypothesis within Maya archaeology more broadly.

CHICH'EN ITZA AT THE DAWN OF THE TWENTIETH CENTURY

Seler's timing resulted from another long backstory. In the 1830s, years after Humboldt had returned from Latin America, after he had seen his *Vue des Cordilleres* receive wide acclaim, and after he had settled into a more conservative diplomatic career, the adventurers John Lloyd Stephens and Frederick Catherwood journeyed through Mayan lands, creating illustrations and narrative descriptions of their encounters (Coe 1999, 92; Stephens and Ackerman 1993). The resulting publications were perhaps just as popular in Europe and the United States as Humboldt's, but few scholars followed in their footsteps through Mayan lands for several decades—in some part because of the military unrest resulting from the region's new political independence from Spain. So it wasn't until after this tumultuous time, by the end of the nineteenth century, that Europeans began trickling back in to the Yucatan Peninsula.

Alfred Maudslay, for one, made his way through the Mayan region carrying Stephens and Catherwood's *Incidents of Travel in Yucatan* in hand (Graham 2002, 79). Maudslay himself epitomized the nineteenth-century adventurer/opportunist, traveling through the South Pacific at a young age and then through Mesoamerica, trying on various positions of privilege before settling in as a successful Maya archaeologist (Graham 2002). Between 1881 and 1902, Maudslay worked for seven seasons at twelve different Mayan sites.

By 1888, Maudslay had spent time at what are now regarded as iconic Classic period sites: Copán and Quirigua in the southern highlands, Tikal in the heart of the Peten, and Yaxchilan on the banks of the Usumacinta River. Maudslay, like Stephens and Catherwood, illustrated numerous monuments and hieroglyphic inscriptions, but he also diligently photographed them. *Biologia Centrali-Americana* (1889) provides powerful scientific records of these monuments, laying photographs side by side with meticulous drawings. Michael Coe writes: "It is impossible to exaggerate the importance to Maya research of Maudslay's published work" (1999, 111).

In turn, this work in the southern Mayan region convinced Maudslay that these sites were mutually contemporary and that they were of some antiquity relative to the first contact of Europeans with Mayans and the communities that Diego de Landa described in his *Relación de las cosas de Yucatán* (Graham 2002, 156). Maudslay's attention was piqued, however, by the fact that Chich'en Itza—which had been illustrated by Catherwood—also was known from the documentary historical record. He realized that this presented an important opportunity: connecting the archaeology of Chich'en Itza to Classic period sites meant connecting the other great cities to the historical record. Maudslay decided it would be worthwhile to attempt to determine the temporal relationship between Chich'en Itza in the north and the sites he had already explored in the south. His plan in 1889 was to spend some time working at Chich'en Itza, considering specifically this problem (Graham 2002, 156).

When Maudslay reached Chich'en Itza at the end of the nineteenth century, it was hardly a "lost city"—in fact, it had never been forgotten. The name of the city itself means "at the edge of the cenote of the Itza," referring on the one hand to the fact that the ceremonial center of the city sits adjacent to a huge natural well or "cenote." On the other hand, the name recalls local histories attributing the city's population by the "Itza" Mayans—an ethnic and linguistic subgroup that appears to have come from the southern Peten (Coe 1999, 71; Tozzer 1941, 20–22). The well itself was the continued target of regional pilgrimages after the depopulation of the city, well into the sixteenth century, including by Landa's consultants, Xiu and Cocom, as well as the Mayans living in neighboring villages throughout the region (Tozzer 1941, 54–55). Xiu himself described pilgrimages of his family members to Chich'en Itza in 1536. In one case, Juan Nachi Cocom's family was in direct conflict with Xiu's family specifically over access to the sacred well (Aldana 2005). In other words, the city's importance had transformed over the centuries, but to local communities it had not diminished.

Thanks to Brasseur de Bourbourg, Maudslay also had Landa's sixteenth-century description of the site to inform his investigations. In his *relación*, Diego de Landa provided a description of a structure in ruins known as both "El Castillo" and the "Temple of Kukulcan," for which he even provided a sketch illustration, attesting to the condition of the building at his time.

> This building has four staircases, which face the four points of the compass. Each of them is thirty-three feet wide and has ninety-one steps, so that it is extremely trying to climb them. The steps are of the same height and depth as we give to ours. Each staircase has two low balustrades on a level with the steps . . . At the time when I saw it, there was at the foot of each balustrade a fierce mouth of a serpent, all of one piece and very (carefully) carved. (Tozzer 1941, 178)

Landa's description went on to treat the stone building at the top of this pyramidal base, noting the enclave in the northern room, which "served as a place to burn . . . incense" (Tozzer 1941, 178). This structure and precisely the features that Landa noted will be of primary importance in the final chapters of this book.

Landa's account went on to address the architectural context for the Temple of K'uk'ulkan:

This building had around it, and still has today, many other well built and large buildings and the ground between it and them is covered with cement, so that there are even traces of the cemented places, so hard is the mortar of which they make them there. At some distance in front of the staircase on the north, there were two small stages of hewn stone, with four staircases, paved on the top, where they say that farces were represented, and comedies for the pleasure of the public. (Tozzer 1941, 178–79)

Based on direct observation, probably accompanied by Juan Cocom, Landa had peered into the government, religion, and theater arts all contributing to and constraining social life at the great city.

As a tourist, then, Landa probably could picture the leaders of the almehen houses consulting with the ajk'ins to arrange the politics and the resources necessary for ritual and everyday life much more easily than tourists can today. Landa's interest in the Great Cenote, however, might strike a modern tourist as brusque: "From the court in front of these stages a wide and handsome causeway runs, as far as a well which is about two stones' throw off" (Tozzer 1941, 178–79).

Two hundred and fifty years later, when Stephens and Catherwood visited the ruins, the principal structures remained standing, but the site was overgrown and primarily populated by plants and animals. Among the overgrowth, Catherwood did find and illustrate the serpent heads at the ends of the balustrades on the north face of the Temple of K'uk'ulkan, but the other architectural subtleties that Landa witnessed were now lost to foliage. On the other hand, Catherwood did spend the time to capture some of what Landa missed. For one, he rendered the "Caracol"—a circular structure that appears to have eluded Landa's acquaintance. (Landa says that he visited the only circular structure in all of Yucatan at Mayapán, even though that one was a small copy of the massive structure at Chich'en Itza.) We will return to the Caracol at Chich'en Itza in chapter 6 and its replica at Mayapán in chapter 10, as they will play significant roles in the interpretation of Tawiskal Uwoojil's Venus Table by the end of this book.

At his arrival sixty years after Stephens and Catherwood and more than three hundred years after Landa, Maudslay initiated his own ambitious project. Maudslay and his crew undertook a huge clearing of trees and brush, as revealed in his now iconic photographs, to facilitate the much more accurate and comprehensive map he aimed to produce (Graham 2002, 163). The work was by no means easy. On the one hand, a henequen boom had taken hold of the Yucatecan economy, drawing local Mayan villagers away from their rural homes and onto hacienda plantations. This made the acquisition of labor for archaeological work very challenging (Graham 2002, 157). Complementing the labor shortage, both Maudslay and his partner Henry Sweet suffered terribly from illness, accompanied by bouts of fever during their stay (Graham 2002, 163).

But the rewards were tremendous. Sweet took what has now become a classic photograph of Maudslay from the temple atop the Monjas Structure—the "office" from which Maudslay poetically described some of the personal reward for his efforts:

To the southward, where no clearing had yet been made, the sea of verdure spread unbroken from our feet. During the lovely tropical nights, when a gentle breeze swayed the tree-tops, and the moonlight rippled over the foliage, it seemed to be a real sea in motion below us, and one almost expected to feel the pulsation of ocean waves against the walls. (Graham 2002, 162)

Maudslay's map rivals the newer ones made with far superior technological advances, and his detailed photographs of the architecture and inscriptions are still treasured today.

For his time, Maudslay's photos and illustrations were critical to the growing study of the inscriptions by Seler and his contemporaries. His clearing work also exposed a close architectural connection between Chich'en Itza and the even more widely known Yucatecan ruins (at the time) of Uxmal. The structures in the southern part of Chich'en Itza—to the south, that is, of the Temple of K'uk'ulkan, and in the vicinity of the Caracol—had been built in the Puuc architectural style, dominant in the Mayan region of the Gulf Coast. This Puuc style is best known today by its representation at Uxmal as well as other sites in northern Yucatan such as Dzibilchaltun or Edzna. Much of what distinguishes the style has to do with a specific architectural feature.

FIGURE 1.20 (*top*) Traditional thatched-roof house; (*middle*) Temple of the Sun at Palenque; (*bottom*) Terminal Classic Puuc architecture at Chich'en Itza

For the stone structures at the Classic period sites of Tikal or Yaxchilan in the southern lowlands, architects generally designed roofs with a slope, mimicking the thatched roofs of commoner houses (Sharer 2006, 215) (figure 1.20). This slope of the roof basically followed the corbel vault it enclosed. In these southern cases, the sloping stone roofs often supported "roof combs"—large, highly decorated stone lattices that looked a bit like modern billboards. Classic period artists decorated these roof combs with historical figures and/or deities important to the ruling dynasty of the city.

In contrast, architects working in the northern Puuc style constructed temples as huge rectangular blocks, maintaining vertical faces all the way to the top (Sharer 2006, 534). The structures still contained corbeled vaults, but the upper portions were "filled in" rather than left to follow the slope of the vault. This feature allowed artists to incorporate the material that would have decorated a roof comb directly into the external façades of the upper levels of the buildings.* At Uxmal, for example, the architects decorated and stylized masks to take on different characters, reflecting the different purposes and meanings of the buildings. The fact that Maudslay found the same architectural styles at Chich'en Itza as that at Uxmal, and the fact that they differed from southern styles, would eventually lead to hypotheses of an "Old Kingdom" in the southern part of the city versus a "New Kingdom" culture of the north centered on the Temple of K'uk'ulkan (Coe 1999, 70–71). The transition was thought to reflect the later presence of Central Mexico—that is, the "Toltec" at Chich'en Itza.

EDUARD SELER AND QUETZALCOATL

Maudslay's connections between old and new, between Postclassic Mayans and Central Mexico, fed directly into Seler's study. On his trip to Uxmal, Seler found what he considered to be Venus symbolism at the site. On the façade of the structure that Stephens and Catherwood called the "Governor's Palace," Seler identified the bar-and-dot representation of the number "8" along with the glyph that formed part of the name of Venus as identified by Förstemann in the Dresden Codex (Aveni and Hartung 1986, 31–32). Both of these symbols were carved into the eyelids of one of the great masks on the corner of the structure, leading Seler to consider this a "Venus structure" (figure 1.21).

For Seler, the number 8 served as a reference to Venus's period of inferior conjunction and *ek'* referred to the planet itself. In this way, Seler was able to make a huge claim. Not only was astronomy to be found in an esoteric Mayan codex; here, he found evidence that the astronomy of Venus itself was part of public art and architecture. At Uxmal, he seemed to be corroborating Förstemann's find of Venus's importance to Mayan culture, extending it even to a specific archaeological site. This led Seler to esteem even higher the place of astronomy in Mayan cultures—a position that some of his later critics suggested he took too far. On the other hand, it proved extremely useful in his work of recognizing common themes between the Borgia Group of codices from Central Mexico and the Dresden Codex Venus Table, which has not been contested since (Coe 1999, 120).

---

*For most of the twentieth century, these masks were thought to represent the face of the Rain God— "Chaak" in Mayan. The iconographic connection seemed very straightforwardly corroborated by ideology. Since Yucatan is dependent on rainfall for fresh water, the Rain God, Chaak, would have been a—if not the— critical deity. While the interpretation definitely appeals to common sense, it hasn't borne out the test of time. As Coe writes: "Long mistakenly identified as faces of the Maya rain god Chaak, they are actually iconographic mountains (witz), the descendants of the corner masks placed on the Classic-period monuments like Copan's Temple 11" (2005, 166).

FIGURE 1.21 Ek' icons on the lower eyelids of the witz' masks on the "Governor's Palace" at Uxmal

By 1904 Seler found himself uniquely positioned to take up Förstemann's findings in the Dresden Codex and examine their relationship to representations of Venus in Central Mexico. Seler's article "Venus Period in the Picture Writings of the Borgian Codex Group" proceeds in wonderfully logical fashion; moving from general descriptions of sixteenth-century Spanish descriptions of Central Mexican astronomy, through mythological records of Venus in particular, he is able to identify deity representations of the planet in the artwork of the codices. These pictorial representations provide him the means to explore astronomical knowledge recorded in the Central Mexican codices, and then extend this to include Förstemann's work on the Dresden Codex. And in his wanderings through historical texts, Seler found further justification for the role of astronomy within Mesoamerican cultures.

One passage in particular, in the *Crónica Mexicana de Tezozomoc*, probably came to life for Seler as he walked through the plazas of Chich'en Itza, which had been cleared by Maudslay. The *Crónica* draws from the speeches given at the Aztec emperor Motecuhzoma Xocoyotzin's accession to the throne; the nobles provided advice to the new ruler (through which they also deftly lobbied for their own interests). In Seler's translation, Motecuhzoma

is exhorted . . . to receive graciously the tributary vassals when they come to the capital and to provide them with all that is necessary for their homeward journey. He is admonished to be valiant against his enemies, but also to employ diplomacy, adulation, and gifts in order to bring them to submission by peaceable means. He should endow the temple and give sustenance to the old people, both men and women. He ought, above all things, to stand well with the nobility, be mindful of their privileges, and daily invite them to be his guests; for his authority and power depend on them. (1904, 355)

Seler, of course, knew that the Aztec Triple Alliance was built on an infrastructure of tribute, so it was relatively easy to transport a similar infrastructure and political organization to Chich'en Itza—even if there was no documentary evidence that the Yucatecan city sat at the top of a formal "empire" (figure 1.22). The plazas he walked across could have hosted marketplaces, and the palaces could have served as the locations for meetings of the nobility. Seler was seeing cultural and political continuity between Aztec Central Mexico and Chich'en Itza and Uxmal in Yucatan.

The *Crónica* guided Seler to appreciate a broader Mesoamerican politics, but its use was even more pointed. The text goes on, enjoining the emperor to be careful in the observance of religious ceremonies and faithful in regard to priestly castigations and to the care of the temples, the sacred places, and the roads leading to them. The emperor himself is urged to rely on some of the practices of priestly timekeepers. Specifically, Motecuhzoma is instructed

especially to make it his duty to rise at midnight (and to look at the stars): at *yohualitqui mamalhuaztli*, as they call "the Keys of Saint Peter" among the stars in the firmament, at the *citlaltlachtli*, the north and its wheel, at the *tianquiztli*, the Pleiades, and at the *colotl*

FIGURE 1.22 Aerial photograph of Chich'en Itza. The Great Ball Court is in the middle left, the Temple of K'uk'ulkan is in the center, and the Caracol is in the lower right corner of the photo. Photo by Linda Schele © David Schele, courtesy Ancient Americas at LACMA (ancientamericas.org)

ixayac, the constellation of the Scorpion, which marks the four cardinal points in the sky. Toward morning he must also carefully observe the constellation *xonecuilli*, the "cross of Saint Jacob", which appears in the southern sky in the direction of India and China; and he must carefully observe the morning star, which appears at dawn and is called *tlauiz-calpan tecuhtli*. (1904, 355)

Seler found in these texts what he took to be justification for his interpretation that astronomy was foundational to Mesoamerican ideology. These were the words that informed his iconographic interpretations at Uxmal and Chich'en Itza. In particular, in this text too Seler understood that the emperor himself was advised to keep track of Venus as Morningstar and that it carried the name Tlahuizcalpantecuhtli.

The core of Seler's argument addressing the interpretation of Venus is still regularly referred to as foundational (Aveni 2001, 186; Bricker and Bricker 2011, 197; Thompson 1972, 62), but the connection to Tlahuizcalpantecuhtli was only one step toward the Mesoamerican Venus common in the literature today. Seler's next move brought the Venus of the Dresden Codex and of Yucatan into conversation with the great legendary hero of Central Mexico, Quetzalcoatl.

Seler's method relied heavily on several other sixteenth-century textual records beyond the *Crónica*. Fortunately, some of them were annotated codices, such as the Codex Mendoza we saw in this book's introduction. These codices he compared against the narrative provided by a Franciscan in Mexico City, Bernardino de Sahagún. Finally, he looked at both of these sources relative to the imagery in several Mixteca-Puebla codices known collectively as the Borgia Group. Seler's first stop in this whole project was with the passage in the *Anales de Cuauhtitlan*, which connected Quetzalcoatl to the planet Venus, as we saw in the introduction—the one noting Quetzalcoatl's death on a raft of serpents and apotheosis as birds and Venus in Morningstar form.

It is worth pausing here to recall that the name "Quetzalcoatl" is a composite of two animals, the serpent and the bird, and so it generates a conceptual metaphor, given that the Mesoamerican cosmos was characterized by three different communities, each largely confined to a cosmic realm. The Underworld was that realm below the surfaces of the earth and bodies of water (cf. Brady 1997, 603). Here lived insects, worms, and rodents along with the souls of human ancestors—their bodies buried or placed in caves—bats, disease, and personifications of death itself. The realm reaching from the tops of the highest floral canopies to the celestial limits was the Upper-world, encompassing the activities and interactions of birds, clouds, the stars, planets, the Sun and Moon, and others. The middle realm was the region occupied by living humans and all of their plant and animal neighbors. Each of these realms was divided further horizontally into the cosmic regions of east, north, west, and south.

The point here is that Quetzalcoatl as a composite creature would have had access to the Underworld as a serpent; to the middle realm as either serpent or quetzal; and to the Upper-world via its quetzal character. This ability to move between all three realms of the cosmos was alluded to by the authors of the *Popol Vuh* in their description of a figure named Cucumatz, which is the K'iche' Mayan term for the same name (*cuc*—quetzal; *cumatz*—serpent). Of this historical figure, the authors of the *Popol Vuh* wrote:

> Cucumatz became a truly enchanted lord. In one transformation he would rise up into the sky, and in another transformation he would go down to Xibalba. In other transformations he would be a serpent, truly becoming a serpent. In another transformation he would make himself into an eagle; and in another transformation into a jaguar. (Christenson 2007, 275)

With Xibalba as the K'iche' name for the frightful region of the Underworld, and with the jaguar transformation for prowling through the middle realm, Cucumatz had powerful influence within each of the cosmic realms.

To this legendary human figure of Cucumatz we will return in the final chapters of this book. For now, we note that Seler drew from the *Anales de Cuauhtitlan*, to connect Quetzalcoatl's eight-day journey through the Underworld to the number 8 that he saw in the imagery of the Governor's Palace at Uxmal and the time interval in the pages of the Dresden Codex Venus Table (1904, 359). Seler therefore now had Quetzalcoatl and his myths to connect to Förstemann's emerging astronomical context. In addition, this passage presented an enigmatic connection to Tlahuizcalpantecuthli, for which Seler's familiarity with the other Central Mexican codices came in handy. These deities—Quetzalcoatl and Tlahuizcalpantecuhtli—were not only referred to in written texts; they were also illustrated.

Seler turned to another Central Mexican manuscript, the Codex Telleriano-Remensis, to corroborate the implications of the *Anales de Cuauhtitlan*. The Codex Telleriano-Remensis was used most certainly for translation purposes during its time of authorship. Like the Codex Mendoza, it contains illustrations of deities, surrounded by commentary handwritten in Spanish. In Florimond's translation (which preserves handwriting differences by using different fonts), Seler confronted the following:

NEUVIÈME TREIZAINE
PATECATLE
tlauizcalpantecutli *o la estrella Venus*
    una caña.

en este dia de vna culebra teniã por malo y si en tal dia alg° caminaua y romperaua un palo o piedra o otra causa q̃ vendria a perder la pierna o morir dello.
    la primera claridad ē q̃ fue criado civaq̃teltona ātes del diluvio dizen q̃ esta lunbre o estrella fue criada ātes q̃ sol.

Este tlauizcalpanteuctli o e.trella venus es el queçalcovatl dizē q̃ es aq̃lla estrella q̃ llamamos luzero de la lus y asi lo pintā con vna caña q era su dia
    cuando se fue o desaparecio tomo su nōbre

Este tlauizcalpantecutli quiere dezir señor de la mañana quando ameneçe y lo mesmo es señor de aquella claridad quando quiere anocheçer. Este es señor destos treze dias, ayunavan los quatro prosteros.

> *propiaṁēte la primera claridad q̄ aparecio en el mundo aqui el ayuno del señor q̄ y lo demas el ciēlo propiaṁete la q̄ c̄ubre sobre las casas o haz de la tierra.*

Whereas the passage from the *Anales* left some ambiguity in the relationships between the named deities and the planets, any subtlety seems to be erased by the commenters on the Telleriano-Remensis.

The *Anales* passage stated that Quetzalcoatl's heart burned and was transformed into Venus as Morningstar such that its author preserved a distinction between Quetzalcoatl becoming a god and the first morning appearance of Venus (as Tlahuizcalpantecuhtli) — they are not identified, but related. In the Telleriano-Remensis, on the other hand, Tlahuizcalpantecuhtli is equated with Venus: "Tlahuizcalpantecuhtli or the star Venus." This is a distinction that will take on specific importance later in this book. For now, however, the important point for Seler was the methodological link between the narrative texts and the codex illustrations of the deities to whom they referred. Now Seler could look for images of the deities within Precontact manuscripts.

In the Codex Telleriano-Remensis, its scribe illustrated Tlahuizcalpantecuhtli, for instance, wearing an elaborate headdress and distinct clothing, but he is perhaps most identifiable by his face decoration. Seler writes:

> The white body with red longitudinal stripes, and with the deep black painting about the eyes, like a domino mask, which is bordered here, but not always, by small white circles is combined with a red painting about the lips, which likewise may be omitted, are characteristic marks of this god (1904, 360).

By inferring that this specific representation provides a secure iconographic characterization of the deity, Seler established the basis for finding representations of Tlahuizcalpantecuhtli throughout other Central Mexican codices (figure 1.23). Recognizing several, he focuses in on one, making clear that his motivation was to elucidate the connection between the work that Förstemann had done on the Dresden Codex Venus Table and a "principal representation, found on pages 53 and 54 of the Borgian Codex" (1904, 373).

While Seler did toy with calendric progressions throughout the Borgia Codex that could be made to fit with some of the Dresden patterns, he found the iconographic connection between Borgia Codex pages 53–54 and Dresden pages 46–50 to be the most convincing. The scribe of Borgia page 53 had included a figure with white-and-red-striped body paint, allowing Seler to link it to Tlahuizcalpantecuhtli in the Codex Telleriano-Remensis and, therefore, to Venus (figure 1.24). This red-striped figure carries a shield in his right hand, and with his left he has launched an atlatl dart into a female victim's ankle. Seler noticed that this was not a singular violent act. On the next page the Borgia scribe illustrated four similar spearing events, with different warriors and different victims in each. These spearings caught Seler's attention, leading him to suggest a comparison between the Borgia warriors and their victims against parallel scenes in the Dresden Venus pages. He went further to suggest equivalences between the first three victims in the Borgia pages and the first three victims in the Dresden Codex Venus Table (figure 1.24).

The female figure, on page 53, Seler considered to be Chalchiutlicue, associated with bodies of water. Correspondingly, the first victim in the Dresden Codex, on page 46, he interpreted as a water deity (1904, 376, also fig. 98). The next two identifications required less speculation. The second victim in the Borgia is readily identifiable as Tezcatlipoca, which was often associated with jaguars. And the victim at the bottom of page 47 of the Dresden Codex is clearly a jaguar (Seler 1904, 379). The third victim in each case is a maize deity (Seler 1904, 379–80). The comparison did not reveal perfect congruence, but the parallels were sufficient for Seler to claim that if the Dresden Codex recorded Venus events, so too did the Borgia.

FIGURE 1.23 Tlahuizcalpantecuhtli as identified by Seler in the Codex Borgia, page 49. LACMA (Codex Borgia). Courtesy Ancient Americas at LACMA (ancientamericas.org)

At this point, however, Seler noted incongruence between his developing interpretation and that proposed by Förstemann. Whereas Förstemann suggested that the warrior illustrations on pages 46–50 of the Dresden Codex represented the Sun, and Venus the victim (Förstemann 1891, 122; Seler 1904, 382–91), Seler identified "Venus . . . as the conquering party" (1904, 384). He speculated that these interactions might be representations of Venus in conjunction with celestial constellations—a supposition that he claimed (without providing details) is "proved by certain reliefs at Chichen Itza" (1904, 384).

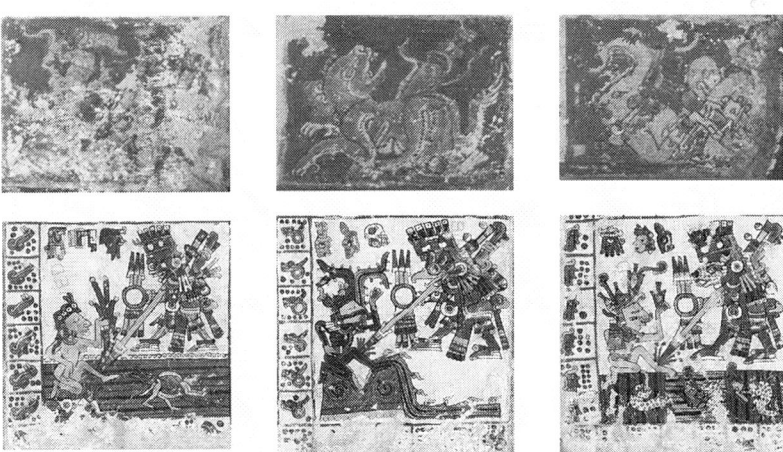

FIGURE 1.24 Parallels identified by Seler across the victims in the Dresden Codex and those being speared in the Codex Borgia. At left is K'awiil—identified by Seler as a "water deity"—and the female deity Chalchiutlicue. In the middle is a jaguar in the Dresden and Tezcatlipoca in the Borgia. At right is a maize deity in both images. Dresden Codex pages 46 and 50, SLUB and Codex Borgia. Courtesy Ancient Americas at LACMA (ancientamericas.org)

It turns out, however, that a second of his speculations was more productive. Seler proposed that the spearing with atlatl darts might be a metaphor for the planet shining its light upon the world and its inhabitants. This he drew from the sixteenth-century *Anales*.

In the Anales de Quauhtitlan, appended to the story of the transformation of Quetzal-coatl into the morning star, there is a more detailed account of these influences ascribed to the light of the planet Venus. It is a remarkable passage, of which I give here the literal translation:

> And as they (the ancients, the forefathers) learned.
> When it appears (rises).
> According to the sign, in which it (rises).
> It strikes different classes of people with its rays.
> Shoots them, casts its light upon them.
> When it appears in the (first) sign, "1 alligator".
> It shoots the old men and women.
> Also in the (second) sign, "1 jaguar".
> In the (third) sign, "1 stag".
> In the (fourth) sign, "1 flower".
> It shoots little children.
> And in the (fifth) sign, "1 reed".
> It shoots the kings.
> Also in the (sixth) sign, "1 death".
> And in the (seventh) sign, "1 rain".
> It shoots the rain.
> It will not rain.
> And in the (thirteenth) sign, "1 movement".
> It shoots the youths and maidens.
> And in the (seventeenth) sign, "1 water".
> There is universal drought. (Seler 1904, 384–85)

Seler thus constructed a cultural triangulation of Venus representations, drawing from sixteenth-century alphabetic texts, Precontact Central Mexican codices, and the Dresden Codex, with the latter anchored to Förstemann's work. The result corroborated Förstemann's proposal that Venus astronomy was important to ancient Mesoamerica and that its periods of visibility were related to effects on different members of society.

Seler, however, was more conservative than those coming after him in his interpretation of a few subtleties—maintaining a distinction between Venus and the deities depicted on pages 46–50. He wrote regarding the Dresden Codex:

> The form armed with the spear thrower and bundle of spears *is not* the deity of the morn-
> ing star repeated five times, as in the representations of the manuscripts of the Borgian

codex group . . . but five different figures. . . . The figures struck with the spear, on the other hand, are clearly the same as in the representations of the Borgian codex group, at least on the first three pages. (Seler 1904, 376, emphasis added)

Seler would only go so far explicitly as to identify the represented deities as "the regents of the five consecutive Venus periods" (1904, 387). Without specifying further what he meant by a "regent," Seler did make it clear that, although the various figures were "divinities," for him only the image on page 47 was an actual representation of a/ the Venus deity (in Morningstar form) (1904, 388). This too is an issue we will return to in the later chapters of this book.

In sum, Seler's contributions to the understanding of the imagery and the oracular role of Venus in Meosamerican cultures fit easily with the basic astronomy Förstemann recovered, so that the vast majority of the interpretation of the Dresden Codex Venus Table was set and would not change significantly for the next one hundred years (Aveni 2001, 184–96; Bricker and Bricker 2007, 106–9; Paxton 2001, 64–81; Seler 1904; Thompson 1972, 62–71). According to this view, Yucatecan astronomers at Chich'en Itza and Uxmal—possibly along with their rulers—closely observed Venus using math and calendar, and they did so in order to prepare for or mediate the influence of Venus as a celestial deity. Seler closed his article accordingly:

The agreement extending to the details that [existed?] between the Mexicans and the Mayas in the system of the calendar and the twenty day signs doubtless corresponded to an agreement in their sacerdotal science in many other particulars. It could hardly be otherwise in view of the active intercourse, which existed between these two great civilized races. I believe the foregoing has, for the first time, furnished conclusive proof of this (1904, 391).

The scope of Seler's work, encompassing the vast majority of what was known from Mesoamerica at his time, was unquestionably impressive. His work was informed by his abilities with Nahuatl and Mayan languages, his trained eye in codex iconography, and his personal visits to Mesoamerican archaeological sites. His erudition brought together the most nuanced Mayan astronomical document with the peerless Quetzalcoatl of Central Mexico. Surprisingly, the next major advance in the modern interpretation of the Dresden Codex came from an individual who possessed none of these traits.

HISTORICITY

One last critical component to the Venus Table's interpretive history, bringing us essentially to its modern interpretation, was awaiting a unique individual. In this case, Mayanist scholarship was impacted by someone without any training in the study of archaeology or Mesoamerica, but with two essential qualifications: keen mathematical talent and a lot of free time to play with numerical patterns. That person was John Teeple, professional chemical engineer and acquaintance of Sylvanus Morley.

In 1927, by which time Förstemann's and Seler's work had become canonical in Maya Studies and new investigations were underway that focused primarily or solely on the recovery of Mayan astronomy, John Teeple received the Perkin Medal in Chemistry. The *Cornell Alumni News* announced the honor, commenting:

> Since receiving his doctor's degree at Cornell Dr. Teeple has been engaged in a chemical consulting practice in New York. . . . He is also an authority on Maya inscriptions, and has contributed numerous articles to *The American Anthropologist.* (1926, 58)

The articles they referred to made specific and lasting impacts on the interpretation of Mayan astronomy. They demonstrated clearly that Teeple had a facility with numbers and scientific reasoning, even if a lack of exposure to Mesoamerica. But it was an accident of history that led to his having the opportunity to work on the Dresden Codex.

Prior to the twentieth century, potash—which had become necessary as an agricultural fertilizer but also was critical to the industrialization of soap production—was obtained by burning wood and then leaching potassium compounds out of the ashes. Lumber produced by the clearing of forests—for example, in Canada—generated the majority of the necessary potash for the international market. By the late nineteenth century, however, large-scale agricultural production around the world depended on much greater amounts, which were primarily available from natural potassium salt deposits within German borders. An "overconfident" German chemist was quoted in the early twentieth century as capturing this circumstance by stating: "With her tremendous deposits of potash it is in Germany's power to dictate which of the nations shall have plenty of food and which shall starve" (*Scientific American* 1917, 282). By the end of the nineteenth century, international demand was being met by mining Germany's natural resources (Brand 1945, 110).

The First World War, of course, disrupted the flow of potash from Germany, and it put pressure on other nations to find alternate sources. Teeple's Perkin Medal came from his clever method for extracting potash from a chemical compound found in Searles Lake, California. Whereas other methods, like that developed in Nebraska, capitalized on the absence of German production, and so folded as too costly when German production recommenced, Teeple's method remained profitable. The *Cornell Alumni News* noted:

> The 1927 Perkin Medal, awarded "annually to the American chemist who has most distinguished himself by his services to applied chemistry," will be given to John Teeple . . .
>
> During the War the acute shortage of potash supplies because of the blockade of German ports led to efforts to secure potash from the brines of Searles Lake. The attempts were at first unsuccessful because of faulty engineering and contamination of the potash with sufficient borax to injure the crops. The British owners, after they had spent large

sums of money, put the whole project in charge of Dr. Teeple, who successfully met the scientific problems involved.

Under his direction potash was gained which compared favorably in cost with the European potash and which was also free from borax. The previously injurious borax was developed into an important output of the plant and is adding largely to the world's supply of this material. (1926, 58)

The point of this digression into World War I economic activity is twofold. For one, it demonstrates that Teeple certainly had a mind for problem-solving, which included the capability of imagining new purposes for old problems. By removing borax in a way that maintained its chemical integrity, its representation as a "problem" became the source of new revenue. Second, Teeple's home was in New York City, but Searles Lake was in California. Through his acquaintance with Sylvanus Morley and Morley's encouragement, Teeple used his time on the train between coasts to puzzle over Mayan hieroglyphic inscriptions (Coe 1999, 134). Without this need to entertain his computationally talented mind on the multiday trip across country, it is hard to imagine who would have come up with the contributions Teeple made, which were indeed formidable.

Through the opportunity provided by his travel time, Teeple's "authority" within Maya Studies was relatively quick in coming. In 1925, Teeple published an astronomical solution to a series of hieroglyphs that were regularly recorded at the beginning of Classic period inscriptions. That is, Classic period scribes carved the texts recorded by Morley, Maudslay, and colleagues in a different manner from those in the codex records treated by Förstemann. The carved stone inscriptions recorded hieroglyphic passages that started with a full Long Count date in extremely formulaic representation (figure 1.25). On Classic period monuments scribes introduced these Long Count dates with a larger-than-normal initial hieroglyph, descriptively referred to as the Initial Series Introductory Glyph, or the ISIG. A given Classic period text, then, would often be transcribed as ISIG 9.15.10.0.0 for its first glyph blocks, providing the intended Long Count date.

The formula followed by these scribes was to append the Calendar Round date to the Long Count, often in a way that split the 260-Day Count date from the 365-Day Count date (figure 1.25). Between these two counts, they consistently included

Initial Series Introductory Glyph	
9 pih	15 winikhaab
10 haab	0 winik
0 k'in	3 Ajaw
Glyphs G & F	Glyph D
Glyph C	Glyph X
Glyph B	Glyph A
3 Mol	tanlam

FIGURE 1.25 Initial Series text on the left side of Piedras Negras Stela 10, including a full Lunar Series as Glyphs A through G.

seven glyphs. The pattern was sufficiently robust to be recognized as variations within a distinct set. Sylvanus Morley called the set the "Supplementary Series" and labeled them generically by letter: Glyph A, Glyph B, . . . Glyph G, although he assigned these names in reverse order. Teeple pooled together "over one hundred rather complete dates with moon series attached" from the inscriptions that Morley had collected from the sites of Piedras Negras, Copán, Naranjo, Yaxchilan, and Quirigua and within them observed a few intriguing patterns (1928, 392).

Teeple's first result was that Glyph A contained two parts: a prefix that was virtually always the same, and a suffix that could either represent the Mayan number 9 (four dots attached to a bar) or 10 (two bars) (figure 1.25). When Teeple picked up the problem of the Supplementary Series in 1925, it had already been "agreed upon" that the prefix had a value of 20 and that Glyph A therefore represented either a twenty-nine- or a thirty-day period (20+9 or 20+10) (Teeple 1925a, 108).

Teeple recognized that the same prefix as a glyph for "20" showed up occasionally in what Morley designated Glyph E, and it was specifically Glyphs C, D, and E that he would tackle in his article for *American Anthropologist*. Again here he was aided by Förstemann's work on the Dresden Codex. That is, besides Förstemann's considerable insight into Venus astronomy in the Dresden Codex, he found that the pages immediately following the Venus pages also concerned astronomical matters. Specifically, Förstemann recognized that the next eight pages recorded lunar periods and may well have been intended to record eclipses of the Sun or Moon (figure 1.26). The key to this recognition was again a set of two rows of numbers in Long Count time interval notation.

As in the Venus Table, Förstemann found that the bottom row of numbers on pages 51–58 represented individual time intervals, while the row above it summed a rolling tally of days. The vast majority of the individual time intervals recorded were of 177 days; the other most common interval was of 148. Förstemann understood these to represent groupings of twenty-nine- or thirty-day lunar months. (Since the synodic period of the Moon is 29.53 days, 29- and 30-day months can be alternated [with a slight imbalance toward 30-day months] to keep visibility in line with calendric records.) Specifically, 177 days is equivalent to $3 \times 30 + 3 \times 29$ and 148 is $3 \times 30 + 2 \times 29$. In the Eclipse pages of the Dresden Codex, long strings of 6-month groupings (time intervals of 177 days) are followed by a single five-month grouping (of 148 days). These are precisely the base intervals useful for tracking eclipse events (Aveni 2001, 173–84).

So the background context to which Teeple was able to appeal was that pages 51–58 of the Dresden Codex showed a Mayan practice of counting moons in twenty-nine- and thirty-day months and that these months were grouped into larger periods of five or six months. This was immediately useful, as Teeple noted that Glyph C had a number attached to it that was never greater than six and so could have been used to record these periods in Classic times parallel to those in the Dresden Codex (1925a, 108).

The fortunate thing about this data set, of course, is that it was attached to the Long Count, which allowed for the computation of the number of days between each Supplementary Series record. Using a set of eighty dates, Teeple showed that Glyph C provided the month within a five- or sixth-month cycle and that Glyphs D and E gave the

Moon Age within the current month (figure 1.25). With this solution to the Supplementary Series, Teeple determined that, for example, the scribes of Piedras Negras Stela 1 recorded: ISIG 9.12.2.0.16 5 Kib Glyph G+F 8E 3C X B A9 19 Yaxk'in. This tells us that on the Long Count date of 9 pih, 12 winikhaab, 2 haab, 0 winik, and 16 k'in, which was 5 Kib 19 Yaxk'in in the Calendar Round, the Moon was near New Moon (Moon Age of 28 days) in its third period and was counted as a month of twenty-nine days.

It was this decipherment—the identification of the Lunar Series—that initiated Teeple's reputation in the field and further promoted the interpretation that the hieroglyphic texts were heavily—if not solely—dedicated to astronomical matters (Coe 1999, 122). The most secure "decipherments" in the field at this point were the cardinal directions, the names of

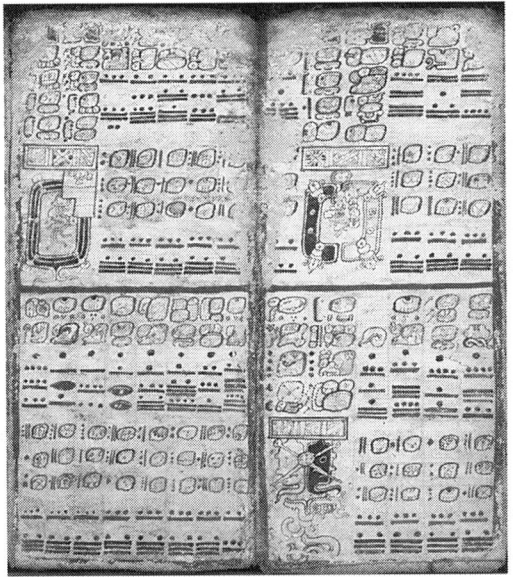

the gods, Förstemann's recognition of Venus and Eclipse Tables in the Dresden Codex, and Teeple's identification of lunar data in virtually every stone inscription from the Classic period. Indeed, Morley could not have been more pleased that his efforts to "bring home the bacon" (which to him was to record hieroglyphic dates and leave the rest undocumented [Coe 1999, 129; Thompson 1975, viii]) were realized through Teeple's problem-solving expertise.

FIGURE 1.26 (*left*) Pages 55 and 56 of the Dresden Codex Eclipse Table, courtesy SLUB; (*below*) translation of dates and time intervals

two columns of text				hieroglyphic text				two columns of text		hieroglyphic text		
		2422	2599	2776	2953	3130	3278			3455	3632	3809
Eclipse Image		2 Muluk 3 Ok 4 Chuwen	10 Kimi 11 Manik 12 Lamat	5 Ak'bal 6 K'an 7 Chikchan	13 Ajaw 1 Imix 2 Ik'	8 Kaban 9 Kawak 10 Etz'nab	13 Chikchan 1 Kimi 2 Manik	Eclipse Image		8 Ik' 9 Ak'bal 10 K'an	3 Kawak 4 Ajaw 5 Imix	11 Kib 12 Kaban 13 Etz'nab
		177	177	177	177	177	148			177	177	177

hieroglyphic text								two columns of text		hieroglyphic text		
8829	9007	9184	9361	9538	9715	9892	10040			10217	10395	10572	10749
13 Kib 1 Kaban 2 Etz'nab	9 Ix 10 Men 11 Kib	4 Chuwen 5 Eb 6 Ben	12 Lamat 13 Muluk 1 Ok	7 Chikchan 8 Kim 9 Manik	2 Ik' 3 Ak'bal 4 K'an	10 Kawak 11 Ajaw 12 Imix	2 Manik 3 Lamat 4 Muluk	Eclipse Image		10 K'an 11 Chikchan 12 Kimi	6 Ik' 7 Ak'bal 8 K'an	1 Kawak 2 Ajaw 3 Imix	9 Kib 10 Kaban 11 Etz'nab
177	177	177	177	177	177	177	148			177	177	177

The chemical engineer's work on the Lunar Series was impressive and important, but it was his next publication that is of most interest to this book. In 1926 Teeple added the final component to the basic interpretation of the Venus Table—all work since his must begin with the framework provided by him, Seler, and Förstemann. Permutations of the building blocks laid by these three scholars—only one of whom actually visited Mayan archaeological sites—delimit the range of published interpretations over the last eighty years.

ACCURACY

Where Seler followed Förstemann by looking broadly for his contributions, Teeple focused on a set of only three numbers for his next major breakthrough. Teeple's work began with a set of time intervals that Förstemann himself had struggled with. By working through the mathematics of the whole manuscript, Förstemann had encountered a connection between the material on pages 46–50 and the content of a page at some remove: page 24. Working through the numbers on page 24, Förstemann found that it started with a table of multiples at the bottom right corner of the page and worked to increasing size from right to left within a row. The scribe had then moved up by rows to fill out a numerical list of Venus time intervals.

Förstemann recognized that the base of all the intervals on page 24 is the number 2,920. This number, 2,920, is the number of days in five Venus Rounds, or 5×584. The time interval is the same, therefore, as the length of time it takes to track one row of dates across all five pages of the Venus Table. Since there are thirteen rows, it would make sense to have a table of time intervals running from $1 \times 2,920$ through $13 \times 2,920$, and that is precisely what page 24 does (figure 1.27). The bottom three rows of numbers in four columns on page 24 give projections of Venus periods from $1 \times 2,920 (= 1 \times 5 \times 584)$ to $12 \times 2,920 (= 12 \times 5 \times 584)$. In a sense, this makes the table on page 24 a numerical summary for pages 46–50. For example, the third multiple, $3 \times 2,920$, gives a 260-Day Count date of 12 Ajaw, and the third row of page 50 ends with 12 Ajaw.

Förstemann thus demonstrated that page 24 was tied very closely to pages 46–50, and so it became clear that the manuscript had been torn at some point and then reassembled in improper fashion—that is, there should have been no gap between page "24" and page "46."

An anomaly did enter the table, though, at the fifth row of Long Count intervals up from the bottom of page 24, or the second row from the top. Here Tawiskal Uwoojil included the only four intervals on the page that are not whole multiples of 2,920 days. These are 1.5.14.4.0 (= 185,120); 9.11.7.0 (= 68,900); 4.12.8.0 (= 33,280); and 1.5.5.0 (= 9,100). Our scribe clearly did not intend to use these intervals to move down set numbers of rows across pages 46–50, as the other intervals in the page 24 table did. At the same time, each interval was written above the date 1 Ajaw—the 260-Day Count anchor to the whole table, and the starting point for all of the calculations on page 24.

This is where Teeple made progress that Förstemann could not. The librarian had speculated that the anomalous sequence of time intervals might have been intended to correlate Venus cycles with periods of Mercury and the Moon. Teeple, on the other hand, discovered a very different use for them. Teeple recognized that each of the anomalous intervals would take a date 1 Ajaw and lead to a later date 1 Ajaw since each was still a multiple of 260 days. His breakthrough was to hypothesize that these intervals might be able to "correct" for the error that would accumulate by using 584-day canonic periods when the actual synodic period of Venus was 583.9214 days (Thompson 1978, 224).

In his 1926 article, Teeple addressed the issue of correction with the recognition of four different 365-Day Count dates, each combined with the 260-Day Count date 1 Ajaw. On page 50, he found 13 Mak and 3 Xul, each corresponding to the 1 Ajaw anchor of the whole table; on page 24, he found two 365-Day Count dates explicitly joined to the 260-Day Count element: 1 Ajaw 18 K'ayab and 1 Ajaw 18 Wo.

Next, through mathematical experiment, he noticed that each of these Calendar Rounds could be generated through the application of an interval derived from the anomalous second row of time intervals. Using a time interval of 4.18.17.0, Teeple could start at 1 Ajaw 3 Yaxk'in and run through 1 Ajaw 18 K'ayab, 1 Ajaw 8 Yax, 1 Ajaw 18 Wo, 1 Ajaw 13 Mak, and then 1 Ajaw 3 Xul. The second, fourth,

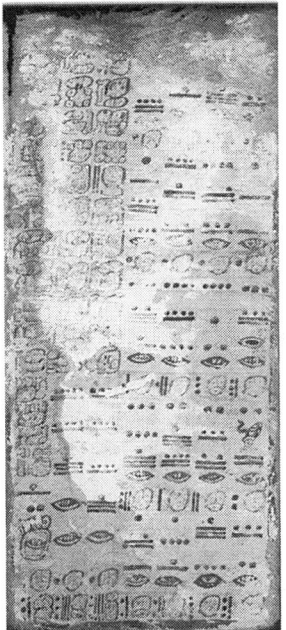

FIGURE 1.27 (*left*) Dresden Codex page 24, courtesy SLUB; (*below*) translation using Arabic numerals in Long Count positional notation and translation recognizing the table's structure based on multiples of 2,920 or 5 Venus Rounds

Left table (positional notation)

column of hieroglyphic text	paired columns of hieroglyphic text		V1	V2	V3	V4
			1	15	10	5
			1	16	10	5
			1	6	16	8
			14	0	0	0
			0			
			1 Ajaw	1 Ajaw	1 Ajaw	1 Ajaw
			1	9	4	1
			5	11	12	5
			14	7	8	5
			0	0	0	0
			1 Ajaw	1 Ajaw	1 Ajaw	1 Ajaw
			4	4	4	3
			17	9	1	13
			6	4	2	0
			0	0	0	0
9	9		6 Ajaw	11 Ajaw	3 Ajaw	8 Ajaw
9	9		3	2	2	2
			4	16	8	0
			16	14	12	10
			0	0	0	0
2	0	16	13 Ajaw	5 Ajaw	10 Ajaw	2 Ajaw
0	0	0	1	1	16	8
			12	4	4	2
			5	6	0	0
			0	0		
4 Ajaw	1 Ajaw	1 Ajaw				
6	16	9				
8 Kumk'u	18 K'ayab	18 Wo	7 Ajaw	12 Ajaw	4 Ajaw	8 Ajaw

Right table (multiples of 2,920 / 5 Venus Rounds)

column of hieroglyphic text	paired columns of hieroglyphic text		R1	R2	R3	R4
			151,840	113,880	75,920	37,960
			4 x	3 x	2 x	13 x
			37,960	37,960	37,960	2,920
			1 Ajaw	1 Ajaw	1 Ajaw	1 Ajaw
			185,120	68,900	33,280	9,100
			1 Ajaw	1 Ajaw	1 Ajaw	1 Ajaw
			35,040	32,120	29,200	26,280
			12 x	11 x	10 x	9 x
			2,920	2,920	2,920	2,920
9	9		6 Ajaw	11 Ajaw	3 Ajaw	8 Ajaw
9	9		23,360	20,440	17,520	14,600
			8 x	7 x 2,920	6 x	5 x
			2,920		2,920	2,920
2	0	16	13 Ajaw	5 Ajaw	10 Ajaw	2 Ajaw
0	0	0	11,680	8,760	5,840	2,920
			4 x	3 x 2,920	2 x	
			2,920		2,920	
4 Ajaw	1 Ajaw	1 Ajaw				
6	16	9				
8 Kumk'u	18 K'ayab	18 Wo	7 Ajaw	12 Ajaw	4 Ajaw	8 Ajaw

fifth, and sixth in this sequence were the four Calendar Rounds he had identified. Teeple showed that although this interval of 35,620 days wasn't actually written on page 24, it could be derived from the difference between the second and third intervals in this row, 68,900 (9.11.7.0) and 33,280 (4.12.8.0). This interval, he noticed, served to provide a four-day correction to match the Venus Table dates with Venus observations while preserving the entry date in the 260-Day Count: 1 Ajaw. It is this difference that Teeple used as the fundamental element in his reconstructed correction of the table (Bricker and Bricker 2007, 102).

Overall, these Correction Intervals suggested that just as the Mayan astronomers of the Classic period used an alternation of twenty-nine and thirty days in the Lunar Series to approximate the lunar synodic period of 29.5315 days, the author of the Dresden Codex made note of the difference between an idealized calendric progression of 584 days and the observable synodic period of Venus at 583.9214 days. Teeple suggested that behind this row of intervals, our scribe had quantified the amount of error that would have accumulated, and on page 24 the scribe recorded the means for correcting the table for future projections. Through Teeple's investigation, page 24 was shown to be a summary of pages 46–50, but also to record the means for correcting the table over long-term use. The author of the Codex would start with one of the recorded "anchors" and use the table to track Venus. Once the visibility of Venus slipped out of line with the table's predictions, the anomalous intervals would be used to "correct" the table.

Teeple's background in engineering proved instrumental in his approach to advancing an academic understanding of the Dresden Codex Venus Table. He took various forms of data and reduced them to a single parsimonious solution, generating a procedure that could be recycled for hundreds of years with very little accumulation of error. We will actually return to this solution in more detail in the final chapter of this book, but not before we see how it was incorporated into other scholarship and put to use for other purposes.

For now it is worth noting that were it not for the impact of World War I on the potash industry, Teeple may not have had the time to explore the math within hieroglyphic texts, and the role of astronomy in later interpretations of Mayan culture may have been very different. In fact, because the field was small, and the scholarship was young, specific people had inordinate impacts on the development of the field. For the Dresden Codex, Förstemann did the heavy lifting, revealing unequivocally its astronomical content; Seler reached broadly to demonstrate that what Förstemann found was part of a rich Mesoamerican tradition; and finally Teeple demonstrated that the Venus astronomy was of impressive complexity and technical sophistication. In the bigger picture, their collective work set the tone with the Lunar Series, the Eclipse Table, and various ensuing studies finding astronomical patterns among the inscribed dates for interpreting Mayan culture as heavily invested in astronomy. On the other hand, two of the three critical contributors to the interpretation of Mayan culture never set foot in Mesoamerica; rather, they encountered the work of Mayan scribes through the logic and coherence of math as language.

CHAPTER 2

B'ahlaj Chan K'awiil and Celestial Warfare in the Late Classic Period

According to the interpretation developed by Förstemann and Teeple, Tawiskal Uwoojil wrote up the Venus Table to closely follow Venus's visibilities in the morning and evening skies. The scientific accuracy that John Teeple found within the Venus Table made the case that Tawiskal Uwoojil had constructed an ephemeris not unlike those developed in ancient Greece or even Mesopotamia. As opposed to the cases east of the Atlantic, though, for Mayan astronomers there was no geometrical component; the calendar allowed for an algebraic-type model that alone, Eric Thompson later would argue, could preserve the match between prediction and observation to "within one day in six thousand years" (1972, 63). This apparent attention to "scientific accuracy," coupled with Seler's demonstration that k'uhulajaws too were invested in astronomy for their statecraft, became a firm anchor for interpretations of ancient Mayan civilization as work continued through the early decades of the twentieth century.

That growing astronomical interpretation was extended even further by archaeological excavation at different sites, suggesting that Mayan architecture had been aligned to mark the movements of celestial bodies. Visiting Chich'en Itza after Seler, for example, Oliver Ricketson identified an equinox alignment to the round structure that Stephens called the Caracol and that Ricketson would dub "the Observatory."

The Caracol is a circular tower with four outer doors facing the cardinal points of the compass. Within is a circular corridor from which four more doors, facing midway between the cardinal points, lead into another circular corridor. This inner circular corridor surrounds a masonry core, inside which a small, spiral staircase leads to the upper part of the building. Unfortunately, the upper terminus of the stairway cannot be ascertained, for, a little more than halfway to the top, the whole northern section of the building has completely fallen. Near the top of the structure is a flat area from which open three rectangular, horizontal shafts known as *ventanas*, or windows. (Ricketson 1928b, 219)

Although much of its roof had crumbled, Ricketson was able to determine that the windows on the upper story were aligned to allow for the accurate viewing of the vernal equinox sunset (1928b, 222) (figure 2.1). This work prompted him to close his article with a call to his colleagues:

> More work of this sort in the field and in the study will undoubtedly prove of value as a means toward the complete decipherment of the Maya hieroglyphs and a better understanding of their relationship to Maya civilization in general. It is sincerely to be hoped that all investigators in this field, whether visiting old sites or exploring new ones, will take accurate and copious bearings whenever the opportunity offers or the faintest suspicion arises that the arrangement of buildings or structural features may have been designed in accordance with astronomical directions. (Ricketson 1928b, 225)

FIGURE 2.1 (*above*) Photograph of the Caracol at Chich'en Itza; (*left*) illustration (after Ricketson 1928a) of the "three windows of the Caracol"

Not only did Ricketson recognize that astronomical records could be found within hieroglyphic inscriptions, but he also believed that finding astronomical evidence in architectural remains could help with the decipherment of hieroglyphic writing beyond numbers and calendric dates.

Part of this perspective was motivated by an earlier trip Ricketson made to an archaeological site near Tikal that Morley named Uaxactun (*uaxac-tun* or "8-stone") after the early Long Count date he found recorded on a stone monument at the site (*tuun* is "stone" in most Mayan languages; Coe 1999, 127; Sharer 2006, 78). Ricketson and colleague Frans Blom followed up on Morley's preliminary work ten years earlier to document a "Group of the Solar Observatory" architectural complex there, which they found to mark sunrise on the solstices and equinoxes (Ricketson 1928a). Specifically, Ricketson found that a pyramidal structure in the middle of a plaza provided a direct view of three "temples" all constructed and evenly spaced, north to south, upon a platform on the eastern edge of the plaza. From the pyramidal structure, a viewer could witness the Sun rising behind the southern edge of the southern structure on the winter solstice, directly behind the central structure on either equinox, and behind the northern edge of the northern structure on the solar equinox (Ricketson 1928a, 436) (figure 2.2). The

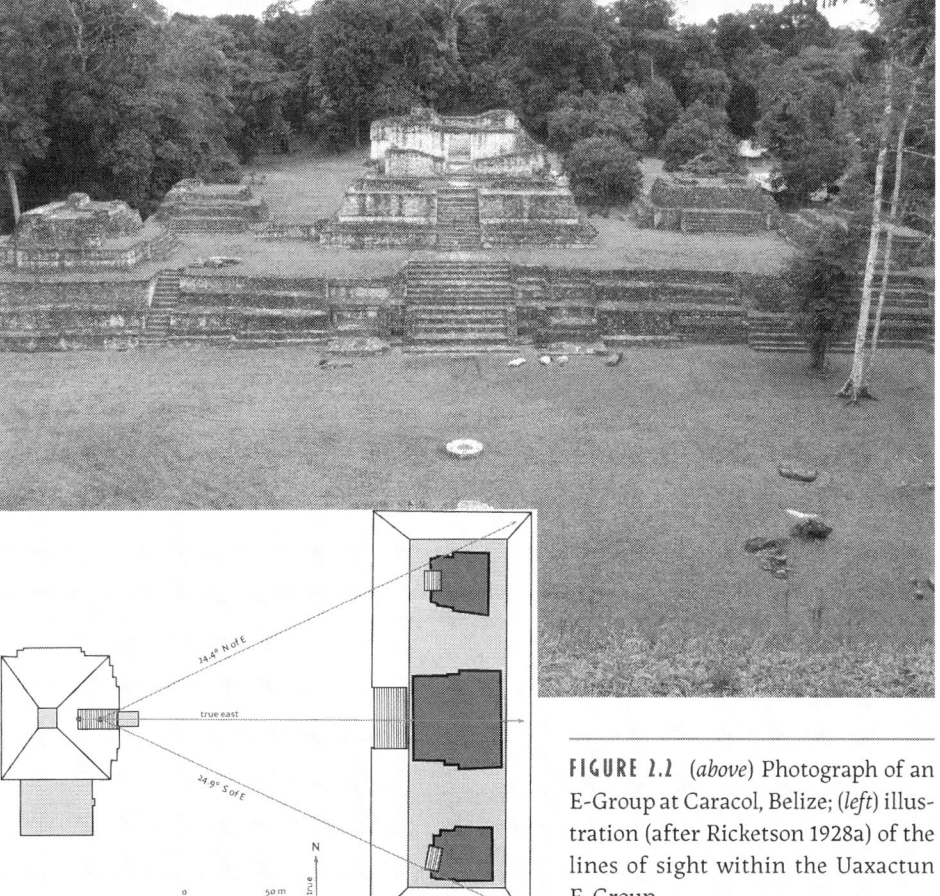

FIGURE 2.2 (*above*) Photograph of an E-Group at Caracol, Belize; (*left*) illustration (after Ricketson 1928a) of the lines of sight within the Uaxactun E-Group

entire architectural complex, therefore, acted like an annual calendar; similar complexes have been found at several other Preclassic and Classic sites since (Aveni 2001, 288–93).

The upshot is that through the growing archaeological work in the Mayan region alongside the computational material we saw in the last chapter, a coherent Mayan astronomy was emerging based on both calendric patterns and architectural features. Unfortunately to this point, little regard was provided for temporal or geographical differences and no substantive input came from noncalendric hieroglyphic textual material.

In spite of the growing coherence, however, there were still puzzles left to be resolved within the Dresden Codex Venus Table itself. In two years of observing the planet, for instance, Förstemann had derived more accurate observable subperiods than those recorded in the Venus Table; so why had Tawiskal Uwoojil used such short-term *in*accuracy when the full table could be made to preserve accuracy over hundreds of years? Why did our scribe mark the subperiods of a Venus cycle at all when the records that Seler had uncovered from Central Mexican codices only emphasized first morning visibility of the planet? And why did page 24 record so many anomalous Correction Intervals when, according to Teeple, only one (which wasn't actually recorded) was necessary?

Questions also arose within other details. As we saw in chapter 1, our scribe recorded two Long Count dates on page 24: one from mythical times, expressed as a "Ring Number" as 6.2.0, the other, 9.9.9.16.0 1 Ajaw 18 K'ayab, as a date within the earlier part of the Middle Classic period. The latter date corresponded to historical times and so immediately raised several questions concerning the use of the table. Was it invented in the Middle Classic period and then recycled by copyist after copyist through the late Postclassic so that Tawiskal Uwoojil was the last in a line of centuries of copyists? Was the table invented even earlier, but accommodated in the Middle Classic to incorporate the numerology of mythic times? Was the Classic period Long Count date a reference point with no real historical bearing on its use for accurate predictions of first morning appearances of Venus at all?

These questions were relatively open for Teeple and his colleagues. But the reference to the Classic period means something different today than it did to them then. As we saw in chapter 1, the early *in*ability to read the hieroglyphic text resulted in interpretations of the inscriptions that focused entirely on astronomy and calendrics, with no substantive historical content (cf. Aveni 2001, 129; Coe 1999, 129; W. Fash 1991, 53–54). The decipherment of the writing system that has matured since the 1980s, however, now has us able to uncover the rich history left by Classic period scribes. And so now we can turn to the content of the inscriptions to guide our responses to the open questions that Teeple confronted. We can now, for example, examine historical trends that provide insight into the changing role of astronomy/astrology from the Middle into the Late Classic period. By doing so, in this chapter we confront examples of Mayan astronomy as it was practiced and maintained in the time period around that Long Count date in the Venus Table, leading up to the time of Tawiskal Uwoojil's copy.

To engage this historical context, we move from the northern Yucatan Peninsula where the Dresden Codex was written and K'uk'ul Ek' Tuyilaj's Chich'en Itza, which Seler and Maudslay visited in ruins, into the southern Peten of what is now Guatemala, where

Morley and Ricketson worked in the 1910s and 1920s. This region was dominated by the network of rivers feeding into a superhighway of Classic period economic activity, the Usumacinta River, which strongly impacted the lives of the k'uhulajaws and almehen presiding over the region during the seventh century AD.

ASTRONOMY IN THE LATE CLASSIC

Although there was never a Mayan empire in the sense of the centuries-later Aztec Triple Alliance, Mayan cities did form political and economic alliances across vast regions—and those alliances were generally not egalitarian. Maudslay's hypothesis that the various cities he visited in the south all participated in a contemporaneous cultural sphere has passed the test of time; the hieroglyphic inscriptions at these cities at least make that clear. As we saw in the last chapter, almost all Long Count dates preserved on monuments from these cities start with a baktun coefficient of 9, placing them within a four-hundred-year period around AD 400 to 800. The use of hierarchical titles and relationships in these inscriptions also attests that there were significant political differences dividing them. Epigraphers Simon Martin and Nikolai Grube write:

> Our own research . . . points to a pervasive and enduring system of 'overkingship' that shaped almost every facet of the Classic landscape (pp. 20–1). Such a scheme accords closely with wider Mesoamerican practice, while seeming to reconcile the most compelling features of the two existing views, namely the overwhelming evidence for multiple small kingdoms and the great disparities in the size of their capitals. (2008, 19)

Martin and Grube go further to argue that the ruling dynasties of two cities in particular, Tikal and Calakmul—two "superpowers"—dominated the Mayan region during the Classic period and that these two kingdoms were generally at odds with each other (figure 2.3). For the purposes of this chapter, we examine the tension at a critical moment during the overall power play between these two dynasties. We find that astronomy played a significant role in the narrative, which in turn will help to inform our reading of the interests Tawiskal Uwoojil may have been serving and the history from which he drew with his Venus Table in the centuries-later Dresden Codex.

Betrayal

At the center of the tension between superpowers during the seventh century AD stood the young Mayan almehen B'ahlaj Chan K'awiil. Member of a ch'ibal that had deep roots in the nobility of the city, he was, however, not the designated heir. Still, it was he who left to posterity a rich historical record providing insight into some fascinating political maneuvering between the two superpowers at a time of the Calakmul dynasty's predominance (Martin and Grube 2000, 42).

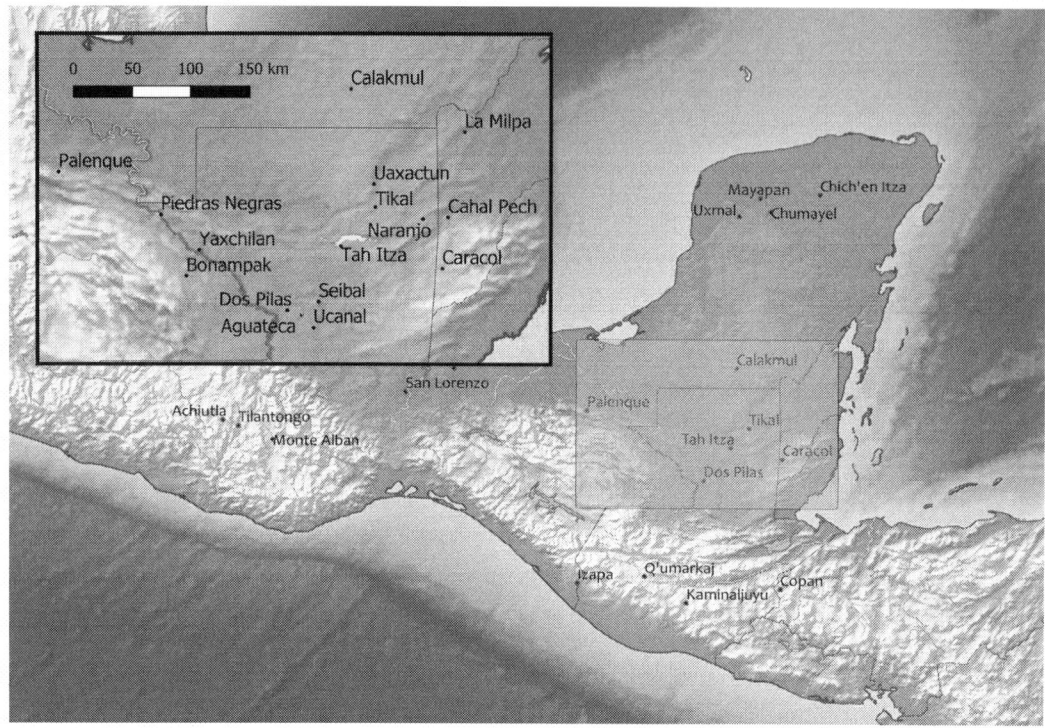

FIGURE 2.3 Tikal, Uaxactun, and other cities in the central Peten

B'ahlaj Chan K'awiil grew up in luxury at the city of Tikal, located midway between the eastern and western watersheds of the Yucatan Peninsula. Agricultural intensification techniques using the swampy *bajos*, invented centuries earlier, generated ample maize-centered nutrition for the population year-round. The location also provided access to navigable rivers to the east and west (Harrison 1999, 46), and since trade in the Peten made substantial use of river-borne canoes, this put Tikal in a very strategic position—one that its inhabitants enjoyed from the Late Preclassic period all the way into the Terminal Classic, or for close to a thousand years. Over that extended time period, B'ahlaj Chan K'awiil's ancestors in the Mutul dynasty had nurtured the population from a somewhat large town in the third century AD into perhaps the most powerful city of the Early Classic Peten (figure 2.4). By the reign of Chak Toh Ich'aak, ninth member of the dynasty in the mid-fourth century AD, its status and location led to an important political and economic shift (Harrison 1999, 76–78).

The growth of the city was spurred during the Early Classic by the intervention of the elite affiliated with the great metropolis of the Central Mexican Basin, Teotihuacan. The hieroglyphic record at Tikal, on stelae throughout the plazas that B'ahlaj Chan K'awiil strolled through in his youth, spells out the arrival of foreigners Sihyaj K'ahk' and Yax Nuun Ayin at the end of Chak Toh Ich'aak's reign at Tikal. This intervention is fraught with mystery even though it has been explored extensively since Proskouriakoff's work on what she referred to as "The Arrival of Strangers" (1993, cf. Harrison 1999, 71–73;

FIGURE 2.4 The civic-ceremonial core of Tikal; photograph of Jasaw Chan K'awiil's Temple I viewed from behind Sihyaj Chan K'awiil's funerary monument, Temple 33 in the northern acropolis

Stuart 2000; Sharer 2006, 310). At its core, the question is concerned with the relationship between the arrival of Yax Nuun Ayiin and the death of Chak Toh Ich'aak. Was this violent overthrow? A planned transition of power? A coup among factions of the nobility? The hieroglyphic record at Tikal does not tell us. We do find archaeologically, though, that the city continued to prosper thereafter.

Six different members of the new "mestizo" dynasty led the city of Tikal's growth in size and political influence for the next century (Aldana 2017; Martin and Grube 2000, 34–39). Near their peak of power, however, in the late sixth century, that growth ceased at the military defeat of their army by their increasingly antagonistic rival, the k'uhulajaw in the dynasty of Kaanul at Calakmul. The conflict between superpowers reached a peak, with victory in this case going to the Kaanul dynasty.

A century of subordinate political stature at Tikal followed their dramatic defeat before B'ahlaj Chan K'awiil and his paramount half brother, Nuun Ujol Chaak, were born. Both were sons of the Mutul k'uhulajaw K'inich Muwan Jol (Martin and Grube 2000, 42). Both learned the stories in their youth that extolled the virtues of the city's growth to great stature and the regional prominence it enjoyed. But the reality they lived was within a politically maintained hiatus of monumental construction. Society was still stratified within the city's population, of course, but relative to the leadership of Calakmul and their allies, Nuun Ujol Chaak, B'ahlaj Chan K'awiil, and members of the Tikal nobility were relegated to subordinate status regionally. During this time, the Kaanul dynasty of Calakmul had

FIGURE 2.5 Yuhknoom Ch'en, Kaanul k'uhula-jaw of Calakmul (after Hieroglyphic Stairway 2 at La Corona)

FIGURE 2.6 (*below*) Dos Pilas within the Petexbatun, near the Usumacinta River; (*right*) B'ahlaj Chan K'awiil, Mutul k'uhulajaw at Dos Pilas (after Stela 8)

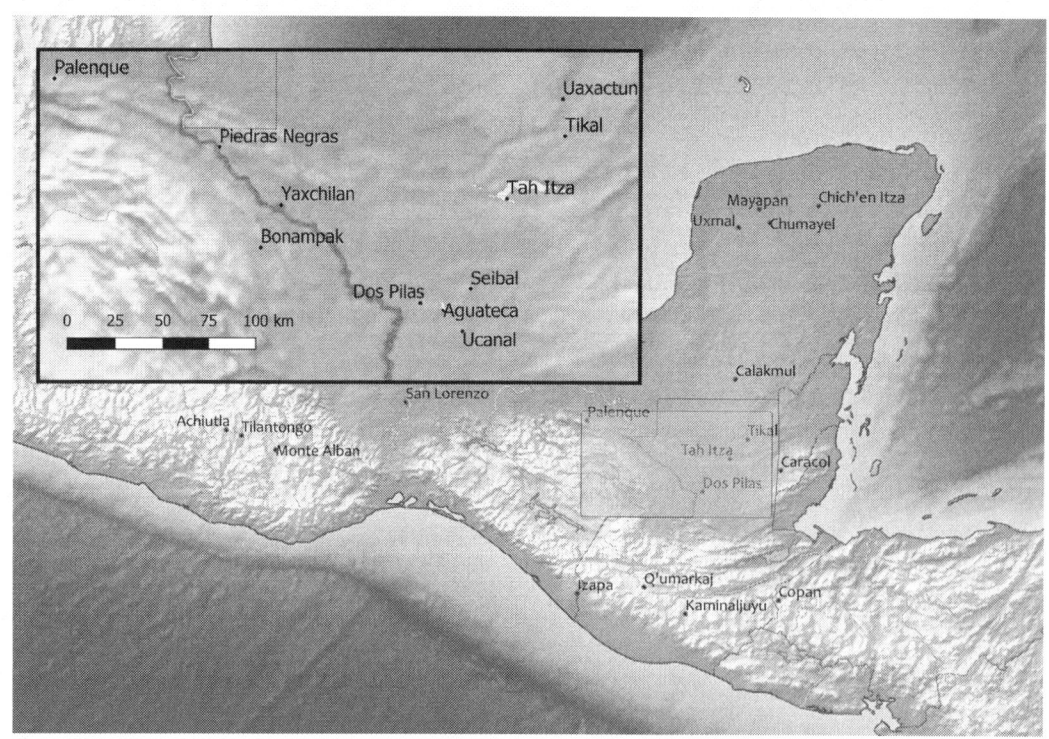

consolidated its strength throughout the Peten, forming alliances with already prestigious dynastic families and supporting new ones; Yuhknoom Ch'en was the Kaanul k'uhulajaw, acceding to his throne at the age of thirty-six on 9.10.3.5.10 8 Ok 18 Sip (figure 2.5).

With just this glimpse of the historical record, we encounter the puzzle that sets the backdrop for the exploration of politics and astronomy during the Classic period. That is, not long into Nuun Ujol Chaak's reign, his younger half brother, B'ahlaj Chan K'awiil, was the head of a new town built in an already densely populated area one hundred kilometers south of Tikal, just off the Pasión River (Martin and Grube 2000, 42) (figure 2.6). B'ahlaj Chan K'awiil had left behind his residence at one superpower and was about to directly engage the political machinations of the other.

Hieroglyphic History in the Petexbatun

In a rich retrospective account patronized very late in his reign, B'ahlaj Chan K'awiil had his scribes record his version of the series of military battles he waged against Nuun Ujol Chaak (cf. Martin and Grube 2000, 42–43 and 56–58). Recalling Landa's comments regarding the content of codices in Precontact Mesoamerica, it is fair to assume that B'ahlaj Chan K'awiil's scribes kept robust historical records of his royal life in their

hand-painted manuscripts. Drawing from what was probably still a modest library of codices at this new city, these scribes highlighted a sequence of martial events within a life history to decorate the stairways of the main structures bordering the central plazas of his adopted home (figure 2.7).* As they crafted it, the story began on the Central Staircase of the largest temple of the site (now referred to as Structure L5-49).

FIGURE 2.7 (a) Hieroglyphic Stairway in the central plaza at Dos Pilas; (b) detail of the stair risers, carved with hieroglyphic text

* There are several excellent treatments of this history (Fahsen 2002; Guenter 2003; Boot 2002; Martin and Grube 2000). Also, note that Sharer (2006, 387) records a substantially different history, suggesting that B'ahlaj Chan K'awiil's father hatched the plan, installing B'ahlaj Chan K'awiil as ruler of Dos Pilas when he was four years old. This interpretation is based on an early reading of the field drawings of the text by Erik Boot (2002). Guenter later argued that the text was sufficiently eroded so that the *hul-iy* reading could not be verified. It looks, therefore, that the text is not referring to B'ahlaj Chan K'awiil acting at Dos Pilas at four years old; instead, this is probably another record of a preaccession ritual performed at Tikal (Guenter 2003, 6–7).

As noted earlier, the corpus of hieroglyphic inscriptions is filled with references of the visits of almehen from one city to their counterparts at another. If B'ahlaj Chan K'awiil himself were to have guided an ally k'uhulajaw through a tour of his town, then, he probably would have started at the Central Staircase of Structure L5-49 with the narrative designed by his scribes. Following conventional practice, they initiated this text with the date recording B'ahlaj Chan K'awiil's birth:

FIGURE 2.8 B'ahlaj Chan K'awiil's birth statement on Dos Pilas Hieroglyphic Stairway 2, Central Stair, Step 6. The last five glyphs read: [On 9.9.9.12.11.2] 8 Ik' 5 Keh, the Mutul k'uhulajaw B'ahlaj Chan K'awiil was born.

This was immediately followed by the preaccession rites he went through at nine years of age (DPHS2, Step 4). As he memorialized his own personal history, then, B'ahlaj Chan K'awiil had been designated (at least a potential) heir in his youth at the great city of Tikal, even though we know now—as would his noble visitor—that the throne would go to his older brother. The narrative, however, makes no mention of his brother, Nuun Ujol Chaak; instead the Central Staircase text of Structure L5-49 continues on unproblematically documenting B'ahlaj Chan K'awiil's ritual activities in his youth through eighteen years of age. In each case, from top to bottom step text, he carries the title of Mutul k'uhulajaw—"ruler of Tikal" (Guenter 2003, 4).

One way to read this abridged account, then, is that he, B'ahlaj Chan K'awiil, had been favored for the throne at a young age, but it was taken from him and given to his half brother, Nuun Ujol Chaak. With such a narrative developing, the k'uhulajaw would have now guided his visitor back to his left, shifting from the Central Staircase of the building to the East Staircase and the riser of the top step there.

The East Staircase narrative picks up the story with the strife that ensued and may have permitted B'ahlaj Chan K'awiil to engage in some theatrical drama at its recounting. At age twenty-one, B'ahlaj Chan K'awiil himself was responsible for the death of a nobleman from Tikal. The inscription itself is that straightforward:

FIGURE 2.9 The death of a Tikal ajaw. The (now eroded) inscription reads: The 6 Ajaw 13 Mak period ending, then it happened on [9.10.15.4.9] 4 Muluk 2 Kumk'u, [eroded]

K'awiil, Mutul ajaw, died at Sakha'al. He was captured, by the authority of B'ahlaj Chan K'awiil, Mutul k'uhulajaw. Dos Pilas Hieroglyphic Stairway 2, East Stair, Step 6

The text gives no other justification, which suggests that, without provocation, B'ahlaj Chan K'awiil fought against his home's own army in military assault.

The death of one Tikal almehen caused by another suggests an attempted coup—at least it may have to B'ahlaj Chan K'awiil's guest. And while it succeeded in dispatching a nobleman of Tikal, it must have ultimately failed as attested by the next hieroglyphic step, from which the visitor would learn that B'ahlaj Chan K'awiil left his home. Whether in haste, driven out by his brother's army, or more slowly, as punishment for a snuffed-out betrayal, the younger brother set out from Tikal—after regrouping, probably atop a palanquin and accompanied by a guard of warriors—through the rainforest to establish a new town near Lake Petexbatun and the Pasión River, 120 kilometers southwest. This was no move out of convenience; a trip of this distance with his guard and supporting noble houses would have taken ten days or more. Nevertheless, it was here that he created the village of Dos Pilas—even if initially it would not be long-lived.

For the next record—on the next step of the East Staircase—B'ahlaj Chan K'awiil's scribes intensified the narrative of tragedy, inscribing his military defeat at the hands of Yuhknoom Ch'en, k'uhulajaw of Calakmul.

FIGURE 2.10 The defeat of B'ahlaj Chan K'awiil at his new home. The inscription reads: Before 12 Ajaw 8 Keh, the eleventh winikhaab [on 9.10.18.2.19] 1 Kawak 17 Muwan, Dos Pilas was defeated under the auspices of Yuhknoom Ch'een, Kaanul k'uhulajaw. The Mutul k'uhulajaw B'ahlaj Chan K'awiil fled to Aguateca. Dos Pilas Hieroglyphic Stairway 2 East Stair, Step 5.

Less than three years after his arrival—and probably just as platforms were being finished and permanent structures were taking shape—B'ahlaj Chan K'awiil's new residence was attacked by Yuhknoom Ch'en, the Tikal dynasty's archrival. According to his own official history, then, it appears that B'ahlaj Chan K'awiil left his home at Tikal and founded a town for his exile. The narrative focuses on his own royal self, but he could not have attempted the coup and founded a new town without a contingent of loyal almehen families. This whole contingent was then attacked by Nuun Ujol Chaak's archenemy, and B'ahlaj Chan Kawiil was forced to find new refuge at the neighboring city of Aguateca. At this point, still only twenty-five years of age, B'ahlaj Chan K'awiil's misfortune had peaked, but his hieroglyphic steps tell us that that was about to change.

Perched on a hillock above the Pasión River, Aguateca was already an old city by the time B'ahlaj Chan K'awiil found refuge there. It was one of several cities that had been established in the Preclassic along these rivers, hundreds of years before Tikal rose to power, to transport quetzal feathers, cacao, jade, and obsidian back and forth by canoe between the Gulf Coast and the southern Mayan region (Aoyama 2008). It was undoubtedly an incredibly lucrative geographical position, and so probably was never very politically stable either. During the seventh century, the ruling families of Aguateca were well established and well positioned to facilitate B'ahlaj Chan K'awiil's arrival in his time of need. Approximately ten kilometers away, he could have reached this place of refuge from Dos Pilas with a few hours travel.

To this point, the personal history that B'ahlaj Chan K'awiil patronized read as one of a prince struggling through the challenges of Classic period politics. But in his own time, we shouldn't fault his older brother, Nuun Ujol Chaak, had he become suspicious of his brother's activities. Once the news reached him that B'ahlaj Chan K'awiil had been defeated by Yuhknoom Ch'en, it may have occurred to him that his stepbrother in fact had set up an alliance with the ruling nobility of Aguateca *before* his treasonous act at home. A covert alliance with a faction of the almehen at Aguateca would have B'ahlaj Chan K'awiil setting up a new city, Dos Pilas, putting only that city and its inhabitants—mostly construction workers, noble allies, and a small army—at risk. Clearly this would have been an easy target for Yuhknoom Ch'en's military force. On the other hand, had B'ahlaj Chan K'awiil fled directly to Aguateca from Tikal, that city would have opened itself up to retaliatory attack by Nuun Ujol Chaak, or possibly worse. By acting only as a sympathetic source of refuge after attack, the almehen families of Aguateca could deflect allegations of complicity and then support B'ahlaj Chan K'awiil in the aftermath. The threat of Yuhknoom Ch'en bringing Calakmul's army into the region must have been on everyone's mind.* So the degree of political strategizing underlying B'ahlaj Chan K'awiil's move to the Petexbatun region was probably open to speculation for Nuun Ujol Chaak then, just as it still is for us now.

It may strike the modern reader as overindulgence to suggest this level of political intrigue behind Classic Mayan politics. With a scant hieroglyphic record, clearly biased in that it only represents one side of the story, can we presume that the politics were this complicated? In fact, we have indigenous histories written during the sixteenth century (such as the Books of Chilam Balam—of which we will encounter more below—and the Central Mexican *Anales de Cuauhtitlan* that we saw in chapter 1) that are filled with historical examples of similar fractured alliances and unstable political relationships (Bierhorst 1998; Roys 1933). Even the *Popol Vuh*, which usually is drawn upon for its rich mythological content, includes political tension and machinations in its culminating, historical narrative—machinations that led up to and in some ways facilitated Spanish conquest (Christenson 2007; D. Tedlock 1985). Given such a context, the rivalries and alliances of the Mayan heartland in the seventh century appear "normal" and, in fact, fall in line with cross-cultural examples from imperial China, Medieval Europe, and even modern U.S. politics.

---

*Itzan reference to Calakmul, Huxte Tuun for 9.11.0.0.0 12 Ajaw 8 Keh P.E. Itzan is 20 km NW of Dos Pilas.

Regardless of its backstory, the outcome was unambiguous. Military defeat by the army of Calakmul meant that B'ahlaj Chan K'awiil now—at best—would be subject to Yuhknoom Ch'en and integrated into the Calakmul sphere of alliances if he wished to maintain any leadership role whatsoever. And if he indeed accepted a vassal status, then he likely would have been able to return to Dos Pilas and build it into a proper Mayan city.

Actually, it appears that B'ahlaj Chan K'awiil did not choose the location of his new home purely out of haste or solely out of an interest in its economic potential. Some planning involving other factors must have been involved. Archaeologists have found that the architects at Dos Pilas constructed the civic-ceremonial center relative to an impressive cave network in the area (Brady 1997). More to the point, these caves preserved evidence of having been used for ritual activity for centuries before B'ahlaj Chan K'awiil and his associates arrived (Brady 1997, 610). While the civic-ceremonial center itself was constructed adjacent to the two natural springs that give the site its contemporary name—and to which the hieroglyphic stairway of Structure L5-49 directly faces—the growth of both ceremonial and residential architecture in the new city followed access points to the Underworld. By placing his new town on top of this regionally important religious landscape, B'ahlaj Chan K'awiil must have had significant support from the almehen families who were making use of the site. It also now provides balance to hypotheses that astronomical considerations were principal in site location and layout. At Dos Pilas, at least, architecting the Underworld and its relationship to the middle realm was just as, if not more, important.

Over the next several years, B'ahlaj Chan K'awiil began to get comfortable with local politics in the middle realm and developed his new city at Dos Pilas both architecturally and economically (Martin and Grube 2000, 57). In part, this involved his marriage to a princess from nearby Itzan (Guenter 2003, 22)—although whether for love or strategy, again we have no record.

A happy and stable home life and entrepreneurial story is not, however, the theme underlying B'ahlaj Chan K'awiil's recorded history—nor the point of our addressing it here. The next step in B'ahlaj Chan K'awiil's stairway history introduces an aspect of Yuhknoom Ch'en's own long-term strategy, and it brings us back to the military theme. Six years after defeating the weaker member of the "house divided," Yuhknoom Ch'en set his sights on the older brother. Step 4 of the structure's east staircase records an attack on Nuun Ujol Chaak at Tikal by Yuhknoom Ch'en of Calakmul.

FIGURE 2.11 Nuun Ujol Chaak at Tikal is attacked by Yuhknoom Ch'en of Calakmul. The inscription reads: The eleventh winikhaab, then it happened on [9.11.4.5.14] 6 Ix 2 Kayab, Tikal was defeated by the authority of Yuhknoom Ch'een, Kaanul k'uhulajaw. Nuun Ujol Chaak was driven out. Dos Pilas Hieroglyphic Stairway 2, East Stair, Step 4

Now this was no trivial battle. With this victory, Nuun Ujol Chaak was driven out of Tikal, further destabilizing the political leadership of that city. Archaeologist and epigrapher Stan Guenter suggests that "much of the damage to the Early Classic stelae and structures" revealed by excavation and still visible at Tikal today "likely can be attributed to this event" (2003, 18). We can imagine that Yuhknoom Ch'en had an interest in marring local histories in order to make room for a new future. Since the other member of the Tikal dynasty was busy with his Dos Pilas project, Yuhknoom Ch'en could now comfortably stand alone as the most powerful k'uhulajaw in all Mayan lands.

The next move by Yuhknoom Ch'en sets the parameters for the next twenty years of interaction between the three main protagonists and the fates of their kingdoms. In the wake of the defeat of both Tikal and Dos Pilas, Yuhknoom Ch'en demanded the attendance of both stepbrothers at a preaccession ritual for his son. The East Stair Step 4 of Hieroglyphic Stairway 2 is heavily eroded, but what remains describes B'ahlaj Chan K'awiil and Nuun Ujol Chaak coming together at the city of Yaxha' to "witness" an event critical to the political legitimacy of the Kaanul dyansty (see figures 2.12 and 2.13). That both were now subject to the city of Calakmul can be inferred from their individual defeats. Their requirement to attend this political event, however awkward, was probably not unexpected (Guenter 2003). It may have been a meeting intended to set things back in some semblance of order—with Calakmul now at the head of a new alliance. Perhaps there was even discussion of a new hegemony with Calakmul atop a hierarchy of empire. Archaeologist and epigrapher Stan Guenter, who has treated these hieroglyphic texts exhaustively, refers to this event as generating a "Yaxha Agreement"—a political summit to construct new relationships (2003, 19). If so, it proved empowering to B'ahlaj Chan K'awiil. The rest of the East Stair continues by establishing B'ahlaj Chan K'awiil's military success, ending with his capture of a warrior named Tajal Mo' (2003, 22).

FIGURE 2.12 The meeting at Yaxha: Yuhknoom Ch'en, Kaanul k'uhulajaw, Nuun Ujol Chaak, B'ahlaj Chan K'awiil, Mutul k'uhulajaw. It happened at Yaxha. Dos Pilas Hieroglyphic Stairway 2 East Stair, Step 4

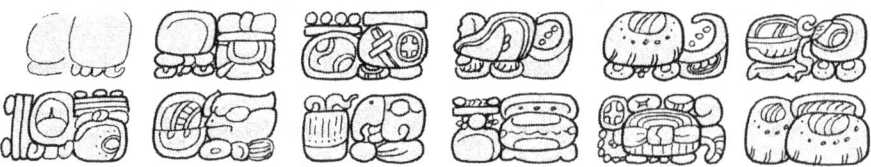

FIGURE 2.13 The half-period end of 12 Ajaw 18 Ch'en, then it happened on 9 Kaban 5 Pohp, Tahal Mo' was captured. He was the captive of B'ahlaj Chan K'awiil, Mutul k'uhulajaw, Bakab. Dos Pilas Hieroglyphic Stairway 2, East Stair, Step 3

Over the next thirty years—actually to his death—B'ahlaj Chan K'awiil becomes a trusted and even powerful and influential ally to Yuhknoom Ch'en and his son. In other words, B'ahlaj Chan K'awiil embraces his position as Calakmul vassal. The West Stair in fact ends with B'ahlaj Chan K'awiil's celebration of the 9.12.10.0.0 period end—dancing with Yuhknoom Ch'en at Calakmul—followed by his sixtieth birthday and his son's birth (figure 2.14). Nuun Ujol Chaak of Tikal, however, decided to take the opposite approach.

FIGURE 2.14 It happened on the 9 Ajaw half-period end, Yuhknoom Ch'en, Kaanul k'uhu-lajaw danced, accompanied by B'ahlaj Chan K'awiil, Mutul k'uhulajaw. Then on 2 Ik' 10 Muwan, B'ahlaj Chan K'awiil, Mutul k'uhulajaw completed his third winikhaab.

Alliances

Nuun Ujol Chaak—the elder brother—still had allies not subject to Yuhknoom Ch'en on whom to rely. While this doesn't show up in the inscriptions of Dos Pilas, we find that in his time of need Nuun Ujol Chaak turned to the famed ruler of Palenque, Janaab' Pakal, Baakal k'uhulajaw. The record in question—which has inspired some debate among modern scholars—was excavated from the now iconic funerary monument for the great Janaab' Pakal—the "Temple of Inscriptions" (figure 2.15). Janaab' Pakal was fifty-three years old and at the peak of his regional influence when Nuun Ujol Chaak was exiled, and while the Baakal scribes undoubtedly were recording in their books the history for which he would become famous. These same scribes eventually summarized some of this history in a long hieroglyphic inscription that takes up three large tablets in the Temple of Inscriptions at Palenque and which is fascinating in its own right (Aldana 2007; Sharer 2006, 453).

The first tablet begins with a formulaic dynastic history, but is interrupted by the critical military defeat of Palenque in the Middle Classic period, more than forty years before the feud began between brothers at Tikal (figure 2.16). When Janaab' Pakal was eight years old, the city was devastated by the army of Calakmul, led by Yuhknoom Ch'en's father (Martin and Grube 2000, 105, 108). Since the inscriptions bearing this history are in the funerary monument to Janaab' Pakal, we shouldn't be surprised to find that the defeat feeds into an historical trope of eventual triumph crafted by the scribes—much like B'ahlaj Chan K'awiil's hieroglyphic stairway. Most of the rest of the narrative over the other two tablets describes Janaab' Pakal's successes through his long sixty-year reign as k'uhulajaw (Aldana 2007; Schele and Mathews 1999). But the last section of the third tablet changes character. It is here that we stumble upon an ancient example of the now common problem of "coauthorship."

Janaab' Pakal lived to eighty-one years of age, so he certainly had the opportunity to consider his own mortality and how it would be represented in local histories. In

FIGURE 2.15 Janaab' Pakal's funerary monument, the Temple of Inscriptions at Palenque

particular, it appears that he "left space" at the end of the three tablets of his funerary monument in order for his successor to include the record of his burial. The last passage in the inscription refers to his interment on 9.12.11.5.18 6 Etz'nab 11 Yax (Martin and Grube 2000, 162). There is, accordingly, a change in the narrative about halfway through this third tablet, which has been interpreted in different ways by modern scholars (Aldana 2007; Schele and Mathews 1999, 101–8; M. Robertson 1983).

In the latter portion of the third tablet of hieroglyphic text in the Temple of Inscriptions, several events are brought together without clear theme. Some do relate important events in Janaab' Pakal's life; others appeal to mythological events in the deep past (Aldana 2007; Schele and Mathews 1999, 106). It is among this jumble of historical and mythological information that we encounter the record of particular importance to the feud at Tikal (cf. Schele and Mathews 1999, 95–97). The hieroglyphic text (figure 2.17) states that "Nuun Ujol Chaak arrived at Palenque."

Why Kan B'ahlam chose to include this reference is not explicit in the rest of the text. It is one vignette among various others. But returning to Dos Pilas, there is a clear importance to the history of the Tikal-Calakmul dynastic rivalry.

B'ahlaj Chan K'awiil's historical stairway noted that Nuun Ujol Chaak was ousted from Tikal by Yuhknoom Ch'en on 9.11.4.5.14. That places it two years before Nuun Ujol Chaak "arrived at Palenque."* If the Yaxha' Agreement occurred on the ritual event

---

*There is still some debate as to the identity of this figure. As noted, the context is so slim that it is hard to be definitive. The text clearly records the name Nuun Ujol Chaak, but the Emblem Glyph is unconventional for Tikal (or Dos Pilas) (cf. Schele and Mathews 1999, 95; Martin and Grube 2008).

FIGURE 2.17 Nuun Ujol Chaak's refuge at Palenque. The inscription reads: 17 k'in, 16 winik, 6 haab after 12 Ajaw 8 Keh, then Nuun Ujol Chaak, Mutul k'uhulajaw, arrived. Palenque Temple of Inscriptions, West Panel

FIGURE 2.16 The defeat of Palenque under Aj Ne Ohl Mat's reign as Baakal k'uhulajaw. The inscription reads: 14 k'in, 6 winik, after 13 Ajaw 18 Mak Lakam Ja' was chopped down on 4 Ix 7 Wo by the authority of the Kaanul k'uhulajaw. Palenque Temple of Inscriptions, East Panel

memorialized on Site Q Panel 6 (Guenter 2003, 19), then the record at Palenque tells us that Nuun Ujol Chaak did not go home to take up service as under-lord to Yuhknoom Ch'en. The hieroglyphic record instead suggests that Nuun Ujol Chaak sought refuge and counsel from Janaab' Pakal and may have remained there for almost a decade, recovering and planning his response.

B'ahlaj Chan K'awiil actually highlighted his brother's eventual response in multiple public places at Dos Pilas (Guenter 2003, 23). On 9.12.0.8.3 4 Ak'bal 11 Muwaan, Nuun Ujol Chaak and his army attacked B'ahlaj Chan K'awiil at Dos Pilas and drove him out of his residence. Fifteen years after he had been defeated by the k'uhulajaw of Calakmul, the disgraced ruler of Tikal mounted an effective military action against his brother, regaining for himself the respect to return to his throne at Tikal. The two brothers remained in conflict for the rest of their lives.

While it leaves many questions open to the imagination, on a historical level we might now reconsider the question of precisely when B'ahlaj Chan K'awiil agreed to ally with Yuhknoom Ch'en. It is possible to posit, for example, that he hatched the plan along with key members of the nobility at Aguateca when he found that he was being passed over for the Tikal throne. That may sound young to be plotting political intrigue on an international level by modern standards, but in Precontact indigenous Mesoamerica and South America, men were often married at fifteen and taking leadership roles long before turning twenty. So maybe B'ahlaj Chan K'awiil was full of bravado at an early age, and he turned that into a complex undercover arrangement that had him building a new base for the Mutul dynasty in a strategically powerful location with key members of the Aguateca dynasty by the time he was twenty. Rather than directly imperil the established polity and nobility of Aguateca, the plan had them create a new town, Dos Pilas, and then arrange for Yuhknoom Ch'en to come and destroy it. It would then only be a matter of time for the betrayal to play out.

Of course there are other possible scenarios. Because the Pasión River was so valuable, it may be that Yuhknoom Ch'en crafted a plan on his own and then waited for the right set of conditions. The Kaanul k'uhulajaw was already thirty-six when he acceded and fifty-six when his army attacked B'ahlaj Chan K'awiil at Dos Pilas. His spies in the royal courts of Tikal and Aguateca may have found the perfect scenario with B'ahlaj Chan K'awiil as an overly ambitious, passed-over prince. Indeed, after Nuun Ujol Chaak's ouster from Tikal and B'ahlaj Chan K'awiil's victory spurred by Yuknoom Ch'en, B'ahlaj Chan K'awiil took on the title of Mutul k'uhulajaw—he was trying to appropriate Nuun Ujol Chaak's legitimacy. Alas, as mentioned above, the record is insufficient to differentiate among these possibilities (cf. Martin and Grube 2000, 56–58; Sharer 2006, 383–87).

But this is not a total loss. And it is in fact neither political allegiance nor allegation of treason that provides the insights necessary for our exploration of the development of Mayan astronomy. Instead, this history provides a view into a specific type of appeal to astronomy that was made throughout this series of military events.

"STAR WARS"

Into the stone risers of hieroglyphic stairways, the scribes of Dos Pilas recorded several of the military events in this history using a provocative verb. The more common hieroglyphic phrase for victory in battle was *jubuy u took' pakal*—"his flint and shield fell," or perhaps metaphorically, "his army fell" (figure 2.18).

FIGURE 2.18 Record of war on Hieroglyphic Stairway 2, West Section, Step 3: 8 Ajaw 13 Tzek is the seventh haab. Then it happened on 13 Kaban 10 Sots' Nuun Ujol Chaak's flint and shield fell. Blood was pooled, skulls were piled up of the 13 Mutul places, by the authority of B'ahlaj Chan K'awiil, Mutul k'uhulajaw, Bakab.

But the main iconographic element of the verb used for several of B'ahlaj Chan K'awiil's military events is **EK'**, the glyph for "star" (Closs 1978, 1981, 1994; Kelley 1977; Lounsbury 1982; Martin 1996; Nahm 1994; Schele and Freidel 1990).

FIGURE 2.19 Record of war on Hieroglyphic Stairway 2, West Section, Step 4: 11 k'in, 1 winik, 5 haab after [unknown] was defeated. 3 Ix 16 Muwan Nuun Ujol Chaak was defeated, burned, and driven out . . . by B'ahlaj Chan K'awiil, captor of Tajal Mo', Mutul k'ujulajaw.

In its fuller form found in inscriptions far beyond the Petexbatun, scribes included what some modern scholars have recognized as water "droplets" or "streams" on either side of this **EK'** glyph (figure 2.19). This imagery and the fact that it consistently showed up in war contexts led scholars of the 1980s to the tongue-in-cheek term: the "Star War verb" (Schele and Freidel 1990, 178).

In the 1980s and 1990s some scholars began looking at collections of "Star War" events, following the practices of Morley, Teeple, Wilson, and their colleagues of the early twentieth century. Without considering the individual historical contexts, researchers hypothesized that the star icon was related to Venus and that there might be an element of ritual timing to Mayan warfare. In fact, some of these studies did claim to find

FIGURE 2.20 Three versions of the Star War verb: Dos Pilas Hieroglyphic Stairway 4; Tikal Temple IV, Lintel 2; Caracol Stela 3

patterns (Closs 1978, 1981, 1994; Kelley 1977; Lounsbury 1982; Nahm 1994). Moreover, these statistical patterns were combined with the warrior imagery of the Venus pages in the Dresden Codex recognized by Seler so that the idea of Venus-regulated warfare made it out of technical publications and into popular literature (Aveni 2001, 167; Schele and Freidel 1990, 178; Sharer 2006, 150).

Unfortunately for its advocates, the hypothesis fell under more rigorous testing (Aldana 2005). The statistical patterns thought to support Venus timing only existed when all records were aggregated and analyzed as a set. When the events were broken up according to the ruler initiating them—and so the interpretations that individual rulers may have held—the patterns fell away. In other words, individual rulers such as B'ahlaj Chan K'awiil or Yuhknoom Ch'en did not time their wars according to specific observations of Venus, but when their events are pooled together with those of their peers, ancestors, and descendants, subsets of those events statistically suggested the possibility (Aldana 2005).

Apart from the mathematical approach, a second complication to interpreting the Star War verb is that, technically considered, it has not yet been completely deciphered. There are strong arguments that it reads *ek'm*, for the verb "to descend" (Aldana 2005, 313; Zender 2005), but a complete phonetic substitution (or even complementation) has yet to be found in the hieroglyphic record that would prove it (cf. Coe 1999). On the other hand, the reading of *ek'm* as "descent" fits the grammatical evidence since the name of the militarily defeated is given immediately after the verb, and then the phrase is often followed by the person *uk'abjiiy*—"by the authority of whom"—the battle took place. In B'ahlaj Chan K'awiil's martial history on Structure L5-49, HS 2, East Staircase, Step 4, which we saw above (figure 2.11), for example, we read that "On 6 Ix 2 K'ayab, Tikal was defeated by the authority of Yuhknoom Ch'en, Kaanul k'uhulajaw." Here "was defeated" is a translation of *EK'M-yi*, or *ek'mey*.

The point here is that without an argument for the timing of war by Venus, we are left to find other reasons for the celestial imagery and/or potentially astronomical association of the verb. While it may have been purely used for its phonetic value, the **EK'** glyph may also have provided other layers of meaning. It may be, for example, that the appeal to astronomy was symbolic. In this case, the glyph would be understood as a logograph: *ek'm* as a term for descent, represented by a star that is "falling like rain"—hence the water droplets (Aldana 2005, 313; cf. Stuart 1995, 313). It is even possible to consider that this may have been a Mayan version of our modern English "meteor shower." Considered broadly throughout Mesoamerica, such a metaphor finds purchase.

Ethnographer Barbara Tedlock, for instance, records that among the twentieth-century K'iche' Maya a "shooting star or meteor is called a *ch'ab'i q'aq'*, 'flaming arrow'" and that "throughout the Maya area, meteors are thought to be evil omens forecasting sickness, war, and death" (B. Tedlock 1992, 28). Similarly, Alan Sandstrom (1991, 248) records that the modern Nahua of Veracruz conceive of meteors as arrows shot by the stars. These associations aren't confined to modern conceptualizations, since Tedlock also has found that "a Colonial Quiché term for meteor was *ch'olanic ch'umil*, 'star that makes war'" (B. Tedlock 1992, 28–29).

Receding further into the past, we find that other Central Mexican records promote the same association between meteors and arrows/war. In Sahagún's *Florentine Codex* (1953), the Franciscan included a star chart identifying several celestial phenomena. Below a comet and above the constellation of a scorpion's tail, Sahagún drew a star with an arrow attached to it and glossed it as *Citlaltlamina* (Aveni 1980, 32, fig. 8). The root, *citlali*, means "star," while *-tlamina* derives from *mina* "herir a alguien, tirarle flechas"—"to injure someone, to shoot arrows at them" (Siméon 1977, 111, 277) (figure 2.21).

Yet another Mexica expression takes the association further. The Nahuatl phrase *auh topan onoc in mitl* was translated literally by dictionary compiler Rémi Siméon as "above us is the arrow" (1977, 277). The meaning, however, Siméon glossed as "estamos sometidos, vencidos por las armas"—"we are subjected, defeated militarily." Now if we go against our initial hunch and accept that the "Star War" glyph itself was intended

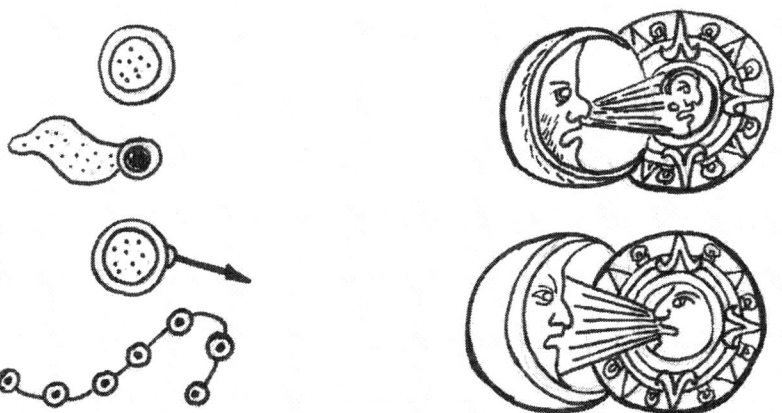

FIGURE 2.21 Sahagún's representations of celestial phenomena on page 282r of the Florentine Codex. A meteor as a celestial arrow is the third image in the left column. The other images are of a star, a comet, a solar eclipse, and a lunar eclipse.

iconographically, then the glyph may be interpreted along this vein as "[the defeated] is under the meteor(/arrow of the gods)." The metaphor thus matches perfectly the context of the Mayan glyph, for it meant that an army was defeated in battle, yet it also held political overtones in that, once vanquished, the losing city was placed under vassal status by the victorious army. Therefore, even though the glyph has not been technically deciphered, both meanings of the Nahuatl phrase are borne out by the history we have of the **EK'**-based verb in Classic period inscriptions.

Star Wars in Context

Bahlaj Chan K'awiil's martial history, therefore, helps to illuminate the role that astronomy played in elite activities during the Middle Classic period. It also encourages the suggestion that we should be willing to find variation in the role of astronomy within Mayan culture over time (Aveni 2001, 166). There may have been a robust presence of observed astronomical phenomena in Middle Preclassic communities, with Ricketson's "E-Groups" providing evidence of public ceremonies tied to sunrise observations at equinoxes and solstices. But by the Early to Middle Classic periods, astronomy may well have become more symbolic or moved into specialized and esoteric forms (cf. Aldana 2006). By this time, for example, the E-Group at Uaxactun had been built over, as was the E-Group at Tikal, where it was left aside for the move to Chak Toh Ich'aak's palace at the city's new center. Accordingly, in his records of battles against his half brother, B'ahlaj Chan K'awiil was appealing to the celestial realm on a metaphoric level, even if there was no quantitative astronomical method underneath its use.

Of course this does not require that astrology be completely absent. For any given event, a particular k'uhulajaw may have appealed to astrological data to augment other sorts of information. But we should be reluctant to appeal to astronomical practices as mechanically triggering social behavior. They may have contributed to part of an issue, but whatever they contributed would have been balanced by other forms of information. The history that B'ahlaj Chan K'awiil left to us alone makes clear at least that various pressures and sources of information had to be accommodated by a k'uhulajaw during the Classic period. We will return to this theme in later chapters. For now we note that the several appeals to the celestial realm in the military history records at Dos Pilas may reflect B'ahlaj Chan K'awiil's adherence to intellectual fashion and not to the mechanical triggering of ritual activity. References to the celestial realm during the Classic period did not have to carry functional astronomical value; they may have been simply rhetorically powerful in describing military history.

With such a hypothesis in hand, it turns out that there is also evidence that a broader shift was occurring during the time that B'ahlaj Chan K'awiil and Nuun Ujol Chaak fought their final battles. Having observed a potential metaphorical use to astronomical knowledge put to political use at Dos Pilas, we now turn to other representations of astronomically related cultural production from the time. Here we find corroboration that the rulers of different Classic Maya cities were changing the character of their appeals to the celestial realm specifically during the Late Classic.

SKY BANDS AT PIEDRAS NEGRAS

If we are now willing to entertain the notion that astronomical concepts functioned symbolically in the self-representation of rulers during the Late Classic, we certainly want to determine how to contextualize this. While we could make cross-cultural appeal to Galileo's naming of Jupiter's moons (Biagioli 1993) or Islamic astrology of the state (King 1993), we actually have other forms of evidence within Classic Mayan royal self-representation. For some of this context, we move downstream from Dos Pilas and the Petexbatun, north along the Usumacinta River past Yaxchilan, to the Classic Mayan site known as Piedras Negras.

In her landmark 1960 publication, thirty years after Teeple cracked page 24 of the Dresden Codex Venus Table, Tatiana Proskouriakoff used the stelae of Piedras Negras to shift the study of Mayan hieroglyphic writing (Coe 1999, 171–76). Having studied architecture as an undergraduate, Proskouriakoff initially impacted Maya archaeology through her artistic reconstructions of sites under active excavation. Initially, she volunteered her efforts in archaeological illustration to the university museum of the University of Pennsylvania while pursuing graduate studies. There, Linton Satterthwaite hired her to join the excavation team at Piedras Negras, and she created her first reconstruction of Mayan architecture (Coe 1999, 168). Eventually, Proskouriakoff would illustrate the architecture of several other sites, and her book *An Album of Maya Architecture* is still an asset to anyone interested in interpreting ancient Mesoamerican cultures. It was only later in her career, though, that she became interested in the writing system, triggering her field-shifting historical hypothesis, as we saw in this book's introduction.

For the purposes of this chapter, we consider a somewhat tangential aspect of her 1960 article. In drawing the monuments of Piedras Negras, Proskouriakoff recognized that the Yok'ib scribes used a consistent iconographic theme on several stelae in their representations of the k'uhulajaws (figure 2.22). The figures on these monuments, we should recall, at her time were not recognized as the historical colleagues of B'ahlaj Chan K'awiil and/or Itzamnaaj B'ahlam. In *Maya Hieroglyphic Writing*, Eric Thompson described the conventional wisdom of his day:

> Usually the front of a stela is carved with the figure of a deity or of a priest impersonating a deity. I believe, although proof escapes me, that the choice of subject for portrayal was directly governed by the dedication date of the monument (Thompson 1978, 19).

To Proskouriakoff's colleagues, the figures represented gods and the dates corresponded to astronomical events related to those gods.

FIGURE 2.22 An example of Proskouriakoff's "cosmic motif" on Piedras Negras Stela 11. The k'uhulajaw is framed on either side by a "sky band."

In the stelae that Proskouriakoff identified, then, the straightforward interpretation she inherited was that gods were portrayed on these monuments, often seated on thrones, which in turn were bordered by bands of symbols. It is the latter bands that occupy our primary concern now. As we saw in the introduction, Proskouriakoff named them "sky bands" as part of her "cosmic motif" (1960, 455). Archaeoastronomers John Carlson and Linda Landis (1985) make the argument that the sky band is actually a representation of the body of the Celestial Dragon (cf. Aldana 2001), a view that comes closest to general consensus support. In any case, Proskouriakoff's paradigm-shifting analysis demonstrated that these scenes were not illustrations of gods, but instead were used by each in a series of seven historical rulers at Piedras Negras to commemorate their accessions to the k'uhulajawlel of the city (1960). This much we have already seen, and it is what ensured Proskouriakoff's place in the history of Maya archaeology. What these monuments also do, on the other hand, is provide us with representations of the sky at different points in time by artists at the same city. And when we look at this series, we find that there is a very clear pattern: the celestial imagery becomes more nuanced over time.

Beginning with a contemporary of Janaab' Pakal and Yuhknoom Ch'en in the seventh century, K'inich Yo'nal Ahk of Piedras Negras patronized the earliest niche with "the cosmic motif." The k'uhulajaw's artists filled his sky band with mostly abstract elements. While a few "standard" items do populate the cells within the band, in this version there are no identifiable celestial bodies. A "crossed bands" icon shows up as a general symbol of the sky, for example, as does the glyph for "brightness" and **CHAN**, the glyph for the "sky" itself (Stone and Zender 2011, 60).

In the next niche within the sequence, artists for the son of K'inich Yo'nal Ahk provided only a partial view of the sky band—only three elements are visible on Stela 33. Carved fifteen years after Stela 25—and only ten years before B'ahlaj Chan K'awiil would leave Tikal—the artists in this case included an **EK'** ("star," "celestial body") glyph as one of the three visible elements. While it has often been mistaken as a Venus glyph, this version does not carry the **CHAK** prefix and so only refers to a generic star, or even less literally marks the band as being celestial (Aldana 2002).

The next monument, however, coming some sixty years after the first, demonstrates much more careful attention to the sky's inhabitants. K'inich Yo'nal Ahk II's accession sky band, in the middle of the "star wars" between Tikal and Dos Pilas, includes glyphs for the sky, **CHAN**, for darkness, **AK'AB**, and **EK'**, along now with a **K'IN** glyph, the last representing the Sun. From this through the last accession niche, dating deep into the Late Classic, the sky bands are populated with significant variation and explicit celestial bodies. Stela 11, for example, carries a Moon glyph, **UJ**, with a rabbit inside. Overall, the clear trend over 140 years is from simple and vague to nuanced and explicit, with the transition occurring within a decade of the final battle between B'ahlaj Chan K'awiil and Nuun Ujol Chaak (table 2.1).

It is also interesting to note that by the fourth ruler in the sequence—now thirty years after the final Dos Pilas–Tikal battle—a new element turns up: the only secure comet glyph from the Classic period, **BUTZ'/K'AHK'-EK'**. It may be that this was one of a series of comets observed and recorded by Chinese astronomers in the mid-eighth century, including Halley's Comet of AD 760 (Aldana 2013, 6; Tsu 1934, 196). Unfortunately because it is only an icon in an artistic motif and not an actual record of an observed event, we cannot pin down a date for it in the Mayan Long Count. We can, however, assume that "Ruler 4" included it on his accession throne because he had some affiliation with it, and that may then suggest that the comet was witnessed right around his birthdate, or between his birth and accession. Such a commemoration should not surprise us; comet appearances are highly influential cross-culturally and across time.

TABLE 2.1 Increasingly complex iconographic representations of the night sky over the Late Classic period

| Monument | Date | # of Cells | Celestial Bodies | | Abstract Icons | |
			#	Elements	#	Elements
St. 25 K'inich Yo'nal Ahk I	9.9.10.6.16	10	0	N/A	10	"brightness," crossed bands, CHAN
St. 33 Ruler 2	9.10.6.5.9	3	1	EK'	2	?
St. 6 K'inich Yo'nal Ahk II	9.12.15.0.0	13	4	EK', K'IN	9	AK'AB, CHAN, brightness, crossed bands
St. 11 Ruler 4	9.15.0.0.0	12	5	EK', K'IN, UJ, BUTZ EK'	7	AK'AB, CHAN
St. 14 Yo'nal Ahk III	9.16.8.0.0	9	2	EK', K'IN	7	crossed bands, CHAN, JA', AK'AB

Notes: K'IN—Sun; UJ—Moon; EK'—star; BUTZ EK'—comet; CHAN—sky; AK'AB—darkness.

The most famous comet in Mesoamerican history must belong to the harbinger of European contact. Archaeoastronomer Anthony Aveni captures the event recorded in a sixteenth-century historical document with reference to the *tlatoani* of Texcoco, Netzahualpilli, in Central Mexico:

> Further evidence of Netzahualpilli's careful attention to the heavens is exemplified in the episode of the great comet which was said to have presaged the fall of the Aztecs (Fig. 3). The chronicler Father Diego Durán tells us that Moctezuma, having observed the comet since midnight, went the next day to Netzahualpilli to seek its meaning. Replied the king of Texcoco, "Your vassals, the astrologers, soothsayers and diviners have been careless! That sign in the heavens has been there for some time and yet you describe it to me now as if it were a new thing. I thought you had already discovered it and that your astrologers had explained it to you. Since you now tell me you have seen it I will answer you that that brilliant star appeared in the heavens many days ago" (Durán 1964, pp. 247–248). He goes on to give details of the frightful omens that soon after befell the unfortunate monarch. (Aveni 2001, 17)

If a similar comet was visible from Piedras Negras, that would put the Long Count boundaries for a relevant sighting at between 9.13.0.0.0 and 9.14.18.3.13;* more broadly, it also may have provided stimulus for more concerted attention to the night sky and its representation in public art.

---

*Using the GMT, this time period corresponds to 701 to 729 AD; with the Aldana Constant, on the other hand, Halley's Comet and two others appear very close to accession as well as the time period for which the stela was raised (Aldana 2011; 2013, 5–6).

At this level of context, then, considering comets and sky bands over time, one trend across generations during the Late Classic is to subtler *public* artistic representations of the night sky and its elements. We might here speculate that the series of comets visible around the world during this time period served to heighten the presence of the celestial realm within everyday public perspectives. That might in turn have pressed rulers to increase the nuance in their artistic representations of the residents of the celestial realm. On the other hand, it may simply have been that the sky as metaphor was gaining use in the political symbolism of authority in the Late Classic—in other words, that the pressure was internal to elite culture. In such case, B'ahlaj Chan K'awiil's appeal to astronomical symbolism in warfare inscribed into his hieroglyphic staircases as we saw above—his "star wars"—would have fit very naturally within this broader developing context.

CREATION ASTRONOMY

The progression of celestial imagery at Piedras Negras is suggestive and is corroborated by other developments transpiring during the same time period beyond the Usumacinta region. Near the end of the thirteenth winikhaab, for instance, the scribes at Coba—far to the northeast across the Peten—picked up a new interest in "Creation." At their lowland city, adjacent to one of the few lakes in northern Yucatan, scribes used a very provocative Long Count placement for their record of a mythological event (figure 2.23). To see that provocation, we first review one specific event within Classic Mayan mythology.

As Förstemann recovered it, and as we have seen in chapter 1, the Long Count is simply a tally of days starting from some "zero point." Scribes throughout Mayan lands recorded the event occurring on that "zero date" consistently within the hieroglyphic record and over time. At Quirigua during the Late Classic, for instance, the artists of K'ak' Tiliw Chan Yopaat were busy with a huge project of monumental aggrandizement. The young k'uhulajaw charged them with constructing massive stelae representing him in traditional style, but on a much larger scale. The content of the hieroglyphic inscriptions on these monuments followed this increase in scale,

FIGURE 2.23 Coba Stela 1, in which the scribe included twenty "extra" periods of the Long Count, each with a coefficient of 13

incorporating mythological events from ancient times into his personal historical narrative. In doing so, they provide us with the record that on the date 4 Ajaw 8 Kumk'u a cosmological ritual was enacted involving three throne stones (figure 2.24). This is followed by the elaboration of the erection of three stone thrones, constituting the "stone-enclosing" ritual.

This short hieroglyphic account speaks to Mayan mythology more broadly. For one, it makes clear that "Creation" was in fact a series of events engaged by multiple actors (cf. D. Tedlock 1985). Several deities worked in concert through travel/procession to enact a ritual of space and time. Just as important as the insight it provides into origin stories, though, is that it sets up templates for Classic Mayan public royal ceremonies. Whereas in this text the actors are deities in a primordial world, the same activities of *k'altuun* (stone-enclosing) and utzapaw tuun (erecting stelae) are among if not the most common activities of the historical k'uhulajaws. The dynastic king list in Janaab' Pakal's Temple of Inscriptions at Palenque, for example, does nothing more than record the accession of each ajaw and then name the period end on which she or he performed a k'altuun event (Aldana 2007, 76). Normalcy, it seems, was represented by rulers continuing to perform the activities first enacted by primordial deities.

We will return to the root of this verb, *k'al*, and associated rituals below. At this point, we focus on one detail of how this specific Creation event was recorded within

---

FIGURE 2.24 Creation narrative on Quirigua Stela C. The inscription reads: [ISIG] *13 pih, 0 winikhaab, 0 haab, 0 winik, 0 k'in 4 Ajaw 8 Kumk'u jehlaj k'oob, oxtuun k'ahlaj. U tzapaw tuun* [Jaguar Paddler] [Stingray Paddler]. *Utihy Na Ho' Chan* [Jaguar Throne Stone]. *Utzapaw tuun Ek Na Chaak?. Utihy kab-ch'en* [Serpent Throne Stone]. *Iwal utihy k'altuun Na Itzamnaaj* [Waterlily Throne Stone]. *Utihy ti' chan, Yax Oxtuunal. U tzutzuy oxlajun pih ukabjiy Wak Chan Ajaw.*

The Jaguar and Stingray Paddlers set up the stone. It happened at the Five Sky House. It is the Jaguar Throne Stone. Ek Na Chaak? set up the stone. It happened at the Sky-Cave. It is the Serpent Throne Stone. And then it happened, the stone enclosing at Itzamnaaj House. It is the Waterlily Throne Stone. It happened at the edge of the sky, the First Hearth Place. Thirteen baktuns were completed by the authority of the Six Sky Ajaw.

hieroglyphic inscriptions of the Classic period. The Long Count date at Quirigua starts with an ISIG and then runs through the five periods of pih, winikhaab, haab, winik, and k'in. It is the lead coefficient that makes clear that this is not a traditional historical date. The pih coefficient is "13," which might suggest a date in the future for the Classic Mayans since we have already seen that most historical dates were led off by pih coefficients of "9." Instead, we find that this text describes a mythological past and a prior "resetting" of the pih coefficient—from a prior accumulation of time to the end of a thirteenth pih as a new "zero date," which initiated the period that would eventually encompass the Preclassic, Classic, and Postclassic historical periods (Aveni 2001, 153; Schele and Freidel 1990, 82; Sharer 2006, 110). That is, the end of the thirteenth pih was followed 144,000 k'in later by the Long Count date 1.0.0.0.0, and so on, until the Mayan historical period around 8.0.0.0.0, 3,000 years after their zero date. The point is that Late Classic scribes recorded their "zero date" of 4 Ajaw 8 Kumk'u as having transpired on the Long Count date of 13.0.0.0.0.

Returning to Coba and our point of departure, the scribes there recorded an abbreviated version of the Creation narrative written by their southern colleagues on Quirigua Stela 1. Where the Coba scribes elaborated was in the representation of the Long Count date itself and not in what transpired on that date. On the back side of the monument, a ruler holding a divination tool is framed by this statement:

> 13.13.13.13.13.13.13.13.13.13.13.13.13.13.13.13.13.13.13.0.0.0.0 4 Ajaw 8 Kumk'u jehlaj k'oob

> On 13.13.13.13.13.13.13.13.13.13.13.13.13.13.13.13.13.13.13.0.0.0.0 4 Ajaw 8 Kumk'u the hearth was replaced.

Rather than just note that "13 pih were completed," as did the Quirigua scribes, at Coba the scribes included 19 additional higher-order periods of time, all with the coefficient of 13. Undoubtedly numerology was important here since 20 periods of 13 symbolically invoked the 260-Day Count, but with this text alone, it is difficult to see other motivation. To provide some context for what these scribes were doing, we turn to the much more elaborate set of Creation texts back at Janaab' Pakal's Palenque.

Palenque Astronumerology

Use of the Calendar Round is much older and far more widespread geographically than the Long Count. This makes it reasonable to assume that some aspects of Mayan mythology were older and so associated with Calendar Round dates, but without Long Count ascription. To some extent, the recognition of this problem is corroborated by painted ceramic vessels from the Classic period. A common style for elite patronage is referred to by modern scholars as the Codex Style, wherein the surface of a cylindrical cup is decorated with a scene accompanied by an explanatory hieroglyphic text (figure 2.25).

 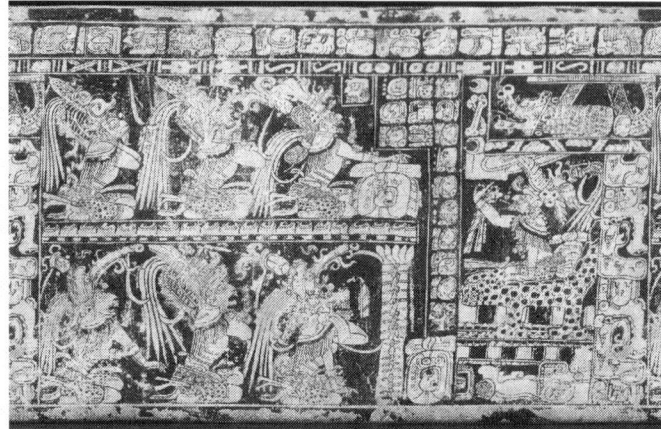

FIGURE 2.25 (*left*) "Codex-style" cylindrical cup (*right*) and photo rollout. This is known as the "Vase of the Seven Gods" and carries the ancient "zero date" of 4 Ajaw 8 Kumk'u (Kerr #2796). Courtesy mayavase.com

What we find, however, is that numerous vessels of this type were painted with an illustration of the same mythological event and described by the same hieroglyphic text, but the Calendar Round date assigned to that event was different on different vessels (figure 2.26; cf. Van Stone 2010). The record provides insufficient consistency to support an interpretation that there was a "correct" date for each event, and a few artists simply "got it wrong." Rather, it appears that the variation reflects local dynastic traditions or local religious interpretations. Such an underlying situation would have presented Classic period scribes with both an opportunity and a challenge. For some of this, we turn to B'ahlaj Chan K'awiil's contemporary at Palenque, Janaab' Pakal, and his son, who dedicated much of his patronage to honoring his father.

Historical records at Palenque show that two nobleman scribes worked with their patron, Kan B'ahlam, to address specifically the issue of Long Count dates for mythological events as part of their efforts to celebrate Janaab' Pakal's memory. Four specific sets of inscriptions leading up to the close of the thirteenth winikhaab, or specifically in preparation for 9.13.0.0.0, include complex calendrical calculations (Aldana 2001, 135–92; 2002; 2007, 83; M. Robertson 1979; Powell 1997, 12–20; Thompson 1975, 212–17; Lounsbury 1978, 773–74). They also introduce a new calendric tool. As I explored in *The Apotheosis of Janaab' Pakal*, the "819-Day Count" was an invention offsprung by mother necessity. Like the Quirigua example we saw above, Kan B'ahlam's project of celebrating his father's success incorporated inscriptional narrative connections between his father's dynastic history and the mythological events of the gods. The scribes of Palenque created the 819-Day Count as an astronumerological tool, allowing them to keep track of astronomical events and calendric counts over huge periods of time. It would therefore facilitate the placement of a floating Calendar Round date within the Long Count (Aldana 2007, 108).

FIGURE 2.26 Different dates assigned to the same mythological event on Codex vessels. The full range of dates for this scene comprises (*top*) 7 Muluk 14 Sak (Kerr #1198), (*middle*) 7 Muluk 15 Yax (Kerr #6754), (*bottom*) 13 Muluk 17 Pax (Kerr #5164); also (not shown) 12 Muluk 14 Sak, 9 Muluk 18 Sak, 7 Muluk 10 Keh, 13 Muluk 1 Pax, 7 Muluk 15 Pax. Courtesy mayavase.com

As we have seen, Mayan computational astronomy is largely algebraic, amounting to arithmetic operations relating astronomical periods to calendric cycles. Venus and Mars would have been easy to track over large periods of time for Mayan astronomers using only the basic calendric system. Venus's synodic period is roughly 584 days, which can be broken down into 8×73. Since the 365-Day Count is made up of 5×73 day periods, Venus Rounds and solar years come into accord with a 5:8 periodicity, as we have seen in the Venus pages of the Dresden Codex. Keeping track of 365-Day Count periods, therefore, makes it easy to extend these to Venus Rounds. Mars is even easier since its synodic period is very close to 780 days, which is just 3×260. Astronomers could then look at the interval of time between an ancient date and a contemporary one to determine the position of Venus or Mars on that date. In fact we know from Palenque that the scribes there did precisely this in their computations of the Lunar Series for mythological events (Aldana 2007; Lounsbury 1978).

The challenging visible planets to keep track of would have been Saturn and Jupiter. It appears that two priests in particular in Kan B'ahlam's court invented the 819-Day Count as a tool specifically to incorporate the synodic periods of Jupiter and Saturn ($399 = 21 \times 19$ and $378 = 21 \times 18$) into their work. This provided Palenque astronomers with a tool to eliminate the ambiguity of the Long Count placement of an ancient/mythological event by constraining it to have an astronomical relationship with historical events. Mythological events would be separated by historical events by integer multiples of the periods of the planets (cf. Lounsbury 1978). This practice has been found throughout the mythological texts at Palenque (Aldana 2007; Powell 1997).

Although they were writing at the same time, these Palenque scribes did not record any Creation events using the Coba scribes' practice of including higher-order periods of thirteens. The Palenque scribes, however, did provide one passage making explicit that they too were interested in this problem of how to numerically and numerologically address the deep past and the distant future. These scribes demonstrated that they did not conceptualize each higher period, or even the pih register, as comprising only thirteen periods. Instead, they showed computationally that twenty pih would be necessary to reach the next higher register, and when that happened, it would have a coefficient of 1 (Aldana 2007, 85; Schele and Mathews 1999, 106). While some modern scholars have worked computationally to rectify the records of Palenque and Coba by making assumptions about how a "universal" Mayan calendar may have worked (cf. Freidel et al. 1993; Stuart 2011), it is just as likely, if not more so, that different royal courts had different interpretations, which changed over time—not unlike the Codex Style pottery records. This becomes even clearer when we consider that there are inscriptions at Tikal, Yaxchilan, Copán, and Quirigua that all explicitly record periods of higher order than the pih, and they cannot all be made to accord with one another computationally (Aldana 2012) (table 2.2). Perhaps more importantly for our purposes here we find that, with the exception of the date from Tikal, all of this astronumerology was undertaken during the Late Classic period.

Stepping back from the details of computation and who followed which practice when, it is worth noting that the kinds of astronumerological projections evidenced in these texts do not require what we would consider modern scientific accuracy. Although they do depend on a scientific tool—that is, one that preserves a mathematical model of a physically observable phenomenon—the mythological connections would have been impossible to verify "objectively." The primary consideration would have been that they be internally consistent computationally and in accord with elite ideology—with mythological records. To this theme we will return in the conclusion to this book. The point is that the science here resembles B'ahlaj Chan K'awiil's "star wars"; Kan B'ahlam and his colleagues were using astronomical knowledge for its symbolic application, as were the scribes at Coba with their string of thirteens.

It is perhaps not surprising at this point to realize that the recently recovered murals from Xultun date to only some decades after the last dates in table 2.1. A wall in the Terminal Classic architecture at the site augments our record of these astronumerological practices. As archaeologist William Saturno found, on the best-lit wall within the structure, scribes not far from Tikal painted a series of numbers in tables, not unlike those in the Dresden Codex (figure 2.27). Some of the intervals are clearly lunar in character; others are more ambiguous (Saturno et al. 2012; Zender and Skidmore 2012). But within

TABLE 1.1 Astronomical and numerological records during the Classic period

Date	Location	Monument	Patron	Record Type
9.3.16.15.14	Tikal	Stela 10	Kaloomte B'ahlam	Piktun (20 baktun) Long Count date
9.9.9.16.0	unknown	Dresden Codex	unknown	Anchor to the Venus Table
9.9.10.6.16	Piedras Negras	Stela 25	Yo'nal Ahk	Sky band
9.10.18.2.19	Dos Pilas	Str. L5-49 HS2, East Stair, Step 5	B'ahlaj Chan K'awiil	"Star War" by Yuhknoom Ch'en against B'ahlaj Chan K'awiil
9.10.10.0.0	Piedras Negras	Stela 33	Ruler 2 (son of Yo'nal Ahk I)	Sky band
9.12.16.16.17	Dos Pilas	HS2, West Stair, Step 3	B'ahlaj Chan K'awiil	Final defeat of Nuun Ujol Chaak
9.12.15.0.0	Piedras Negras	Stela 6	K'inich Yo'nal Ahk II	Sky band
9.12.10.5.12	Coba	Stela 1	Ruler X	13 pih numerology
9.12.11.15.18	Palenque	T1-W	Janaab' Pakal	1 piktun = 20 pih
9.13.0.0.0	Palenque	Cross Group Tablets	Kan B'ahlam	Invention of 819-Day Count
9.14.0.0.0	Copán	Stela C	Waxaklajun Ub'aah K'awiil	13 kalabtun completed
9.15.0.0.0	Piedras Negras	Stela 11	Ruler 4	Sky band (comet)
9.15.13.6.9	Yaxchilan	Str. 33, HS2, Step VII	Yaxun B'ahlam IV	13 pih numerology
9.16.8.0.0	Piedras Negras	Stela 14	Yo'nal Ahk III	Sky band
9.16.10.0.0	Quirigua	Stela F	K'ahk' Tiliw Chan Yopaat	19×20^7 haab
9.17.5.0.0	Quirigua	Stela C	K'ahk' Tiliw Chan Yopaat	13.0.0.0.0 Creation event

(glyph table)

FIGURE 2.27 Table of Long Count time intervals painted on the walls of Xultun Structure 10K-2. The first entry is 8.17 or 177, which is the base interval of the Dresden Codex Eclipse Table (see above). The intervals increase as multiples of 29- and 30-day lunar periods to a value of 13.5.4 or 4,784. Accounting for 162 moons, this gives an approximation to the synodic period of 29.53086, the same approximation used by Kan B'ahlam's scribes at Palenque.

these tables, we find intervals that appear to be built upon the 819-Day Count—the astronumerological tool invented at Palenque a few decades earlier. It is fascinating to recognize this room as an astronomer's computational workshop for addressing the kinds of problems tackled earlier by the scribes at Copán, Palenque, and Coba. Regardless, this Xultun record punctuates the pattern that by the Late Classic such astronomical computations must have been of considerable importance, if they were allowed on walls within the elite architecture of the city.

A UNIQUE VENUS RECORD

One final example pointing to an intensifying interest in astrology/astronomy during the Late Classic arises in the patronage of Yax Pahsaj Chan Yopaat, the sixteenth ruler of Copán. There scribes had adopted the 819-Day Count developed at Palenque and included it within a set of wall inscriptions in Structure 10L-11. Midway through the narrative carried by these inscriptions, immediately before this 819-Day Count record, the scribes carved an explicit historical record of a Venus event (Aldana 2005, 309; Schele and Freidel 1990, 328, 490). Notably, with this record Yax Pahsaj Chan Yopaat was not including his own observations of the celestial body. Instead, he was appealing

to his predecessor of thirty years—the thirteenth ruler of Copán, Waxaklahun Ub'aah K'awiil (figure 2.28). The text reads *5 Kib 10 Pohp k'alwaniiy Ajaw Chak Ek' u muuk(?) Jun Ajaw Winik.*

As treated elsewhere (Aldana 2005; Fuls 2008), this Calendar Round of 5 Kib 10 Pohp corresponds to the Long Count position of 9.15.15.12.16. Of interest at this point is that it also represents a direct link to the hieroglyphic text in the Venus Table. We can see this link by recalling that on every page of the Venus Table, there is a repeating phrase of four glyphs, which in a post-Proskouriakoff context we can now read. The first column on page 46, for example, is as follows:

FIGURE 2.28 (*left*) Venus record in Copán Str. 10L-11, using the verb **K'AL**; (*right*) Venus record from the Dresden Codex Venus Table

3 Kib 4 Yaxk'in k'ahlaj [Cosmic Region] [Deity] *Chak Ek'*

The Venus record at Copán is remarkably similar:

> *5 Kib 10 Pohp k'alwaniiy Ajaw Chak Ek'* [Deity]

Both events are described with the same verb root, *k'al*, and both pair Venus (Chak Ek') with a partner, leading at least to the implication that the astronomy behind each should be relatively similar.

The case becomes even more provocative when we look at the specific Long Count dates recorded. Here we find continuity between the Copán scribal records and those in the Dresden Codex. Starting with the Middle Classic period anchor recorded on page 24 of Tawiskal Uwoojil's manuscript and running forward Venus Rounds to Yax Pahsaj Chan Yopaat's inscription, we find that both are perfectly consistent with observable events. If the 9.9.9.16.0 1 Ajaw 18 K'ayab anchor date corresponded to a first morning visibility of Venus, then the 9.15.15.12.16 5 Kib 10 Pohp date would have corresponded to a first evening visibility of the planet. In other words, if a similar table of dates were available to Yax Pahsaj Chan Yopaat's astronomers, they would have started with the base date on the thirteenth row of the last column on page 50 and moved through the table to the third row of the second column on page 48 to arrive at the first evening visibility they witnessed. The Classic period anchor to the Venus Table in the Dresden Codex is consistent historically with the record in Temple 11 at Copán.

There is much to explore in this Venus record, and we will return to it in chapter 5. For now, what is nice about Yax Pahsaj Chan Yopaat's record is that at the very least it tells us that, by the onset of the Late Classic, rulers were working with astronomical phenomena for political metaphors, with abstracted, idealized synodic planetary periods for calendric computations, and at the same time they were keeping track of planetary events—official records existed giving the relative positions of the planets. The record at Copán at least adds the case of Venus to go along with the Moon, which had been duly recorded in the Lunar Series since the late Preclassic, as we saw in the last chapter. It is sufficient to note for now that the above trends within numerology, astronomy, and the 819-Day Count all came together in Structure 10L-11 at Late Classic Copán.

ASTRONOMY IN THE LATE CLASSIC

In this chapter, we have seen what appears to be a coherent historical narrative. At some point during the early part of the Late Classic, appeals to the celestial realm became fashionable. Under the pressure of intersite political competition, an interest in combining astrology and numerology went beyond metaphor, expanding the sophistication of each. When we consider the military strife between B'ahlaj Chan K'awiil and Nuun Ujol Chaak, we find that it occurred during the early period of royal interest in astronomical self-fashioning such that B'ahlaj Chan K'awiil's penchant for recording his victories over his half brother using the Star War verb matches the convention of his day. It does not challenge the imagination to suggest that this interest in astronomy may have been corroborated and even spurred by a high incidence of comet visibilities as well as the numerological implications of the end of 13 winikhaab. Once developed, these astronu-merological tools became established mechanisms for those with the interest. And here again it should be pointed out that not everyone was interested. The 819-Day Count only shows up at Palenque, Copán, and Quirigua. Likewise, the enigmatic Glyphs Y and Z only show up at Yaxchilan (Andrews 1938). As with any other fashion, some dynasties adopted it wholeheartedly, while others looked the other way.

Given this exploration of astronomy and numerology in the Late Classic, we return to the question that initiated this chapter. What are we to make of the "historical" date in the Dresden Codex? Did Tawiskal Uwoojil or K'uk'ul Ek' Tuyilaj have access to their own historical records within which the 9.9.9.16.0 1 Ajaw 18 K'ayab date had been preserved since Middle Classic times? Consistency with Yax Pahsaj Chan Yopaat's record at Copán and in line with astronomical symbolism used by B'ahlaj Chan K'awiil at Dos Pilas suggests that it very well may have been. It could easily have been recorded during this time of rising interest in representing the celestial realm and used for the political purposes of the almehen—if not just the k'uhulajaw.

So it is with this backdrop that we may return to the Dresden Codex with a perspective that is radically different from the one that Teeple or his colleagues had access to. If Teeple and his colleagues enjoyed the fruits of the script's decipherment, they might have gone another way with their interpretations. They might have recognized that the time period of the anchor to Tawiskal Uwoojil's Venus Table was appropriate for the recording of planetary observations and that the specific record we do have is internally consistent with an explicit Venus record at Copán. But they didn't. And in fact the standard interpretation of the Dresden Codex Venus Table that coalesced after Teeple's work goes *against* its historical interpretation.

The prevailing interpretation today is that the base date of the Venus Table, 9.9.9.16.0 1 Ajaw 18 K'ayab, was not historical—it was *in*accurate and primarily a numerological fiction. Within most of modern scholarship, the base date is understood to have facilitated the connection of mythological times to much later historical observations of the planet, but it did not match what was observable in the sky (Aveni 2001, 191; Bricker and Bricker 2011; Lounsbury 1983; Thompson 1972). This leads us, of course, to address how such a distinction came about.

To do so, we now return to a consideration of the interpretation of the Dresden Codex Venus Table beginning with Eric Thompson's revision of John Teeple's work to find that a bias was introduced that remains within treatments even today.

CHAPTER 3

The Books of Chilam Balam and the Quest to Correlate Calendars

T he indigenous Yucatec Mayan hand of Juan Josef Hoil was the last to copy the full manuscript that modern scholars refer to as *The Book of Chilam Balam of Chumayel.* The vast majority of the 107 pages of the book are in the same handwriting, which matches the personal observations he recorded at the end. Hoil notes, for example: "On this 18th day of August, 1766, occurred a hurricane. I have made a record of it in order that it may be seen how many years it will be before another one will occur" (Roys 1933, 80). A hurricane tearing through the village of Chumayel—not far from Merida, the capital of Yucatan—obviously would have been worthy of note. Rebuilding after the destruction would have been a communal effort, and as a community leader and keeper of the book, Hoil very likely played a key role in the organization of reconstruction efforts. We will return to some of the implications of this specific historical record in the later chapters of this book. Here, though, we note that Hoil's final statements are of disease and drought in 1781 and 1782. While speaking to the hardships endured under colonial rule, these events also securely date for us the final copying of his manuscript to the end of the eighteenth century.

Written in his native language, but using the Latin alphabet, Hoil's book appears to be far more narrative in content than the codices of K'uk'ul Ek' Tuyilaj or Tawiskal Uwoojil. Grids and tables of numbers and dates found in the Precontact codices no longer show up in the Books of Chilam Balam. In its content, though, Hoil's *Book of Chilam Balam of Chumayel* matches well many indigenous histories—oral and written—treating mythology, history, and prophecy. Such traditions do not appeal to Western European "objective" perspectives, either in the sense of being "unbiased" or in that of adhering strictly to a scientific contextualization of historical facts. The tradition is other and, in that sense, has presented numerous complications for modern scholars interested in interpreting the manuscript.

Scholars since Morley have recognized that despite its different form, there is *potential* for tying together Hoil's calendric records to the Long Count records of the early Postclassic period. As we will see, the Long Count component of the winikhaab is mathematically identical to the "katun" of Hoil's manuscript. According to Morley, scribes after Tawiskal Uwoojil tracked a "Katun Count," which simply took over from the Long Count without skipping a beat. For modern scholars, this constitutes an essential component of the "continuity assumption" (Aveni 2001, 208).

While the continuity assumption does provide a convenience, and it seems reasonable enough, when we turn to the historical context such an assumption looks forced. It does not consider that there may have been internal political motivations and even external political forces to shift from a linear to a cyclical count (cf. Thompson 1978). And if there were a political motivation for a change in practice, there very well also may have been an adjustment of the anchor for the new cyclical count. Moreover, Hoil included suggestive references, for example, to *u uodz katun*, or "the folds/wrinkles of katuns" (Roys 1933, 74). If these corresponded to calendric shifts of any kinds—motivated by politics or ideology—the continuity assumption is rendered untenable.

In this chapter, therefore, we encounter the complication arising from the fact that besides the Dresden Codex, there are other books written by Yucatec Mayan hands that have played important roles in the interpretation of Precontact Mayan astronomy. Without such a set of manuscripts, a relatively hard break would have been left to modern scholars between the traditions of Precontact scribes and those living in the vastly different contexts of the nineteenth century, witnessed by Alexander von Humboldt, Charles Etienne Brasseur de Bourbourg, and Eduard Seler. More pointedly, without these other books, Tawiskal Uwoojil's Venus pages could have been studied primarily within their own historical context relative to other hieroglyphic codices—the Madrid Codex, the Paris Codex, and those from Central Mexico—along with Classic period inscriptions.

On the other hand, with these other Postcontact Mayan language documents, scholars encountered the opportunity to find a new role for the Venus Table within modern scholarship and interpretations. Through the possibility of using the Dresden Codex as a bridge between the earlier hieroglyphic inscriptions from the Classic period and later Colonial period historical events, the Venus Table itself was transformed. With these other books, modern scholars were able to convert the Venus Table into a subordinate mathematical instrument useful in fine-tuning the solution of another interpretive problem—the Mayan Calendar Correlation Problem. With this development, the Dresden Codex would not be studied only for the insights it provided into Mayan culture or the lives of its authors; now it would be used as one piece in the larger project of translating chronologies.

In this chapter, we take up this shift in the interpretation of the Dresden Codex Venus Table, which was made possible by John Teeple's work on page 24 of the manuscript, the "preface" to the table. To get at the first part of this shift, we first examine the role played by these other Mayan books, about which—unlike the Dresden Codex—we have quite a bit more direct historical information. In the process, we are challenged to develop an understanding of the approaches indigenous Mayan scribes used in these documents, along with how they may have changed over time. For the second part of the shift, we realize that apart from subordinating the interpretation of the Dresden Codex Venus pages to secondary status, scholars were motivated by a fundamental interest within the field of archaeology: relating cultural developments in one geographic region to those in another.

The latter problem became particularly important for the interpretation of Chich'en Itza, as scholars began to wonder whether the hegemony of the Postclassic originated in the Mayan region and moved west, or if it was created in Central Mexico and moved into the Mayan region. A handle on the problem of chronology could shed light on this issue. Returning to the mural containing K'uk'ul Ek' Tuyilaj's image in the Great Ball Court, for example, we are led to ask, What temporal relationship existed between K'uk'ulkan of Chich'en Itza and Quetzalcoatl of Tula? In these chapters, we find that the same problem impacting the interpretation of the Dresden Codex also related to the interpretation of cultural developments across Mesoamerica and with the rest of the world. In this chapter, we review the documentary context for the interpretation of the Dresden Codex and its conversion into a practical tool within the Calendar Correlation debate. This sets up chapter 4's treatment of the role that tool played in the innovation of the now ubiquitously used archaeological tool of radiocarbon dating, at the expense of the Venus Table's own nuanced appreciation.

JUAN JOSEF HOIL, A MAYAN SCRIBE

In his copying efforts for *The Book of Chilam Balam of Chumayel*, it is clear that Hoil did not attempt to edit substantively the previous versions he worked from—he did little to no reorganization, leaving duplications of material that otherwise could have been tightened up. Instead, it appears that he simply copied what those before him had copied, and then added a little more (cf. Farriss 1984, 247). That preservation of previous voices, without subjecting them to a single narrative, appears to have been the consistent practice from one generation to the next.

It is this lack of editing that provides the material of particular relevance in this chapter. Hoil's copying efforts generated what appears to be a consistent narrative, but it belies the fact that the text is much more like an archaeological excavation unit. Each "historical account," each individual record, sits within its own cultural and historical context, sedimented and distinct from those before and after it. Attempts to fit together narrative pieces from one layer into pieces from others without taking the changing contexts into account run the risk of finding patterns where none was intentional. This issue constitutes a central factor in the efforts of late nineteenth-century and early twentieth-century scholars, as we will see below.

When it comes to the dated material in the book, still another complication entered the record. Hoil and his predecessors had moved away from Tawiskal Uwoojil's Long Count and had begun relying on a Katun Count to record progressions through time.

That is, we have seen that the scribes of earlier periods used the linear Long Count to anchor dates uniquely in time with precision to the day; but Hoil and his contemporaries shifted to a practice that embraced a cyclical calendric count.

Here it is important to recognize that Hoil's "katuns" are mathematically identical to the winikhaab of the Long Count, comprising twenty 360-day periods (or 20 haab), and so close to—but just short of—twenty years.* The key difference, however, is that scribes of the Classic period tallied the winikhaab of the Long Count linearly, subordinate to pih (or baktuns), as we have seen in chapter 1. Scribes of the late Postclassic and Colonial periods, on the other hand, recycled the katuns in the Katun Count endlessly in groups of thirteen, as represented in figure 3.1. So winikhaab can be identified uniquely in the Long Count, but within the Katun Count there are numerous possible assignments of each Katun. While Hoil recorded, for example, that the founding of Chich'en Itza occurred in a Katun 13 Ajaw, which Katun 13 Ajaw this may have corresponded to in the Long Count cannot be determined unambiguously without additional information (Roys 1933, 28).

The new structure introduced ambiguity, but it also facilitated a specific aspect of Mayan record keeping and it followed more common Mesoamerican practice. The Katun Count emphasized a form of prophetic history in which the character of a time period may have been more important in the narrative of a historical record than the specific number of days that separated it from another event (Jones 1998, 13). One Katun 13 Ajaw set expectations for the next Katun 13 Ajaw. The latter recycling of time periods or "eras" was followed throughout the rest of Mesoamerican with their reliance only on the Calendar Round, which repeated every fifty-two years.

Parts of the Katun Count passages are similar to previous Mayan practices, as notable in the content of pages 72 and 73 of *The Book of Chilam Balam of Chumayel* (Roys 1933, 72–73):

> Katun 11 Ajaw is established at Ichcaanzihoo. Yax-haal Chac is its face. . . . The drum and rattle of Ah Bolon-yocte shall resound. . . . they shall find their food among the rocks, those who have lost their crops in Katun 11 Ajaw.

> The katun is established at Uuc-yab-nal in Katun 4 Ajaw. At the mouth of the well, Uuc-yab-nal, it is established . . . It shall dawn in the south. . . . There is mourning for water; there is mourning for bread.

> The katun is established at Maylu, Zaci, Mayapan in Katun 2 Ajaw. The katun is on its own base. The ramon is the bread of 2 Ajaw. It shall be half famine and half abundance. This is the charge of Katun 2 Ajaw.

> The Katun is established at Kinchil Coba, Maya Cuzamil, in Katun 13 Ajaw. Itzamna, Itzam-tzab, is his face during its reign. The ramon shall be eaten. . . . The sun shall be eclipsed. Double is the charge of the katun: men without offspring, chiefs without suc-

---

*To be precise, the interval is of 20 × 360 days = 7200 days = 19.71 years.

FIGURE 3.1 Comparison of Classic period winikhaab to Colonial period katuns. The text from the West Tablet of the Temple of Inscriptions at Palenque states that after 8 k'in, 5 winik, 10 haab, 11 winikhaab, and 10 pih it will occur, 5 Lamat 1 Mol on the Long Count date of 1.0.0.0.0.8, after the seating of the tuun on 10 Ajaw 13 Yaxk'in. It makes clear that at Palenque, the Long Count is not cyclical but proceeds in higher periods above pih by powers of 20. The Katun Count, on the other hand, repeats cyclically as depicted in the wheel. Each "Ajaw" symbol is accompanied by a Roman numeral in the sequence: XI (Buluc), IX (Bolon), VII (Wuk), V (Ho), III (Ox), I (Jun), XII (Lahka), X (Lajun), VIII (Waxak), VI (Wak), IV (Kan), II (Ka), XIII (Oxlajun). This follows the repeated addition of 7,200 days (or one "katun") to each Ajaw date.

cessors. For five days the sun shall be eclipsed, then it shall be seen. This is the charge of Katun 13 Ajaw.

These records are what Sylvanus Morley referred to as *u kahlay katunob*—"the sequence of the katuns" (1975, 79). In each, the katuns as time periods are established at specific geographical places, ritual accoutrements to the katun are described, and the passage ends with a prophecy. The ceremonial practice of marking time, therefore, did not differ drastically from previous periods.

Hoil's Book of Chilam Balam does testify to a degree of cultural continuity, then, and numerous scholars have relied on forms of it for academic interpretation. Morley, for example, noticed that Terminal Classic period scribes at Tikal included what he took to be a new dating convention within their very late monuments. Royal scribes there initiated

the inscription on Stela 22 with *13 Ajaw 18 Kumk'u. U 17 winikhaab*—"13 Ajaw 18 Kumk'u. It is the 17th Winikhaab" (figure 3.2). Instead of the traditional format of ISIG 9.17.0.0.0 13 Ajaw 18 Kumk'u, the Tikal scribes replaced the Long Count periods with a narrative statement that provided the winikhaab coefficient. Morley recognized this as similar to Hoil's Katun Count convention, which he suggested would have reduced it further to simply Katun 13 Ajaw. Morley hypothesized that this abbreviation is what elicited an eventual abandonment of the Long Count and its connection to Classic period mythology built into its "zero date." Brevity superseded precision.

The scholarly interest in continuity, however, runs the risk of eliding what was otherwise a protracted period of tumult and transformation. Mayan communities throughout the Yucatan Peninsula experienced waves of influence and direct impact between Tawiskal Uwoojil's copying of his manuscript and the time that *The Book of Chilam Balam of Chumayel* made its way into Mayanist scholarly endeavors. These changes provide important context for the very attempts at appealing to forms of "cultural continuity" by Morley and his colleagues during the mid-twentieth century.

THE BOOK LEAVES CHUMAYEL

A later local administrator—an assistant curate named Justo Balam—decided that he didn't have the authority or the responsibility to recopy the entire manuscript. Instead, he simply appended to Juan Josef Hoil's manuscript, adding birth records to the final pages for the children of a relative—Andres Balam and wife, Maria Juana Xicum—on April 4, 1832, and December 22, 1833. Five years later, the book changed hands again, leaving Justo Balam's, now almost sixty years after Hoil finished his copy. Pedro de Alcántara Briceño added the latest content to the manuscript, writing:

FIGURE 3.2 "Short Count" on Tikal Stela 22. The opening text reads: "13 Ajaw 18 Kumk'u. It is the seventeenth winikhaab." The event is a "k'altuun" performed by the Mutul k'uhulajaw, which will be addressed in the next chapter.

This is the day on which I purchased the book: July 1st, 1838. It cost me one peso in <my> poverty. This was the price <I paid> to the Señor Padre: <one> peso. This is the year of the purchase ... I have recorded it in order that it might be known that at this time it passed into my hands by purchase.

I, Pedro de Alcantara Briceño, resident of San Antonio. (Roys 1933, 82)

If nothing else, these final records make clear that this book was changing hands precisely during another time of transformation within the history of Yucatan.

By the end of the eighteenth century, the region had drifted far from the internationalism that K'uk'ul Ek' Tuyilaj or Tawiskal Uwoojil experienced during the Mesoamerican Postclassic period. Contact with Europe in the sixteenth and seventeenth centuries generated a "reduction" of the Mayan population (Spanish *reducción*)—both deliberate and accidental (Farriss 1984; Jones 1998). Village populations throughout the peninsula substantially decreased due to war and disease, but many Mayans also were moved into the colonial centers of Merida, Izamal, and Campeche or into freshly "urban-planned" villages spread out across the peninsula to suit colonial administrative needs (figure 3.3). Of course some Mayan communities dissented and left the Spanish Crown's reach entirely—the largest such independent city flourished until the end of the seventeenth century on an island in the central Peten at Tah Itza, supported by towns on the lakeshores and by nearby allied communities. Positioned between colonial centers in Yucatan to the north and Antigua Guatemala to the south, Tah Itza and surrounding villages became a region of refuge. As we saw in the introduction to this book, Kan Ek' was visited by Andrés de Avendaño there—the latter claiming to have read Mayan hieroglyphic books written by local scribes as late as the 1690s (Jones 1998, 3–6; Means 1917, 146).

But even those indigenous Mayan communities directly under colonial authority during this time maintained a significant degree of autonomy. Yes, there were taxes and labor to be paid to the Spanish Crown, and the titles held by local officials intercalated Spanish offices with Mayan ones, but in many towns contact with Europeans and even Spanish creoles was minimal, if existent at all (Farriss 1987, 87). While Hoil was copying the Book of Chilam Balam, for instance, rural towns such as Chumayel were still very much under the management of indigenous Mayans. In most cases, local Precontact administrative structures were adapted to Spanish forms, such as the *cofradía*, which brought together political and social roles to enable community religious activities (Carlsen 1997; Farriss 1987, 319).

Historian Nancy Farriss writes that in Spain cofradías were "joined on a voluntary basis and supported with alms and dues." She continues:

Spanish *cofradías* were primarily urban in character; they served to define social identities in a complex society in addition to promoting personal piety and offering mutual aid in the form of burial insurance and sometimes benefits for the sick, widows and orphans. (Farriss 1987, 265)

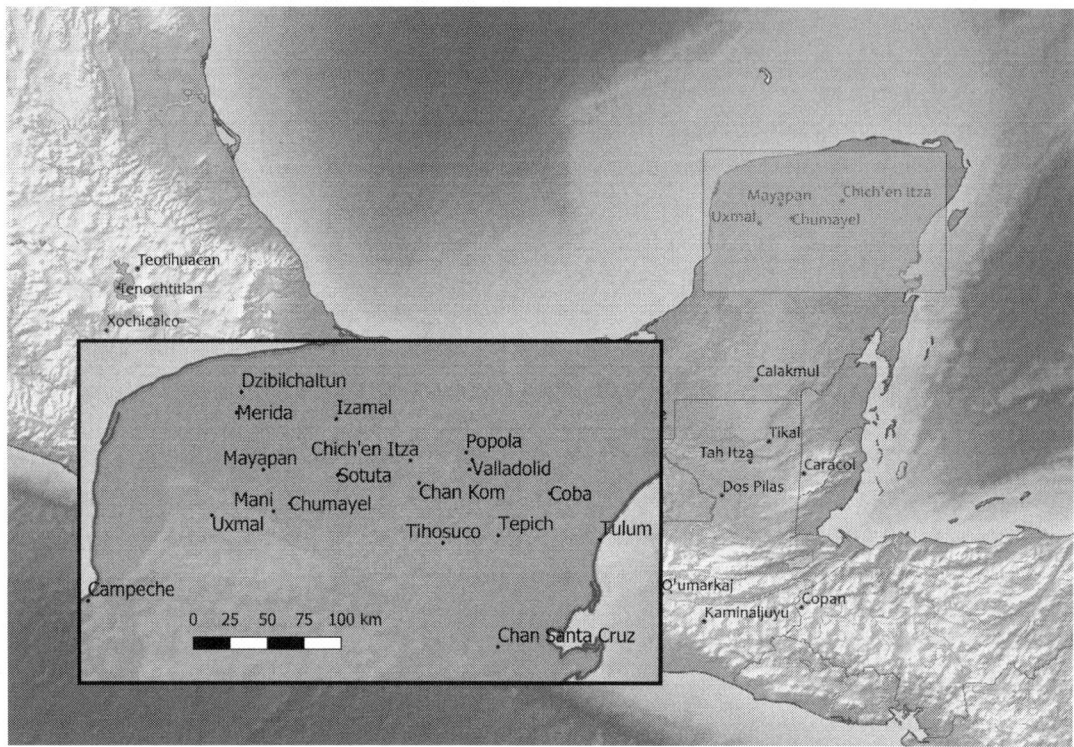

FIGURE 3.3 Cities of the Late Postclassic, towns of the Colonial period, such as Chumayel and villages of the Caste War, including Chan Santa Cruz

With the cofradía, the Catholic Church thereby provided Yucatecans a structure in which "each pueblo might nominally have several cofradias dedicated to different saints, [but] they were all integrated into a single cofradia organization, with a single set of officers who also served either simultaneously or alternately in the parallel civil posts" (Farriss 1987, 266). European hegemony was certainly in place, but indigenous Mayans managed everyday life.

Along with their religious organization, Mayan officials were managing the municipal governments for their own families and neighbors. Hoil, no doubt, was a member of this local Mayan "upper class" as a literate keeper of the Book of Chilam Balam. In fact he was very likely a *maestro cantor* for his community, trained both in indigenous religious traditions—maintained in part by indigenous codices—and the Catholic mass (Collins 1977; cf. Farriss 1987, 313). And this would have put Hoil into a tradition that reached back to the first encounters between Mayans and Spaniards in Yucatan more than two hundred years earlier, as well as to the authors of the words he copied.

As we saw in chapter 1, Gaspar Antonio Xiu and Juan Nachi Cocom maintained a scribal tradition that Landa tapped into in order to write his *Relación de las cosas de Yucatán*. During the sixteenth century, this must have been an especially challenging position for either Xiu or Cocom. On the one hand, they were attempting to preserve the welfare of their local indigenous communities; on the other, they were being trained by and within a new hegemony centered in Europe. Part of the tension resulted in scribes like Xiu shifting to write manuscripts on new types of paper in hybrid forms. While they kept their native language, they now wrote on imported European paper with the Latin alphabet to record the myths, histories, calendars, and prophecies that would eventually be compiled into the Books of Chilam Balam (Farriss 1987, 247).

Other forms of tension, however, were far from productive. These are evident in the passages that could have been authored by any of the copyists between Xiu and Hoil:

> It was only because these priests of ours were to come to an end when misery was intro-duced, when Christianity was introduced by the real Christians. Then with the true God, the true *Dios*, came the beginning of our misery. It was the beginning of tribute, the beginning of church dues, the beginning of strife with purse-snatching, the beginning of strife with blow-guns, the beginning of strife by trampling on people, the beginning of robbery with violence, the beginning of forced debts, the beginning of debts enforced by false testimony, the beginning of individual strife, a beginning of vexation, a beginning of robbery with violence. This was the origin of service to the Spaniards and priests, of service to the local chiefs, of service to the teachers, of service to the public prosecutors by the boys, the youths of the town, while the poor people were harassed. These were the very poor people who did not depart when oppression was put upon them. (Roys 1933, 31)

The author refers to "real Christians" and their deity as "the true God," yet she or he attests that what they bring is far from the salvation that Christian evangelists have been promising in Latin America for the last five centuries. While this tension was determined to be worthy of recopying by scribe after scribe over three centuries, it boiled over during the nineteenth century.

GUERRA DE CASTAS

Where Hoil maintained at least some semblance of the community charge held by Xiu, exhibiting sociological continuity between Hoil and Alcántara Briceño, a new insta-bility was introduced into the political context. At the turn of the century, Charles II, king of Spain, was unable to father an heir to his throne, capping a period of significant economic decline for Spain and its colonies (Hamilton 1938, 174). Charles, therefore, left his throne in 1724 to his grandnephew, Philip of Anjou, who was also grandson of Louis XIV, king of France. Discontent of the other European kings with the potential unification of France and Spain led to war in Europe and an eventual treaty, maintaining a separation of the kingdoms (Hamilton 1938, 175). The result, however, was a move of the crown's governance from the Hapsburgs to the House of Bourbon, which led to a shift in approach to empire, initiating the Bourbon Reforms in Spain as well as in the colonies. Specifically, Charles III in 1786 shifted to the French model of the intendancy (Farriss 1987, 367; Kuethe and Blaisdell 1991).

For indigenous Mayans in Yucatan, the political unrest of Europe may have appeared somewhat innocuous at first: officials were given new titles but performed basically the same functions. As Farriss teases out of the colonial record, though, the major impact was not structural, but came from the fact that creoles and/or Spaniards took over the positions of authority from local Mayans (1987, 356). Instead of money, resources, and judicial decisions being handled by Mayans for Mayan communities, as in Hoil's day, by the time Briceño held his copy of the Book of Chilam Balam, Spaniards or creoles were moved into governmental positions, and resources were no longer managed according to local Mayan interests.

By itself, then, this "reform" changed Mayan life significantly. What compounded the problem was the boom in international trade dependent upon production directly from the land in Yucatan, resulting from Charles III's "Decree of Free Trade" (Farriss 1987, 367, 371). The shift from local, sustainable economies serving local populations to the production of cotton, sugar, and then overwhelmingly henequen for export generated a huge pressure to actively cultivate more and more land. Had local Mayans still been in resource management positions, the impact may not have been so great. With non-Mayans in these positions, however, the impact was dramatic. Entire cofradía lands (considered "gifts to the saints" and charged to the Mayan cofradía) were reappropriated from indigenous needs and put to profit for the Bourbon Empire (Farriss 1987, 372–74). By the time Alexander von Humboldt traveled through Mexico and published illustrations of the Dresden Codex in 1810, then, and while the Book of Chilam Balam went from Balam's to Briceño's possession—now purchased and not inherited—the economic boom in henequen and sugar was in full swing.

The intendancy of the Bourbon Reforms in Yucatan, however, did not last long—independence from Spain was initiated by *el grito* in Mexico City in 1810, followed in Yucatan specifically by extensive political strife and then the Caste War. Initially, the Yucatec response to independence was led by criollos—three different men attempted to lead coalition armies made up of Mayans, creoles, and mestizos. The criollo Santiago Iman first led the opposition to governmental authority centralizing in Mexico City, but then two governors, Miguel Barbachano and Santiago Mendez, traded leadership of the Republic of Yucatan, differing in the proposed relationship between Yucatan and the new Mexican state (Reed 2001, 29–30).

By the late 1840s, however, several Mayan leaders recognized an opportunity by taking the creoles at their word (Rugeley 1995, 486). The creoles had first inspired Mayans to join their armies by promising freedom from the oppressive economics to which they had been subjected and which we have seen attested in the Books of Chilam Balam. One Mayan soldier, Cecilio Chi, left his home of Tepich in eastern Yucatan to fight with mixed-population armies in Campeche against Antonio Lopez de Santa Anna, who had been sent by the Mexican government. But Chi then returned to the east and allied with Jacinto Pat, Batab of Tihosuco, and Manuel Antonio Ay, Batab of Chichimila. Chi explicitly addressed opposition to economic oppression: "We are prepared to proceed through fire and blood to liberate ourselves from the payment of any contribution as long as we live" (Rugeley 1995, 486). These three indigenous Mayans laid the foundation for the Caste War, which aimed to establish an independent Mayan nation. The governor Santiago Mendez sent armed forces to the east in 1847 to prevent a suspected rebellion, but instead unleashed a chain of responses that resulted in Mayan forces taking over the entire peninsula with the exception of walled defenses at Merida and Campeche (Reed 2001, 84–107).

Unfortunately for the last independent Mayan nation since Tah Itza, their victories had spread their forces thin, and new troops from Mexico pushed Mayan armies back to their stronghold in the southeast. The two regions each consolidated strength, with Mayans creating an independent nation and a new capital at Chan Santa Cruz, which maintained its independence politically and economically until well after Förstemann picked up the Dresden Codex in his library across the Atlantic (Reed 2001, 186) (figure 3.4).

The point of this historical digression is that in 1838 when Pedro de Alcántara Briceño recorded in the back of the book that he purchased it from "a priest" for a single peso, the political infrastructure of local Mayan communities was in turmoil, and military activity throughout Yucatan was on the rise. What Briceño meant by writing that he "made a loan" twenty years later (Roys 1933, 80) is thus clouded by uncertainty, as is his purchasing it for "a peso." We might also recognize that, to a degree, this condition had become

FIGURE 3.4 Temple of the holy cross at Xocen

"the new normal" in Yucatan since the time of Contact with Europe. Political instability had become the rule, and authority was taken further and further out of Mayan hands.

And it is at this point that Hoil's book was taken over by Mexican hands. Like Tawiskal Uwoojil's "Dresden Codex," therefore, *The Book of Chilam Balam of Chumayel* moved from indigenous hands to European hands during a time of political unrest and military conflict. While we know less about the specific movement of Tawiskal Uwoojil's manuscript, we get some sense of its possible itinerary by comparison with that of his professional descendent Juan Josef Hoil.

The Book of Chilam Balam of Chumayel passed through local hands without official annotation until 1868, when Hermann Berendt came across it in the possession of the bishop of Yucatan, Crescencio Carrillo y Ancona (Roys 1933, 14). This bishop, though, did not follow Landa's lead of burning indigenous books, nor did he gift it to reside in a private library far from its homeland; rather, Carrillo y Ancona lent it to Berendt. The latter was a German medical doctor who moved to Central America and shifted to the amateur study of languages and ethnography. Berendt's first effort accordingly was to hand copy the manuscript, which he completed in 1868. He then had the opportunity in 1887 to have the irascible Teobert Maler make a photographic copy of it (Roys 1933, 13–14). Maler first saw it in Merida, then, between his photographic work at Palenque and that at Chich'en Itza, while battles of the Caste War were still common in eastern parts of the Yucatan Peninsula.

It was through Maler's photographs and then American historian Daniel Brinton's copies that the content of the Books of Chilam Balam found their way into the hands of early Mayanists and the burgeoning scholarship to which Förstemann, across the Atlantic, was contributing. While old forms of monarchy were being transformed into new governmental forms in Europe and North America, Mayanists were beginning to access the material culture that would inform the basis of archaeological interpretation for the next century. In the early 1900s, through the efforts of early Mayanists, the handwritten text of the Books of Chilam Balam came into interpretive contact with the carved inscriptions of the Classic period.

So while we now have been able to reconstruct the long-term connections between traditions, communities, and content, this has only become possible after a century of scholarship. When Förstemann's colleagues encountered the books at the end of the nineteenth century and the beginning of the twentieth, they investigated a completely different context—one that focused on thin patterns of connections between the Dresden Codex and the Books of Chilam Balam. During this time, as we have seen, the success of mining Mayan hieroglyphic texts for their astronomical content without the ability to read the language spilled over into the treatments of the Books of Chilam Balam. Of critical importance to the interpretation of the Dresden Codex, a number of these early Mayanists picked up the books to address the calendrical records within them relative to the growing presence of Mayan astronomy in the more ancient material. These provided the context for the new layer of interpretation to which the Dresden Codex Venus Table would be subjected.

The issue of continuity and the Calendar Correlation between and among these documents—that is, among the different versions of the books themselves, and between them and the Dresden Codex—has generated a direct impact on our ability to understand the Dresden Codex Venus Table. Its treatment, therefore, is essential to the goals of this book.

TEEPLE'S CAVALIER PROPOSAL

By 1926, with Yucatan in the wake of the Mexican Revolution, Teeple's cross-country computation sessions had taken him beyond recovery of the Lunar Series and into the more complicated material in the Dresden Codex. While passing the hours on the train—no doubt quite comfortably with the luxury accommodations available to those who possessed the financial means—Teeple hit on a possible "solution" to the anomalous intervals on page 24 of the Dresden Codex. His physical comfort was probably matched by his confidence by this time in his grasp of Mayan astronomy. The Perkin Medal announcement titled "Chemists Honor Cornellian" in the *Cornell Alumni News* of October 28, 1926, for example, capped the list of Teeple's chemical engineering successes with recognition of his contributions to *The American Anthropologist* as noted above.

It may well be, then, that this comfort level pushed Teeple to "go out on a limb" within the field of Maya Studies. Specifically, his new insights into the computations recorded on page 24 allowed him to pursue "another" correlation between the Mayan and Christian chronologies. Through his latest work on Venus combined with his previous work recovering the Mayan Lunar Series, Teeple believed that he had discovered a new way to translate Mayan Long Count dates into their corresponding Julian dates.

The problem he was tackling here—known as the Maya Calendar Correlation Problem—is itself an issue with a fascinating history that continues to play out even today (Aldana 2009; 2016b; Coe 1999; Fuls 2008; Kennett et al. 2013; Thompson 1978). The problem itself is relatively simple, originating in the observation reviewed in chapter 1 that the Long Count uniquely identifies single days in historical time relative to each other. For the purpose of cultural translation, the Long Count can be understood as a prior invention of the Julian Day Number (JDN), which is used commonly within modern astronomical practice.

The JDN itself originated with Joseph Justus Scaliger in 1583—a French historian and contemporary of Landa and Xiu—who brought together three independent calendric cycles (the "indiction cycle" of 15 years [used in Medieval Europe for tax purposes], the Metonic cycle of 19 years, and the solar cycle of 28 years) to create a period of 7,980 years. Scaliger "zeroed" his cycle on January 1, 4713 BC in order to capture all known historical calendric records within a single period. His work and intent was primarily historical, incorporating Persian, Babylonian, and Egyptian records into accounts of ancient Rome and Greece, so for him there was no need to deviate from the chronological base of the Julian year of 365¼ days.

The ramping up of astronomical projects three centuries later as the European Scientific Revolution raged on, however, had an English astronomer adopting, and repurposing, Scaliger's count. While serving as the president of the Royal Astronomical Society, John Herschel published *Outlines of Astronomy* (1861), in which he shifted Scaliger's count of years to a count of days only, avoiding the complications of accounting for leap years. The JDN was therefore formally established as an integer count of days, zeroed at January 1, 4713 BC.* Within this system, for example, Scaliger was born on 2,283,759 JDN, which corresponds to August 5, 1540 AD, and Herschel himself died on 2,404,558 JDN, corresponding to May 11, 1871 AD.

Across the Atlantic in Mesoamerica, some two thousand years prior to Herschel's work, calendric specialists of flourishing Late Preclassic cities chose an arbitrary starting point roughly three thousand years in *their* past and used it to count the number of days elapsed since. These scribes, some at Tres Zapotes near the Gulf Coast in Mexico at the western base of the Tuxtla Mountains, and their colleagues at Chiapa de Corzo in the heart of the Isthmus of Tehuantepec made use of their modified vigesimal numerical system to invent an earlier version of Herschel's count (figure 3.5). As we saw in chapter 1, the dating system they invented is represented differently, but now we can also appreciate that conceptually it is identical to the JDN (Bricker and Bricker 2011; Satterthwaite and Ralph 1960, 180).† One of the Preclassic scribes at Tres Zapotes—four hundred years before Chak Toh Ich'aak built his palace at Tikal—recorded the Long Count of 7.16.6.16.18,‡ which in modern decimal notation would be 1,125,698 days after their zero date.

Although the two systems did not start on the same zero date in absolute time, the calendric opportunity they afforded to Teeple and his colleagues was to correlate one Long Count date with one JDN unambiguously. This in turn would provide a correlation between Mayan and Christian chronologies, making possible the alignment of Meso-american histories with others around the world—and so following up on Scaliger's historical agenda while bringing it across the Atlantic. Translating hieroglyphically recorded events from Mayan texts into JDNs, for instance, would allow scholars to place Mayan events relative to those of the fall of the Roman Empire in Europe, the rise of Islam in the Mediterranean, or the rule of the Tang Dynasty in China. On a regional level, a solution might contribute to the alignment of Aztec calendrics with the Mayan calendric system. Of more interest to the early twentieth-century researcher, though, was that a solution to the Calendar Correlation Problem would allow for the reconstruction of the visible night sky for any Long Count event.

---

*December 31, 4713 BC, therefore, was 365 JDN; January 1, 1 AD was 1,721,424 JDN; and July 23, 2015 was 2,457,227 JDN.

†The 260 DC seems to maintain continuity between the Late Preclassic dates we have and the Classic period inscriptions; the 365 DC definitely shifts. See, for example, Justeson 1988, in which he demonstrates continuity in the 260-Day Count between the Epi-Olmec Long Count dates, but a clear shift in the associated 365-Day Count.

‡on the monument we now refer to as Stela C

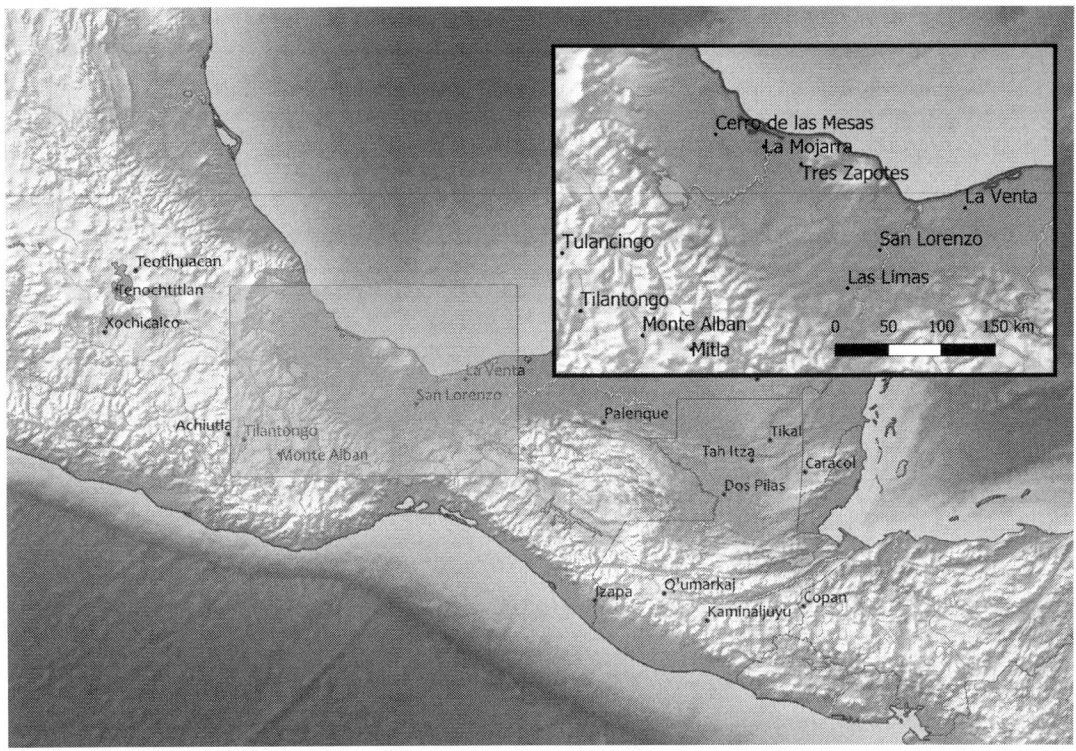

FIGURE 3.5 Early sites in the Gulf Coast region with Long Count date inscriptions

With a solution to the Calendar Correlation Problem, researchers could reconstruct the positions of the Sun, Moon, planets, and stars for any date in the hieroglyphic record—to "see what the Ancient Maya saw" in the dates recorded on their monuments. Early twentieth-century Harvard astronomer Robert Willson noted that a Calendar Correlation Constant would only have to be a simple time interval moving from the JDN zero point to the Long Count zero date (Aveni 2001, 207; Willson 1924).

LC + A = JDN
Where
LC = decimal equivalent of a Long Count record
A = "the Ahau Equation"
and
JDN = the Julian Day Number for the Long Count record. (Willson 1924)

With an appropriate choice of A, any Long Count date could be translated easily into its Julian or Gregorian equivalent. For example, using Teeple's proposed Correlation Constant of 492,621, the Tres Zapotes scribe's Long Count date noted above would correspond to 1,125,698 + 492,621 = 1,618,319 JDN, or September 19, 283 BC.

And here we confront the problem's visceral attraction to the computationally minded. A single historical event identifiable in both systems would provide a solution to the Calendar Correlation Problem as a simple integer constant. In his 1926 article, Teeple found what he believed to be just such an opportunity.

To develop his Correlation Constant, Teeple leaned on his engineering sensibility: he stripped the problem down to its simplest form, using as little data as necessary in order to derive an analytical solution. In taking such an approach, however, Teeple was going against the consensus in the field and so made possible a controversy that has waxed and waned ever since. The material Teeple wished to "strip away" was precisely what "insider" Mayanists had recovered from Juan Josef Hoil's work, along with the other Books of Chilam Balam. Teeple pitted a new astronomical data set against an archive of historical records, setting up a tug-of-war contest that has yet to achieve unanimous resolution by Mayanist scholars today (Aldana 2009; Kelley 1983; Thompson 1978).

THE "HARVARD SCHOOL"

The favored Calendar Correlation Constant for which Teeple provided "another" possible solution was largely assembled by Sylvanus Morley. Morley's charisma and extensive travel agenda—as both archaeologist and U.S. spy—put him in a position to gain access in his day to most of the data that is still used by modern scholars to correlate the Mayan Long Count with the Christian calendar. Moreover, his interest in the problem was perhaps matched by his eloquence in describing his goal:

> One of the most important problems in American Archaeology is the correlation of the Maya system of counting time with our own. Long before the first appearance of the white man in the Western World, the Maya race of Central America and Southern Mexico had developed an accurate system of reckoning time and recording events. So accurate indeed is this aboriginal chronology that were it possible to translate a single Maya date into the corresponding notation of our own calendar, the age of all the great cities of the Maya culture would be known, probably more exactly than the age of Nineveh, Babylon, or even Rome. (1910, 193)

For years, Morley has been portrayed as a paragon of the American archaeologist, although he came to the field a little late by modern standards—after an undergraduate education in civil engineering at the Pennsylvania Military College. As related by colleagues and obituarists, Morley broke from his father's interests in engineering to pursue the study of ancient Egypt at Harvard College (Barrera Vásquez 1948; Thompson 1949). In 1907, he traveled to Yucatan, visiting multiple sites and even witnessing the dredging of the Great Cenote at Chich'en Itza by the archaeologist Edward Thompson. These travels and the contacts he made would contribute critically to the material Morley eventually applied to three major projects: (1) his master's degree, which he took in

1908; (2) his dissertation, which would never officially be completed; and (3) two related publications on the Mayan calendric system. The latter two publications are of greatest concern to this chapter.

Morley's plan was to complete his PhD at Harvard based on a dissertation that included the (calendrical) interpretation of a set of hieroglyphic inscriptions from the Classic Mayan site known as Naranjo in Guatemala, near its border with Belize. This probably would have left Morley just behind his classmate Herbert Spinden, who graduated with his PhD on "Maya Art" from Harvard in 1909. Morley had "jumped the gun, though," by attempting to use the data from Naranjo for his doctoral research (Thompson 1975, vi). In 1909, Morley published the article "The Inscriptions of Naranjo, Northern Guatemala" with the title footnoted: "Work done in partial fulfillment of the requirements for the degree of Doctor of Philosophy in the Department of Anthropology, Harvard University." The problem was that those inscriptions had been photographed by the abovementioned Teobert Maler via an expedition of the Peabody Museum, which in turn was funded by an influential Bostonian, Charles Bowditch, who was himself an amateur Mayanist. Morley apparently had not obtained permission to use the photos that were still "in course of preparation" to be published as volume IV, number 3, of the *Memoirs of the Peabody Museum of American Archaeology and Ethnology, Harvard University*. According to Eric Thompson, Bowditch took offense and blocked Morley's thesis, leaving him with only a master's degree (Thompson 1975, vi). Stepping back, we might recognize that the politics of publication in early twentieth-century academia resonated with the Classic period politics among the k'uhulajaws they aimed to study.

The lack of a doctorate in archaeology did not slow Morley down (Coe 1999, 127). By 1915, Morley published a book that is still printed in paperback form today. In *An Introduction to the Study of the Maya Hieroglyphs*, Morley wrote that his intent was to "enable the general reader, without previous knowledge of the science, to understand" the current state of the field (1975, xix). While he noted that there were already a considerable number of publications on the topic—including Charles Bowditch's—Morley claimed that they were either insufficiently comprehensive or too technical for the public. His project was warranted, therefore, and it is in this text that the perspective of the early twentieth-century Mayanist is laid clear. Morley writes:

> The classification followed herein is based on the general meaning of the glyphs, and therefore has the advantage of being at least self-explanatory. It divides the glyphs into two groups: (1) Astronomical, calendary, and numerical signs, that is, glyphs used in counting time; and (2) glyphs accompanying the preceding, which have an explanatory function of some sort, probably describing the nature of the occasions which the first group of glyphs designate. (Morley 1975, 26)

Here we confront the explicit foregrounding of astronomy within the interpretation of ancient Mayan culture. Without the ability to read the glyphs, Morley rendered the noncalendric portions of text to be strictly subordinate to the astronomical patterns that he and his colleagues would look for within the calendric material.

In preparation for *An Introduction to the Study of the Maya Hieroglyphs*, but after his troubled Naranjo inscriptions article, Morley took on the Calendar Correlation Problem. In 1910, he wrote an article that pursued the opportunity presented by his personal familiarity with Yucatan and the work done by the historian Daniel Brinton, who had translated Colonial period Yucatec Mayan language documents. The latter included the Books of Chilam Balam (others besides the one from Chumayel had surfaced in the intervening years). Morley himself saw the Chumayel manuscript in 1913 after it had been returned to Yucatan, and he acknowledged in a footnote to *An Introduction* and in his 1910 article that he had "freely consulted" Brinton's translation "in [his] discussion" (1975, 3n1).

The particular opportunity that this afforded for the Calendar Correlation Problem is that Morley now could include the Katun Count histories found in the books, which went so far as to reference events transpiring hundreds of years before cross-Atlantic contact at the major cities of Postclassic Yucatan. Within these historical records Morley recognized indigenous Mayan histories, including the rise and fall of Chich'en Itza. Here Morley was able to follow up on a specific goal set by his predecessor, Alfred Maudslay. As we saw in chapter 1, Maudslay's experience at the Classic period sites in the south and his interest in Chich'en Itza in the north led him to an interest in exploiting the potential overlap between ethnohistory and the hieroglyphic history (Graham 2002, 156). If he could find overlap, or if the two sets of records could be made to align, then he would have calendric continuity from Chich'en Itza at the Terminal Classic to the inscriptions of Palenque, Tikal, and Yaxchilan during the Classic period. Now Morley would take this even further to extend Maudslay's proposed chronology into Colonial Yucatan. For all his ascription of importance to Mayan astronomical records, the Books of Chilam Balam provided Morley the possibility of a *historical* solution to the Calendar Correlation Problem.

Fortunately, he did not have to start from scratch with his calendric treatment of the Books of Chilam Balam. For one, he could at least find inspiration in the work of a man he publicly admired. In his obituary for Joseph Goodman,* Sylvanus Morley wrote that he was starstruck at meeting the self-made scholar at the young West Coast institution, the University of California. Morley was thirty-three years old at the time (1916), not yet excavating at Chich'en Itza. The meeting made a significant impression.

The writer, on the single occasion he was so fortunate as to have the privilege of meeting the great Maya scholar scarcely a year before the latter's death, well remembers in this connection the assurance and vigor with which Mr. Goodman defended this thesis.

It was at a lunch at the Faculty Club, Berkeley, in September, 1916; and the writer, because of his studies along the same line, had the honor of sitting next to Mr. Good-

---

*In *Breaking the Maya Code*, Coe mistakenly attributes this quote to Alfred Tozzer.

man, then just seventy-eight. It was a personal moment long anticipated and never to be forgotten. (Morley 1919, 444–45)

The meeting was certainly an amiable one, but Goodman himself seemed to regard his own scholarly efforts as distinct from those of the "Harvard school" (including Morley), as he referred to his compatriot-competitors in the debate over Mayan chronology (Goodman [1905] 1974, 647).

Joseph Goodman had been introduced to Mayan hieroglyphic writing by Alfred Maudslay. Maudslay shared his photographs and illustrations of hieroglyphic inscriptions with Goodman, who—having made a fortune in gold prospecting in California—initiated his own study of these inscriptions during his retirement in the 1880s (Coe 1999, 112). Historians have since pointed out, unfortunately, that Goodman's work paralleled Förstemann's, and he may have claimed discoveries that were actually made by the German librarian. Michael Coe writes: "There can be little doubt that Goodman was quite aware of what Förstemann had already published on the Dresden" (1999, 112).

But Goodman did have some unprecedented and unchallenged epigraphic success; for instance, he identified "head variant" representations of the far more common bar-and-dot notation for numbers (figure 3.6). More important for our work in this chapter, Goodman combined his familiarity with Maudslay's hieroglyphic texts with his own considerable aplomb, and he went on to try his hand at a project he openly considered to be unresolvable: the attempt to correlate the Mayan and Christian calendars. Five years before Morley published his own more confident attempt, Goodman wrote:

It is not certain the thing can be done even by the use of proper data; but as our only present hope of coordinating the Archaic dates with ours lies in such a correlation, I have deemed it worth while to make one as correct as possible ([1905] 1974, 642).

Goodman provided some healthy perspective on the overall project, but in the end he valued the ability to correlate civilizations across geographic regions above the fact that there was insufficient data for him to solve the problem adequately. This view that the Calendar Correlation could be understood simply as a hypothesis and that it can be overturned with additional evidence is one that persists to this day.

FIGURE 3.6 Scribes often used head variants to represent Mayan numbers; occasionally they would go so far as to carve "full-figure" representations as on Stela D of Quirigua. The image represents the date 8 Kaban, with a personified number 8 holding the Day Sign in his lap.

It is notable, however, that even though it may be taken as a hypothesis, the interpretations of astronomical data that have resulted from its application regularly enjoy broad acceptance.

From Goodman's view, there were two fundamental "obstacles" to any correlation. Each he handled by making coordinated, strategic assumptions. For the first, Goodman stumbled with the lack of continuous records between Katun Counts from the Postcontact period and Long Counts from Precontact times (Goodman [1905] 1974, 642). As noted at the beginning of this chapter, the Long Count had fallen out of use by the time of the writing of the Books of Chilam Balam, so the Mayan practices transitioning from one system to the next went undocumented. More importantly, the Katun Count records that could be found were internally inconsistent—many of the dates had certainly slipped, such that the dates from one section to the next *in the same document* and certainly across documents could not be made to fit into a single continuous progression (Kelley 1983; Thompson 1978). An example of the complications shows up in what Roys designates chapter 18 quoted above, where the sequence progressed from Katun 11 Ajaw to Katun 4 Ajaw to Katun 2 Ajaw to end with Katun 13 Ajaw. (A complete, continuous sequence would have been Katun 11 Ajaw, to Katun 9 Ajaw, through 7, 5, 3, 1, 12, 10, 8, 6, 4, and to Katun 2 Ajaw, to end with Katun 13 Ajaw.) While the last three are in sequence, there is a gap of two hundred years between Katun 11 Ajaw and Katun 4 Ajaw. What had the original scribe intended with this sequence? Had any later scribe elided the text of his predecessor? These were critical issues that any modern interpreter would have to face.

First, then, to handle the complication, Goodman assumed that there was an underlying continuity to calendric practices over time and across regions—he advocated the continuity assumption. As Morley would also do—along with the vast majority of scholars treating the subject throughout the twentieth century—Goodman found the purported key to his solution by making the assumption of strict calendric continuity. For the above sequence, the "missing" katuns were simply not included by the original author or by Hoil as copyist. They were left out, but they didn't reflect calendric revision or accommodation.

Second, Goodman assumed that Gaspar Antonio Xiu's lineage was a privileged one—and not just economically. This reverence allowed Goodman to hypothesize a specific line of calendric integrity:

> That no other change was made is certain from the facts that the Xius did not align their katun count with that of the Itzas, Cocoms, and Chels, and that its character remained unaltered and its continuity unbroken from the time they left their mother-country. ([1905] 1974, 644)

For Goodman, all the calendric inconsistencies found in the Books of Chilam Balam were the result of other (i.e., non-Xiu family) copyists deviating from the one true and continuous count. The one ideal count was kept by the Xiu family, and so only their records made a Calendar Correlation possible.

Unfortunately for his efforts, Goodman was unable to find a clear "double-date"—that is, an event recorded in both Christian and Mayan chronologies by the same scribe. Instead, he put together a series of events dated by the Mayan year in which they occurred and cross-referenced them with records penned by Spanish chroniclers. The anchor to his proposal was the recorded death of one Napot Xiu on 9 Imix 18 Sip within a year beginning on 4 Kan (Goodman [1905] 1974, 644).

Goodman's proposed solution was not perfect. It left him still, for example, to "correct" scribal "confusions." With these corrections, though, he was able to compute the position of this date within the Katun Count, concluding that "it could be only 13 Ajaw-7 Xul, October 30, 1539" (Goodman [1905] 1974, 645). An example of the problems he encountered was that the 260-Day Count portion, 13 Ajaw, was fine, but it should not have corresponded to a coefficient of 7 for the month Xul. During the Classic period, the convention would have had 13 Ajaw fall on the date 8 Xul. To address this, Goodman didn't change the 260-Day Count; he simply adjusted the 365-Day Count portion such that October 30, 1539, fell on 13 Ajaw *8 Xul* and not on 13 Ajaw 7 Xul as the scribe recorded. He introduced a one-day adjustment for a presumed slippage between Xiu's Colonial period dates and those of the Classic period. The revised Calendar Round date he was able to assign to 11.16.0.0.0 in the Long Count, with the assumption of continuity between the Terminal Classic period and the early Colonial period. This connection further generated a Calendar Correlation constant of 584,280, which led Goodman to write:

> The result shows that Copan, Quirigua, Tikal, Menche [Palenque], Piedras Negras, and the other more modern capitals, flourished from the sixth to the ninth century of our era, speaking in round terms. ([1905] 1974, 646)

The high esteem that Morley held for Goodman probably led the junior scholar to take much of his approach from Goodman, while at the same time emphasizing a different data set within it. Whereas Goodman chose one Yucatec family and privileged its records, Morley emphasized individual records from different contexts (and different families) that fit into a consistent pattern. Of course no modern archaeologist would be comfortable picking and choosing data from different material contexts to construct a meaningful archaeological interpretation, but Morley did just that with the calendric data he found in the Books of Chilam Balam.

Morley also stood fast with Goodman's assumption that one calendric component was inviolate and had not skipped a beat, nor been revised by even one day from the Early Classic through his own twentieth-century times and beyond. Even though other components may have been revised at different times or in different regions, Morley took the 260-Day Count as an "intertribal tzolkin," maintained by all parts of Mesoamerica since its origin (Morley 1910). Again, there was no documented evidence for this, but it served as a useful assumption because it meant that selected dates in the Katun Counts of the Books of Chilam Balam could be dropped into an unbroken progression of the Long Count from the Classic period. Morley, like Goodman, took as his anchor for this assumption the statement noting that Napot Xiu died in 1536 AD, which fell in a Katun 13 Ajaw (1910, 196).

And so with this assumption in hand, Morley too was able to constrain the mathematical relationship between the cyclical Postcontact Katun Count and the linear Classic period Long Count. Rather than a near infinite number of possible Calendar Correlation Constants, now he had only a handful. Since Katun 13 Ajaws repeated every 260 years, he needed a context that would allow him to select among 260-year interval options. For Morley the key was to incorporate the data from Chich'en Itza.

In his pre–Carnegie Institution travels through Yucatan, Morley was able to view the "Initial Series Lintel" that Edward Thompson excavated from the site. The date on that lintel was legible, and Morley transcribed it as "10-2-9-1-9 9 Muluc 7 Zac." In order to bring this into line with the Katun Count data, he converted the date to the later convention. The end of the winikhaab in the Long Count prior to the inscribed date would have been 10.2.0.0.0 3 Ajaw 3 Keh, which, then, was a Katun 3 Ajaw (1910, 201). That provided Morley with his objective: "Our next problem is to find in the [Books of Chilam Balam] a Katun 3 Ahau in which there is a stated occupation of Chichen Itza" (1910, 201). Here Morley turned to the histories recorded in the books to find three different Katun 3 Ajaws as candidates, separated by a total of 780 years.

Without access to more precise dating technologies, which would be developed after his time (as we will see below), Morley weighed the historical implications of the three candidates against one another. He eventually settled on his first proposal, putting the earliest historical dates in the Books of Chilam Balam at 9.0.0.0.0 8 Ajaw 13 Keh in the Long Count. Morley found this reasonable given that the vast majority of hieroglyphic inscriptions recovered from the "Golden Age" dated between 9.0.0.0.0 and 10.0.0.0.0. The result, though, placed Katun 13 Ajaw not on 11.16.0.0.0 13 Ajaw 8 Xul, as Goodman argued, but 260 years later on 12.9.0.0.0 13 Ajaw 8 K'ank'in.

> The ancient Maya, with whom this volume deals, emerged from barbarism probably during the first or second century of the Christian Era; at least their earliest dated monument can not be ascribed with safety to a more remote period. . . . This period of development, which lasted upward of 400 years, or until about the close of the sixth century, may be called perhaps the "Golden Age of the Maya"; at least it was the first great epoch in their history, and so far as sculpture is concerned, the one best comparable to the classic period of Greek art. (Morley 1975, 2–3)

Morley's work to link the two calendars was certainly a feat and demonstrated his command over a broad range of material. It did leave some gaps, though, and he acknowledged that it was only one among a collection of viable alternatives (Morley 1975, 2). Indeed, we have seen that Goodman's correlation relied on the same historical anchor as Morley's—both started with the death of Napot Xiu occurring during a Katun 13 Ajaw. The difference was simply where in the Long Count that Katun 13 Ajaw would be placed.

Whereas Morley assigned it to 12.9.0.0.0, Goodman had it at 11.16.0.0.0—these differed by thirteen katuns or almost 260 years. The only way to distinguish between the two was reliance on different interpretations of the Katun histories in the Books of Chilam Balam in all their complexity. And while it was certainly arbitrarily selective in culling data points from several disconnected sources, Morley's approach went further to rely on multiple types of data, making it challenging for other scholars to adequately contest his argument. It is worth noting that, at the time, few scholars could effectively work with the hieroglyphic calendar, the excavation data, and the historical record to develop a convincing counterargument. During the first decades of the twentieth century, though, Morley's former classmate Herbert Spinden took up the Calendar Correlation Problem in an effort to address those gaps.

THE SPINDEN CORRELATION

Sylvanus Morley and Herbert Spinden were classmates at Harvard College, with the former taking his AB in 1908 and the latter two years earlier in 1906. While Morley came from the Pennsylvania Military College, Spinden had come straight from adventure in the Alaska Gold Rush (Sorensen 2018). They did travel to the Mayan region together in 1914, though by this time Morley was already working for the Office of Naval Intelligence, and on Morley's recommendation Spinden soon joined him (Browman 2011, 11). Besides their military work, Spinden and Morley also found common interest in working on the Calendar Correlation Problem.

Spinden's article "The Reduction of Maya Dates" came out fourteen years after Morley's work, while Spinden was in his mid-forties and Morley was getting to work at Chich'en Itza. In his position as a curator for the Peabody Museum at Harvard, Spinden's primary contribution was to finesse Morley's results through a new appeal to Landa's *Relación de las cosas de Yucatán*, adding a key element. As we saw above, Förstemann already made use of Landa's discussion of Mayan calendrics in his treatise, helping him even to crack the Long Count itself. Spinden went back to some of the same material, though with a different eye.

The extensive list of dates that Landa had recorded using both Mayan and Christian elements is what specifically drew Spinden's attention. This list comprised the Mayan Calendar Round—both the 260-Day Count and the 365-Day Count—along with the Julian month and day within the Christian calendar. The great tragedy for the growing twentieth-century interest in correlating the Mayan and the Christian calendars, though, is that Landa did not include the Julian year of his record. And this oversight specifically turns out to be absolutely critical. All modern scholars since Spinden making use of historical data construct arguments that rely on Landa's record as the foundation. All also take its year as AD 1553—this is the nexus upon which all assumptions of continuity have been laid (Lounsbury 1992b). All ultimately rely on Spinden's intervention, which we will see is flawed.

The proposal that Spinden made came from one additional set of data alongside the Christian dates. In line with the 260-Day Count dates, Spinden recognized the Christian practice of including the dominical letter, assigned to the Julian portion of the date (see figure 1.5a). In his own words:

> In connection with the orderly presentation of the days of the Mayan tzolkin occupying stated positions in Mayan and European months, Landa gives a cycle of seven letters which correspond to the days of the week, Sunday being marked by a capital A and the other days by the lower case letters b-g. The year-bearer 12 Kan has the letter A and therefore corresponds to Sunday, July 16, 1553. (Spinden 1924, 85–86)

Spinden here refers to 12 Kan as the "year-bearer," since it corresponds to the first day of the Mayan year; that is, 12 Kan lands on the first date in the 365-Day Count year, which is 1 Pohp. According to his understanding of how the Christian dominical system worked, Spinden took the dominical letters to correspond to the days of the week, wherein Sunday was designated with the letter "a," and "b" through "g" corresponded to Monday through Saturday. In Landa's list of dates, January 1 is assigned the dominical letter "a," so for Spinden, Landa intended to note that the year of this record transpired within a Christian year beginning on a Sunday. Thus, even though Landa didn't record the calendar year of his list of dates, Spinden could at least narrow it down to only those years that started on a Sunday (1924, 86). He logically concludes that 12 Kan 1 Pohp transpired during the year 1553 because that was the only year that began on a Sunday during Landa's time in Yucatan (Spinden 1924, 84–86; Tozzer 1941, 151).

His hypothesized reconstruction helped Spinden make sense of what otherwise appeared to be a mistake in Landa's list. The tradition throughout Mesoamerica was to consider the last five days of a 365-Day Count year as a period of rest and reflection. In the Yucatec Mayan calendar, these five days were known as the Wayeb; in Nahuatl, they were referred to as the *nemontemi*—or the "nameless days." The 260-Day Count dates corresponding to these five days are not included in Landa's list, literally reflecting the Nahuatl tradition. If we back up, though, the sixth to last date before 12 Kan 1 Pohp New Year date gives us 7 Ak'bal, occurring on July 10 in the Julian calendar. Spinden realized that this presented a problem.

If Spinden started on 7 Ak'bal as the date for July 10, and then counted through the Wayeb, the result would be 12 Lamat as the last day of the Mayan year on July 15. That would put July 16 on 13 Muluk. Instead, as we have seen, 12 Kan is explicitly recorded as the first day of the next Mayan year on July 16 in the Julian calendar. Clearly something was amiss.

Spinden proposed that this riddle could be solved by hypothesizing that Landa started his sequence with the Mayan New Year date of 12 Kan 1 Pohp on July 16 of one year, and then followed through a complete record of 365 days, ending on July 15 of the next calendar year. But in Spinden's view, Landa considered it objectionable to start the list on the Mayan New Year's Day instead of the Christian New Year's Day, and so Landa must have reordered the list to start with January 1. Not considering the impact on the sequence of dates (or not caring), he simply cut his list starting on January 1 of the second year and pasted before those of the first year. As Spinden put it:

> [T]he year 12 Kan ran from July 16, 1553, to July 15, 1554, the part corresponding to 1554 being placed before the part corresponding to 1553. This cutting and patching process gave a record of Mayan days equivalent to the European year except that the last day of Uayeb was a day 12 Lamat which is followed in the table by 12 Kan, *an impossible situation*. If we turn from 11 Eb on December 31 at the end of Landa's table back to 12 Ben on January 1 at the beginning we have two days which are really in sequence. For 12 Lamat as the last day of Uayeb terminates a year beginning with 12 Kan. Now the count of Landa's European year is continuous from January 1 to December 31 in the week day letters as in the days of the months. (Spinden 1924, 86, emphasis added)

In the end, this led to Spinden's claim that he had largely corroborated his former class-mate's work on the Calendar Correlation Problem. With 12 Kan on July 16, 1553, Spinden also found that he could keep Morley's primary criterion that 1536 occurred within a Katun 13 Ajaw.

The academic advisor to both Morley and Spinden also jumped into the debate. In his heavily annotated translation of Landa's manuscript, Alfred Tozzer supported the import of Spinden's appeal to the awkward list of dates and the role of the dominical letters in it. Tozzer's footnote included the quotation from a letter by Juan Martinez Hernandez, a Mexican historian and translator of Yuctatec Mayan texts. Martinez Hernandez summarized and endorsed Spinden's work:

> Landa, in editing his typical year, began it January 1 with the Christian dominical letter A which means Sunday. At the time of the conquest the years 1525, 1553 and 1581 alone could have begun with Sunday. The Christian solar cycle is composed of twenty-eight years. During the first Landa was not in Yucatan and in the last he was already dead. It is, then, the year 1553 which he had in mind when he drew it up. . . . The year 12 Kan began on Sunday, July 16, 1553. Having run through to the end of this year he returned to the beginning of the same year of 1553 instead of continuing with the year 1554. January 1, 1554 fell on Monday, Christian dominical letter G. (Tozzer 1941, 151n748)

Spinden, therefore, had fortified a division between the would-be solvers of the Calendar Correlation Problem. On one side was Goodman with his preference for the Xiu lineage records and his point of articulation that would place the Colonial period date of November 22, 1539, on 11.16.0.0.0 13 Ajaw 8 Xul in the Long Count; on the other was the "Harvard school"—Morley and Spinden put the date of April 22, 1536, on 12.9.0.0.0 13 Ajaw 8 K'ank'in in the Long Count. Both sides agreed that revisions had occurred within various components (including Spanish adherence to the Gregorian calendar reform in 1582), but they both also subscribed to the key assumption of continuity in the 260-Day

Count, which they believed to be reflected in Landa's historical anchor. This assumption of an uninterrupted progression over thousands of years is what constrained the overall interpretation to allow only one of these two solutions to be correct. The archaeologist and epigrapher Eric Thompson wrote as late as 1978 that: "If a new correlation is needed, we may have to discard the katun count, but at all costs keep the equation 12 Kan 1 Pop = July 25, 1553 [Gregorian]" (1978, 309).*

The bulk of the work going into the mid-twentieth century, then, relied on subtle differences in the interpretation of the historical material written by Yucatec hands, while leaning heavily on a weird list of dates provided by Landa. The resulting difference in interpretations was substantial, amounting to a shift of 260 years such that the Classic Mayan period ran from the third to the ninth century AD, or from the first to the sixth. It was this division of heavyweights—Goodman on one side and the "Harvard school" on the other—that Teeple would find himself addressing with his "alternate" solution to the problem.

HISTORICAL COMPLICATIONS

Teeple's primary objection to the solutions available by the 1930s was that they relied on inconsistent historical records. Given his focus on and training in mathematical computation, this may have been more of a hunch than a well-informed conclusion, but we know now that his concerns had merit and that there are significant challenges to the base assumptions made by all who had been addressing the problem, on both sides of the divide.

In the first place, the assumption of calendric continuity in the 260-Day Count is challenged by comparable data elsewhere in Mesoamerica. Addressing this fundamental prerequisite to prior solutions, historian Howard Cline has shown that calendric *revision* is more evidenced as a rule in Mesoamerica than calendric continuity (1973). Aztec and Zapotec calendric records over time demonstrate internal inconsistencies that appear similar to what we find in the Mayan Books of Chilam Balam. But scholars addressing these Central Mexican records have understood them as different counts maintained by different communities (Aldana 2009, 111; cf. Justeson and Tavárez 2007; Cline 1973). In fact, there is some practical sense here. During the Classic Mayan period, the Long Count constrained continuity, and it did so with observational phenomena. Teeple's work on the Lunar Series demonstrated this relationship. He was able to verify that scribes from city to city throughout the Classic period recorded the Moon Age with Glyphs C, D, and E in the Supplementary Series with minimal deviation. This was possible because the Long Count itself was not reset or adjusted during the Classic period.

Without the Long Count, however, and having dropped the practice of regularly recording observable lunar data, scribes of the Late Postclassic and early Colonial peri-

---

*See below for the explanation for "July 25, 1553" instead of "July 16, 1553."

ods were free to adjust all parts of the calendar relative to solar time. Perhaps more importantly, societal forces were also challenging continuity. During the sixteenth century, indigenous calendric use was persecuted by Spanish colonial authorities, fracturing and forcing underground the social networking necessary to keep calendric counts continuous (Farriss 1984, 290). The kinds of ritual activities that symbiotically coexisted with the indigenous calendar were also persecuted by priests of the Catholic Church. Moreover, movements such as rallying around the Speaking Cross in Yucatan during the nineteenth century opened the door to revision for ideological purposes. All of these pressures and a lack of phenomenological constraints make it more likely that the variation in the dates in the Books of Chilam Balam reflects the revisions made by scribes serving the needs of local communities over time, and not computational mistakes made by one copyist or another.

Beyond the cultural complications, there are technical philological problems that modern scholars face with the reliance on Landa's text as a reliable source for dated material. The reader may have noticed already the discrepancy between Thompson's subscription to 12 Kan 1 Pop = July 25, 1553 (Gregorian) = July 15, 1553 (Julian) and the placement everywhere else of 12 Kan 1 Pop = July 16, 1553, or one day later in the Christian calendar. Such details have plagued all researchers working with the problem—they claim strict continuity, but then require slippage of one or two days for their arguments to hold (Aldana 2009, 151–52). For example, Thompson's proposal for this discrepancy was that Landa received the information of 12 Kan 1 Pop = July 16 in some year prior to 1552; then in 1552 the leap year would have pushed the correlation to 12 Kan 1 Pop = July 15, but Landa didn't take this into account and instead used the old correlation in 1553 of 12 Kan 1 Pop = July 16, even though it wasn't valid for that year (Thompson 1978, 304). While Thompson's argument can be made to work, such proposals also require the recognition that Landa's data may be simply inaccurate.

In fact doubts about the use of Landa's list of dates already had been expressed by 1935. A colleague of Eric Thompson, Richard Long, noted that the practice of assigning dominical letters reflected in Landa's list differed from Spinden's (and Thompson's and Martinez Hernandez's) understanding. Concerning Spinden's proposal, there were different Christian traditions for assigning dominical letters in use during the Colonial period, and in New Spain in particular. Within Landa's historical context, Long argued that the practice was actually the one in which the dominical letter "a" was assigned to the first day of the year, January 1, and not specifying Sunday (Long 1935, 97). In such case, the dominical letters didn't tell us anything about the specific year the list was intended to reference—every year started with the letter "a" regardless of the day of the week (cf. Justeson and Tavárez 2007, 28). If it didn't correspond to the tradition Spinden assumed, then it could not be used to constrain the year Landa had recorded it at all (Long 1935, 97). The dominical letters no longer constrained the list to correspond necessarily to AD 1553.

More recently, Matthew Restall and John Chuchiak introduced a yet more concerning complication. They emphasize that scholarly perspectives have insufficiently accounted for the fact that the manuscript that Brasseur de Bourbourg recovered was not Landa's—it was a copy of Landa's work for use in a compilation along with other manuscripts (Restall and Chuchiak 2002). They explicitly warn that

> even if the *Relación* is viewed not as a whole but as a source on specific and isolated topics, scholars cannot take for granted the authorship and dating of particular passages—let alone the reliability of published editions. . . . [T]he *Relación* [is] a complex and messy compilation, one that should be handled as gingerly as, say, one of the colonial Maya compilations known as the *Books of Chilam Balam*. (2002, 664)

It may be, therefore, that the list of dates, for one, is not entirely Landa's work. Such a proposal would make sense of the complication that Spinden focused on: the sequence started on January 1, but the Mayan dates didn't properly sequence. Perhaps this wasn't Landa who was offended by a list starting on a Mayan New Year's Day, but a later copyist in Spain who was more interested in setting it appropriately according the Christian calendar. This too may suggest that the dominical letters themselves were not even written by Landa's hand. They too may have been included by a copyist after editing the sequence to make better sense to a European audience. The point is that for all that scholars have taken Landa's "double-date" to be secure, it is far from it.

The upshot is that we know now that there are significant and potentially debilitating problems with the assumptions underlying the historical interpretations used by Goodman, Morley, and Spinden to solve the Calendar Correlation Problem. And whether or not Teeple was prescient in this regard, his approach was to avoid them entirely when he took up page 24 of the Dresden Codex in 1925. Much more compelling, Teeple thought, would be an analytic solution that depended only on astronomical data. And this he found to be possible through his new analysis of the Venus Table.

RELYING ON ASTRONOMY

Teeple turned his back on the Katun Count and the records left by Hoil and Landa to follow the hints within the math itself. Specifically, he realized that combinations of the right kinds of data would allow him to construct a solution to the Calendar Correlation Problem without having to make assumptions about history, or having to borrow simplifying assumptions. Teeple noted instead that if the same astronomical events could be identified uniquely in both calendars, then that would provide the unequivocal connection leading to a secure Calendar Correlation Constant. After such a solution was found, only then would historical records such as those in the Books of Chilam Balam or Landa's manuscript be checked or accommodated accordingly.

At the end of chapter 1, we saw Teeple's proposal that page 24 could have been used to correct the table against observation. He recognized that a strict calendric progression of 584-day periods through the calendar would slip over time relative to the positions of Venus in the morning and evening skies. He also found that the intervals recorded on page 24 allowed for the dates in the table to shift in accommodation of that slippage. These correction factors, in turn, led Teeple to the realization that he could now match up historical Venus events with corrected records from the table. In other words, his corrected table was able to track observable Venus events over hundreds of years for any historical period.

Unfortunately, however, Teeple's table corrections would only give him a sliding match. Once he could accurately track Venus events, he could place that accurate tracking in the first century or the tenth without knowing which was intended. A parallel problem would be to have a tracking of the phases of the Moon, but not having the historical records to know if a given record corresponded to a Full Moon in January or July. The Venus pattern, therefore, provided an important constraint, but it was insufficient by itself for a solution to the Calendar Correlation Problem. What Teeple needed was at least one other astronomical event—preferably a relatively rare one, like an eclipse or a comet. He would then be in a position to look for a date on which he knew *both* the position of Venus *and* the date of a rare astronomical event. This combination would generate a solution to the correlation problem far stronger in Teeple's view than one built on slippery historical dates. He could ignore the continuity assumption entirely, for example, and just match up records of natural phenomena observed and recorded in both Mayan and Christian documents.

Teeple thus went back to the inscriptions he had consulted for his work on the Lunar Series, including the illustrations published by Maudslay. He first looked for corroboration that his corrections of the Venus Table were evidenced in these Precontact hieroglyphic inscriptional records. In his searches, Teeple looked for texts that included the "Venus glyph" and that could be assigned to Long Count dates. Although he found only a few such records, his treatment of them again gives us a view into the prominence of the astronomical perspective in the early twentieth-century practice of "reading" Mayan hieroglyphic texts.

> Turning again to the inscriptions, we have one other piece of evidence for the position in the long count. The wooden lintel of Temple C at Tikal (Maudslay drawing volume III, No. 78) gives a date 11 Ik 15 Chen which is usually and apparently safely considered to be 9.15.12.2.2,11 Ik 15 Chen, and in the immediately following Glyphs is a statement that "the Venus year ended in Kayab 24 days from a new moon day." Now the 10th year of our 1 Ahau 18 Kayab Venus calendar would have ended on 9.15.11.10.0 1 Ahau 18 Kayab, and the actual appearance of Venus might have been a day or two before at 16 or 17 Kayab. There was a new moon about 9.15.11.11.3, just twenty-four days after 17 Kayab, all of which at least is in agreement with our long count dates. (Teeple 1926, 405)

Never mind that the writing system was still decades away from decipherment, Teeple was sufficiently confident to read the hieroglyphic text and find support for his interpretation that page 24 not only held the key to correcting the Venus Table over the long run

FIGURE 3.7 (*above*) View from Tikal Temple IV; (*left*) Teeple "reads" the inscription in this structure fifty years before the writing system has been deciphered.

within the Dresden Codex, but that in fact may have been used to do so during the Classic period (figure 3.7).

Teeple's reading set him up to go one step further. His earlier work on the Lunar Series, which in turn built upon Förstemann's work on the Dresden Codex Eclipse Table that we saw in chapter 1, yielded the proposal that an eclipse was visible and recorded in the Mayan region on 9.16.4.10.8 12 Lamat 1 Muwaan. Here was his rare combination.

Our problem then is to find an eclipse of the sun which occurred between 0 and two days after the sun passed the moon's node and between nineteen and twenty-four days after an inferior conjunction of Venus. I am not an astronomer, but it seems likely that not more than five or ten dates in a thousand years would meet these conditions. At any rate, an examination of all eclipse dates between 462 and 517 A.D., (the date commonly assumed for 9.16.4.10.8 is from 480 to 500), showed only a single one that even remotely met the conditions imposed; this was the eclipse of November 22, 504 Julian calendar. (1926, 406–7)

Teeple therefore used his Dresden Codex–derived Venus records in concert with a hypothesized solar eclipse record in the Eclipse Table to generate a rare combination of celestial events. He then looked to astronomical records produced during modern times to find a reconstructed match, giving him an elegant and efficient solution to the Calendar Correlation Problem. The result put Mayan hieroglyphically recorded events near those generated by Spinden's correlation.

With a bit of skepticism, we might question whether the connection to Spinden's correlation reflected Teeple's friendship with Morley and Morley's continued intelligence work with Spinden. That Spinden was an influence shows up in Teeple's extension of his solution into the twentieth century:

> Until further evidence is brought forward I think we may say that in the time of the Maya Empire 9.16.4.10.8 was November 22, 504. Brought forward to Spanish times this would make **12.9.0.0.0** 13 Ahau 8 Kankin occur on February 22, 1545, O.S., or March 4, 1545, Gregorian. (Teeple 1926, 407, emphasis added)

Teeple, therefore, chose to demonstrate the impact on Spinden's "12.9" correlation and not Goodman's "11.16" version. His results did generate an "alternate" Calendar Correlation Constant; at the same time, his solution deviated from Spinden's by less than ten years. One might expect, therefore, that it held the potential to gather significant support—particularly since Teeple was enjoying considerable success in the interpretation of Mayan astronomy. Although his solution did have to give up calendric continuity, his Calendar Correlation depended only on astronomical data.

Of course, with work done so early in the history of the field, and coming from one with no training in history or archaeology, there were bound to be problems with Teeple's results. The first, very obviously, was that the writing system had not been deciphered yet, so there was no certainty that Teeple had appropriately identified and "translated" Venus events in the hieroglyphic record. Complicating matters further, Teeple adopted a practice that would plague would-be interpreters of the hieroglyphic record over the next century.

We have already seen that the Dresden Codex Venus Table includes the hieroglyphic compound that Förstemann recognized as the Mayan name for Venus, **CHAK EK'**. In one instance within the table, however, the prefix *chak* was left off, and the text recorded only the *ek'* element. Knowing now that *chak* means "great" and *ek'* means "celestial body," it becomes straightforward to see that the scribe, Tawiskal Uwoojil, may have simply written "the celestial body" in one of the columns, with the context removing any ambiguity that he was referring to the celestial body known as Chak Ek'. Twentieth-century scholars, however, often have understood this differently. Teeple among them, these scholars take any/all instances of **EK'** as referring to Chak Ek' (Bricker and Bricker 2011; Thompson 1972).

Teeple's "Venus event" in his proposed solution exemplifies this practice, coming from what has since been labeled Lintel 3 in Tikal Temple IV. Rather than recording the end of a Venus year in the month K'ayab related to the previous New Moon, the text is actually an extension of the history reviewed in chapter 2 of this book (see figure 3.7). Teeple's "Venus event" is the *ek'm* or "Star War verb," his "K'ayab" glyph is just the phoneme "'a" in the name of the defeated city, and his "Moon glyph" is the verbal suffix "ja." We can now read that the lintel records the military victory of Yik'in Chan K'awiil over a local adversary, where Yik'in Chan Kawiil is the son of Jasaw Chan K'awiil—the Tikal k'uhulajaw who finally found revenge against the dynasty of Calakmul (Martin and Grube 2000, 48). In other words, Teeple's "Venus" record anchoring his solution to the Calendar Correlation Problem had nothing to do with Venus at all.

Beyond his translation effort, Teeple also ran into a complication by substantially limiting his interpretation of eclipses in the Dresden Codex Eclipse Table. First of all, scholars are still debating whether the Long Count anchors are meant to record a lunar eclipse or a solar eclipse (Aldana 2016a; Aveni 2001, 173–84), but Teeple only considers the latter. Additionally, Teeple restricts the window of possible solar eclipse records to within a fifty-year time span centered on the expectation generated by applying Spinden's correlation. Given that Goodman's correlation differed from Spinden's by more than 250 years and Seler's even more, Teeple's approach was dramatically favoring a solution close to that of his friend's colleague.

Aside from all its complications, Teeple's article did generate one very substantial impact, though it was methodological, not a result per se. Part of this may have been foreshadowed in his closing statements within his 1926 text. By implementing his proposed correction method for the Venus Table and his new correlation constant, Teeple noted that

> we should at present be using a Venus calendar whose zero date was 13.8.0.0.0 1 Ahau 13 Yax, and the third year of the calendar would have ended on 13.8.4.15.12 11 Eb 0 Yaxkin, and the Venus conjunction would have been on 13.8.4.15.8 which was July 3, 1924, in our calendar. The actual Venus conjunction occurred July 1, 1924, which is not bad agreement for a calendar that was in use over 1500 years ago. (1926, 407)

This statement, however casually intended, did precede an eventual focus on the role of accurate prediction by the Venus Table. And even with his implored final caveat—"This, of course, shows the accuracy of the calendar and not the correctness of the correlation" (Teeple 1926, 407)—later scholars consistently attempted to argue that the accuracy of their Venus event predictions served as testament to the correctness of their Calendar

Correlation Constant. Showing that the Venus Table could be made to work with incredible precision, however, did not mean that the accompanying interpretation was correct. Moreover, on an academic level, the accuracy of the table's predictions should be considered independently of any proposed solution to the Calendar Correlation Problem.

It is the exact opposite method from his recommendation, though, that has been implemented over the next eighty years. And while the decipherment would bring politics and history to the fore in the interpretation of Mayan records, it was because of this role of accuracy that a reliance on astronomical records would ensure that astronomical expertise would always have a back door into cultural interpretations.

It was not Teeple's work, however, to fan the flames of controversy around the Calendar Correlation Problem; that task was taken up by one who would become one of the most influential Mayanists of the twentieth century, Eric Thompson.

THOMPSON AND THE GMT

It's not hard to see the "young whippersnapper" in Eric Thompson when he entered the field of Maya Studies. Wounded as he fought for the British in World War I, then working on a family cattle ranch in Argentina after returning to good health, he entered the academic world after a youth of exploration (Coe 1999, 124; Hammond 1977, 180). Some of that spirit also appears to have translated into an intellectual bravado, and in part that is what took the emerging Calendar Correlation debate to its next level. Thompson's work would eventually result in what has become known as the "GMT," which in Maya Studies refers somewhat confusingly not only to Greenwich Mean Time but also to (Joseph) **G**oodman, (Juan) **M**artinez (Hernandez), and (Eric) **T**hompson. By the end of the twentieth century, the GMT had become the primary Calendar Correlation Constant in use by scholars, as polished and championed by Eric Thompson. And it was utterly dependent on a new and very specific interpretation of Teeple's work on the Venus Table.

After his participation in World War I and his time in Argentina, Thompson enrolled in anthropology at Cambridge University, to be mentored by Alfred Cort Haddon (Coe 1999, 125). Haddon himself was trained as a biologist, but his expeditions to the Torres Strait inclined his interests toward indigenous cultures. Like much of the early twentieth-century work into indigenous cultures, it is fairly clear that Thompson's training in the study of Mesoamerica per se was largely self-guided. On the other hand, he clearly possessed specific talents. After completing his certificate at twenty-seven years of age, he landed his first job offer based on his ability to work with the Mayan calendar. Also key here was his academic network, resulting in an interview in London with Oliver Ricketson—Morley's colleague who was on his way to excavate the E-Group at Uaxactun, as we saw in chapter 2 (Coe 1999, 126; Hammond 1977, 180; Thompson 1963, 5–6). A key factor in his hire was that Thompson taught himself Mayan calendrics using Morley's *Introduction to Maya Hieroglyphs* (Coe 1999, 125), which in fact rather quickly led to an interest in the Calendar Correlation Problem. In 1927, with "A Correlation of the Mayan and European Calendars" (published by the Field Museum of Natural History), Thompson specifically took up the problem, which as we have just seen already had demonstrated to be of considerable computational and calendric complexity.

For this first foray into the material, Thompson reviewed the problem and the data impinging on it through a relatively balanced survey, exhibiting some humility and even recognizing the delicateness of the continuity assumption.

> The present contribution is offered, not in the sure conviction that the correlation is correct, but in the sincere belief that it bears more evidence of being the true correlation than the others yet published. There is always the possibility that the Maya time machine had broken down before the arrival of the Spaniards, in which case a day for day correlation based entirely on astronomical evidence may eventually be accepted. (1927, 3)

Thompson, therefore, acknowledged the possibility of Teeple's position on the historical data, without committing to it. Neither, though, did he end up supporting Morley's (his boss) or Spinden's proposed solutions. Instead, Thompson side-skirted the "Harvard school" to set up a new framework in his approach to the problem, initiating a new debate.

Either because he was politically astute, or because he was still getting comfortable with all of the different forms of data bearing on the problem, Thompson's first treatment of the Calendar Correlation Problem took a somewhat neutral "bigger picture" approach. Rather than completely disregard the historical data as Teeple had, Thompson suggested that any given solution should be able to account for *both* the historical data so thoroughly reviewed by Spinden and Morley *and* the astronomical records newly elucidated by Teeple:

> [T]he historical evidence is to a large extent contradictory, and the astronomical data have been translated in two different ways . . . The present correlation is an attempt to reconcile the historical and astronomical data. (1927, 5)

So while Thompson approached Teeple as an ally, he also recruited further afield. Thompson's lasting contribution would come from his appeal to the earlier proposal by Joseph Goodman and his newly acquired ally, Juan Martinez Hernandez.

The Mexican historian Juan Martinez Hernandez, whom we have seen above, was the source of the support for Spinden's accounting of the dominical letters in Landa's manuscript. Martinez Hernandez himself was born in Yucatan, went to Georgetown University for his BA, and worked in the conservation of monuments in Yucatan for the Mexican government during the 1910s. Also an historian, he would go on to edit the *Diccionario de Motul*, an important dictionary of Yucatec Mayan, still useful today in the continuing decipherment of the ancient hieroglyphic writing system. In the *Diccionario de Motul*, for instance, Martinez Hernandez's work identified a sixteenth-century entry for *butz' ek'* or "smoky star"—the Mayan name for comet. The dictionary representation explicitly referenced a comet observed in colonial Mexico: "BUTS' EK' 1: cometa crinito como el que apareció el año 1577" (Barrera Vásquez 1980, 72). Martinez Hernandez's historical detective work went as far as to include an extensive genealogy of Juan Josef Hoil himself, the last copyist of *The Book of Chilam Balam of Chumayel*.

Through his friendship with Goodman, Martinez Hernandez was the one to introduce a critical document into the Calendar Correlation debate as a companion to the Chilam Balam material; this was the so-called Chronicle of Oxkutzcab. Morley too had access to this handwritten text, described by Eric Thompson as having been contained within

> a collection of titles, family papers, births, etc., of the Xiu family, who prior to the Spanish conquest were one of the ruling families of Yucatan, and in all probability the most important family in the land. (1927, 5–6)

Thompson here, we can see, was following Goodman's perspective on the Xiu family, stating that the Chronicle of Oxkutzcab contained "the most reliable information" of the histories relating to the Calendar Correlation Problem. Rather than emphasize the record of the death of Napot Xiu in the Katun Chronicles of *The Book of Chilam Balam of Chumayel*, Thompson turned to the record in the Chronicle of Oxkutzcab that 13 Ajaw 8 Xul "occurred" during the year of AD 1539.

The argument was a stretch and appears to have plagued Thompson for decades. He struggled with it extensively, arguing for and against his own position even as late as *Maya Hieroglyphic Writing* through its second and third editions in the 1950s and 1960s. The problem that Thompson faced—and that Morley did not—is that he took this one "most reliable" record as critical, but it was off by a full year.

The line in question comes from page 66 of the Chronicle of Oxkutzcab. Juan Xiu had recorded a list of Christian years, each followed by the Mayan year bearers, and the Calendar Round date for the "end of the tun" (or 360-day period). These ran from 1532 to 1545. Morley had identified some scribal errors in the list (e.g., a 13 written in place of a 3), but those were inconsequential. The two lines of interest to Thompson (1927, 7) were these:

> 1539 10 Muluc on Pop 1. 4 Ahau the tun on 11 Xul.
> 1540 11 Ix on Pop 1. 13 Ahau the tun on 7 Xul.

To interpret these dates, Thompson turned to Landa's manuscript, which recorded that 1 Pop—the Mayan New Year's Day—started in July. This meant that each Mayan year would straddle two Christian years, as Spinden had noted in his proposal for the use of the dominical letters. The problem for Thompson arose, though, with his argument for the Mayan "tun ending on 13 Ahau" actually occurring in 1539, not 1540 as *explicitly recorded*. Thompson didn't resort to calling this a scribal mistake—that would have invalidated his whole attempt to privilege the accuracy of the manuscript. Here he was claiming that we should shift the 13 Ajaw 8 Xul "katun end" to the year 1539 from an explicit record placing it in 1540 because he claimed that the years given are those of the ends of the Mayan years. In other words, Thompson is claiming that the author of the text, Juan Xiu, knew that 11 Ix 1 Pop and 13 Ajaw 8 Xul occurred during the year 1539, but the year of the year bearer ended in 1540, so Xiu defied normal practice and assigned them both to 1540.

The opposite situation, in fact, makes much more sense. If the Mayan New Year began in July, then the month Xul would have transpired entirely within the same year—putting 1 Pohp on July 16, for example, would put 1 Xul on October 24 and 19 Xul on November 12. It would therefore make perfect sense to record the Christian year that corresponded to both the year bearer and the tun end. Interpreted in that way, Juan Xiu's records meant that 10 Muluk was the year bearer and 4 Ajaw was the tun end in 1539 and 11 Ix was the year bearer with 13 Ajaw on the tun end in 1540. That would have made sense and would have required no tweaking of the recorded information.

The problem for Thompson, however, was that if 13 Ajaw 8 Xul did occur in 1540, taking Juan Xiu's words as written, and 11 Ix was the year bearer at the time, then that would disrupt his other critical connection. If 13 Ajaw 8 Xul occurred in 1540, then Landa's date of 12 K'an 1 Pop would have occurred in July of 1554, not 1553. The only way Thompson could preserve Spinden's use of Landa's dominical letters—the only way to keep the assumption of continuity—was to follow an otherwise unattested interpretation of Juan Xiu's "Chronicle of Oxkutzcab" (Thompson 1927, 7).

One might assume that this interpretation could not have held up for long—it appeared specious and went contrary to all other calendric scholarship. As we have seen, though, it was well acknowledged that the Calendar Correlation Problem itself was underconstrained; because there were insufficient quantitative constraints, assumptions had to be made to make the problem resolvable at all. Perhaps as important, the field of scholars addressing the problem was relatively small, such that Thompson had to consider both the political implications of following the specific assumptions made by specific colleagues along with the technical arguments put forth in each case. And while the technical side was amenable to innovation and more data, the assumptions were still reflective of individual bias and political inclination.

So while complications in his interpretation of the Oxkutzcab manuscript and in other technical details were rife throughout his proposed solution, Thompson's approach was also ingenious. He pushed both the historical and the astronomical data sets to incorporate greater subtlety in each. This allowed him to draw from—and so valorize—each of the previous contributors while not exactly following any of them. Thompson incorporated his boss's (Morley's) work, some of Spinden's results, and Goodman's basic approach and balanced it all, as we will see, with Teeple's work on the Venus Table. At the same time, he created an argument so complex, its details would vex scholars for generations (Aldana 2011b, 2016b; Kennett et al. 2013).

IMPACT ON THE DRESDEN CODEX VENUS TABLE

Comfortable with his modified use of Juan Xiu's Chronicle of Oxkutzcab, Spinden's treatment of Landa, and careful appropriations of other historical records, Thompson now turned to the astronomical record. Here Thompson found that championing the GMT left him in yet another awkward position. Teeple had taken the anchor of the Venus Table as a historical record. As we have seen, he took 9.9.9.16.0 1 Ajaw 18 K'ayab on page 24 as a historically observed first morning visibility of Venus, which he then used to inform his exploration of other "Venus events" in the hieroglyphic inscriptions from Classic times. Thompson, however, didn't have that luxury. Because he followed the historical treatments' assumption of calendric continuity, he had to simply take whatever the GMT produced and see if the astronomy could be worked out. And when he did this, he found that indeed he could preserve the Moon Ages of the Classic period. He only needed to adjust Goodman's correlation by three days (or Martinez Hernandez's by 2) to have the Lunar Series data and the Dresden Codex Eclipse Table accord with reconstructed lunar phases.

The anchor to the Venus Table, however, did not work out.

We have seen by now that Förstemann and all following his work view the Venus Table as having been constructed with an anchor on the first morning visibility of Venus; the Venus Table also ties all progressions and corrections of the first morning visibility to the date 1 Ajaw. Using Thompson's proposed solution, the GMT, however, the base date of the Venus Table not only misses a first morning visibility event; it does not correspond to any of the Venus events recorded in it. This was a significant problem. With the GMT, the table's first morning visibility base date actually places Venus on the wrong side of the sky, near the end of its evening star phase in the west, weeks away from first visibility in the east. According to the GMT, the 9.9.9.16.0 1 Ajaw 18 K'ayab anchor to the Venus Table deviated sixteen days from the first morning appearance it should have recorded.

Rather than admit the possibility that his correlation was wrong and then revisit the various complications he had worked around with the historical dates, Thompson found a way to circumvent his Venus problem. His work-around shows up in his 1935 revisitation of the Calendar Correlation Problem—this time as a Carnegie publication. Here Thompson went back to Teeple's methods.

As we saw at the end of chapter 1, Teeple applied his inferred Correction Interval serially to a 1 Ajaw 18 K'ayab base date. Doing so generated the list of Calendar Rounds in the Venus pages, but it also produced one "extra" Calendar Round date, 1 Ajaw 8 Yax (1926, 403–4). For Teeple, this was important as a demonstration of the author's intent. Tawiskal Uwoojil would have left this out to vary the quantity of correction necessary for specific periods of observation. Thompson, however, ignored the specifics of Teeple's proposal and focused on the base date. He knew he had to address the sixteen days of difference between the prediction of the GMT and the base date, and he knew that "uncorrected" runs of the full Venus Table introduced "errors" of five days each round.

This provided Thompson with his puzzle pieces—it was now just a matter of fitting them together.

Thompson proposed that the scribes writing up the Venus Table were working sometime in the Postclassic and so were fitting dates to their own observed Venus cycles as well as to future predictions. From this perspective, Thompson worked to place the table's use according to when it would most accurately track reconstructed Venus observations. Here Thompson picked up on what Teeple had implied. Now Thompson claimed that the accuracy with which he would be able to make the Venus Table fit with hypothetically observed Venus events determined the strength of the interpretation. Accuracy was a critical element in interpreting the Dresden Codex Venus Table—not because Tawiskal Uwoojil or another Mayan scribe ever had written such, but because it had become a useful tool in the twentieth-century attempt to solve the Calendar Correlation Problem (Aldana 2016a).

Thompson tinkered with the details over the years, but his method eventually settled on the following (Aldana 2011d, 42; Bricker and Bricker 2007, 102–4). He started with the application of the GMT, which he knew put the base date in error of sixteen days. He then applied the time interval at the far left of the top row on page 24 made up of four complete Venus Tables ($4 \times 5.5.8.0 = 4 \times 11{,}960$) to give him a second base:

9.9.9.16.0 1 Ajaw 18 K'ayab

+ 1.1.1.15.4.0

= 10.10.11.12.0 1 Ajaw 18 K'ayab

These amount to four "uncorrected" runs of the full Venus Tables so that now Thompson has made up for the sixteen days of "error," bringing the table's first morning visibility prediction closer to the GMT's prediction.

Next, Thompson applied a time interval from the second row—one of Teeple's "Correction Intervals"—to shift the base date from 1 Ajaw 18 K'ayab to 1 Ajaw 18 Wo and to generate a closer match between the Venus Table and the GMT.

10.10.11.12.0 1 Ajaw 18 K'ayab

+ 4.12.8.0

= 10.15.4.2.0 1 Ajaw 18 Wo

Finally, he introduced the "implied" interval derived by Teeple from there on to preserve accuracy in the long run (Bricker and Bricker 2007, 102–4) (figure 3.8). Teeple created this implied interval by subtracting the two middle Correction Intervals of the second row:

9.11.7.0

− 4.12.8.0

= 4.18.17.0

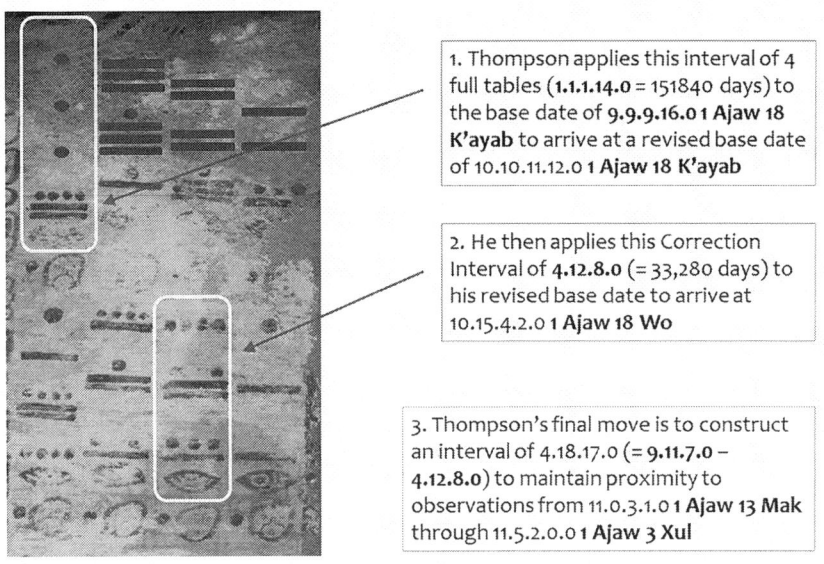

1. Thompson applies this interval of 4 full tables (**1.1.1.14.0** = 151840 days) to the base date of **9.9.9.16.0** 1 Ajaw 18 **K'ayab** to arrive at a revised base date of 10.10.11.12.0 1 **Ajaw 18 K'ayab**

2. He then applies this Correction Interval of **4.12.8.0** (= 33,280 days) to his revised base date to arrive at 10.15.4.2.0 1 **Ajaw 18 Wo**

3. Thompson's final move is to construct an interval of 4.18.17.0 (= **9.11.7.0** – **4.12.8.0**) to maintain proximity to observations from 11.0.3.1.0 1 **Ajaw 13 Mak** through 11.5.2.0.0 1 **Ajaw 3 Xul**

FIGURE 3.8 Thompson's method for using the deciphered dates and numbers of page 24 in order to place the Venus Table into historical times. This method allows him to use the Venus Table as a tool in his attempt to solve the Calendar Correlation Problem. Courtesy SLUB

So that:

10.15.4.2.0 1 Ajaw 18 Wo

+ 4.18.17.0

= 11.0.3.1.0 1 Ajaw 13 Mak

+ 4.18.17.0

= 11.5.2.0.0 1 Ajaw 3 Xul

It is worth emphasizing that this interval does not show up anywhere in the Venus pages, but because it was "implied" by two other Correction Intervals and because it accurately preserved alignment between the GMT and the Venus Table predictions, it has been used by virtually all researchers since (Bricker and Bricker 2007, 104).

And this is what now gave Thompson an advantage over the other proposed correlation constants: Thompson didn't just have a "better" solution than that provided by the Spinden correlation; he now also had a rationale to explain the complications of correcting the Venus Table. To do so, Thompson proposed—contrary to Teeple's position—that the original date wasn't supposed to record an observed Venus event. Instead, he claimed that it was crafted by Postclassic scribes as a numerological anchor for the connection of their own observations to the mythological past.

Whereas Teeple assumed that the dates recorded actually reflected observations and historical records of them, Thompson assumed a combination of numerologically inspired dates with later observed phenomena. And this changes what we infer about

Tawiskal Uwoojil's intent or that of any other scribes behind the manuscript. For Thompson, the scribes' historical past was irrelevant; the important data were the future predictions made possible by the computations on page 24. From this perspective, with an emphasis on the accuracy of prediction, Thompson was arguing that the scribes were creating an ephemeris for Venus. Accuracy was important for the future, but apparently not for the historical past.

LEGACY

Interestingly enough, the Calendar Correlation problem remained unresolved even after Thompson's early work on the Venus Table. Not everyone was comfortable with Thompson's interpretation of the sixteen days of error, and there was an even bigger picture that introduced another realm of ambiguity. That is, Spinden's and Thompson's correlation constants depended on continuity in the 260-Day Count, but the Katun Count allowed for distinct solutions that differed by 260 years. The situation as understood by many in the field was thoroughly reviewed and summarized by Lawrence Roys:

1. Two answers to the problem are before the public. Spinden places the typical Old Empire date 9.15.0.0.0 at Oct. 22nd 471 A.D.; while the advocates of the Goodman correlation place it within a day or two of Aug. 22nd 731 AD.
2. Both these solutions were originally computed from post-Conquest material. If any breaks occurred in the Maya time counts between Old Empire and Spanish times, the computations for these solutions do not hold good; and there is considerable chance that some such interruptions occurred.

    . . .
5. After comparing the demonstrations of the correlation solutions with the demonstrations of the solutions of the accepted calendrical and astronomical systems, it seems evident that the correlation problem is as yet unsolved. (1933, 416–17)

More important than Roys's caveat here is that Thompson's work ensured that the Dresden Codex Venus Table would be subject to its role in the Calendar Correlation Problem and not explored on its own. Yet the story does not end with the formulation of the GMT and its contested state, primarily because of events transpiring completely unrelated to Mesoamerican calendrics.

 During the late nineteenth and early twentieth centuries, without calendars anchored to modern ones, ancient civilizations were understood within a soup of time—tools to approach the passage of absolute time were as yet undiscovered. As Danish antiquarian Rasmus Nyerup put it as early as 1802:

Everything which has come down to us from heathendom is wrapped in a thick fog; it belongs to a space of time we cannot measure. We know that it is older than Christendom, but whether by a couple of years or a couple of centuries, or even by more than a millennium, we can do no more than guess. (Trigger 1989, 71)

We can see some of the same constraint in Joseph Goodman's work, which actually went further than just historical projection, since he had access to the very early dates in the Palenque inscriptions that we saw in chapter 2. Handling them without the benefit of the decipherment of the writing system, however, Goodman "stuck to his guns."

I am aware that the older Palenquean dates are so remote that it has been commonly agreed to discredit their historical value. There is no warrant for this. They stand on exactly the same footing as the dates assumed to be historical, and all must be accepted or rejected alike . . . and that Palenque was in existence 3,143 years before Christ. ([1905] 1974, 646).

In fact, Goodman went so far as to justify the early Palenque dates on stylistic grounds, not allowing for any leeway in deviating from his interpretation. In it each city defined its own history, and Palenque was simply far more ancient than many others (Goodman [1905] 1974, 647).

With ambiguities amounting to thousands of years—at least in Goodman's read—the fact that the two options that Thompson and Spinden presented were separated by 260 years probably did not seem too significant. But it does to us today. From a twenty-first-century perspective, 260 years is tremendous, and it is so in large part because of a revolution in dating technology that occurred just as the Calendar Correlation debate settled into a comfortable stalemate.

If we have attributed some of the advancement of the understanding of the Dresden Codex Venus Table to World War I and Teeple's role in the potash industry, then we also find a critical intervention from World War II on the development of the Calendar Correlation Problem and so indirectly on its interpretation. The "revolution" that transpired for the study of the past occurred through the birth of "molecular biology" during the 1930s and 1940s. And one of the first applications of this revolution was to a consideration of the Calendar Correlation Problem by the Nobel Prize–winning physicist Willard Libby.

CHAPTER 4

Molecular Biology, Jasaw Chan K'awiil, and Radiocarbon Dating

I n the aftermath of the Late Classic internecine warfare at Tikal that we saw in chapter 2, Nuun Ujol Chaak's son took the throne. Jasaw Chan K'awiil, the son, cemented his place in Mayan history by avenging his father, leading his army against and defeating Yich'aak K'ahk', k'uhulajaw of Calakmul (Martin and Grube 2008, 44; Sharer 2006, 393). He therefore completely turned around regional politics, revived old alliances, and patronized neo-classical artistic forms celebrating his ancestral links to Teotihuacan (Harrison 1999, 125–46; Martin and Grube 2000, 47).

Some of his Teotihuacan-related ancestors had been buried in the now well-developed North Acropolis, the buildings from which Jasaw Chan K'awiil eventually drew inspiration for the construction of what we now call Tikal Temple II. Built as a terraced mountain, with corbel-vaulted rooms in the structure at the top, he dedicated it to his wife, Lajchan Unen Mo' (Martin and Grube 2000, 46; Sharer 2006, 395). The large open doorway of the structure at the top was made possible by a long and hefty wooden lintel hewn from the trunk of a sapodilla tree (*zapote*). This man-made "mountain cave" or *chan-ch'en* would make possible the appropriate ritual communication with the soul of Lajchan Unen Mo' in her afterlife within the Underworld.

Across the central plaza, adjacent to the North Acropolis, the masons of Tikal constructed Jasaw Chan K'awiil's funerary monument to mimic and to directly face Temple II. Taller and thinner than Temple II, Temple I too housed within it a set of large lintels to support the doorway. Upon one of these, Tikal artists carved a hieroglyphic text commemorating his victory over Yich'aak K'ahk', alongside a representation of the Tikal k'uhulajaw sitting on his jaguar palanquin (figure 4.1). This text would remind members of the Mutul dynasty of Jasaw Chan K'awiil's historical achievements for another century, as well as provide access to the space connected to his tomb underneath the base of

FIGURE 4.1 Temple I at Tikal, Jasaw Chan K'awiil's funerary monument. Heavy wooden lintels spanning the doorway were carved with images and hieroglyphic inscriptions.

this temple. Centuries later, with Maudslay's and Morley's archaeological projects, Tikal Temples I and II would become symbolic of all Mayan civilization (Martin and Grube 2000, 47). And not long after that, but long after Jasaw Chan K'awiil's stories and the ability to read the text were forgotten, this wooden lintel would command the interest of the world's leading nuclear physicists, with direct impact on the Mayan Calendar Correlation and the interpretation of the Dresden Codex.

In this chapter, we confront a scientific tool of such formidable scope that it was able to contain and restrain the historical exploration of the Dresden Codex and its contents. Radiocarbon dating has found application far beyond Maya archaeology, even though Jasaw Chan K'awiil's patronage served as some of the very first material to be addressed with it. More importantly, the technology resulting from World War II efforts added fuel to the fire of a problem that took on much greater presence within Maya Studies. The Calendar Correlation Problem would now dwarf the details of Mayan astronomy such that, like Jasaw Chan K'awiil's lintels, the pages that Tawiskal Uwoojil painted would be subordinated to a modern intellectual agenda. As we will see by the end of this chapter, much of this scenario was driven by the personal agenda of a scholar who has often been portrayed as something of a villain within the history of the decipherment of Mayan hieroglyphic writing. Here we confront the legacy of Eric Thompson within the subfield of Mayan astronomy.

MOLECULAR BIOLOGY AND JASAW CHAN K'AWIIL

As Teeple traveled across the United States by train to study the chemistry of Searles Lake, the PhD physicist Willard Libby was becoming interested in cosmic-ray activity in the upper atmosphere. The buzz in the scientific community around the phenomenon that Marie Curie dubbed "radioactivity," along with her discoveries of radium and polonium, had scholars looking for new forms of radiation as well as their application during the early twentieth century (Latour 1987, 88). The field was so new and the interest so intense that the French physicist Prosper-René Blondlot discovered a fictitious form of radioactivity, the N-Ray, in 1903 (Latour 1987, 75). While proper scientific vetting eventually proved Blondlot's mistake, the international fascination and investment of resources in radioactivity was set.

Thirty years after Curie had coined the term, the labs at the University of California at Berkeley in the United States were pushing theoretical and application research in the field. Key to this progress was the Nobel Laureate Ernest Lawrence. By 1928, Lawrence had completed his doctorate at Yale and moved on to a tenured professorship at the University of California; "two years later he became Professor, being the youngest professor at Berkeley" (Nobel Media 2014) (figure 4.2).

Although he did not originally conceive of the idea of particle acceleration, Lawrence developed and eventually patented the first operating cyclotron in 1931 (Lawrence and Livingston 1932). Notably, Lawrence's cyclotron differed from the linear accelerators already invented in an extremely productive way. For his device, atomic particles were introduced into the center of the cyclotron; then a magnetic field accelerated them to the desired speed and emitted them through a port, allowing for their capture or their collision into other particles (figure 4.3). Although it was his cyclotron at Berkeley that contributed to a critical revolution in archaeological dating, it was not Lawrence himself who made the specific contribution of our interest here.

Sam Ruben was a brilliant young scientist who was encouraged by his advisor Willard Libby while he was still an undergraduate at Berkeley "to study theoretical physics and advanced theoretical math because he feels I can do it and therefore profit tremendously" (Sam's letter of May 3, 1935) (Johnston 2003,

FIGURE 4.2 Campanile at the University of California, Berkeley, overlooking the "Rat House" and Strawberry Creek

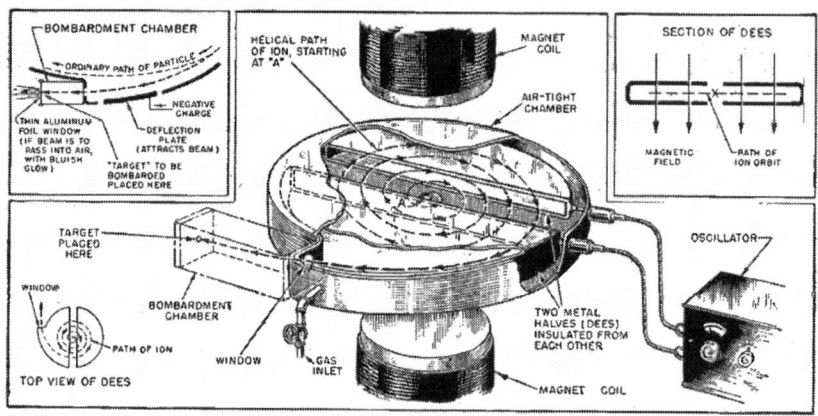

FIGURE 4.3 Schematic illustration of the Lawrence cyclotron at Berkeley used by Ruben and his team to generate tracers. Courtesy Wikipedia

107). Ruben went on to complete his dissertation "Studies in Artificial Radioactivity" and received his PhD at Berkeley in 1938 (Johnston 2003, 107).

The bulk of Ruben's dissertation research used radioactive elements produced by Lawrence's cyclotron as tracers within biological activities. For example, Ruben injected radioactive phosphorous into the stomachs of rats and then dissected them to track how it traveled through the body (Johnston 2003, 107). While these experiments met some success, in 1940 Ruben shifted his postdoctoral research to take on a "hot topic" in biology. Ruben initiated an investigation into photosynthesis. While Eric Thompson was at the Carnegie Institution writing up articles on the ceramic record of Belize and working through revisions of his Calendar Correlation proposal, Ruben was across the country, working on radioactivity in California. In some fifteen years, the results of their independent work would meet.

Working on photosynthesis, Ruben almost immediately ran into a significant complication: the tracer he had been using on rats, ^{11}C, had a very short half-life, which required experiments to be carried out within a two-hour period (Johnston 2003, 109). That was sufficient for many processes, but it wouldn't be enough for his photosynthesis work. Ruben needed a more stable radioactive element. Here he looked to his advisor Libby's earlier work on tritium, which led to the hypothesis that a different isotope of carbon, ^{14}C, might be generated in the upper atmosphere and possess a half-life on the order of years (Gest 2004, 78). Such a longer half-life would provide Ruben with the time necessary and allow for less rushed experimentation. But it was still hypothetical since there was no known readily available source for or process of generating ^{14}C.

Libby and Ruben turned to Lawrence's cyclotron, but according to Libby's retrospective account for the Department of Energy, their first efforts met with failure.

The project "was undertaken by Samuel Ruben for his doctoral thesis under the direction of the author in the Chemistry Department at the University of California at Berkeley. However, it was mistakenly supposed that the number of atoms needed would correspond to something like the case of sulpher-35, with which experiments had been made. . . . Actually, of course, the lifetime is 8300 yr on average, so Ruben and the author made only 1/25 000 of what was needed; therefore there was failure to detect any radiocarbon produced in this first experiment." (Libby 1967, 4)

But this was only a minor setback. The next part of the story is very brief in Libby's account:

Ruben went on to take his degree with the author on other subjects in physical radiochemistry and, after having finished, joined forces with Martin Kamen, who had come to the Lawrence Radiation Laboratory from Chicago a couple of years before to have another try at radiocarbon. This time they succeeded [3]. By bombarding graphite with a strong deuteron beam from the cyclotron, the (d, p) reaction on the ^{13}C present in the natural mixture gave enough radiocarbon for them to detect. On the basis of this they gave the tentative value of 25 000 yr as the lifetime. This was in 1940. (Libby 1967, 4)

The creation of ^{14}C, or what is now commonly referred to by archaeologists as "radiocarbon," therefore, actually belonged to Ruben and his partner, Martin Kamen.

Their results met immediate fame:

On February 29, 1940, there was a momentous ceremony in Wheeler Hall [University of California at Berkeley], where the Swedish Consul presented the Nobel Prize to E. O. Lawrence. The chairman of the Physics Department, R. T. Birge, gave the presentation address. He spoke of the great importance of radioactive isotopes as tracers in biology and possibly as therapeutic agents. Then in a dramatic gesture wholly atypical of him, he stepped back, raised his arm, and portentously announced, "I now have the privilege of making the first announcement of very great importance. This news is less than twenty-four hours old and hence is real news. Now, Dr S. Ruben, instructor in chemistry, and Dr M. D. Kamen, research associate in the Radiation Laboratory, have found by means of the cyclotron, a new radioactive form of carbon, probably of mass fourteen and average life of the order of magnitude of several years. On the basis of its potential usefulness, this is certainly much the most important radioactive substance that has yet been created." (Gest 2004, 79)

Inventing a process to create ^{14}C was a brilliant success. It was, however, only half of the work needed for their ultimate impact beyond the field of chemistry.

The work that would bring Ruben and Kamen fame was in some ways the result of a gentlemanly bet. Some years earlier, Ernest Lawrence had what he thought to be interesting results of an experiment run using his cyclotron. He showed these to yet another Nobel Laureate in Physics, Niels Bohr, during the latter's visit to Berkeley, and Bohr responded with incredulity (Gest 2004, 78). Rather than argue the point, Lawrence recruited a colleague at Berkeley to "review the problem." That physicist suggested that

Kamen pursue the project experimentally. Lawrence then went to the dean of the College of Chemistry to find an appropriate partner; Dean Wendell Latimer recommended Sam Ruben while he was still a graduate student. Although they were unable to champion Lawrence's claims—Bohr was right—Kamen and Ruben found that they worked well together and decided to continue collaborating on other projects (Gest 2004, 78).

In his much later memoir, Kamen described his very hands-on and even frenetic work with Ruben at Berkeley (Johnston 2003, 108; Kamen 1985). Ruben's group referred to their lab as the "Rat House" since it had previously been used to house the rats used for experimentation:

> At the Rat House, Sam and Zev [W. Z. Hassid] would be waiting for me like sprinters at the starting gate. Beakers would be filled with boiling water or other solvents and pipettes ready to suck up measured volumes of radioactive solutions onto absorbent blotters, which would be held by tongs over hot plates and dried. All the necessary reagents and apparatus would be in place. The counter would be ticking away establishing the background activity. Each experiment had to be planned ahead in every detail so that no time was lost in confusion or delay deciding what procedure to follow. Anyone looking in on the Rat House when an experiment was in progress would have the impression of three madmen hopping about in an insane asylum. (Johnston 2003, 108; Kamen 1985)

Their frantic work paid off. Kamen had been trying to create ^{14}C at the University of Chicago since 1935; with Ruben, he was successful.

As presaged by R. T. Birge, Ruben and Kamen became academic celebrities and decided to collaborate on an appropriately challenging new project. Their work on photosynthesis, to which they would first apply ^{14}C, was equally successful. Ruben and Kamen were featured within an article in *Life Magazine* in 1941, titled "Science and the Future: Search for Knowledge Enters an Era of Great Conclusions," for overturning the previous model of photosynthesis (Johnston 2003, 109; *Life* 1941, 80–81). The text for their two-page photo spread, bringing science to the public, was titled "Photosynthesis: Plants Support All Life." The text hyped the new findings, mentioning Ruben and Kamen explicitly.

> How photosynthesis works is a mystery that is now yielding to attack by a new potent weapon of science, the tagged atom. These are artificially radioactive atoms which can be introduced into plants and animals and traced throughout the process of absorption into tissues. In exploring photosynthesis, Drs. Samuel Ruben and Martin Kamen employ carbon atoms tagged by the Lawrence cyclotron at the University of California. These they have traced from their inorganic partnership with oxygen in carbon dioxide gas (*above*) to their organic union with hydrogen in carbohydrates in bean and sunflower leaves (*right*). (*Life* 1941, 80–81)

At the peak of this success, however, World War II intruded into the scientific community at Berkeley. Libby was pulled away to the Manhattan Project. Kamen slipped into an ambiguous identity, accused of ties to Russia and possibly espionage (Johnston 2003; Kamen 1985). Ruben went to work for the National Defense Research Committee (NDRC) and became intrigued by a riddle in the toxicity of phosgene, Cl_2CO—a chemical weapon used extensively in World War I (Johnston 2003, 110). The current state of knowledge for the project was recorded in an "Unpublished Report Written by T. N. Norris, October 22, 1943, Chemistry Department, UC":

> Research on the physiological effects of phosgene has yielded only an incomplete understanding of the reason for the extreme toxicity of the substance. It has been generally supposed that phosgene because of its ready solubility in organic materials, was able to penetrate the cell walls in the lung tissue, there to undergo hydrolysis to CO_2 and HCl, and that the HCl so formed was the cause of the severe edema observed. Death has been attributed to the edema, which fills the lungs with fluid thereby interfering drastically with oxygen exchange.
>
> Experiments are projected to test this hypothesis and to characterize better the nature of the substance or substances formed in the reaction of phosgene with organic matter of the organism. Such knowledge could make possible the deriving of a more fruitful method of medical treatment for cases of phosgene poisoning than has been available hitherto. (Johnston 2003, 112–13)

Ruben took up the effort to find a cure for phosgene poisoning.

Andrew Benson recalls being hired to support Ruben, "a young and rising star in the Berkeley faculty . . . who was on the verge of being advanced to Assistant Professor" (Johnston 2003, 111). Benson recalled being assigned a desk in the lab and a "mass of dirty glassware freshly deserted by Henry Taube (Nobel Laureate, 1983)" (Johnston 2003, 111). Working as a team, Benson and Ruben developed a method for creating phosgene samples. Benson's contribution had been to generate the radioactive material and then contain it for transport to Ruben, who would run trials on it.

> Andy's method required cooling the source glass ampoule in ice water, scratching the side of the narrow glass tube close to its upper end, fitting a rubber tube over the glass tube past its scratch mark, clamping the rubber tube tightly with wire, connecting the rubber tube to a special expansion bulb, crushing the glass tube at the scratch mark, and passing the liquid phosgene through an open metal valve into the ice cold metal tank. The operation was a complex and dangerous one. (Johnston 2003, 120)

With the war intensifying and their research in full swing, though, new pressures impacted the lab's work. Benson avoided the draft by registering as a conscientious objector and so was forced to leave the lab, sent off to fight wildfires in the forestry service. The way he tells it, Latimer would not renew his teaching contract, which would have allowed him to stay on campus and continue his research (Johnston 2003; Kamen

1985). In any case, Benson ended up on fire lines in Nevada. More importantly, when he left, there was no one to replicate his delicate method for safely handling phosgene.

Undeterred and committed to furthering his research, Ruben tried to work with phosgene samples on his own, relying primarily on graduate student assistants. Unfortunately, the complexity of the process caught up with them. On September 27, 1943, disaster struck.

> First Sam slowly carefully dipped one end of the phosgene ampoule into a Dewar flask of liquid air in order to lower its vapor pressure and make it safe to transfer. Instantly the soft glass container cracked and split open. Still at room temperature and under pressure, liquid phosgene squirted into the liquid air. The leading slug froze solid, but in doing so, it boiled off a large amount of liquid air. This sprayed liquid phosgene high into the air which fell back as a dense cloud of cold air and vaporized phosgene. Sam, Kent, and Robert got out of the room as fast as they could and rushed away from the building. As Andy Benson said, "They lay down on the grass by Strawberry Creek." (Johnston 2003, 120–21)

Sam Ruben was studying the mechanism of death by phosgene in hopes of finding ways to treat those poisoned by it. The coroner pronounced his death by phosgene poisoning.

> We the jury do find that the name of the deceased one was Samuel Ruben, a native of California, aged 29, and that he came to his death on Sept. 28, 1943 at Memorial Hospital, Berkeley, Alameda County, California, and we further find that the death was caused by pulmonary edema, massive and diffuse, due to inhalation of poison gas, dilation of right ventricle, suffered accidentally in his regular line of duty in the Chemical Laboratory on the U. C. Campus, Berkeley, when a glass vial of poison gas accidentally broke. We find death was accidental. (Johnston 2003, 122).

The entire tragic story of Ruben's contribution to the history of radiocarbon dating was curtailed, however, in Libby's account.

> World War II came, Ruben died, and after the war Kamen went on to biochemical work. However, before his death, Ruben published a most distinguished series of papers on the use of radiocarbon in the study of photosynthesis. (Libby 1967, 4)

Perhaps it was sensitivity to Ruben's death, or perhaps it had to do with the technicality that Ruben's work on ^{14}C was limited to photosynthesis as process, and not in its implications, that his mentor, Libby, elided Ruben's role in reflecting on his post–World War II work with the material.

In any case, by the time Libby returned from the Manhattan Project, it was clear from Ruben's work at least that plants would absorb radioactive carbon dioxide along with nonradioactive CO_2 in the process of photosynthesis. Libby took this observation further, realizing that the half-life of ^{14}C could be used to date plant matter of considerable age. Since plants would cease absorbing CO_2 from the atmosphere at death, the amount of ^{14}C in them would only decrease due to decay over time. The percentage of ^{14}C left in a specimen, therefore, would correspond to its age. It was another of Libby's associates, though, who would first bring the latest physics into contact with ancient Mayan artifacts. This was the work of J. Laurence Kulp.

Kulp received his PhD from Princeton in 1945 and also went to work on the Manhattan Project "at the Nash Building at 130th and Broadway" while he worked at Columbia University. As an assistant professor at Columbia, he set up the Lamont Laboratory for Geophysics in 1949, inspired by Libby. In an interview for the American Institute of Physics, Kulp stated:

> The method, for which he [Libby] received the Nobel Prize totally revolutionized archaeology and Pleistocene geology. Well I had heard of this development through one of my friends Dr. Jim [James] Arnold who was working in Libby's lab as a post-doc. Jim and I had been at Princeton at the same time working on the Manhattan Project. And so I called Jim and asked him if Libby might be willing to let me come out and learn the method if I were willing to take three or four months and be a slave in his lab. Libby said fine. So as soon as I could, I went out to Chicago, got a room, went to his lab, and worked seven days a week for about four months and as an assistant. (Kulp 1996)

Part of Kulp's interest was nurtured by the impression that his lab could get substantial funding from the government and from oil companies. In fact the acknowledgments to his laboratory's first publication proved that his hunch for funding was spot on:

> The research reported in this paper has been made possible through support and sponsorship extended by the Geophysical Research Directorate of the Air Force Cambridge Research Center, under contract No. AF19-(122)-214. The support and cooperation of the American Museum of Natural History, the Anthropology Department of Columbia University, and the U. S. Geological Survey are also gratefully acknowledged.
>
> The comparatively rapid development of this laboratory was made possible by the wholehearted support of W. F. Libby, Institute for Nuclear Studies, University of Chicago, who permitted the senior author to spend a month in his laboratory. (Kulp et al. 1951, 568)

Before long, Kulp was able to reflect on the early success of his lab:

> We got the money to build our system, which was an improvement over the one at Chicago and we began the dating process at Lamont. And of course virtually [every] sample produced an exciting result as the age of these objects were almost entirely unknown. (Kulp et al. 1951, 568)

Through Kulp's and Libby's interventions, the soup of time that had been lamented by archaeologists for decades now began to reveal its structure. Materials themselves could now contribute to civilizational chronologies, not historical records alone.

With a laboratory at Columbia University in New York City, Kulp had only to look nearby for new material to test. One of the first applications of the new technology was to a wooden lintel, housed in the collections of the American Museum of Natural History. This lintel had been taken by Herbert Spinden in 1914 from Tikal Structure 5D-52 (aka Structure 10), and it bore a hieroglyphic date (Coe and Shook 1986, 21). In fact the date was somewhat eroded, with only the Calendar Round of 3 Ajaw 3 Mol preserved. Most of the short inscription was also illegible. What secured its place in the Long Count was that the Calendar Round was followed immediately by the glyph for a "half-period end," meaning that the haab coefficient would be 10. Fortunately, another inscription corroborated the date and its patron. The date 9.15.10.0.0 3 Ajaw 3 Mol as a half-period end was inscribed on the truly majestic Lintel 3 of Temple IV—the largest structure at Tikal. The lintel's patron was Yik'in Chan K'awiil, son of Jasaw Chan K'awiil and grandson of Nuun Ujol Chaak. Yik'in Chan K'awiil was now enjoying the fruits of his predecessors' struggles.

Although he didn't know anything about Yik'in Chan K'awiil and he misspelled "Tikal" in his sample description, by accident or intent, Kulp had the Calendar Correlation Problem right in his crosshairs.

> The discovery of the Carbon 14 method of age determination by W. F. Libby and co-workers (1–4) has so many potential applications in geology, anthropology, archaeology, oceanography and meteorology that the development of several laboratories equipped to make the necessary measurements is imperative. About a year ago the construction of the requisite facilities was undertaken at the Lamont Geological Observatory. (Kulp et al. 1951, 565)

With Libby's new technology, made possible by Sam Ruben's research, Kulp stoked the Calendar Correlation Problem debate (table 4.1).

This last line in Kulp's comment is explicit: "Looks like Spinden date." Nuclear physics and the technologies of World War II were now being marshalled in the battle for the Mayan Calendar Correlation. Despite Eric Thompson's massive efforts to move the Calendar Correlation problem by rationalizing numerous calendric anomalies, now Kulp glibly corroborated Spinden's constant with a single line.

But this too was far from the end of the story. For one, chemists would go on to refine the measurement of the half-life of ^{14}C, eventually revising it from 5568 to 5730. Just as important, Elizabeth Ralph set up her own radiocarbon lab at the University of Pennsylvania.

> In this joint report, Elizabeth K. Ralph, research associate at the Department of Physics, in charge of the University of Pennsylvania Radiocarbon Laboratory (which is sponsored

TABLE 4.1 Archaeological samples

Sample No.	Description	Age (yr)
113	*Mayan lintel*: Wood (Zapote) taken from carved Mayan lintel from structure 10, Tekal, Guatemala. Am. Mus. Nat. Hist., submitted by Junius B. Bird.	1600 ± 200 1400 ± 150 Av 1470 ± 120
	Carved date on lintel: 9.15.10.00 in the Mayan calendar which, according to the Goodman-Thompson correlation, would be June 30, A.D. 741. According to the Spinden correlation, this would be Aug. 30, 481. Wood has about 30 years' growth. Hence expected date is (1210–1240) +, Goodman-Thompson or (1470–1500) +, Spinden. Comment: Looks like Spinden date.	

Source: Kulp et al. 1951, 566

jointly by the Museum and the Department of Physics), is entirely responsible for radio-carbon data and analyses. William E. Stephens of the Department of Physics contributed invaluable advice in the important matter of statistical analysis. Robert Stuckenrath, research assistant at the Radiocarbon Laboratory, processed the samples for dating. Linton Satterthwaite, curator of the American section of the museum, supplied information on Maya dates and correlation hypotheses to be tested. (Satterthwaite and Ralph 1960, 166)

The new results produced by the latest technology and more careful attention paid to sampling were now beginning to favor the GMT—not the Spinden correlation.

Ralph's new numbers—and along with them, her lab—were vindicated by Libby himself, three years later.

The measurements reported in this list have been made in the Isotope Laboratory at the Institute of Geophysics, UCLA during 1962. Dates have been calculated on the C^{14} half life of 5568 years. (Fergusson and Libby 1963, 1)

And with the heading "Tikal series, Guatemala" Libby continued:

Wood samples from Mayan temples at Tikal (17° 13.3' N Lat, 89° 38.5' W Long), Petén, Guatemala to check on correlations of the Mayan and Christian calendars. Samples were from lintel and vault beams and, where possible, precautions were taken to avoid "post-sample growth" errors. Subm. by Linton Satterthwaite, University Mus., Philadelphia. (Fergusson and Libby 1963, 13)

General Comment: the C14 dates on Temple IV support the Goodman-Thompson correlation while those on Structure 10 fall between the Goodman-Thompson and Spinden correlations. These results can be compared with the Tikal series dated by the Univ. of Pennsylvania (Satterthwaite and Ralph, 1960). . . . The agreement between the two laboratories is good, especially considering that the Pennsylvania ages were computed using pre-1900 Oak wood as the modern standard and the UCLA ages were computed on basis of 0.95 NBS oxalic acid. The agreement is also additional support for the validity of 0.95 NBS oxalic acid as a C14 standard for the contemporary activity of the biosphere. (Fergusson and Libby 1963, 14)

While the first numbers reported by Kulp supported Spinden's correlation, continued work shifted support to the GMT.

Now it was Thompson's turn to appeal to the latest technology, and his response was far from restrained. In the "Preface to the Third Edition" of *Maya Hieroglyphic Writing*, ten years after Satterthwaite and Ralph published their results, Thompson wrote with flair:

The few sporadic first readings supported the Spinden (12.9.0.0.0) correlation a matter which filled me with dismay. However, I felt that all other lines of evidence so strongly supported the Goodman-Martinez-Thompson (11.16.0.0.0) correlation, as is made clear in Appendix II, that one must not act hastily, but wait to see how this new technique developed. New processes or new machines, from the latest detergent to the slickest computer, have their teething troubles—and how consoling to be able to envisage a computer so human and defenseless as to suffer the pangs of teething. . . .

It is pleasant that Carbon-14 readings now fall into line with the other factors which so strongly support the 11.16.0.0.0 correlation. In my opinion there is now no serious doubt that it is the correct answer. (1978, vi–vii)

Thompson's attempt to relish victory did not go completely unchecked. For a volume of *Science*, Frederick Johnson provided a more nuanced review of the state of the problem, presenting a view that endured for decades.

The Maya calendar has not yet been successfully correlated with the Gregorian. It was natural to use radiocarbon in an attempt to confirm some one of the at least nine different attempts at correlation. The various estimates differ by some 260 years. Over many years the problem of correlation developed a seemingly endless orgy of arithmetic and astronomy . . . Libby and the Lamont Laboratory reawakened interest by deriving dates from a beam from structure 10 at the site of Tikal, which in a spectacular way appeared to confirm Spinden's correlation. . . . The University of Pennsylvania, which was excavating at Tikal, commenced a project to date beams found there in the temples . . . If certain basic assumptions are correct, the results of this and of independent analyses favor the Goodman-Thompson-Martinez correlation (27). However, some believe that the distribution and comparisons of certain types of Maya pottery favor the Spinden correlation; the controversy continues. Nevertheless, radiocarbon dating has introduced a new element that may eventually help to resolve the problem. (Johnson 1967, 168)

There is no doubt that the new data provided by radiocarbon dating supported the GMT. When we look back on it now, though, we should recall that it only did so relative to the other solutions being considered. The subtle complication is that, by itself, radiocarbon dates cannot verify a day-for-day correlation, which the GMT purports to be. Satterthwaite and Ralph represented this factor well in 1960:

> It must be remembered, however, that they are all day-for-day correlations founded on various types of evidence. A radiocarbon spread of years may in principle render untenable some or all proposed day-for-day correlations, or impose limits for possible new ones. *If one is satisfied that he has a list which includes the correct solution*, and also that radiocarbon results eliminate all but one of them, logically he will accept the latter as correct. (Satterthwaite and Ralph 1960, 179, emphasis added)

Satterthwaite and Ralph's assessment is spot on: if one has all possibilities, and if they are separated by more than the tolerance of radiocarbon dating methods, then, logically, radiocarbon dates will determine the correct correlation. But it is not clear that all possibilities were considered in these late twentieth-century studies. The only ones that were used in Satterthwaite and Ralph's study all subscribed to the contested and contestable conditions and assumptions coming from the historical data. They all also depended on the assumption of continuity, which, if only this one constraint is released, introduces $365 \times 50 = 18{,}250$ alternative correlation constants.

In the same "Preface to the Third Edition," Thompson himself snuck in an admission of new counterevidence to his adherence to continuity. The very last paragraph of the preface read:

> One final matter before leaving the present subject. On page vi of the Preface to the Second Edition, I briefly mentioned the discovery of a Mixe 260-day almanac which I said was synchronous with surviving highland Maya almanacs. Later information makes clear that the almanacs are not synchronized; the Mixe count, in fact, **varies from village to village**. In one center it differs by five days from the highland Maya almanacs. (1978, vi–viii, emphasis added)

Yes, radiocarbon dates did show that choosing from the correlations previously considered, the GMT is closest to a correct correlation; but they also made clear that alone they cannot prove it correct. While the GMT fits with ^{14}C data, the latter also allow for hundreds of other possibilities given the analytic tolerances. Moreover, they heighten the complications of the historical and astronomical data, which are all that can resolve the day-for-day problem. Nevertheless, Thompson and the GMT seemed on the precipice of winning his colleagues over.

The point now is that before Libby's work, 260 years was a reasonably short time period for the relative dating of ancient civilizations. Through ^{14}C dating, 260 years of difference between correlations between the Mayan and Christian calendars became resolvable.

One of the key factors in this revision was the process of obtaining a sample. The original Mayan loggers at Tikal who felled the tree for use as a lintel also shaped it accordingly. What Kulp had not considered in taking his sample was how much material the final artists had taken off from the outer growth of the tree. Had they stripped away much material, preserving only the core? Or had they used virtually the full diameter of the tree trunk? The difference corresponded to different ages of the tree and so different times relative to the date carved on the lintel. Ralph and Stuckenrath (1965) addressed, though could not definitively solve, this "postsample growth" by taking samples from the outer rings. The same issue resurfaced very recently with the advent of yet newer technology. In 2013, Douglas Kennett teamed up with a brigade of collaborators to introduce a new technology into the approach. Now utilizing calcium concentrations within tree cross sections as proxies for visual tree rings and the latest Accelerated Mass Spectroscopy equipment, Kennett and his colleagues claimed to champion the GMT (Kennett et al. 2013, 2).

This work by Kennett and his colleagues certainly contributed substantively to the consideration of the Calendar Correlation Problem, but in fact it did not go as far as claimed. For one, Kennett and his colleagues made the same assumption as Satterthwaite and Ralph, believing that the list of possible correlations in their study was complete (Kennett et al. 2013, 2). In fact, they used the same list, which was already fifty years old and mostly obsolete. Furthermore, they used Satterthwaite's reading of the dates in the hieroglyphic inscription, which compromised their results (Aldana 2016b, 13; Kennett et al. 2013, 2). At the time of Satterthwaite's reading, the decipherment of the writing system was just getting underway—Proskouriakoff had just published her historical hypothesis in 1960, and Sattherthwaite and Ralph were working on the Calendar Correlation Problem in 1965.

We have seen at the beginning of this chapter, though, that the dates inscribed on Lintel 3 of Temple I do not correspond to the dedication of the structure in which they were found. Instead, the structure is unequivocally the funerary monument for Jasaw Chan K'awiil (Coe 1990; Harrison 1999; Martin and Grube 2008), and the text on the lintel in it records the most important military victory of his life (Martin and Grube 2008, 44–45). These two events, in turn, are separated by some forty years (9.13.3.7.18 and 9.15.3.6.8), not the maximum seventeen years suggested by Satterthwaite and Ralph. The latest possible "dedication date" included by Satterthwaite and Ralph is therefore twenty-three years short of the earliest possible construction of the monument. Because Kennett and his colleagues with their new methods were working with much smaller tolerances than Satterthwaite and Ralph—on the order of twenty-five years—a twenty-three-year error is significant.

Spurred by Kennett and his colleagues' work, I recently reran an analysis of the dates from Tikal, using the data generated over the latter half of the twentieth century along with new data produced by Kennett and his colleagues in 2013 and using the same Bayesian statistical software (Aldana 2016b). Rather than use calcium concentrations, my study used the historical information in the hieroglyphic inscriptions to assign Long

Count dates to radiocarbon samples. The carved lintel in Temple I is now associated with the Long Count date of that building's construction—not the historical events it records (Aldana 2016b, 7). Since we can now read the inscriptions, we can use their Long Count dates as intervals between building construction dates, analogous to tree ring data in the American Southwest (Aldana 2016b, 4). The result shows that the GMT is definitely close to an appropriate solution, but that if we allow for the possibility of variation and revision among scribal communities during post–Long Count times, then a better solution to the Calendar Correlation Problem is statistically suggested to be about twenty to forty years later than the GMT.

The problem is still not solved and ^{14}C data alone cannot resolve its day-for-day formulation, but it should be clear that there are enough open questions within the evidence used to construct the GMT that scholars should avoid using it to constrain interpretations of the Dresden Codex Venus Table. That is not, however, how the modern literature generally refers to the GMT or the Venus Table. In *Breaking the Maya Code*, for example, Coe writes:

> In spite of oceans of ink that have been spilled on the subject, there now is not the slightest chance that these three scholars (conflated to GMT when talking about the correlation) were not right; and that when we say, for instance, that Yax Pasah, King of Copan, died on 10 February 820 in the Julian Calendar, he did just that. (1999, 114)

The difference between the modern reassessment and Coe's perspective has everything to do with an emphasis on authority over evidence and one last intervention by Eric Thompson.

THOMPSON'S LEGACY

A Commentary on the Dresden Codex (1972) was a testament to an era. The historian of science David Pingree reviewed it for the *Journal for the History of Astronomy*, writing:

> Thompson's commentary, in conjunction with his many other contributions to Mayan studies, some of which have been mentioned, provides a secure basis in fact for the interpretation of the Dresden Codex, and for the understanding of Mayan culture in general. His is indeed an admirable achievement. (1974, 138)

At the time, however, Thompson was on fragile scholarly ground. He continued to press the idea as late as 1978 that the writing system was a sort of ideographic writing (1978, vi), and he suggested further that the ideographic approach had been the source of greatest epigraphic success during the century. But he was now writing after Proskouriakoff published her historical hypothesis based on Piedras Negras and after Thompson's own concession that she was right (Coe 1999, 176). In Coe's words, this amounted to "the demolition of one of the pillars supporting his general views about the Maya" (1999, 176). And by the late 1960s and early 1970s, work was beginning to demonstrate the productivity of Proksouriakoff's historical hypothesis along with a revised approach to Knorosov's phonetic hypothesis.

On the other hand, no Mayanists contested the interpretation that the Dresden Codex contained astronomy, so it may be that as Thompson saw his epigraphic work slip away, he saw the Venus Table as a secure retreat to his original strength in calendrics and astronomy for his final opus. This would allow him to produce a volume in which he could display his intellectual prowess with much greater potential for long-term saliency. Whatever the primary motivation, Thompson used this opportunity near the end of his career to "read" the whole document. In *A Commentary*, each almanac in the Dresden Codex, each table, is fully commented and interpreted and is accompanied by a new facsimile of the manuscript.

For his colleagues familiar with his publications, Thompson's 1972 treatment of the Venus Table presented no surprises; it solidified the interpretation he had built based on Teeple's work. In some cases, though, he did introduce some nuance. For instance, Thompson changed his interpretation of the illustrated figures in the Venus pages. Now he offered a revision of the spearthrower-holding figures relative to Seler's 1904 proposal (1972, 65). Thompson argued that the warrior figures were not "regents" for Venus periods, as Seler argued more than fifty years earlier; they were representations of Venus itself as a warrior god (1972, 67).

> On the right of each page of the Venus revolutions are three pictures. . . . In the middle . . . manifestations of the Venus god(s) at heliacal rising after inferior conjunction with spears and spearthrower menace the victim who is shown with dart-pierced body in the bottom . . . set of pictures. (1972, 67)

Thompson overrode Seler's early, more conservative interpretation to suggest that Venus had multiple identities, and each of these was depicted as the spearthrower figure on the right-hand side of each of these pages (figure 4.4). This warrior interpretation of Venus took hold in the academic and in the popular literature. Eventually, it led even to the "Venus-timed warfare" hypotheses and "Star Wars" interpretations that we saw in chapter 2.

Thompson also took one more step with the Venus pages in *A Commentary*, coming from his new confidence in the ^{14}C evidence. What he had recognized as a complication, he would now recast as a constraint on any proposed interpretation. We have seen that Thompson's work with Teeple's corrections had the GMT correlation placing the anchor of the Venus Table sixteen days earlier than the first morning visibility it was supposed to capture. Rather than admit that the sixteen days of error resulted specifically from his application of the GMT, Thompson argued that the error reflected the scribe's original intent. Now the error resulted from the implementation of Correction Intervals—as though they had actually been scripted within page 24—and therefore were independent of any proposed Calendar Correlation Constant.

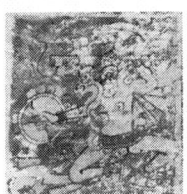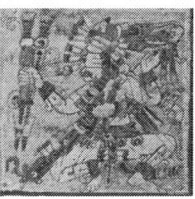

FIGURE 4.4 Thompson's "Venus Warriors" in the Venus pages. Courtesy SLUB

However one looks at the data, if the last three entries of the table correctly register heliacal risings of Venus, as they do in terms of the Goodman-Martínez-Thompson correlation, the 9.9.9.16.0 entry must lie some sixteen days before heliacal rising; both sets of dates cannot be correct. My suggestion that it was an accumulated error of 16 days at the first date which was corrected by the addition of 16VR+16 days may be wrong, but whatever the circumstances there is no doubt that the first entry is 16 days wrong in terms of the later entries, *whatever correlation is used,* and that the Maya amended their 61 VR − 4 days and 57 VR − 8 days groupings to include this extraneous 16 VR + 16 days. (1972, 63−64, emphasis added)

Thompson notes that the correction mechanisms that he and Teeple derived for the Venus Table's operation do not allow for both the early "base date" of the preface to correspond to an observable Venus event *and* for the much later corrected dates also to correspond to observable events. Rather, one of the sets must be erroneous if the other is presumed to be accurate. He has therefore ignored that the sixteen days only arise as a result of both his subscription to the GMT and his interpretation of the method of correction. No matter, Thompson had now solidified his interpretation of page 24 and set forth the challenge that if one Calendar Correlation could explain the sixteen-day error better than others, then that correlation had a better chance at being the right one.

And that is where the problem moved from interpretation on a technical level to one involving rhetorical manipulation. Starting in 1935, Thompson referred to Teeple's work as though it unequivocally supported his own position on the GMT (Thompson 1935, 63; 1972, 62−63; 1978, 224−25). Thompson wrote, for example:

There is little to add to the deductions made by Teeple with reference to the bearing of the Venus data on the correlation question. He has shown that the only correlations based on Sixteenth Century data which will agree with 9.9.9.16.0, 1 Ahau 18 Kayab as the approximate date of an heliacal rising of the planet Venus are those which make the Katun 13 Ahau of the Conquest fall on 10.10.0.0.0 or 11.16.0.0.0. (1935, 63)

As late as 1931, though, in his last publication on the subject, Teeple himself wrote:

There is not yet sufficient data available from the inscriptions either alone or with the *Dresden Codex* to determine a correlation. . . . In the meantime if one must use a correlation, then [the GMT] seems to meet fewest objections, although I am far from being convinced yet that it is the correct one. (1931, 109)

Even though Teeple expressed a lack of confidence in the GMT and the specific interpretation of the Venus Table that would support it, Thompson's representation of Teeple's work led to the acceptance of both.

With *A Commentary*, then, Thompson consolidated the overall interpretation of the Venus pages and secured at least the calendric and astronomical components of his legacy. With the GMT and his modified implementation of Teeple's corrections, Thompson claimed that in the Venus Table, Postclassic Mayan astronomers had access to an ephemeris for the planet accurate to "within one day in six thousand years" (1972, 63). This claim consolidated the long-recognized interpretation of Mayan astronomical acumen at the same time that it lent credence to the GMT. His own scholarly weight was now behind three aspects of the Venus Table's interpretation: (1) Venus was a warrior god, (2) the Venus Table supported the GMT's correctness, and (3) the accuracy with which a Venus ephemeris could be constructed out of the Venus Table was the metric by which all interpretations should be judged. Despite the counterevidence to each of these claims, Thompson's reputation firmly entrenched them in the literature for the next several decades.

ONE MORE MATHEMATICIAN'S INTERVENTION

By all accounts Floyd Lounsbury was a scholar and a gentleman; colleagues who knew him personally describe him with affection and respect (pers. comm. with the following: Michael Coe [2011], Marianne Mithun [2013], David Stuart [2014], George Stuart [2014], Bruce Love [2014]). Like Förstemann and Teeple, Lounsbury found himself with the right talents at the right time to make important contributions to the growing field of Maya Studies. But also, like both of them, his interests initially led him professionally some distance from Mayan cultures, hieroglyphics, and astronomy. At the time that he entered the field of Native American linguistics—just a few years after Ruben created ^{14}C at Berkeley—the intellectual movement in linguistics was in the analysis of spoken language, not on writing systems. Lounsbury found an interest in the Oneida family of languages, working on a project under Morris Swadesh at the University of Wisconsin, both as an undergraduate and as a master's student (Chafe 2000, 226). Also like Ruben, Lounsbury's career was impacted by World War II, though it only sent him to Brazil with the Air Force, where he picked up an interest in meteorology (Chafe 2000, 226). After his return, he enrolled in the doctoral program in anthropology at Yale, which he completed with a return to the Oneida language and to verb morphology—one of the more challenging languages to master for those whose first language is Indo-European in origin.

By 1949 at thirty-five years old, Lounsbury was teaching at Yale with faint interest in Mayan writing. Seven years later, though, he picked up Yuri Knorosov's phonetic hypothesis for the Mayan hieroglyphic writing system. More out of duty to the profession than academic interest, Lounsbury read Knorosov's work via invitation to review it for an academic journal. What he did not expect was to encounter an interest that took him back to his bachelor's degree in mathematics. "What really got me hooked was not the decipherment, but the mathematical puzzles of the Dresden Codex" (Coe 1999, 198). Those puzzles and his respect for Knorosov's linguistic proposal served as an entryway into the writing system more broadly.

By the early 1960s, Lounsbury began teaching graduate seminars on the Dresden Codex and the Mayan hieroglyphic writing system. This preparation matured into his first paper on the subject for a Dumbarton Oaks conference in 1971 (at age fifty-seven), titled "On the Derivation and Reading of the Ben-Ich Affix." With meticulous logic, Lounsbury broke down and reinterpreted with Knorosov's phonetic hypothesis what Thompson had interpreted ideographically as the "Ben-Ich" glyph. For his approach, Thompson had taken the glyph in question and reduced it into two familiar elements: one looked like the 260-Day Count Day Sign "Ben"; the other looked like an eye, which in Yucatec Mayan is "ich" (Coe 1999, 199–200). Drawing from both Classic period inscriptions and examples from the Dresden Codex, Lounsbury instead read the affix as "ahpo" or alternately "ajaw" (1973, 141). Although the latter has turned out to be correct, it was primarily his reliance on highland rather than lowland Mayan languages that led him to favor the former (Coe 1999, 200). Regardless, Lounsbury was thus heavily investing linguistic analysis into the approach to the hieroglyphic writing system, joining the methodological departure away from Thompson.

Lounsbury's next move within Mayan Studies was to attend the now fabled Palenque Round Table, hosted by art historian Merle Greene Robertson in the sleepy town adjacent to the archaeological site. As has been well told many times now, Lounsbury found himself there with the art historian Linda Schele and the graduate student working with David Kelley, Peter Mathews (Coe 1999, 201; Schele and Freidel 1990, 466). Lounsbury could not have been better prepared for that meeting. With his familiarity with titles of rulership and associated texts, and with his mastery of Mayan mathematics, calendrics, and numerology, he played a crucial role as the group assembled a dynastic list for the rulers of Palenque that spanned more than two hundred years (Coe 1999, 193–217). In turn, Lounsbury found himself in an optimal position in 1978 to contribute a masterful entry in the *Dictionary of Scientific Biography* entitled "Maya Numeration, Computation and Calendrical Astronomy." This came out one year before he retired from the professoriate at Yale. Although at the tail end of his career, Lounsbury had definitely left his mark on the decipherment of Mayan hieroglyphic writing.

Invigorated by success at Palenque and mastery of Mayan calendrics, Lounsbury kicked off his retirement by tackling the Calendar Correlation Problem. Whereas in his earlier work Lounsbury seemed to retain a clear and critical eye, he revealed a heavy bias within his work on this problem.

The Thompson correlation *will be assumed* (in its original value: Julian day number = Maya day number + 584285), since I am convinced that it represents the truth; but it can as well be understood merely as a working hypothesis, about to be put to a test. (Lounsbury 1983, 5–6, emphasis added)

For whatever reason, Lounsbury took it on himself to champion the GMT, recognizing that it was intimately related to the Dresden Codex Venus Table.

The longcount date that appears in the preface to the Venus table of the Dresden Codex poses a problem that up to the present has resisted satisfactory solution. So also does one of the intervals tabulated in the fourth tier of numbers on that same page. These problems, which are related, bear on the astronomical circumstances and the chronology of the Venus table, and indeed on Mayan chronology in general; for the interpretation accorded to them affects crucially the solution of the Maya-to-European calendar-correlation problem. A resolution of the Venus problems is now at hand; and this opportunity is taken to present it. (1983, 1)

Within his 1983 article on the Calendar Correlation, Lounsbury clarified his perception of the problem he wished to address, and he resurrected Thompson's test in so doing. Regarding page 24 of the codex, Lounsbury reviews what we have already seen: that it

highlights a day 1 Ahau 18 Kayab which it places at 9.9.9.16.0. The most natural assumption would be—and has usually been—that this must designate the primary base of the main table. The trouble with that, as Eric Thompson expressed it (1950: 226), is that "in no correlation so far suggested, which is not derived solely from astronomical data, does 9.9.9.16.0 coincide with a heliacal rising of Venus after inferior conjunction." Correlations can of course be coined ad hoc to accomplish this end when other pertinent evidence is ignored; but those that have taken historical evidence into account are not able to achieve it. Thompson noted that by his correlation (which is derived from post-Conquest historical data) the prediction is about sixteen days too early for the event. (1983, 4)

With the latter portion of this statement, Lounsbury has taken up Thompson's implicit test. Through it, Lounsbury has set up a problem in which he characterizes the GMT as a correlation derived by history but one that ambiguously meets astronomical constraints. Lounsbury has disregarded all of the complications with the historical "evidence" and picked up the problem as it was rhetorically contrived. Lounsbury saw two possibilities—the GMT and the Spinden correlation, set against each other by radiocarbon data. He then saw the Venus Table as the means of choosing a winner.

Following Thompson in *A Commentary on the Dresden Codex*, Lounsbury turned away from the various issues behind each of the historical dates used in the GMT and focused on the astronomy. Lounsbury goes on to lay out the desideratum for his work, asking,

> why was the date 9.9.9.16.0 given such prominence and made the focus of the preface if it was not the intended base of the main table or even a historical heliacal-rising date? Its discrepancy by the Thompson correlation has posed for many a serious obstacle in the way of their acceptance of that correlation. I hope now to remove that obstacle. (1983, 5)

For Lounsbury it is no longer a question of why a discrepancy of sixteen days might exist; it is now a question of *what kind of base date* 9.9.9.16.0 must have been *given that* it was sixteen days from the event it was supposed to record. Lounsbury has now transformed Thompson's implicit test into a formal problem for which he intends to provide a solution.

In order to proceed, Lounsbury borrowed from his early academic training to build a very useful little mathematical instrument intended to test Thompson's correction mechanisms. Lounsbury graphed the difference between historically reconstructed first morning visibilities of Venus and their prediction by the Dresden ephemeris, having invoked the GMT. He plotted the averaged error beginning in 9.9.9.16.0, accruing linearly over periods of 104 haab, against the saw-tooth function generated by Teeple's and Thompson's correction mechanism (figure 4.5). For Lounsbury, this produced a very provocative result: a "zero" error occurs on 10.5.6.4.0 1 Ajaw 18 K'ayab.

> Thus, if there was ever a heliacal rising of Venus on a day 1 Ahau 18 Kayab that motivated the ascription that is made in the Dresden Codex, and if the Thompson correlation is

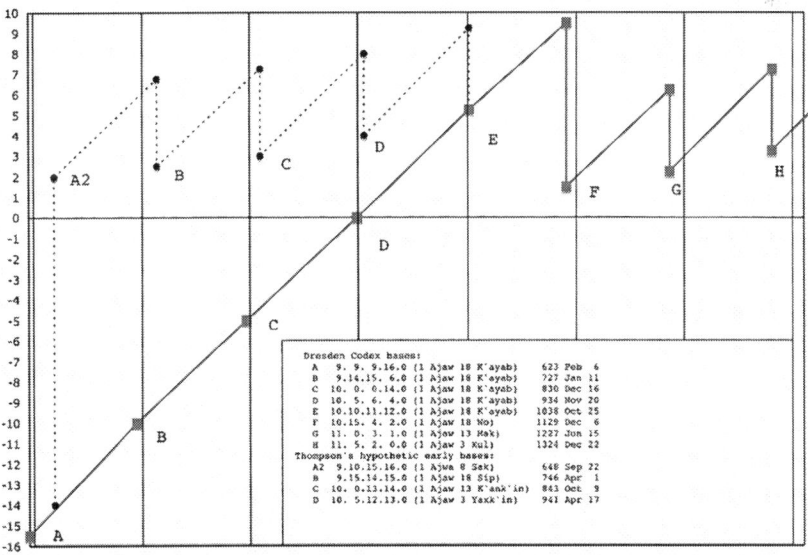

FIGURE 4.5 Reconstruction of Lounsbury's graph of Venus Rounds and his error-correction mechanism

correct (or any correlation that respects Landa's equation), then the one of base D was it. Its date was A.D. 934 November 20 (Julian), equal to Maya 10.5.6.4.0 by the Thompson correlation. There can have been no other. (1983, 11)

Considering the natural variability of Venus's period along with the steadily accruing error generated by the approximated synodic period, the heliacal rise of Venus occurs on the day predicted, with zero error: 10.5.6.4.0 1 Ajaw 18 K'ayab on November 20, 934. Precisely following Thompson, Lounsbury makes the claim that any Calendar Correlation must meet this constraint. That is, any correlation must put a 1 Ajaw 18 K'ayab date on the heliacal rise of Venus on AD 934, November 20. Since the GMT is the only one that does so, and because it does so so well, to Lounsbury it must be correct.

With this article, Lounsbury consolidated Thompson's claims and fundamentally shifted the interpretation of Tawiskal Uwoojil's hieroglyphic manuscript. Within Mayan Studies, it could no longer be considered on its own. It was now a component of the Calendar Correlation Problem, with the aspect of an instrument used to fine-tune the solution.

Perhaps more importantly, Lounsbury's "proof" set off a larger movement in the field. Now scholars began to take the GMT as correct by assumption at precisely the time that the field of archaeoastronomy began to grow. Many scholars even pushed the interpretation of Venus further by using the GMT to record its position on every historical date recorded in the inscriptions—even when the hieroglyphic textual content had nothing to do with matters astronomical (Schele and Freidel 1990, footnotes). Others invoked the GMT and then looked for astronomical events on the postdicted dates to argue for causal relationships between ritual events and astronomical configurations (Closs 1978, 1994; Dütting and Schramm 1985; Tate 1992; Bricker and Bricker 1986a). The subtler points were avoided, and instead scholars began taking it on Lounsbury's authority—in conjunction with Thompson's reputation—that the GMT was the definitive solution to the Calendar Correlation problem. Evidence of the movement of the field here in response to Lounsbury's work can be found in a book that might be considered the bible of Mesoamerican archaeoastronosmy. Whereas Anthony Aveni was still cautious in his first edition of *Skywatchers of Ancient Mexico* in 1980, his hesitancy was gone by the late 1980s (Aldana 2011b, 142; cf. Aveni 1980, 204; 2001, 207).

There are still very real problems with the GMT, but the inertia of convenience and authority maintain its popularity. More importantly for our purposes in this book, it is possible to see the Dresden Codex Venus Table as having been held hostage by Thompson's use of its predictive accuracy and dependence on the GMT. One of the latest treatments exemplifies this. In 2007, Harvey Bricker and Victoria Bricker started with the work of Thompson and Lousbury, but introduced a new way to handle the now "relatively minor complication" of how to interpret the short-term accuracy of the table's predictions. They shifted the accuracy argument away from exact prediction to one of strict anticipation. That is, they see the Venus Table as a warning table (Bricker and Bricker 2007, 106). As such, it would be most important for all dates in the table to *anticipate* observable events, not predict their occurrence exactly.* Although Lounsbury's emphasis on exact prediction is altered by Bricker and Bricker's interpretation, the role of accuracy remains essentially the same, preserving the interpretation of the Venus Table as an ephemeris. In fact, they appeal to an improved accuracy to the table as support for their interpretation (Aldana 2011b; Bricker and Bricker 2007, 113).

In sum, the words of Joseph Goodman recorded by Morley may be taken metaphorically as describing the twentieth-century approach to the Dresden Codex Venus Table.

> The veteran scholar discussed the Maya texts for upwards of an hour, always emphasizing more and more the importance of the numerical elements, and finally in conclusion stating as his belief that it was not history of which they treated, but of arithmetic and the science of numbers; and that the only promising method of approach to the meaning of the yet undeciphered characters—the method by which he had made all his great advance, he added—was the mathematical, and not the phonetic, indeed he rejected the latter with some show of impatience. (Morley 1919, 445)

Common to all contemporary approaches is the assumption that the first Calendar Round date anchor in the table is fictitious. Thompson considered this an "error" of which the scribes were aware and for which they used the intervals on page 24 to construct a later period when the table would become accurate. Lounsbury (1983) took the argument further to claim that the base was intentionally wrong. He suggested that the base date was constructed for numerological reasons, which took precedence over accuracy, and then had to be corrected to match observational reality. These are not completely unbelievable interpretations, but they also appear to be attempts to circumvent the inconvenience of the possibility that the GMT is simply incorrect.

There is also an internal, very concerning inconsistency. Interpretations rest on the accuracy of the table as an ephemeris for first morning visibility, and accuracy is generally used as a test of the quality of any given interpretation, but everyone acknowledges that two of the four events corresponding to subperiods are far from accurate on a cycle-to-cycle basis. And these inaccuracies are not made up for with long-term corrections. In fact, there is little room for the kind of event-to-event accuracy that scholars suggest for the Venus Table. The only place for real accuracy is in the long-term insurance that one observable event does not drift too far from expectation.

For all these complications and internal inconsistencies, it is worth noting that demonstrating that the Calendar Correlation Problem has artificially constrained the interpretation of the Dresden Codex Venus Table is not the same as showing that the resulting interpretation is wrong. Assuming, for example, that the Long Count anchor is a numerological fiction (as opposed to an observed event, preserved in historical records) may

---

*This move slightly altered Bricker and Bricker's inherited notion of accuracy since now they want a table that strictly anticipates first morning visibility; the constraint, of course, is that it cannot come so early that the prediction overlaps with last evening visibility—a rather small range of possibility over the long run given that inferior conjunction varies and can be as short as half a day.

be useful for modern computations, but it also may have actually been used that way to facilitate computations in ancient times. Similarly, assuming continuity of the Long Count through the Katun Count today may be convenient as Aveni characterizes it (1980, 205), but it also may have resulted from concerted political or ideological maintenance in ancient times. In other words, it may simply be a happy accident that the assumptions scholars have taken up only constrain the solution to the Calendar Correlation Problem in ways that conform to its actual solution.

On the other hand, there are numerous unanswered questions in the current interpretation of the Venus Table. Why was the method for correction so awkwardly written on page 24? What is the meaning of the hieroglyphic text within the table? Did Förstemann correctly infer it a hundred years before it was deciphered? We will find that there are even more questions regarding where the Dresden Codex may have been written and what motivated it in chapters 7 and 8. At this point, we proceed somewhat intellectually conservatively. Let us recognize that the current interpretation of the Venus Table is not the last word—but let us neither deny it nor embrace it. Instead, let us look to the manuscript itself within its own context and see where it leads us. I suggest that to do so we can follow the suggestion provided by the record from Copán that we saw at the end of chapter 2. We find that that connection goes much further than simply computational consistency between Tawiskal Uwoojil's manuscript and Classic period astronomical practices; there is also a robust ideological continuity between the two. The ideology and politics that Tawiskal Uwoojil worked within also shaped the concerns of the advisors to Jasaw Chan K'awiil or his uncle B'ahlaj Chan K'awiil. To see this continuity, though, as well as where it might lead, we return to the now legible historical records of ninth-century Copán.

CHAPTER 5

Processions and Precontact Time-Space Ritual Activity

S itting on his bench, waiting for his summons, we may wonder if Mak Chanil could hear the Copán River gurgle by, or if the activities of his neighborhood drowned out such natural sounds. Perhaps instead, shouts of merchants directing canoe traffic on that river dominated the local soundscape. Either way, in the late eighth century at his residence to the east of the civic-ceremonial center, Mak Chanil's thoughts would have been consumed by the colossal charge he was responsible for. As ajk'uhuun for the Copán k'uhulajaw, he would have to transform his ruler's political agenda into artistic and architectural construction. Of course he wouldn't have to do this alone, and he also had an extremely rich body of examples all around him. Copán was renowned for its public sculpture; stelae, altars, and buildings were all testaments to the talent of Copán artists over centuries.

Mak Chanil's own family held a prestigious position within the community of Copán artists. His ancestors since the notorious thirteenth member of the dynasty, Waxaklajun Ub'aah K'awiil, served as scribes to the royal court, and his own residence was a large and impressive architectural compound (figure 5.1). Once he had completed his charge and gone on to glorious reputation, he would rebuild his home and include within it an elaborately carved bench, worthy of a king's residence. Centuries later, the building in which he sat would be referred to by archaeologists as "the Scribe's Palace" or "the House of the Bacabs" (W. Fash 1991, 120; Schele and Freidel 1990, 329). But at the turn of the seventeenth winikhaab of the tenth pih, he was still contemplating the shape of his work. At any moment, a page to the Copán k'uhulajaw would arrive to summon him to meet with Yax Pahsaj Chan Yopaat, sixteenth member of the dynasty.

In chapter 2, we saw that B'ahlaj Chan K'awiil's history of treason and the Piedras Negras dynasty's iconographic depictions of the sky provided a window into some provocative astronomical practices from the Late Classic period. They also foster support for an interpretation of the Venus Table pulled away from the needs of the Calendar Correlation

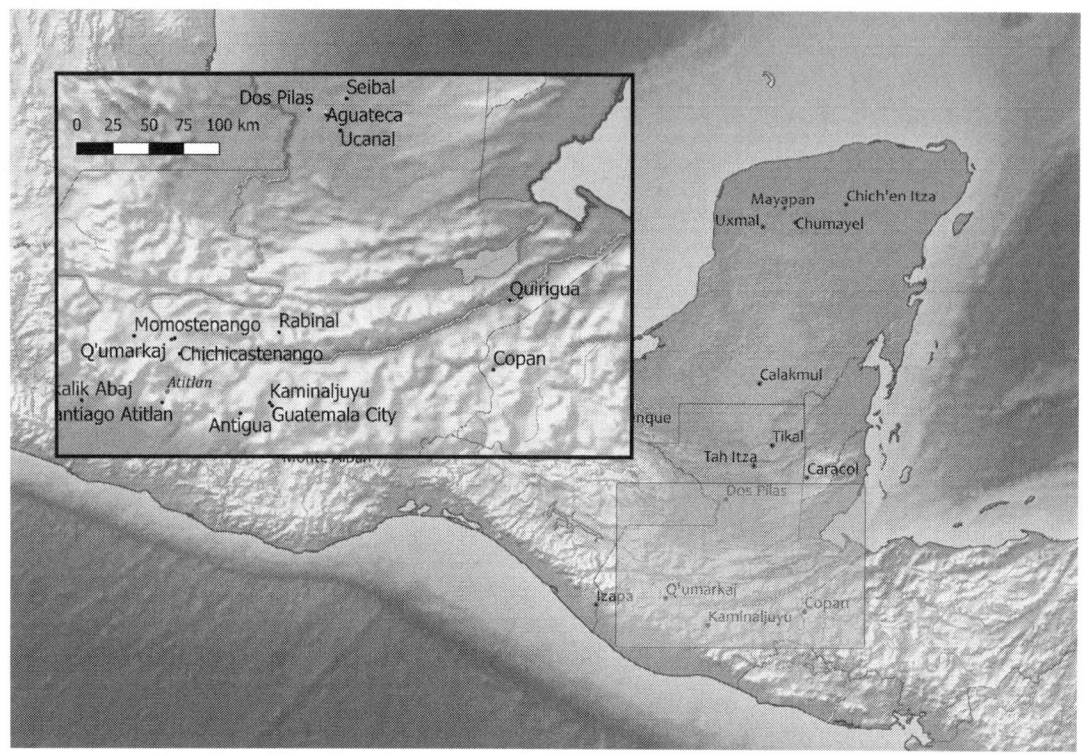

FIGURE 5.1 (*top*) Southern highlands; (*bottom*) Mak Chanil's residence, otherwise known as the "Scribe's Palace" at Copán, in the architectural group known as Las Sepulturas. Inside is a hieroglyphic bench, carved with the name of the scribe and his relationship to the k'uhulajaw, Yax Pahsaj Chan Yopat. LACMA (Schele Drawing Archive #63078), photo by Linda Schele © David Schele, photo courtesy Ancient Americas at LACMA (ancientamericas.org)

problem and instead aligned with Classic Mayan scribes' own words and practices. Here we turn to colleagues of the scribes at Dos Pilas, Tikal, and Palenque, who lived and worked at the southern edge of the Mayan realm, at Copán during the Late Classic. There, as we saw at the end of chapter 2, Yax Pahsaj Chan Yopaat's scribes—led by Mak Chanil— recorded a Venus event almost identically to Tawiskal Uwoojil's records in the Dresden Codex. By taking a much closer look at this record, we encounter a broad Mesoamerican conceptualization of ritual space and time through hieroglyphic text, which in turn makes better sense of the Venus Table than has thus far been accounted for. Rather than rely predominantly on computational methods as have scholars interpreting these pages for the majority of the twentieth century, we now take advantage of the decipherment of the writing system as it informs the interpretation of the astronomy in the Dresden Codex. This opens up the opportunity to rely on a form of ideological continuity in place of an assumed calendric continuity.

VENUS AT COPÁN

The Copán that Yax Pahsaj Chan Yopaat inherited in the eighth century as the sixteenth ruler in the dynasty was not the Copán of his twenty-five-year predecessor, Waxaklajun Ub'aah K'awiil. By the late eighth century, many of the political alliances that dominated the economic luxury of the Classic period had been realigned or destroyed completely (Martin and Grube 2000, 63). The Pasión River region where B'ahlaj Chan K'awiil once had enjoyed regional dominance was falling into chaos with wall fortifications built to protect city centers and even eventual abandonments (Houston 1993, 47; Martin and Grube 2000, 65–67). Copán's northern neighbors on the east coast of the Yucatan Peninsula, on the other hand, were thriving. The dynasty at Caracol (in what is now southern Belize), for example, enjoyed a building renaissance at the end of the eighth century that would last for decades (Chase and Chase 2001, 280; Martin and Grube 2000, 98–99). Chich'en Itza in the northern lowlands would begin its transitional stage from its Puuc-style architecture in "Old Chich'en" to its more international style in the northern part of the city (Schele and Mathews 1999, 199).

The regional political environment was certainly changing, and Copán's dynasty would go through its own crisis. In the most ambitious building effort of his reign, Yax Pahsaj Chan Yopaat aimed to reach across it (figure 5.2). In consultation with Mak Chanil, the Copán k'uhulajaw conceptualized Structure 10L-11 ("Temple 11") as massive and central to the civic-ceremonial center of Copán—prominently visible from both the public spaces and private elite spaces of the city (W. Fash 1991, 168). In the primary hieroglyphic text within this monument, Mak Chanil recorded Yax Pahsaj Chan Yopaat's personal history, but he included one exception. The narrative was broken in the middle to reach back to record an event occurring within the reign of one of Yax Pahsaj Chan Yopaat's predecessors, and he did so using astronomy. Here Mak Chanil included a Venus record just like the ones in the Dresden Codex, but this record came from the

FIGURE 5.2 The south face of Copán Temple 11. At the top is visible the open doorway to the south, with inscriptions on either wall.

reign of the thirteenth member of the dynasty, Waxaklajun Ub'aah K'awiil (figure 5.6). This reference was not motivated by caprice.

Waxaklajun Ub'aah K'awiil, as thirteenth k'uhulajaw, had taken over from his wildly successful father, who in the Middle Classic period advanced the art and science of his city and extended powerful political influence over regional cities. While the brothers at Tikal and Dos Pilas were battling each other to the north, a large cylindrical monument at Quirigua attested to the Copán k'uhulajaw's diplomacy and regional prominence, recording the visit that Waxaklajun Ub'aah K'awiil's father made to the smaller subordinate city (W. Fash 1991, 104). The hieroglyphic text around the rim of that Quirigua altar announced that "K'ahk' Uti' Witz K'awiil arrived." In the center of the altar, artists depicted the Copán k'uhulajaw sitting upon the glyphs that informed us that during his visit "he danced" (Aldana 2002, S29; Martin and Grube 2000, 201; Schele and Freidel 1990, 314). As noted in the introduction to this book, the Classic Mayan hieroglyphic record is full of visits such as this, from one noble to another, in the formation, consolidation, or maintenance of alliances.

The inscribed cylindrical monument has been identified by archaeologists as a "katun altar,"* reflecting the tradition of carving the 260-Day Count date into the top of a cylindrical stone in commemoration of the completion of the winikhaab; in this case, "the eleventh winikhaab has ended"—that is, 9.11.0.0.0, with the date here as 12 Ajaw 8 Keh

---

* Quirigua Altar L, see Looper 2003, Schele and Freidel 1990.

in the Calendar Round. The image on the monument, accordingly, was a representation of the date 12 Ajaw. What is clever here is that the local Quirigua scribe included a visual metaphor by using the k'uhulajaw of Copán himself as the representation of the Day Sign "Ajaw" in the 260-Day Count date. We can see the bottom of the "Day Sign cartouche" under the ruler's feet and part of the bars of the number 12 above his head (figure 5.3). As explored elsewhere (Aldana 2002; W. Fash 1991; Schele and Freidel 1990; von Schwerin 2011, 277), this monument reflects the Copán k'uhulajaw's specific agenda concerning regional prominence as well as the more broadly growing economy of the Mayan world at the start of the Late Classic.

In any case, when he took over from K'ahk' Uti' Witz K'awiil in the seventh century, Waxaklajun Ub'aah K'awiil undoubtedly looked to maintain his father's success.

FIGURE 5.3 Quirigua Altar L. The circular form facilitates the Surface representing the 260-Day Count date 12 Ajaw, with the coefficient at the top, and the seated k'uhulajaw in the center of the Day Sign cartouche. Courtesy Wikipedia

FIGURE 5.4 Traditional rectangular relief-carved Stela P from Copán, compared to Waxaklajun Ub'aah K'awiil's full-round sculpture on Stela H (a–c)

The architectural record makes clear that he took up this charge by initiating major building projects and patronizing artistic innovations in the stelae representing himself throughout the public spaces of his own city (W. Fash 1991, 113; Schele and Freidel 1990, 315; Schele and Mathews 1999, 133). Before Waxaklajun Ub'aah K'awiil's reign, stelae throughout the Mayan region were basically rectangular blocks with a profile representation of the ruler on one side and hieroglyphic text on the others (figure 5.4). But the artists of Copán had been forced early on to develop a more sophisticated sculptural tradition than those of other cities. With little access to limestone, but substantial quarries of volcanic tuff, sculptors at Copán gave up the decoration of their buildings with stucco and began carving figures in the full round out of very durable stone (W. Fash 1991, 113). Waxaklajun Ub'aah K'awiil carried this tradition further by having Mak Chanil's predecessors sculpt representations of the Copán k'uhulajaw in deep relief to fill the Great Plaza. Certainly he was pushing an artistic agenda and so too the public representation of power (Newsome 2001, 11; von Schwerin 2011).

FIGURE 5.5 Temple 22 at Copán: external decoration (*top*), inner doorway (*bottom left*), and ek' symbol in the Celestial Dragon's ear (*bottom right*)

There is also evidence that Waxaklajun Ub'aah K'awiil was developing a complementary new form of astronomy in his elite architectural space. Whereas his artistic innovations were intended for a public audience, though, his astronomical innovation was reserved for his eyes and those of his closest advisors only (Aldana 2005; Aveni 2001, 257–59). Under his tenure—at the same time that Kan B'ahlam's nephew propagated the tradition of the 819-Day Count at Palenque—the masons of Copán built Temple 22, a structure at the northeastern edge of the acropolis. The sculpted entrance of Temple 22 enhances the "mountain" characteristic of the temple: artists carved *witz* masks for its external corners and a great serpent's open mouth for the front doorway. Walking through these reptilian jaws today, visitors confront yet another of Waxaklajun Ub'aah K'awiil's artistic extravagances. Here an elaborate cosmogram frames the entry to an inner room (figure 5.5).

Modern scholars refer to the Temple 22 inner-doorway image as a "cosmogram," reflecting the three levels of the Mesoamerican cosmos (Aveni 2001, 148; W. Fash 1991, 122; Schele and Freidel 1990, 325; von Schwerin 2011, 275). The base of the cosmogram was carved into the lower step of the inner room. Here skulls punctuated a hieroglyphic text broken up into four passages, representing the Underworld under the feet of anyone occupying the inner room. The text itself is unique, representing the only known quote from a historical k'uhulajaw. Providing information on the dedication of the building, the text starts with *ti 5 Lamat ni winikhaab*, or "on 5 Lamat, it is my winikhaab."* Here, Waxaklajun Ub'aah K'awiil followed the common practice of commemorating his first winikhaab (about 20 years) on the throne, but he brought even more attention to himself by including a first-person narrative. This is in line with his public artistic innovation of representing himself on stelae and perhaps betrays something of a narcissistic trait within his personality.

On either side of the doorframe, directly on top of large skulls—and so representing the middle realm of the cosmogram—at the beginning and end of the hieroglyphic text, two seated men raise their hands to hold up a creature spanning the lintel. The two figures wear headdresses that mark them as elders and wise men, as Pahuatuns, and they hold up the sky (W. Fash 1991, 122; Schele and Freidel 1990, 325). Finally, the creature they hold up is the Celestial Dragon representing the Milky Way as it arches overhead across the sky. This representation reflects a Creation narrative present throughout Mesoamerica, describing the formation of the world (Aldana 2001; Stuart 2005; Velasquez 2006).

For the exterior of the structure, Waxaklajun Ub'aah K'awiil's artists continued the broadly religious theme. Recent excavation has revealed twenty Maize God sculptures adorning the façade, contributing to what Jennifer von Schwerin states represented "the Yax Hal Witz ('First True Mountain' or Creation Mountain) of Maya mythology" (2011, 272; cf. Freidel et al. 1993, 149). In representation to the exterior world and to those privileged to step across its threshold to witness the inner-doorway cosmogram, Waxaklajun Ub'aah K'awiil's new building spoke of his cosmic import.

Four distinct spaces within the building each appear to have served a different purpose. The first space, into which Waxaklajun Ub'aah K'awiil entered by walking into the serpent's mouth, served principally as a passage providing access to the three inner spaces: a central chamber and a room on each side. The central chamber, entered through the cosmogram doorway, was likely a ritual space where Waxaklajun Ub'aah K'awiil would have meditated or communicated with ancestors (W. Fash 1991, 122; Schele and Freidel 1990, 325; von Schwerin 2011, 275). The east room is relatively featureless, rendering ambiguous the activities conducted there, perhaps opening up the possibility that it was occupied by attendants to the k'uhulajaw during ceremonies. Our interest here is in the room to the west of the cosmogram.

---

*Cf. "Temple 22," Peabody Museum of Archaelogy and Ethnology at Harvard University, accessed June 6, 2021, https://www.peabody.harvard.edu/node/2234.

The west room, to his left as Waxaklajun Ub'aah K'awiil entered the building, had an odd-shaped window built into its west wall. Scholars over the last forty years have suggested that three characteristics of this window constitute the argument that the Copán royal architects designed it for the observation of Venus (Aldana 2002, S46; Aveni 1980, 245; Closs et al. 1984). The first is its shape. The masons at Copán do not appear to have constructed this window for purely utilitarian reasons—for example, for ventilation or for lighting. From the inside the opening was relatively large, but halfway through the wall the opening was reduced to a small slit, a fraction of its original width, but of the same height. Second, the Venus phenomenon visible from this window was a regularly repeating one, and so once discovered it could be verified several times over the course of a single observer's lifetime and used for design purposes (Aldana 2002; Closs et al. 1984). The third aspect is that Venus's visibility through the window is restricted to a very provocative span of time. Each time the planet becomes visible within it, it remains within the range of visibility through the window for twenty days.

All three of these characteristics could have been intentionally developed without challenging the observational technologies of Classic Mayan astronomers. It is even possible that the observation was Waxaklajun Ub'aah K'awiil's own doing, representing a personal investment in astronomy. From a much later era, historical records inform us that the ruler of Texcoco—one city of the Aztec Triple Alliance—engaged in his own astronomical observations. According to a Spanish chronicler, Netzahualpilli

> was a great astrologer and prided himself much on his knowledge of the motions of the celestial bodies; and being attached to this study, that he caused inquiries to be made throughout the entire extent of his dominions, for all such persons as were at all conversant with it, whom he brought to his court, and imparted to them whatever he knew, and ascending by night on the terraced roof of his palace, he thence considered the stars, and disputed with them on all different questions connected with them (1969, Vol. 1, Book 2, Chapter 44, p. 188). (Aveni 2001, 16)

Whether designed by Waxaklajun Ub'aah K'awiil or commissioned of his court astronomers, the alignment of the window to Venus would not have been very difficult.

Viewed from a specific geographic location, Venus makes observable repeating geometric patterns in the twilight sky, as we saw in chapter 1. The planet's motion has it drift relative to the setting Sun, the background stars, and the horizon, with some portions of this motion recognizably more regular than others (see figure 1.18). For Waxaklajun Ub'aah K'awiil, an observation post at the top of the previous building (Str. 10L-22-Sub), for example (or atop an adjacent structure), would have allowed him or his astronomers to fix markers aligned with Venus's setting points on the western horizon for different parts of its eight-year cycle (Aveni 2001, 65). Repeated observations would easily allow for the selection of an appropriate gap between two markers, delimiting a span of twenty days of visibility. Those markers would then have been preserved and used to demarcate the size of the window as the final structure was built. It would take a maximum of six-teen years to confirm this pattern or as few as eight years to identify it. As we have seen, the hieroglyphic text tells us that Waxaklajun Ub'aah K'awiil dedicated the structure one winikhaab (or almost twenty years) after his accession, which would have given him plenty of time to confirm the celestial pattern and his masons sufficient time to build it.

Waxaklajun Ub'aah K'awiil's motivation for capturing a twenty-day span resonates with basic calendrics. Visibility for strictly twenty days means that all Day Signs within the 260-Day Count would be accounted for in a single passage through the window. In other words, the Day Sign of the first day that Venus is visible will return the day after the planet's last visibility through the window. In that way, the window captures one unique, discrete sequence within the 260-Day Count Day Signs, characterizable by the first or last date of visibility. This is directly analogous to the katun altar commemorating K'ahk' Uti' Witz K'awiil's dance, noted at the beginning of this chapter. That altar too was built to record the completion of one full run of a period of 7,200 days, marked by a single 260-Day Count Day Sign (cf. Thompson 1978, 181). For Quirigua Altar L, the commemorated date was the end of the eleventh winikhaab, or 9.11.0.0.0. In a Mayan sense, then, this wasn't the first day of a new winikhaab, but the "completion" of the previous winikhaab on an Ajaw date in the 260-Day Count. To extend the analogy, the visibility of Venus in Waxaklajun U'baah K'awiil's window would have been completed on its last day, so the last day would have provided its symbolic value.

Given these characteristics, Waxaklajun Ub'aah K'awiil may have been using his Venus window for astrological prognostication. If we anchor this speculation to the ti'huun astrology reviewed in this book's introduction, then Venus's first appearance in the window would have it "meet" the Day Sign current at the time. Over the next nineteen days, Venus would "meet" each of the other Day Signs, constituting a specific twenty-day period of Venus omens. We will return to indigenous Mesoamerican divination practices in chapter 6, and so provide context for what Waxaklajun Ub'aah K'awiil may have been seeking. Unfortunately for him, though, any oracular signs he may have obtained did not provide him sufficient warning of at least one critical political matter. In fact it may be that Waxaklajun Ub'aah K'awiil was so distracted by his feats of artistic and scientific innovation that he let slip his attention to regional politics.

According to Stela J at Quirigua, K'ahk' Tiliw Chan Yopaat—the ruler of that city—attacked and defeated Waxaklajun Ub'aah K'awiil during the latter's forty-third year of reign (Looper 2003, 76; Martin and Grube 2000, 219; Sharer 2006). While Waxakla-jun Ub'aah K'awiil was stargazing, the ajaw of Quirigua—installed by K'ahk' Uti' Witz K'awiil—attacked and defeated him. K'ahk' Tiliw Chan Yopaat actually ended up record-ing this event three times on the public stelae of his own town, which he began immedi-ately transforming architecturally into a proper Classic Mayan city, including those stelae examined in chapter 2 (Looper 2003, 77). At Copán, the defeat was recorded only once, near the bottom of Structure 10L-26, on its "Hieroglyphic Stairway." One carved step-riser reads *i-KAY-yi u-SAK-IK'-li tu-TOOK' tu-pa-ka-la 18-u-B'AAH-K'AWIIL-la*—"then he died with his flint and shield, Waxaklajun Ub'aah K'awiil" (Looper 2003, 78).

According to Barbara Fash, William Fash, and colleagues (B. Fash et al. 1992), the defeat of Waxaklajun Ub'aah K'awiil by the Quirigua army meant that the Copán dynasty was forced to kowtow to the local nobility and consolidate their support in part by constructing a new temple in the ritual center of town, dedicated to a council made up of the leaders of the almehen houses. Rather than proliferate monuments aggrandizing the next k'uhulajaw or their dynastic history, the next monument celebrated a level of "shared governance." Barbara Fash and colleagues call Structure 10L-22A a "Popol Nah" or "Council House," reflecting a temporary broadening of political power, backfilling against the possibility of an even deeper crisis (Fash et al. 1992; cf. Schele and Mathews 1999 379). While it certainly may have served this political purpose, the Popol Nah, Temple 10L-22A, was built immediately west of Temple 22, or directly in line with the "Venus window." The view Waxaklajun U'baah K'awiil's successors would have had of the planet from the western room was now cut off. It appears, therefore, that at least one instrument for Waxaklajun Ub'aah K'awiil's esoteric astrological pursuits was sacrificed for the survival of the polity.

Relative to the broad developments in astronomy reviewed in chapter 2, we find that even if astronomy/astrology had become ever more fashionable within elite representation into the Late Classic at Dos Pilas, Palenque, Piedras Negras, and elsewhere, at Copán, with the defeat of Waxaklajun Ub'aah K'awiil, the local almehen relegated celestial observation to greatly inferior priority in favor of seeking to restore the city's political and economic dominance of the region.

REBUILDING

Waxaklajun Ub'aah K'awiil's engagement with astronomy was almost certainly on Mak Chanil's mind as he contemplated his work for Yax Pahsaj Chan Yopaat. By the reign of the sixteenth ruler, Yax Pahsaj Chan Yopat, some of the astronomical rhetoric seems to have made a comeback. The sixteenth member of the dynasty began a building program of massive proportions, heavily emphasizing cosmological and astronomical themes. At the center of this building program, at the center of the civic-ceremonial portion of the city, and in line with the cut-off view of Venus from Temple 22, Yax Pahsaj Chan Yopaat put up a powerful new construction that we now call Structure 10L-11, or prosaically, Temple 11.

The sixteenth Copán k'uhulajaw certainly intended the building to serve as a mediator between realms. With her reconstructive eye, Tatiana Proskouriakoff saw the same "cosmogram" in the exterior decoration of Temple 11 as Waxaklajun Ub'aah K'awiil built within Temple 22. On its south side at its base, Temple 11 possessed Underworld imagery of shells, caymans, and ballcourt markers with Chaaks at each end (Schele and Freidel 1990, 322–23). On its north side, Proskouriakoff reconstructed the fallen sculptural remains to reflect the placement of the great Celestial Dragon supported by two Pahuatuns (Newsome 2001, 50; Schele and Freidel 1990, 489). The southern face of Temple 11 looked out onto a private, elite plaza, while the northern face opened out onto the

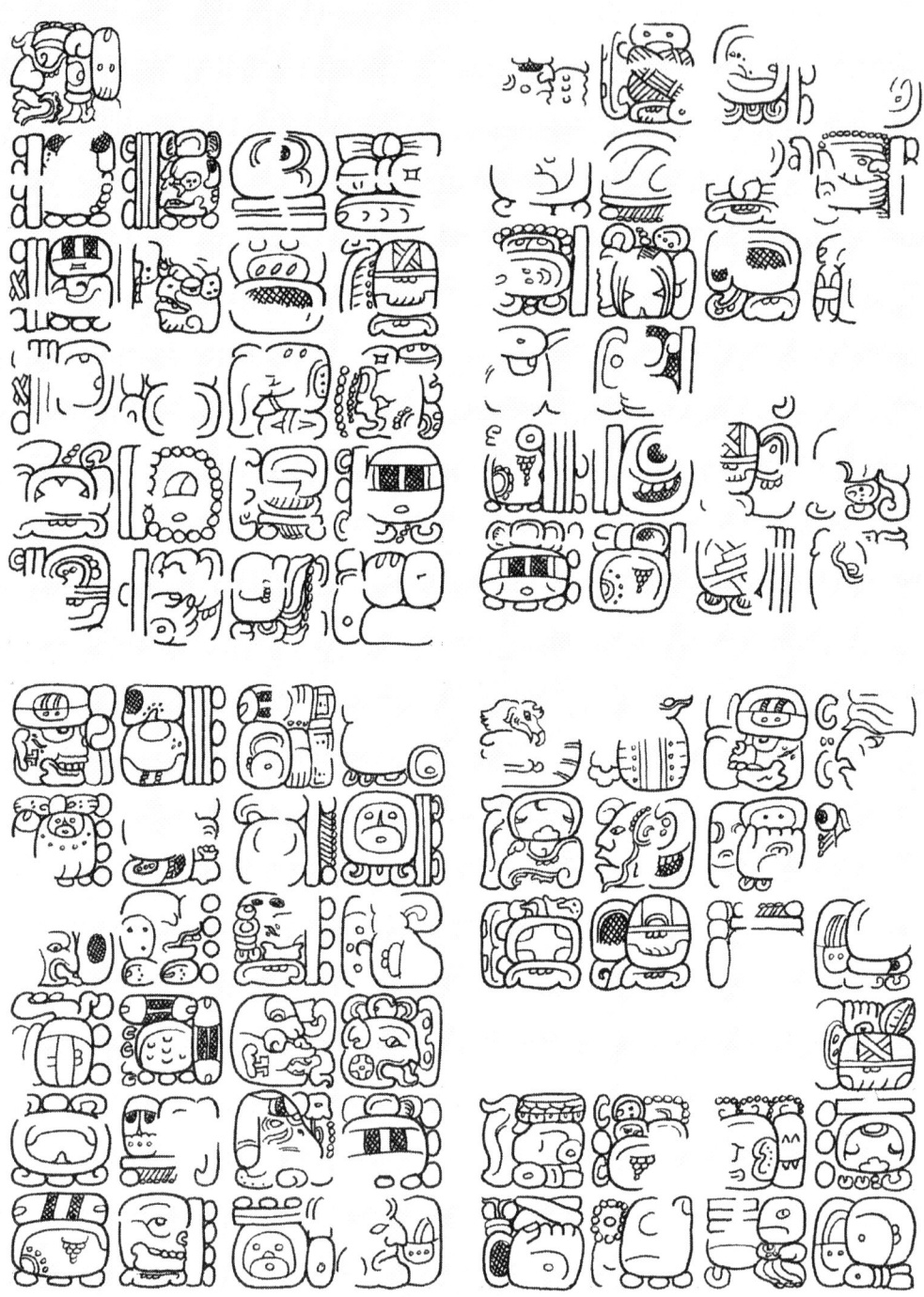

FIGURE 5.6 The inscriptions on the walls in the four doorways of Temple 11. The first two are NE and NW (*above, top, left and right*), corresponding to the north doorway and the inscriptions on the east and west walls accordingly. These are followed by SE and SW (*above, bottom, left and right*) for the south doorway, EN and ES (*facing page, top, left and right*) for the east doorway, and WN and WS (*facing page, bottom, left and right*) for the west doorway.

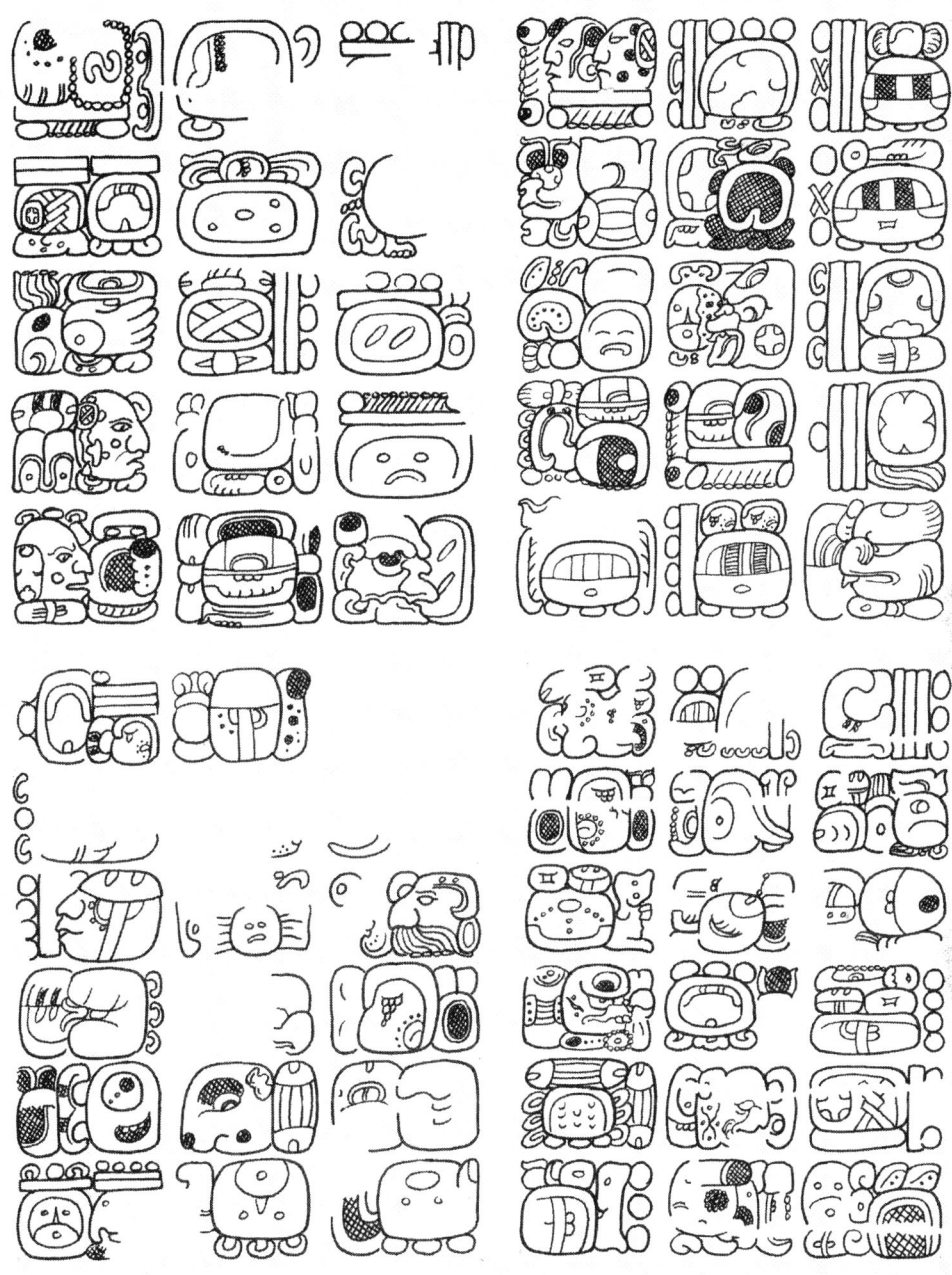

public plaza containing its magnificent Ball Court and probably the open space of a great market. The temple, therefore, sat right in between public and elite private spaces, and in between the Underworld and the celestial realm. It sat at and manifested the center of the city.

Yax Pahsaj Chan Yopaat and Mak Chanil took the design further to incorporate cosmology into the building's interior. Two corbel-vaulted hallways running the length and width of the building intersected at its center. This created eight walls, each of which was decorated with hieroglyphic text at its entryway (figures 5.6; 5.7). The specific story that Mak Chanil composed in these inscriptions mirrored the masons' work, weaving together sky and earth but now also old and new (W. Fash 1991, 168; Fash and Fash 1996, 138; Schele and Freidel 1990, 322–28). It is specifically within this temple and these texts that we return to the reference by Yax Pahsaj Chan Yopaat to his predecessor Waxaklajun Ub'aah K'awiil. It is also in the hieroglyphic text within this building that we encounter a window into the Venus Table of the much later Dresden Codex.

Following convention, Mak Chanil and Yax Pahsaj Chan Yopaat's royal artists started the long textual narrative on the north exit east wall (NE) with a statement of their patron's accession before moving on to other contemporary events. In outline, the content of the rest of the inscription follows an oscillation between "old and new" (figure 5.6 and table 5.1).

Immediately following his accession statement on the first wall (NE), Mak Chanil reached into the historic past, three winikhaab earlier, to record an event that may refer to the death of the thirteenth ruler, Waxaklajun Ub'aah K'awiil.* Next, the narrative jumps forward to record the first major period end after Yax Pahsaj Chan Yopaat's accession: 9.17.0.0.0 13 Ajaw 18 Kumk'u. This set the stage for the Calendar Round 1 Kib 19 Keh, corresponding to the Long Count date of 9.17.2.12.16—the dedication of Temple 11 itself. That sequence constituted the historical theme within the text.

For the astronomical narrative, Mak Chanil began the EN wall with the phrase *u tz'akaj* ("to arrange," "to put in order" [Mathews and Bíró 2008]), which is a phrase commonly used in Classic period inscriptions and specifically here throughout the eight tablets to break up the narrative according to events separate in time. A Calendar Round date follows; here it is 5 Kib 10 Pop. On this date, Mak Chanil recorded the verb root we see throughout the Dresden Codex, *k'al*, followed by the titled name of Venus: **AJAW-wa CHAK EK'** (see figure 5.6). Here lies the connection between Yax Pahsaj Chan Yopaat and the concept underlying the Dresden Codex Venus Table. A critical link is the understanding of the verb *k'al*, which we take up below.

The Venus record on the EN wall is followed immediately by another *u tz'akaj* phrase, this one introducing the 819-Day Count station of ES. As we saw in chapter 2, the

---

*The text is eroded, but appears to provide a verb taking up two glyph blocks immediately preceding the readily legible name of 18 Ub'aah K'awiil. The first block appears to be *aj-ne-?* and the second *mu-chu-bu* or perhaps *mu-MUK-bu*. The latter might refer to the burial of Waxaklajun Ub'aah K'awiil.

FIGURE 5.7 The WN (*above*) and WS (*left*) inscriptions of Copán Temple 11

TABLE 5.1 Parallel narratives in the inscription of Copán Structure 10L-11

	Past	Intermediate	Contemporary
Historical	Waxaklajun Ub'aah K'awiil	9.17.0.0.0 period end	Dedication of Structure 10L-11
Astronomical	Venus	819-Day Count station	New Moon

819-Day Count was one of the fruits of the astronomical fad of the Late Classic and so propagates an astronomical theme here at Copán. What is intriguing is how the use of the count may have come to Copán from Palenque's royal court.

In chapter 2 we saw that Janaab' Pakal's son Kan B'ahlam wrote the first texts for the commemoration of his Triad Group just before the 9.13.0.0.0 period end, which depended heavily on computations with the 819-Day Count (Aldana 2007, 113). By the middle of the fifteenth winikhaab, thirty years later, Kan B'ahlam's nephew, Ahkal Mo' Naab, adapted the practice to his own monuments. Ahkal Mo' Naab used astronumer-ology based on the 819-Day Count to tie his own history to the Creation narrative of the formation of the cosmos from a great primordial dragon—the same entity repreenting the Milky Way in Waxaklajun Ub'aah K'awiil's Temple 22 at Copán (Aldana 2007, 185; Stuart 2005, 177). These texts he included in what is now referred to as Palenque Temple XIX, a structure in the tradition of Kan B'ahlam's Triad Group structures, but just outside that plaza to its south. Some twenty years after the completion of Temple XIX, a member of the Palenque almehen and relative to Ahkal Mo' Naab, Chak Nik Ye' Xook, left her hometown and moved to Copán to marry a high-ranking member of the local nobility. Her child was Yax Pahsaj Chan Yopaat, who we just saw took the throne on 9.16.12.5.17 6 Kaban 10 Mol (W. Fash 1991, 153; Martin and Grube 2000, 209). It may well have been from Chak Nik Ye' Xook, Yax Pahsaj Chan Yopaat's mother, that Mak Chanil learned of the 819-Day Count and then included it in Temple 11.

Returning to the narrative in Temple 11, Mak Chanil used this 819-Day Count record as a stop on the way to an unorthodoxly placed full Long Count record: 9.17.2.12.16. The most common practice in the inscriptional record was to use Long Counts at the beginning of inscriptions, and then lead the reader through historical events using time intervals and Calendar Round dates. To have included a full Long Count date in the middle of a narrative, therefore, would have caught the attention of any Classic period reader. It appears that Mak Chanil broke with convention here in order to continue the astronomical theme by including a provocative Lunar Series record. Here Mak Chanil took advantage of the opportunity to poetically describe a "New Moon" (figure 5.6).

> **i-IL-ji** [eroded glyph block] **tu- UH NAH ch'o-ko**, or *i[h]l-[a]j ? t(i) Uh nah. Ch'ok*. See-
> [PASS]-3SA PREP Moon House. Sprout-[INTR]-3SA.

> "The sprouting moon was seen in/from the Moon House."

By itself this is a very nice lunar record; relative to the rest of the text, the passage serves a larger purpose. We may now recognize the narrative of the EN, ES, and WN walls as recording an explicit Venus event, an 819-Day Count station, and then an explicit lunar event—that is, a self-contained set of astronomical statements.*

The final record of the whole set of inscriptions describes an event transpiring on 9.17.0.0.16, recorded by the Calendar Round 3 Kib 9 Pop. The interplay of this historical record with its astronomical reflection occurs in the recognition that the date 9.17.0.0.16

1. was exactly 15 Venus Rounds after the explicit Venus record from Waxaklajun Ub'aah K'awiil's tenure (9.17.0.0.16 − 9.15.15.12.16 = 8,760 = 15 × 584);
2. was the first Full Moon of the eighteenth winikhaab; *and*
3. was the twenty-fourth solar year anniversary of the Venus event.

All three points demonstrate the extended use of the 819-Day Count within astronumerological texts. The third point pushes the practice further. A twenty-fourth anniversary means that it captured the same Venus event as the one recorded from Waxaklajun Ub'aah K'awiil's time. Since Venus events repeat on an eight-year cycle, whatever phase they captured in the early record was commemorated by Yax Pahsaj Chan Yopaat's record. Moreover, we saw in chapter 2 that if we take the Dresden Codex historical records and project them back to the time of Copán, we find that both Waxaklajun Ub'aah K'awiil's and Yax Pahsaj Chan Yopaat's records correspond to first evening visibility events over the western horizon.

In moving through Yax Pahsaj Chan Yopaat's narrative in Copán Temple 11, we return to the larger point raised by the last chapter. The argument within the traditional scholarship on the Dresden Codex requires that the first morning visibility anchor of 9.9.9.16.0 1 Ajaw 18 K'ayab was contrived and did not correspond to what was observable during the Classic period. Mak Chanil's hieroglyphic inscription at Copán, though, stands against that interpretation. Whereas Thompson and Lounsbury argued that the anchor was a numerological fiction useful for connecting a fixed mythological date to observations of the Late Postclassic, the fact that the Copán record and the Dresden Codex record both accord with first evening visibility events suggests that they were both intended to maintain actual observations for posterity. Unless we claim some kind of collusion between Mak Chanil of Copán and Tawiskal Uwoojil in Yucatan across hundreds of years in pushing a numerological agenda, it is much easier to see both as historical records of observed Venus events. Each may have been recorded for political or religious purposes, but each recorded what was visible to observers in the Mayan region.

---

*Unlike other arguments for planetary references in hieroglyphic inscriptions (Aveni 2001, 167–69; Dütting et al. 1982; Dütting 1985; Tate 1985; Closs 1994; Aldana 2007), we do not need to infer astronomy behind a cloak of patterns among dates here. These are explicit.

Furthermore, if we recall that Waxaklajun Ub'aah K'awiil's Venus window opened out onto the west, then we witness a strong metaphor. Yax Pahsaj Chan Yopaat built Temple 11 with cosmic themes that mimicked those of Temple 22, but took the imagery from the inside and put it on the outside. Temple 11, like Temple 22, housed an opening out onto the western horizon, but now as a doorway, not a small window. And in this Temple, Mak Chanil explicitly included a record of Venus's visibility in the west. These come together to suggest that when confronted with Yax Pahsaj Chan Yopaat's political context in the Late Classic, Mak Chanil conceptualized a means of reconnecting with the past. It looks as though the architectural alignment and imagery implied that in some ways the new architecture and its royal patron were meant to "repair" the astronomical/ cosmological link that was broken through dynastic tragedy, inflicted by the defeat of Waxaklajun Ub'aah K'awiil by K'ahk' Tiliw Chan Yopat of Quirigua. In other words, Mak Chanil's efforts do not represent "pure science," but science as applied to religious and political needs (cf. Aldana 2007).

What is very nice about this context is that Yax Pahsaj Chan Yopaat's inscriptions in Temple 11 provide even more information by giving us a subtler understanding of the Venus reference. To see it, though, requires some digging. Here we consider a number of visual, calendric, and distributional abnormalities in the Temple 11 inscriptions to better understand the operative verb in the Dresden Codex Venus Table, *k'al*.

COSMOLOGICAL SYMBOLISM

The artists carving the glyphic narrative into the walls of Temple 11 introduced an abnormality: Mak Chanil designed one-half of the glyphs to be written "backward." That is, Classic Mayan scribes in general worked with a writing system in which most hieroglyphs were asymmetrical and possessed a conventional orientation. For four of the eight texts in Structure 10L-11 (NW, SE, EN, and WS), the standard orientation is reversed (W. Fash 1991, 168; Schele and Freidel 1990, 326–27). Hands usually with fingers pointing to the right now pointed to the left; faces generally looking left now looked right. Some modern scholars have suggested that Mak Chanil created this as a visual pun intended for appreciation by "the gods" (Schele and Freidel 1990, 327).

Yax Pahsaj Chan Yopaat's Temple 11 glyph orientations are odd, but they are not without precedent in the corpus of Mayan hieroglyphic writing. Such a conventional break was used, for example, by Itzamnaaj B'ahlam, Late Classic k'uhulajaw of Yaxchilan. Fixed into the entry doorway to Yaxchilan Structure 23, Lintel 25 was dated 9.12.9.8.1 5 Imix 4 Mak and possessed "mirrored" image glyphs (W. Fash 1991, 168; Tate 1992, 119–21, 204–8). Here at Copán, though, the rest of the architecture provides us with a rationale for this break with convention through yet another anomaly. In fact here there is a reasonable suggestion, as we will see next, that Mak Chanil oriented the glyphs in this way so that they could "face out" of the structure and "look" out of the doorways they

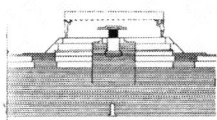

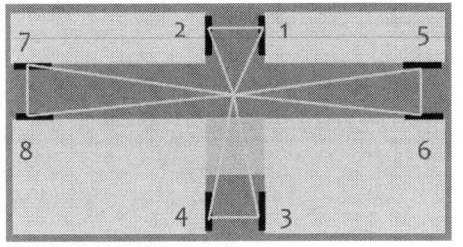

FIGURE 5.8 Reading order of texts in Copán Temple 11

were carved into, out onto the four cosmic regions with which the walls were aligned.

A second unorthodoxy Mak Chanil introduced into the Temple 11 narrative is that the reading order from one wall to the next is not "linear." In order to read these texts "in order," "in succession" (as the glyphs themselves require [*u tz'akaj*]), the narrative forces the reader to follow an interesting path.* Specifically, the narrative covers its historical theme by starting in the NE, then moving to the NW before crossing through the center of the temple, continuing with the SE wall, and concluding with the SW wall (Aldana 2011d; cf. Schele and Freidel 1990, 325–27). The astronomical theme is picked up on the EN wall, and then continued in mirrored fashion, proceeding to the ES before passing through the center of the temple, and then finishing with the WN and then the WS walls (see figure 5.8). The narrative arrangement thus moves the reader through intersecting loops, symmetric about a NE/SW axis. Although it may seem eccentric, there is in fact good reason for this pattern. Anyone wishing to read the narrative as composed would find themselves tracing out a pattern that bears resemblance to a hieroglyph used by other scribes at Copán and elsewhere in the Mayan region. And the layered meanings of this particular glyph have not gone unnoticed.

In his landmark publication *Biologia Centrali-Americana: Archaeology*, published in 1890, Alfred Maudslay drew explicit comparison between the shapes of a specific Mayan hieroglyph and the "cosmograms" built into calendric representations within two Late Postclassic books—the Féjérvary-Mayer and Madrid Codices. Referring to his Plate XXXII (figure 5.9), Maudslay writes:

The Maya calendar occurs in the MS. known as the Codex Cortesianus, and the Mexican calendar is taken from plate 44 of the

FIGURE 5.9 "20" as completion after Maudslay's Plate XXXII

---

*Schele and Freidel state that to read the text, one "moves through the 4 regions . . ." However, they do not point out the specific path through these regions discussed here.

Fejervary Codex . . . it will be seen that the arrangement of the calendar is in exactly the same shape as the sign which I suppose to be the numeral twenty.

If I am right in my suggestion the likeness in the arrangement of the calendar to the numeral twenty would suggest a numeration in scores, which is in accordance with what is stated by the early Spanish writers. (1974, 41–42)

Maudslay argued that the Classic period hieroglyph for "20" is the same shape as graphic representations of the calendar in Postclassic codices from the Aztec and Mayan regions. A closer look at Maudslay's illustrations sheds new light on the concept that informs the verb used in both the Dresden Codex Venus Table and Copán Temple 11. As he points out, there is a superficial connection. Page 1 of the Féjérvary-Mayer Codex contains a picture at the center that bears strong resemblance to the warrior figure on page 49 of the Dresden Codex Venus pages. We saw this figure in this book's introduction, exemplifying Tawiskal Uwoojil's mixing of Central Mexican and Mayan styles. Here, though, we focus on the shape of the overall illustration into another type of "cosmogram" (Aveni 2001, 150–52).

The artist of the Féjérvary-Mayer Codex frontispiece framed the central warrior with four trees, each topped by a distinctive bird and each anchored to one side of a central square (figure 5.10). The Aztec artist depicted these frames in trapezoidal forms, painted in red, yellow, black, and blue, which in turn surrounded each tree and two figures attending it. The colors and symbols at the base of each tree associate each assemblage of imagery with a specific cosmic region: east, north, west, and south.

The Aztec scribe used the trapezoidal frames,* augmented with intercardinal lobes, to create a continuous path around the ritual activity. This path is marked by twelve dots on each line segment with each vertex carrying a Day Sign from the 260-Day Count. Archaeoastronomer Meredith Paxton notes that for an analogous image in the Madrid Codex, the path ends where it begins, with the implication that it should start over again (2001, 33).

For example, the last Day Sign in the upper right intercardinal lobe is the eighth Day Sign of the *tonalpohualli*, the rabbit, Tochtli. With twelve dots counted counterclockwise we return to the first Day Sign, at the base of the upper arm of the cross, Cipactli, where the count began. The straightforward pattern is that any dot (representing a date within the 260-day Count) is its own beginning and end in this cycle. The imagery tells us, therefore, that the 260-Day Count encircles space as it progresses in time. This cosmogram itself depicts a ritual "completion" or "making whole" of space and time. It gives the horizontal counterpart to the vertical cosmogram provided by the inner doorway of Copán Temple 22.

---

*This form is commonly referred to as a Maltese cross in the literature (Aveni 2001), but for those who study such things, this is an incorrect classification. Rather than having flat ends, the cross members of a Maltese cross have "v's" taken out of them, producing a cross with eight very distinct points (and each point bearing symbolic meaning). The cross pattée is a more general form in which the cross members are wider at their ends than at their centers.

FIGURE 5.10 The Codex Féjérvary-Mayer frontispiece. Maudslay recognizes that the path of the 260-Day Count follows the shape of the glyph for 20 as completion. Los Angeles County Museum of Art, courtesy Ancient Americas at LACMA (ancientamericas.org)

As Maudslay pointed out, though, the scribes' depiction of the specific path of this "completion" or "space-time enclosing" is not unique to the Féjérvary-Mayer and Madrid representations. Paxton (2001, 31) extends Maudslay's case to note that it was not just a concept from Classic Mayan communities into Postclassic Mesoamerica; the shape is ubiquitous in Mesoamerica across time and space (Aveni 1980, 155–56; León-Portilla 1988, 65–66; Villa Rojas 1945, 127–34).

In its hieroglyphic form, Maudslay also noted that the "flower" symbol suggested an equivalence between the number 20 and a sense of completion. In a Long Count context, this flower would signify that the period of time was completed. We find at Quirigua and Copán that another glyph substitutes for this flower: a glyph that Thompson descriptively called the "flat-hand" glyph. And it is a version of this "flat-hand" glyph that

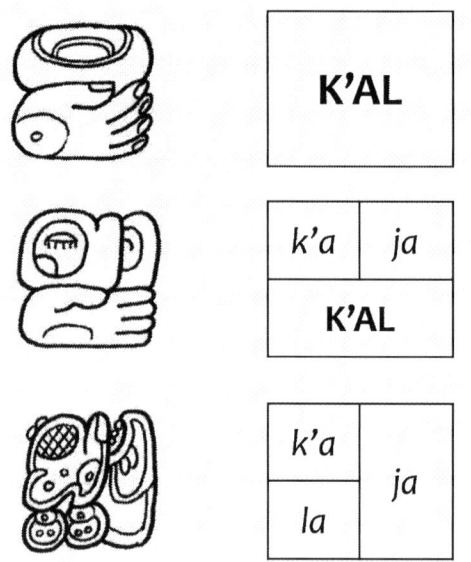

will bring us back to the controlling verb in the Dresden Codex Venus Table (figure 5.11). What we find here is that there is an overall coherence to the symbolism of Yax Pahsaj Chan Yopaat's Temple 11 and the ritual it preserved. Were Yax Pahsaj Chan Yopaat to lead a member of the Copán nobility through a reading of the text, they would have walked out a path mimicking the shape used to represent ritual enclosings of space-time. Each reading, then, constituted a ritual tracing of the cosmic regions from the temple, and each completed reading implied the beginning of another.

It would appear, then, that there was symbolic performative activity here, involving the cosmic regions, given that each doorway opens up onto the cardinal directions. Appropriately, the pattern defined by the reading of the inscriptional narrative engaged the architecture of the building to complement the cosmic

FIGURE 5.11 Hieroglyphic versions of *k'al*: K'AL; k'a-K'AL-ja for k'ahlaj; k'a-la-ja for k'ahlaj, after Stuart 1996, figure 9, p. 156

themes long identified within its decoration (W. Fash 1991, 168; Miller 1988; Schele and Freidel 1990, 323–28; Newsome 2001, 50–51). Whereas the size, placement, and imagery of the celestial and Underworld realms of Temple 11 connected cosmological layers vertically, the reading of the text on the walls in the doorways encompassed the four directions horizontally (figure 5.12).

The two anomalies incorporated into the hieroglyphic inscriptions of the building therefore held meaning. The unorthodox path that Mak Chanil and Yax Pahsaj Chan

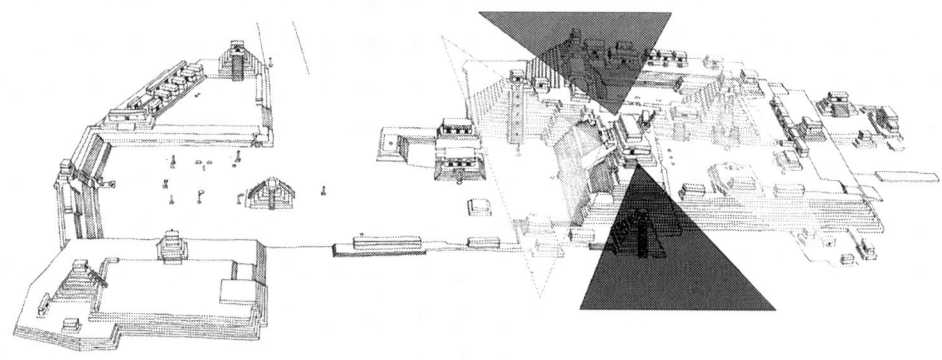

FIGURE 5.12 Temple 11 functioned as the "center" of Copán, both relative to the *k'al* enclosing of the horizontal plane, but also vertically with celestial imagery on its north-facing roof and Underworld imagery on it south-facing stairway. LACMA Schele Drawing Archive Photo by Linda Schele © David Schele, photo courtesy Ancient Americas at LACMA (ancientamericas.org)

Yopaat created to read the narrative in Temple 11 implies the notion of an enclosure of space and time. To read the texts "in order," a reader had to follow the path that also was used to symbolize the cosmic enclosure of space and time. Emphasizing the point, half of the texts were positioned so that the faces within the glyphs look out onto the cosmic regions as the reader encompasses them. The pattern is significantly deeper, though. We have seen that the Féjérvary-Mayer frontispiece implies that the cycles repeat, and that in the Temple 11 narrative the Venus Round of 584 days is repeated numerous times within a time interval to close off / complete the narrative. It turns out that this is also what the text explicitly tells us, providing us with a new, much richer interpretation of both the Copán record and the Dresden Codex Venus Table (Aldana 2011d).

THE HANDS THAT BIND

Recognizing that Mak Chanil included a resonance between the reading order of the text in Temple 11 and graphic depictions of ritual "enclosings" helps us resolve a puzzle that has quietly formed within the interpretation of the Dresden Codex Venus Table over the last one hundred years. This puzzle concerns the verb controlling the Venus record in Temple 11—the same one in the Dresden Codex Venus Table, as noted above. Given our work of the last few chapters, however, we can now move away from approaching the issue as a puzzle missing pieces; now we can work toward considering the question of what Tawiskal Uwoojil intended with his use of a specific verb. While the Copán version has not received much attention, Tawiskal Uwoojil's use of the verb has a long history of interpretation.

As we saw earlier, although they both contributed tremendously to the interpreta-tion of the Venus Table, neither Förstemann nor Seler gave much consideration to the actual hieroglyphic text in it because at their time the writing system was still undeci-phered. Almost seventy years later, though, Thompson's *Commentary on the Dresden Codex* attempted to build on Seler's iconographic inferences through hieroglyphic textual support—that is, Thompson did concern himself with the glyphs.

When it came to the Venus Table, Thompson applied his approach to the verb he catalogued as "713." Glyphs are now still occasionally referred to by their "Thompson numbers," so in this case "T713" was alternately described as the "*il*-hand" verb or the "flat-hand" glyph. But Thompson's efforts were largely for naught; by utilizing a "reading" method that has not withstood the test of time, Thompson inferred the meaning of indi-vidual hieroglyphs rather than linguistically analyzing them (Thompson 1978, 289–96; see also Coe 1999, 140–42). As noted in the last chapter, he proposed numerous "read-ings" for textual passages throughout the Dresden Codex that have not held up against the advancing decipherment. In this case, in his *Commentary on the Dresden Codex*, Thompson wrote: "The *il*-hand of line 15 should have some meaning such as appears, is visible, or influences. The clause would then read 'God X appears? in the East [with?] great Venus'" (1972, 65).

Archaeoastronomers taking over the subject after Thompson have come to similar if not identical conclusions (Aveni 2001, 194; Bricker and Bricker 2007, 109; Paxton 2001, 80) for the interpretation of Tawiskal Uwoojil's Venus records.* Certainly, this is due at least in part to the recognition that the calendric structure along with the elements of the hieroglyphic text makes it intuitive to follow Thompson's logic. The astronomical pattern strongly implies an "appears" reading for T713. By the 1980s and 1990s, however, Tatiana Proskouriakoff's decipherment intervention blossomed, generating countless new attempts to address old interpretations and to revise them drawing from linguistic evidence.

In 1983 Linda Schele and Jeffrey Miller suggested two semantic values for T713. Schele was just at the time making a name for herself, having attended the first Palenque Round Table almost ten years earlier, which resulted in her collaboration with Peter Mathews and Floyd Lounsbury on the reconstruction of the Palenque dynasty, as we saw in the last chapter. She was active in the small cadre of scholars concentrating on moving the decipherment forward using dictionaries, living Mayan languages, and even (a relative novelty at the time) computers.[†] As a practicing artist retrained as an art historian at the doctoral level, Schele has contributed a tremendous repository of illustrations that already have influenced generations of scholars.

By the early 1980s, for their first attempt at it, Schele and Miller suggested that the primary value for T713 came directly from the back of the hand, which they took to be *pach* in Yucatec (or in Cholan and Tzeltalan languages *pat*), meaning both "the back or end of something" and "to form/make something" (1983, 36). They recognized that while the hand repeated in numerous different instances of the glyph, an object set atop that hand varied significantly. Their approach allowed them to adapt the reading according to the variable element "held" in the glyphic hand (figure 5.13). In so doing they also brought the astronomical cycles of the Moon into the consideration.

As the main sign in Glyph C of the Lunar Series and as a period-ending glyph for various parts of the katun, T713 appears to have meant "the end" of the particular cycle as marked by the glyph appearing over (or in) the hand. In accession phrases, it appears to have been read as the verb "to make." Paraphrased, T713

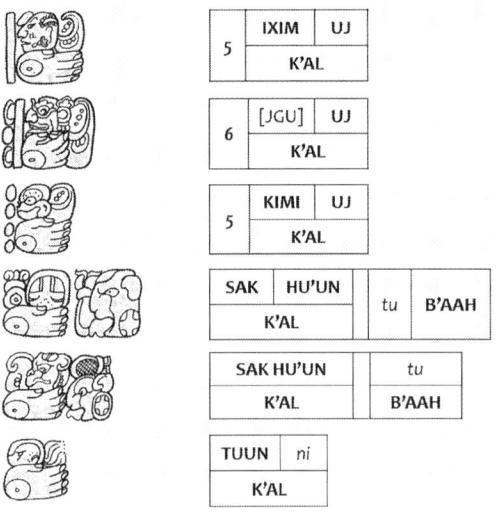

| 5 | IXIM | UJ |
| | K'AL | |

| 6 | [JGU] | UJ |
| | K'AL | |

| 5 | KIMI | UJ |
| | K'AL | |

| SAK | HU'UN | | tu | B'AAH |
| K'AL | | | | |

| SAK HU'UN | | tu |
| K'AL | | B'AAH |

| TUUN | ni |
| K'AL | |

FIGURE 5.13 *K'al* within various phrases: In the Lunar Series: *k'al 5 Ixim Uj; k'al 6 JGU Uj; k'al 4 Kimi Uj.* In k'uhulajaw accessions: *k'al sak hu'un tu b'aah; k'al sak hu'un tu b'aah.* In period ends: *k'altuun*

---

*Paxton writes: "These dates are associated in lines 15–19 with a verb that evidently means 'appear' (or something similar)" (2001, 80).

[†]Peter Mathews notes that this use was not always felicitous.

would read "(he) was made . . ." with the glyph for the title assumed placed over (or into) the hand. (Schele and Miller 1983, 36)

Schele and Miller were here appealing to an aspect of the glyphs that would eventually become familiar: polyvalence. It is now well attested that Classic Mayan scribes put some glyphs to multiple purposes (Coe 1999, 234). This may seem odd to the reader, but it frequently occurs in English as well. For one, the word "can" can represent a verb or a noun. Even more interesting, in English the letter "e" is sometimes pronounced as part of a word, and sometimes it only functions to change the value of another vowel in the same word. In hiero-glyphic writing, when used as a phoneme, T528 held the value **KU**, but it could also hold the logographic value **TUUN** (figure 5.14). For Schele and Miller, the principle of polyvalence allowed them to argue that T713, too, held different meanings depending on its context.

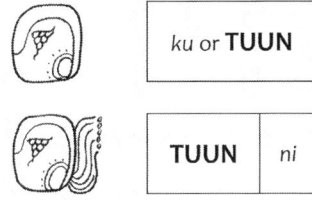

FIGURE 5.14 **KU** vs. **TUUN**

Schele took a similar approach in her collaboration with Yale art historian Mary Miller not long after her work with Jeffrey Miller, suggesting that in the Dresden Codex Venus pages the T713 hand held a "mirror." They went on to combine an argument for the mirror as *nen* with an interpretation of the role it played to suggest that now the glyph read *nentah*, "to shine" (1983, 19). This reading fell in line with the traditional archaeoastronomical interpretations of the text: "shining" versus Thompson's "appearing." Yet Schele expressed clear reservations, reflecting a concern with the specific semantic value and not just general interpretation:

> The primary meaning of the *nen* compound on the Venus pages is perhaps "shine," but since Venus is invisible in two of its four phases, the glyph is more likely to mean "succeeds in office," referring to the successive initiation of each of the four phases. (Schele and Miller 1983, 20)

Schele's assessment here presaged the eventual decipherment of T713, not through its intuitive connection in the Dresden Codex, but through its tendency toward more emphasis on linguistic coherence over interpretive convenience.

Schele stepped up her analysis over time, later turning to the verb's other contexts. In a number of phrases, Barbara MacLeod found that in context the verb could be followed by a reflexive clause, *tu b'aah*, "on oneself" (Schele 1999, 36) (see figure 5.13). MacLeod was a student of Schele's, contributing to the lively epigraphic scene in Austin, Texas, through her expertise in the linguistic study of contemporary Yucatec Mayan along with her fascination with exploring caves. In her work on the T713 verb, MacLeod appealed to the numerous instances of the verb on ceramic drinking vessels within the "Primary Standard Sequence," which led her to suggest a reading of "to raise" or "to present" (Stuart 1995, 107). These readings, however, also were based on interpretations of actions, not on phonetic substitutions; therefore, they still did not constitute a secure decipherment.

According to Schele, it was not until "1995 [that] all the parts came together and yielded a reading" (1999, 36). Schele attributed the actual decipherment to physicist and Mayan astronomy enthusiast Werner Nahm's identification of a phonetic substitution (**k'a-la-ja**) at Chich'en Itza (1999, 36). The same year saw the completion of David Stuart's dissertation, "A Study of Maya Inscriptions," culminating a critical step along his trajectory under an international spotlight generated by his MacArthur "Genius Grant" fellowship, which he took to Princeton University. He was seventeen when he received the award.

By the time he completed his dissertation, Stuart had already presented at international conferences among the leading scholars in Maya archaeology—himself at the forefront of epigraphy. With his father, George Stuart, a photographer and editor for the magazine *National Geographic*, Stuart grew up steeped in records left by ancient Mayan scribes and even lived at the archaeological site of Coba in the northeastern Yucatan Peninsula during his youth. David Stuart combined his lived experiences with his powerful intellect to produce a landmark dissertation in the field (Coe 1999, 231–44). In it, he treated the verb in question here, T713, in its accession context:

> *K'al* means "to bind, fasten" in Yucatec, which would be appropriate in this particular context. The entire accession statement, with the prepositional phrase **tu-ba-hi** (*t-u-bah*, "to, on himself") may read something similar to "the paper headband was fasten[ed] on himself." (Stuart 1995, 204)

Unfortunately, the published literature lacks clarity on the issue of primacy, since Stuart's monumental 1996 article "Kings of Stone" claims that it is the first to formally publish the decipherment of a use of *k'al*. Referring to the interpretation of a glyphic compound made up of T713 and the glyph for *tuun* or "stone," Stuart wrote:

> The *k'altun* reading deserves some comment, as it has not been published before now. In the inscriptions of Chichen Itza and northern Campeche, the hand-mirror verb is replaced by the syllabic sequence **k'a-la**, suggesting the logographic value **K'AL** (fig. 9a). In another much earlier inscription from the Bonampak region, the hand 'holds' the syllable **k'a**, where it presumably serves as a phonetic complement in **k'a-K'AL**. (1996, 155)

Here, we leave the determination of scholarly precedence between Nahm and Stuart to the fog of history, but it is worth noting that this example attests to the rapid pace of decipherment rippling through the epigraphic community during the 1980s and 1990s. It also speaks to some of the difficulty that modern epigraphers encounter in working toward a glyph's syllabic decipherment and an argument for its "meaning."

TABLE 5.2 Definitions of *k'al*

Yucatecan (Barrera Vásquez 1995, 367–68)

[1]K'AL	veinte; veintena, cuenta de veinte
[2]K'AL	cerrar con cerradura y encerrar y atrancar y detener encerrado
[2]K'ALAH	clausurar; encerrar, encajonar
[4]K'AL	abrochar cosa que encaja como botones o corchetes
[6]K'AL	armar lazo, abrochar como jubones

Ch'orti' (Wisdom 1950, 83, 145)

k'ar	"getting, holding on, retention (as of a bad sickness)"
k'ar u ya'bich	"retention of urine (lack of flow)"
k'ar	"opening fissure, interior of neck, throat"

Ch'orti' (Hull 2005, 68)

-k'ar	clasificador numeral para contar un veinte de algo [palabra Antigua que casi no se usa actualmente]

TABLE 5.2 Data used by Kaufman (2003) to reconstruct proto-Mayan *k'al*

Awakateko	**k'al**	lo amarro
Itzaj	**k'a'"a"li**	se cerro
Mopan	**k'aali**	se cerro
Teko	**k'aalo7**	amarrado
Mam	**k'alo7n**	amarrado
Mam	**k'alb'il**	pañuelo
Ixil	**k'alb'al**	cinturón

Note: There is considerable time depth to some of these glosses as evidenced by Terrence Kaufman's linguistic reconstruction work.

In any case, agreement among scholars that *k'al* is the verb root of T713 does not by itself settle its semantic value. Tables 5.2 and 5.3 show that we are still left with numerous possible glosses.*

Several of these glosses refer to "tying," which Stuart (1996) picked up in his "Kings of Stone" article, yielding the intuitive appeal of *k'al sak huun* as "tying on a white headband." Stuart took this further to persuasively suggest a "stone-binding" reading of *k'al-tuun* (1996), which in turn has led to a general acceptance of *k'al* as "to tie, to bind" (Lacadena 2004, 174; J. Robertson et al. 2004, 266; Stuart 2007). On the other hand, when it comes to the verb's association with Venus in the Dresden Codex, these analyses

---

*Here we focus on Yucatecan and Ch'orti' because the latter is the closest extant language to "Classic Ch'olti'an" and the former is recognized as having crept into the Dresden Codex (Lacadena 2004; Wald 2004). Also, by following the practice of recognizing Classic Ch'olti'an as the language of the inscriptions, I am making a statement about the linguistic camp I am associating with, as there is still nontrivial debate on the issue. In this matter, I am most influenced by the work of John Robertson, Stephen Houston, and David Stuart, complemented by that of Alfonso Lacadena and Marc Zender.

leave us with the relatively straightforward question of whether or not "being tied in the East" is the same thing as "appearing in the East."

There are other problems with each of the above readings when extended to the verb in the Venus Table. For *k'al* as "to appear," in the first instance, we would have to acknowledge some level of conceptual inconsistency. As Schele noted, two out of four *k'al* records on each page correspond to the periods when Venus is in conjunction with the Sun and so not visible at all. As Schele noted in 1983, referenced above, it is difficult to accept that the scribe referred to Venus's periods of *in*visibility as "shining" in the north or south. She felt that there must have been a more abstract notion of *k'al* that was opaque to us, perhaps connected to rulership. Similarly, the "tying" or "binding" meaning of *k'al* relative to Venus would have to refer to some obscure relationship since it does not seem intuitive for Venus to be tied to each of the cosmic regions. As such, it does not have much to recommend for or against it.*

In his treatment of the Venus pages within the canonical archaeoastronomy text, *Skywatchers*, Anthony Aveni provides an illustration that gives the "reading" of the Venus Table text in operation. Here Aveni translates the lower section of hieroglyphs as "On 7 Kan in the East reappears Venus from the South having been absent 8 days" (Aveni 2001, 194). In his narrative description within the same book, however, Aveni follows the translation of Schele and Nikolai Grube (1997) for the upper text. In this case, Aveni gives the first part of the text as "tied to the east is Kaktonal Great Star" (2001, 193). In the same treatment, therefore, he has captured both readings in the literature of his time—"Venus appearing" and "Venus tied"—without considering them to be in conflict.

Recently I have shown that resolution comes from the same source material, interpreted through a slightly different perspective. When I first started working on this verb, I discussed it with a Yucatecan colleague, Pedro Pablo Chuc Pech, in Popolá—a village not five kilometers from the larger city of Valladolid (figure 5.15). Chuc worked for decades as a bilingual (Mayan–Spanish) instructor in Yucatan and eventually as an assessor of bilingual instruction. He and I came into contact while I was in graduate school through a common friend and our mutual interest in Yucatec Mayan literacy.

Over the course of his career, Chuc was disappointed that the Mexican educational system used the Mayan language only as a crutch. It was used in elementary school to transition students into speaking Spanish, which then would be used in the rest of their education (pers. comm., 1997). Chuc became invested in preserving literacy in Yucatec Mayan and worked with the Mexican national organization for arts and culture,

---

*The problem arises because Venus is not just "tied to the east"; it is tied sequentially to the north, west, and south as well. Each page has Venus associated with a *k'al* event in each of the four cosmological regions, and in each one, a different deity name is paired with Chak Ek'. This much is evident from Figure 5.17, and was recognized by Förstemann and Seler over a hundred years ago. If we follow the logic of the traditional model, this would identify Venus with different deities in each cosmic region. As noted above, though, a complication arises in that there is only one column on each page dedicated to the other directions, yet there are two *k'ahlaj lak'in* statements describing Venus in the east. And the planet Chak Ek' is paired with different deities in each of these statements.

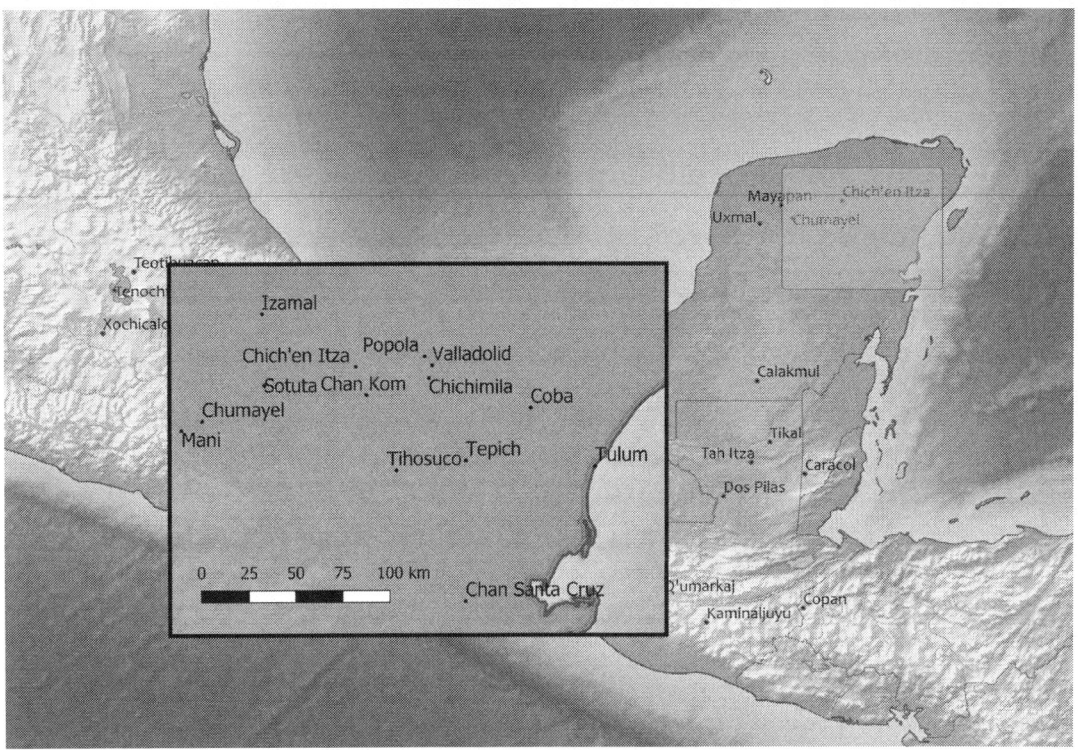

FIGURE 5.15 Popolá, Valladolid, and Chich'en Itza in northern Yucatan

CONACYT, to publish books of local legends in both Spanish and Yucatec toward that end. Recognizing our common interests, he and I began putting on workshops for the Popolá community to augment Yucatec literacy with Mayan hieroglyphic writing, but we also used our time together to delve into all matters relating to ancient and modern Mayan cultures.

During the summer of 2009, on a lunch break from one of our workshops, Chuc and I sat with his son, Bolay, in the shade of a tree at a wooden table in his household patio—an open space framed by three household structures, as traditional in the region. On the patio's northern side was Bolay's room—a thatched roof structure, with cement block walls. On the patio's western side was the kitchen—also a thatched roof, but with walls made of sticks and branches. On its southern side was a large cement block structure with a tiled floor. During our wide-ranging conversation in that patio, I asked them both what they might use the verb *k'al* for and if they would use it in a sentence for me—for comparison against the records in the Cordemex Dictionary. Bolay responded first: "Es como un corral, ¿no, Papá?" to which his father replied: "Si: tu k'alaj, es decir 'fué encerrado'" (field notes 2009).

Chuc's response, not surprisingly, was directly in line with one of the definitions of *k'al* in the Cordemex Dictionary. The full entry reads:

K'AL 3: armar lazos, ballestas 4: armar lazo 2. **K'ALAH** 3: V. **k'al** 3. **K'AL CHULUL** 2, 5: armar lazo para iguanas 4: k'al a chulul tu hol aktun: arma tu lazo en la boca de la madriguera 6: armar lazos a pájaros 8: vn armar el arco de la flecha, el lazo de la iguana 4. **K'AL K'AAN** 1: lazo que está armado ya; ti' biní u naabte u k'al k'aan: es ido a requerir sus lazos. (Barrera Vásquez 1995, 368)

My conversations with Chuc along with definitions coming from the Colonial period *Motul Dictionary* and *Diccionario de San Francisco* began to pull the verb's meaning away from "tying" or "binding" per se, although maintaining a connection. *K'al chulul*, for example, was glossed as "to arm a lasso-for-iguanas"—in other words, to set a snare or trap. That would mean that we now have *k'al* as "to arm" in addition to the other meanings, although further consideration also suggested an important subtlety.

The example sentence garnered by Juan Pio Perez from a native speaker around 1850 is itself provocative: *K'al a chulul tu hol aktun*—"arm your snare around the entrance of the iguana's burrow." Arming here is very clearly setting a snare—forming a loop with a cord, "encircling" the entrance. This example actually brought a new perspective to the meaning of the verb. Certainly it still resonated with the notion of "tying," but "encircling" shifted the emphasis. Furthermore, from this definition we find that there is a sense of "arming weapons" with bearing on our understanding of *k'al*.* Closer attention to the dictionary demonstrates that the specific examples given for *k'al* as "arming" refer to weapons that are "armed" by a particular means (see table 5.4).

The specifics of the weapons make it clear that we do not have to reach far for a productive interpretation. Each of them is "armed" by closing a loop with a cord of some kind. The bow and the crossbow both are "completed" by closing off an open loop with a cord. Arming these weapons constitutes the forming of a closed loop. These also seemed well connected to Chuc's suggestion that *k'al* was used to corral or "encircle" animals, not tie them up.

The parallel between the Cordemex example phrase and the hieroglyphic phrase that MacLeod and Schele first worked with—the "tying of the headband" phrase—is also straightforward and extends the argument (table 5.5).

Under this comparison, then, *k'al* is not a *tying* of the white headband; *k'al* is the encircling of the white headband around the ruler's head. In its iconographic forms, the knot serves as a symbol for the closing of a loop—not for the process of tying. More importantly, the resulting reading of *k'al* as "to encircle" or "to enclose" now becomes very productive relative to its other contexts, including those cosmological. The meaning that emerges is that *k'al* is the encircling or enclosing that holds important things together as exemplified by the glosses in tables 5.2 and 5.3. This meaning works on both everyday practical levels as well as abstract cosmological ones.

---

*Recently, however, Harvey and Victoria Bricker along with Gabrielle Vail have added a new proposal. They appeal to a Yucatec gloss of *k'al* as "to arm" and suggest that it refers to Venus as a warrior deity being armed for the attacking of its victims. It is also worth noting that another possibility is that we might conclude that these are simply homophonic terms of significantly different meaning.

TABLE 5.4 *K'al* and weapons

Term	Definition	Translation	Source
K'AL	armar lazos	to arm lassos	Vocabulario de Viena (17th century)
K'AL	armar ballestas	to arm crossbows	Vocabulario de Viena (17th century)
K'AL	armar lazo	to arm a lasso	Diccionario de San Francisco (maya-español) (17th century)
K'AL CHULUL	armar lazo para iguanas	to arm a lasso for iguanas	Motul II (español-maya) (16th century); Diccionario de San Francisco (español-maya) (17th century)

TABLE 5.5 Parallel sentence constructions based on *k'al*

Verb	Absolutive	Preposition	Possessed Object
k'al	*a chulul*	*t(i)*	*u hol aktun*
encircle	*your lasso*	*around*	*its burrow opening*
k'al	*sak huun*	*t(i)*	*u b'aah*
encircle	*the white headband*	*around*	*his head*

K'AL AND MAYAN ASTRONOMY

As mentioned above, Classic period scribes used the verb *k'al* in the Lunar Series of Initial Series dates. While Teeple had cracked the numbers to reveal the lunar records, he did so without reading the glyphs in the Supplementary Series. We can now go back to Teeple's treatment of its most common form. Within the Supplementary Series, we saw in the introduction that the *ti'huun* glyphs accounted for Glyphs G and F for time-keeping and astrological purposes. In chapter 2, we saw that the Lunar Series comprised Glyphs E, D, C, B, X, and A. Glyphs D and E refer to the Moon Age, giving *X huliiy*—"X days ago" and "it arrived." The last component, comprising Glyphs X, B, and A, gives X_n *u k'aab'a 29*, or "X_n is the name of the 29 [days]," and where X_n varies accordingly. Of primary import here, though, is Glyph C.

As we saw in chapter 1, Teeple showed that from a computational perspective, Glyph C counts periods of up to six moons each, with each lunar deity governing (five or) six moons before passing on duties to the next lunar deity (Schele et al. 1992; Linden 1991; Aldana 2006; see also Teeple 1931). We also saw that in the Dresden Codex Tawiskal Uwoojil wrote out day tallies of 148 (= 3 × 30 + 2 × 29 = 5 moons) or 177 (= 3 × 30 + 3 × 29 = 6 moons) in his Eclipse Table. From a strictly computational perspective, then, Glyph C corresponds to the tracking of the completions of twenty-nine- or thirty-day lunar periods.

When we turn to the hieroglyphic text, we can now transcribe the glyph block as **u**-[#]-**K'AL**-[patron]-**UH**. Note that here the **K'AL** hand "holds" the head of a lunar deity or a

26 days ago	arrived
the completion of the 4th Kimi Moon	X-vi
is the sprout name of	the 30 days

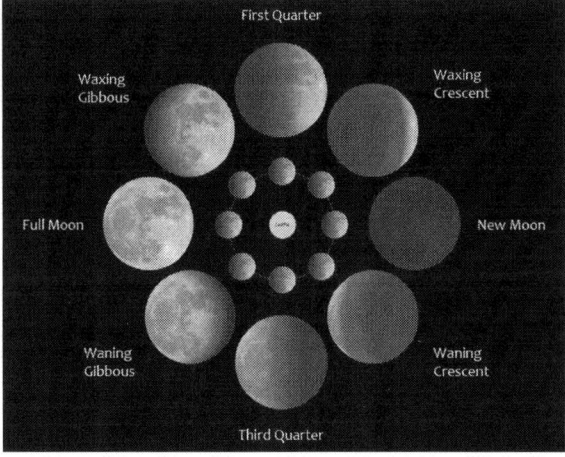

FIGURE 5.16 (*top*) Lunar series translation as a couplet; (*bottom*) lunar phases. Twenty-six days after New Moon would be waning crescent, near the next New Moon. Photo courtesy Nancy Aldana

deity patron instead of a stone, a white headband, or a "celt"* (Schele et al. 1992; Linden 1991; Aldana 2006). With the *u* prefix, this can be read as the completion of the nth Patron Moon, or in the example of figure 5.16, the completion of the 4th Death Moon. That would yield an overall translation of the Lunar Series as a couplet (Hopkins, pers. comm., 2018).

26 days ago, the completion of the 4th Death Moon arrived. X-vi is the name of the 30 days.

Accordingly, the Lunar Series provides compelling evidence that *k'al* was used to refer to the completion of (sub-)periods of larger cycles (i.e., each of six *k'al* events transpires before the next deity's *k'al* events begin).

The upshot of this exploration of lunar records is that the "enclosing" referred to by *k'al* could be physical *or* temporal. Moreover, we find that scribes referred to the completion of time periods beyond temporal completion of Moon Ages, to address the Sun as well. In Yucatec, for example, we encounter the term *k'al k'in* "9: todo el dia 13ddp: todo el dia completo" (Barrera Vásquez 1995, 370). A secondary definition, **K'ALAB K'IN**, helps with the analysis. According to Barrera Vásquez (1995, 1), *-ab* is a verbal suffix, generating the passive form in the incompletive aspect. *K'alab* in modern Yucatec, therefore, plays the same role as *k'ahlaj* in Classic Ch'olti'an. Here, though, the absolutive of the verb is *k'in*, which is itself a bit ambiguous. *K'in* could refer to the Sun; it could stand for day; or it could refer to time itself. The Yucatec definition of *k'alab k'in* provides a solution: Barrera Vásquez gives "todo el dia completo"—"all of the completed day" (1995, 370), so that the referent is the time period, day, and *k'alab* here is "completing" it.

The point of this contextualization, then, is to suggest that "tying" or "binding" are too specific as translations of *k'al*. A more abstract notion of "encircling," "completing," or even "enclosing" would actually fit scribal use of both the astronomical contexts and the royal ritual events. Moreover, a more abstract interpretation would resonate with the codex records, illustrating the calendric enclosing of both time and space.

---

*This celt is what Schele and Miller identified as *nen*, "mirror" (T614) (1983, 20).

BACK TO THE DRESDEN CODEX VENUS TABLE

This all brings us back to Tawiskal Uwoojil's work on Venus periods in the Dresden Codex with a new opportunity. As noted by Förstemann, in each column of the Dresden Codex Venus Table the "flat-hand" verb is followed by one of the cosmic regions/cardinal directions. The sequence starts in the north with what we can now read as *k'ahlaj nohol*, *k'ahlaj ochk'in*, *k'ahlaj xaman*, and *k'ahlaj lak'in*. For most of the twentieth century, this was read as "appears in the north," "appears in the west," "appears in the south," and "appears in the east." The reading presented here suggests otherwise. Now we have the passive verbal phrases: "the north is enclosed . . . the west is enclosed . . . the south is enclosed . . . the east is enclosed." And this reading of hieroglyphic text appears to be precisely what is depicted graphically on the Féjérvary-Mayer frontispiece (figure 5.17). Such a reading pushes our new interpretation even further. The second part of each hieroglyphic sentence gives the two entities that enclose it—the name of a deity and Chak Ek', the name of Venus, paralleling the two deities in each of the four lobes of the Féjérvary-Mayer cross. Finally, the dates at the top of each of the Venus pages play the role of the Day Signs on the Féjérvary-Mayer frontispiece.

Through this interpretation, the Féjérvary-Mayer frontispiece does with an image what the Dresden Codex Venus Table does with hieroglyphic text. Both documents say the same thing.

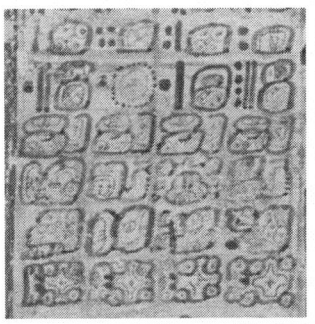

5 Lamat	4 Etz'nab	7 Lamat	2 Kib
11 Sip	1 Mol	6 Wo	14 Wo
is enclosed	is enclosed	is enclosed	is enclosed
North	West	South	East
Te Ajaw	Kimil	K'awiil	Jun Ajaw
Chak Ek'	Chak Ek'	Chak Ek'	Chak Ek'

FIGURE 5.17 In the Dresden Codex Venus Table, the text describes *k'al* enclosings of the cosmic directions, whereas the time-space enclosing is depicted visually in the Codex Féjérvary-Mayer. Courtesy SLUB

It is now worth stepping back to note that a coherent and consistent method has emerged concerning Mesoamerican astrological practices. The underlying motivation is the interpretation of a meeting of two entities, one of which is celestial and the other "cosmological." For Glyphs G and F—that is, the ti'huun practice of telling time—the "star of the night" meets each of nine different deities, with the zenith meeting generating the omen of the night. For the Lunar Series, the Moon in lunar patron deity form meets the deity who is the "name of the month." For Waxaklajun Ub'aah K'awiil's Venus window, Venus meets each of the Day Signs of the 260-Day Count specifically in the west. And finally here, in the Dresden Codex, Venus meets each of twenty deities in cycles through the four cosmic regions. We will see in chapter 6 that each enclosing also generates a set of omens, listed on the right-hand side of each page in the Venus Table. The point is that Mayan astronomy appears to have been the quantitative arm of the interest in predicting meetings between celestial bodies and cosmological deities. As concerned meetings of k'uhulajaws from different cities, or meetings of other authority figures, such meetings had layered meanings to be prepared for and interpreted. In astronomical cases, the specific celestial body provided the periods of interest—and so the opportunity for scientific investigation—while the cosmological deities provided the oracular meaning.

Furthermore, with this perspective in hand, and following up on Seler's work, we can even go further with this reading of k'al to explain the fact that Tawiskal Uwoojil provided two grammatically equivalent statements with different deity names for the same cosmic region on each page of the Dresden Codex Venus Table—an inconsistency not addressed by other scholars.* Here, we focus on a comparison of the Féjérvary-Mayer frontispiece with the Dresden Venus pages.

(OSMI(VIOLEN(E

We have already examined the center of the Féjérvary-Mayer frontispiece, which depicts a warrior deity holding a spearthrower along with a pair of darts. Here we also note that the Aztec scribe illustrated this warrior's victim ripped into pieces, with body parts in the intercardinal regions (Boone 2010, 115–16; V. Bricker 2010, 318). A close look reveals that, in fact, it is the victim's blood that follows the dots of the 260-Day Count to provide the pattée cross pattern.

The Mayan version of this figure, painted by Tawiskal Uwoojil on page 49 of the Dresden Codex, also holds a spearthrower and darts in the same position as the Féjérvary-Mayer warrior. We saw this identification in chapter 1 within Seler's comparison of Tawiskal Uwoojil's illustrations to those in the Borgia Codex. Also, the figure below

---

*There is a complication here in the Brickers' interpretation. If we take k'al as simply "to arm," then each of the four columns should describe an "arming." But they don't depict that. We have seen that the upper register is the deity of the east column and is described with Chak Ek' with the same verb, k'al, but there is no hint that this figure is a warrior or arming a weapon.

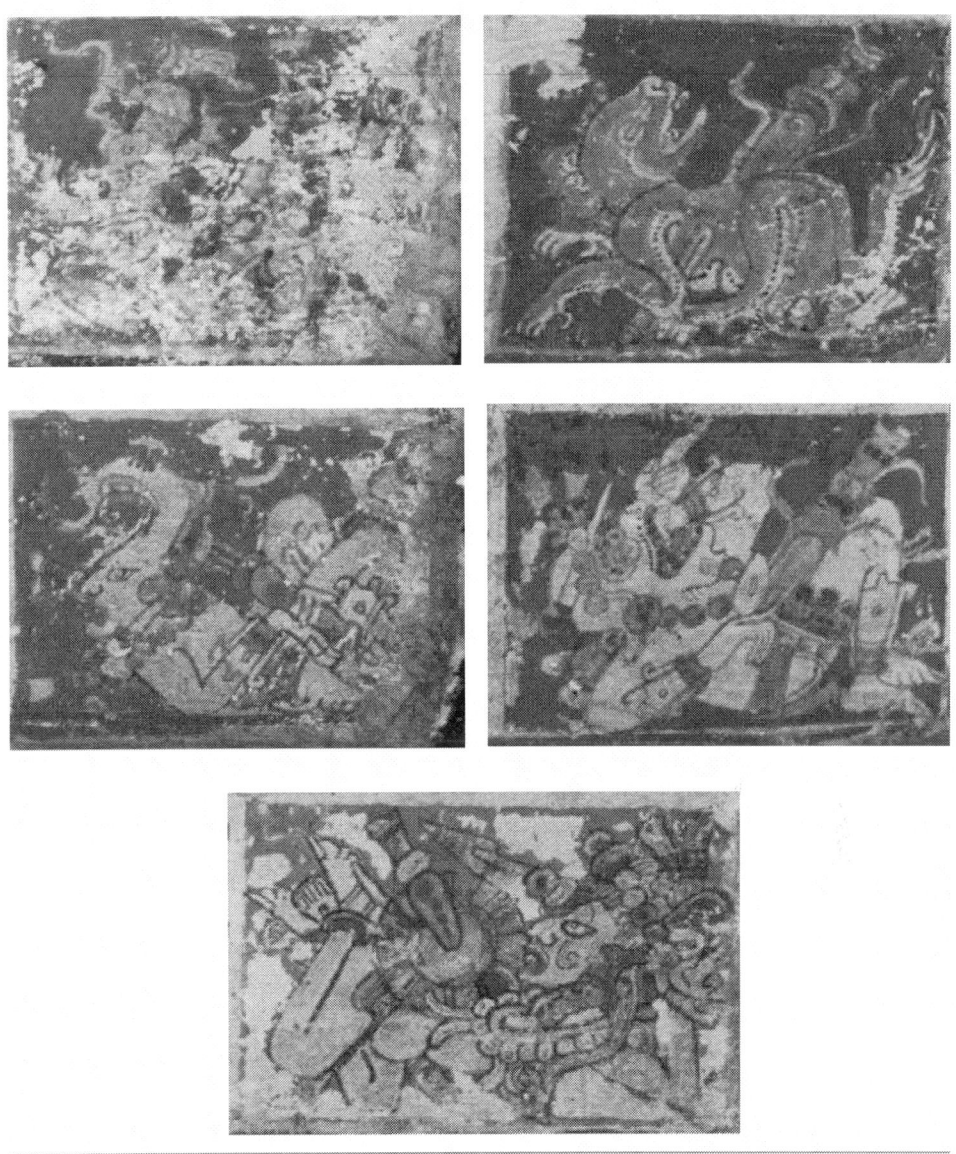

FIGURE 5.18 The five victims that are speared in the Dresden Venus pages. Courtesy SLUB

each warrior figure in the Dresden Venus pages is identifiable as his victim since it has a dart lodged in its body (figure 5.18). This complex is what Thompson reinterpreted as a representation of Venus itself as a warrior. It also provides the inspiration for the Brickers' reading of *k'al* as "nocking an atlatl," corresponding to one version of the dictionary definition of "to arm" (Bricker and Bricker 2011, 197). What we note here, however—and what neither Seler nor Thompson had access to—is that the hieroglyphic text accompanying this warrior and actually naming the figure with the atlatl provides a second instance of the phrase *k'ahlaj lak'in Y Chak Ek'*. Each of the Venus pages, then, possesses the statement "k'ahlaj lak'in Y Chak Ek'" twice, but each with two different values for Y for the same Venus Round.

If we stick with Y as the name of Venus (Bricker and Bricker 2011, 197; Thompson 1972, 67), then we are left with Venus being two different deities at once, which is challenging to make sense of. On the other hand, the sentence makes ready sense if we see *k'al* as "to enclose." The enclosing of the east within a Venus Round completes the enclosing of both that specific cosmic region and the entire pattern. Two things are enclosed when the entire pattern is completed; two different deities are involved. The interpretation is sound for either the Féjérvary-Mayer frontispiece or the pages of the Venus Table. For the Féjérvary-Mayer frontispiece, the spearing of a victim occurs when the entire pattern is completed, and that happens when the east is enclosed. Completing the east is also completing the entire sequence around the cosmic regions. For the hieroglyphic version, two statements are required: one referring to the enclosing of the east and one referring to the enclosing of all four cosmic regions.

When we put these lines of investigation together, we confront a coherent new reading of the Dresden Codex Venus Table that doesn't have to differentiate between Venus "appearing" or Venus "being tied." Now we can see each statement as the driving force behind the whole table, which is perhaps why it also shows up in the preface to the table, on page 24. Each *k'al* statement corresponds to Venus completing one of the well-defined parts of its synodic cycle, and it tells us which deity was responsible for maintaining that period with Venus. Our reading of the verb also becomes grammatically consistent. First, each of the cosmic regions (east, north, west, and south) "is enclosed" by Venus. Then, the return to first morning appearance constitutes an enclosing of all four regions. The enclosings of each individual region correspond to Venus's association with one deity; the completion of the full cycle corresponds to an association with another. The whole table, then, prescribes the tracking of these cosmic enclosing events through the continuous progression of Venus subperiods.

This new perspective of Mayan astronomy brings us back to the beginning of the chapter. We now find continuity between the description of Chak Ek''s movement by Tawiskal Uwoojil in the Venus pages and that by Mak Chanil in Copán Temple 11. Mak Chanil included a historical Venus record using *k'al* along with dates marking multiple completed Venus periods in the hieroglyphic text of Temple 11. In order to read this text, the viewer had to follow a prescribed path through the building, following the shape of the hieroglyph that substitutes in meaning for the *k'al* hand glyph. Each of the doorways containing the inscriptions opened onto a different cosmic region, which would also have corresponded to different audiences. The north opened up onto the huge public plaza, the south onto the smaller elite plaza, and the east and west each faced Venus in its distinct realms of visibility as Morningstar and Eveningstar. Because he was also specifically referencing the Venus observations of his troubled predecessor, it is possible to read this as an attempt to re-create a new cosmic order at Copán—one that repaired the damage done to the dynasty by the defeat of Waxaklajun Ub'aah K'awiil. And one that required a repeating maintenance of a cosmological contract between Yax Pahsaj Chan Yopaat as k'uhulajaw and the entities of the local cosmos.

If we take the hint from Copán, then, the emphasis of the Dresden Venus pages was not on creating a scientific ephemeris—it was not an attempt to isolate a set of "pure" scientific records. The attempt was to enable a set of ritual events guided by the movements of Venus. This verb, *k'al*, brings us back to procession and community activity, celebration and ritual contracts. Mayan astronomy is pulled away from being a project abstracted from culture and is placed back in contact with daily lives and experiences. Moreover, textual procession through Temple 11 and the verb in these Venus pages suggest that there is much greater continuity between the astronomy of the Late Classic period and that of the Terminal Classic and that the apparent discontinuity between the astronomy in the Dresden Codex and that of the Late Classic is an artifact of modern scholarship on the manuscript. Working our way through the architecture at Copán also suggests that the Long Count anchor to the Venus Table was most likely not a purely numerological convenience. The use of historical observations at Copán hints that we may want to consider seriously that the Venus Table's anchor was also the result of historical observation.

It may be, then, that the use of the Dresden Codex Venus Table within the battle over the Calendar Correlation Problem skewed its interpretation to create an illusion of discontinuity between the Classic and Postclassic. In this chapter, we have turned to the hieroglyphic record itself to find that—as Thompson noted emphatically—the primary interest was in bringing together observation with the regular progression of the calendar. Breaking Venus's movements up into regular subcycles and then coordinating these with deities and dates was the driving force—not a proto-scientific "accuracy" of predicting first morning visibilities.

It is important to emphasize, then, that according to this interpretation neither the accuracy of prediction nor the warning of an event was the centerpiece of Tawiskal Uwoojil's Venus Table. Certainly large disparities would have been detrimental and potentially could thwart the whole project, so a degree of accuracy was important. On the other hand, small discrepancies would have been challenging to predict on a case-by-case basis and so would have been expected. What the table needed to be effective was a long-term accuracy, but short-term regularity that didn't violate the basic parameters of observation. What, then, was the purpose of subperiods if it wasn't highly accurate prediction? Here again we find that the clues are in the manuscript itself.

CHAPTER 6

Incensarios and the Public Life of the Venus Table

With the results of the last chapter, we should now be able to see that in the text of the Venus pages of the Dresden Codex Tawiskal Uwoojil was recording ceremonies connected to a long tradition of space-time enclosing religious practices. These types of practice certainly go back to Mak Chanil in the Late Classic at Copán, but also as far back as the origins of the Lunar Series at the end of the Preclassic. The Venus record we reviewed at Copán, then, was not a bare "data point," but provided information about Chak Ek' within ritual and historical contexts. To have used the same verb and grammatical structure in the Venus pages, Tawiskal Uwoojil would have been familiar with such ceremonial activity and accordingly would have held other responsibilities in line with Landa's description of indigenous priests:

> The office of the priest was to discuss and to teach their sciences, to make known their needs and the remedies for them, to preach and to publish the festival days, to offer sacrifices and to administer their sacraments. The duty of the *Chilans* was to give the replies of the gods to the people and so much respect was shown to them that they carried them on their shoulders. (Tozzer 1941, 111–12)

This may have us speculating whether or not Tawiskal Uwoojil—as the author of numerous pages in the Dresden Codex and likely the supervisor of the ceremonial activity it recorded—was sufficiently experienced and well regarded so as to have been carried on the shoulders of local citizens. Certainly he would have been a close witness to a multitude of ceremonies even when he was not the priest in charge of their proper performance. In these duties it should not be hard for us to imagine him walking within processions through the city, surrounded by festivity.

We are not left to our imagination alone, however, in reconstructing these processions, or to projections back in time based on modern indigenous ceremonies. To get at a better sense of these activities, we may turn to the origins of Tawiskal Uwoojil's text, now as narrative and not just as interpretation or "explanation" of calendric and astronomical patterns. We therefore take the next natural step from having focused on the operative verb in the last chapter to seek perspective on those rituals for which he was in charge. Our first step in seeing the lived experience of the Venus Table starts with this reading as narrative.

THE VENUS TABLE AS NARRATIVE

To begin his reading on page 46, Ajk'uhuun Tawiskal Uwoojil first turned to the left-hand side to assemble the Calendar Round date. Addressing the text in vertical columns, he combined the first element in the column of 260-Day Count dates with the 365-Day Count date that follows. Here the first such date is 3 Kib 4 Yaxk'in (figure 6.1). Next our scribe encountered the sentence of primary concern in chapter 4. On this date "the North is enclosed by Ulum and Chak Ek'." The time interval at the end of this string he read, then, as a statement: "236," which we may translate as "It is 236 days."* Through Förstemann's early work, we know that this is the time that has passed since the end of the planet's previous first morning appearance; it also refers to the length of time it takes for the north to be enclosed by the motion of Venus. Its meaning, therefore, is more fully translated as "It is 236 days since the last Venus *k'al* event." Since each Venus Round is completed at first morning appearance—or, as we have seen, first morning visibility serves as the anchor to the whole Table—this sentence could also be understood as "It is 236 days since the previous first morning appearance of Chak Ek'."

The remaining text in the first column provides an alternate phrase to be used when the whole table is "corrected." We will explore this further when we return to the corrections on page 24 in chapter 10. For now, we can take this whole lower portion to function like the table of 260-Day Count dates at the top—the reader uses only the upper one or the lower one, depending on what is appropriate at the time (see figure 6.1). Tawiskal Uwoojil's entry into the table therefore read:

> On 3 Kib 4 Yaxk'in, the North is enclosed by Ulum and Chak Ek'. It is 236 days since the previous first morning appearance of Chak Ek'.

Our scribe's attention would then move through the next three columns in sequence, culminating with the east:

---

*Stand-alone nouns in Mayan sometimes may be translated as "stative" phrases or sentences in English with the prefatory phrase "it is."

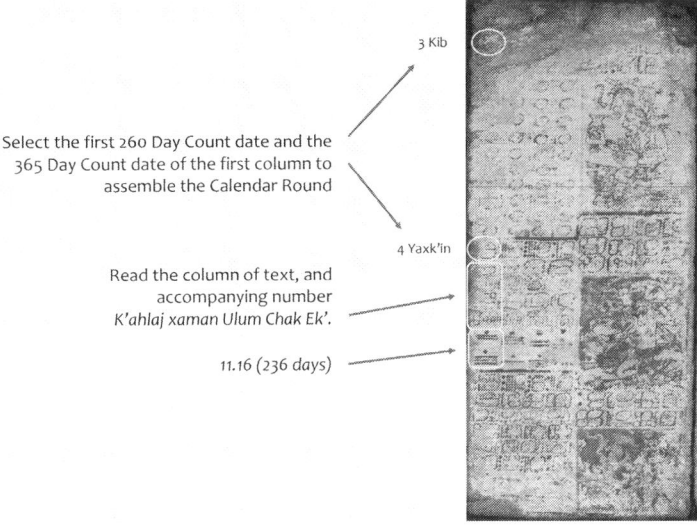

3 Kib

Select the first 260 Day Count date and the 365 Day Count date of the first column to assemble the Calendar Round

4 Yaxk'in

Read the column of text, and accompanying number *K'ahlaj xaman Ulum Chak Ek'.*

11.16 (236 days)

FIGURE 6.1 Reading the Venus Table as narrative, Tawiskal Uwoojil pulls from the dates to assemble: "On 3 Kib 4 Yaxk'in the North is enclosed by Ulum and Chak Ek'. It is 236 days since first morning appearance." Courtesy SLUB

On 2 Kimi 14 Sak, the West is enclosed by Sinaan and Chak Ek'. It is [now] 326 days. On 5 Kib 19 Tsek, the South is enclosed by Chak Tul and Chak Ek'. It is 576 days. On 13 K'an 7 Xul, the East is enclosed by Kimil and Chak Ek'. It is 584 days.

With the decipherment of the hieroglyphic script and with more attention to the text itself, we can now see that the left-hand side of the page narratively describes each of the four cosmic regions being enclosed in progression, taking a full Venus Round period of 584 days to go from one first morning appearance on one page to the next on the following page.

This reading of the left-hand side of page 46 results from only a slight revision in the interpretation of its grammar, but it also presents new possibilities for the overall interpretation of the Venus Table. As we have seen in previous chapters, by the end of the twentieth century, accuracy of prediction and relatively esoteric alignments of structures formed the core of the consolidated modern scholarly interpretation of the Venus Table and its purpose. These attributes would have fit within the activities of the elite—possibly even inaccessible to the public at large. Such interpretations might have been aligned better, for example, with Waxaklajun Ub'aah K'awiil's private consultation of his Venus window than the public performances of Yax Pahsaj Chan Yopaat's at Temple 11. In this chapter, we will find that by bringing the new reading of the left-hand sides of the Venus pages into dialogue with the imagery and text of the right-hand sides we encounter a significant role of the public in the commemoration of Chak Ek', along with what they witnessed relative to this narrative.

In order to consider the inclusion of the public, at the most basic level, certain characteristics would have been necessary. The public's engagement in ceremony would have required that the table be of sufficient accuracy to prevent the rituals of each subperiod from slipping out of phase. If someone were to notice that an evening visibility were being commemorated when Chak Ek' was already Morningstar, that would be a problem. On the other hand if Chak Ek' were visible for a day or two in the east at sunrise before the ritual commemoration of Venus as Morningstar, that would not present a problem and most likely would not even be noticed by members of the public. This latter point is worth emphasizing.

Precise first and last morning visibilities of Venus would be restricted to those members of society who had access to views of the horizon from the tops of the structures that pierced the floral canopy. The vast majority of the population would have had views of the sky from the open plazas down below. Accordingly, first and last morning visibilities would occur several days after or before (respectively) these events as they were visible from the highest temples. Accommodation for this differential visibility would therefore have been necessary at least at some level in order to engage the public.

Returning to the text with these considerations in mind, we move on to the material suggesting public participation within events associated with the Venus Table. Examining both sides of each page, we find that Ajk'uhuun Tawiskal Uwoojil organized the narrative on the right-hand sides of pages 46–50 differently from the table structure of the left. The top of the right-hand side of the page, above the first illustration, he filled with hieroglyphs, organized in conventional double columns and zig-zag reading order rather than the straight columns of the left-hand side (figure 6.2). Unfortunately, the top of each page is heavily damaged—in part from normal wear and tear, but also significantly from the bombing of Dresden in World War II (Thompson 1972, 19). Almost completely illegible on page 46, the glyphs that are visible suggest that its content was very similar to the lower sections of text, with an emphasis on the phrase *u muuk*, which we will address below.

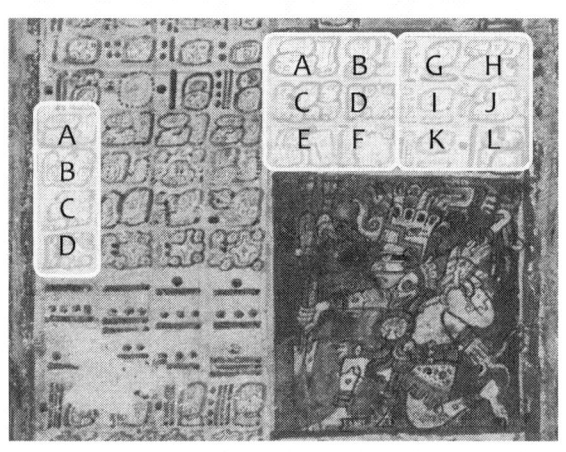

FIGURE 6.2 Reading order of right-hand side vs. left-hand side of pages 46–50. Courtesy SLUB

Interspersed within the text, Tawiskal Uwoojil painted three illustrations on the right-hand side of the page, each taking up the space of a grid of four columns and three to four rows of text each. We have already seen that, in the bottom two images, Tawiskal Uwoojil painted warrior figures along with their victims. The uppermost illustrations on the right-hand sides of each page provide the core interpretive material for this chapter. Exploring them extends the connections between the Dresden Codex Venus Table and the Féjérvary-Mayer frontispiece recognized in chapter 4; they also introduce a connection between the Venus Table and the material culture of the Postclassic, through which we encounter the public's participation in Venus rituals and aspects of Tawiskal Uwoojil's daily life.

CERAMIC EFFIGIES AND THE UPPER IMAGE

In his *Commentary on the Dresden Codex*, Eric Thompson recognized sky bands within the imagery of the upper right-hand illustrations of the Venus pages (Thompson 1972, 67). This imagery we have already seen in chapter 2 as a border for the Piedras Negras rulers' accession thrones. As the Piedras Negras sculptors did for their rulers, Tawiskal Uwoojil placed all of the figures in his illustrations seated on sky-band thrones (figure 6.3). Scholars since Thompson have taken this depiction as evidence that they are celestial deities, or deities actually located in the celestial realm (Bricker and Bricker 2011, 189). Thompson also noted that the fact that they are interacting with Venus on a periodic basis suggests that these figures represent "constellations" or asterisms in the night sky (Thompson 1972, 67). Each *k'al* event, then, would put Venus in conjunction with the recorded deity as constellation. While appealing on an intuitive level, the calendric shifts specified by the lower text (as we will see in chapter 7) work against such an interpretation (Aldana 2011d).

It is also worth noting that Tawiskal Uwoojil did not choose unique deities for this set of illustrations; we find the same ones showing up in other almanacs within the Dresden Codex. Within these other records, there have been no arguments for their representations as constellations (figure 6.4). Like the Aztec Lords of the Night, then, the deities named in the Venus pages appear not to be directly tied to observable celestial phenomena, but instead to play symbolic cosmological roles. We saw in this book's introduction that the Lords of the Night were both well-recognized deities, and they were non-physically tied to the seats they occupied as part of the ti'huun astrological project of keeping time at night. Rather than look for an observation-based mechanism behind the records, we here may allow for symbolic roles and interactions as well. And for this, we return to further details within the illustrations themselves.

We noted in the last chapter that Tawiskal Uwoojil recorded that five different deities somehow participate in the *k'al* rituals on each of the Venus pages. He associated one deity for each of the north, west, and south directions and then two different deities for the east. In the last chapter, we concluded that the two for the east corresponded to the fact that the enclosing of the east also completes the full Venus Round. A comparison with the Féjérvary-Mayer would suggest that each of the first four figures should be in the trapezoidal "petals" of the *k'al* "flower," with the fifth as the central warrior figure.

FIGURE 6.3 Uppermost illustrations on Dresden Codex pages 46–50. Courtesy SLUB

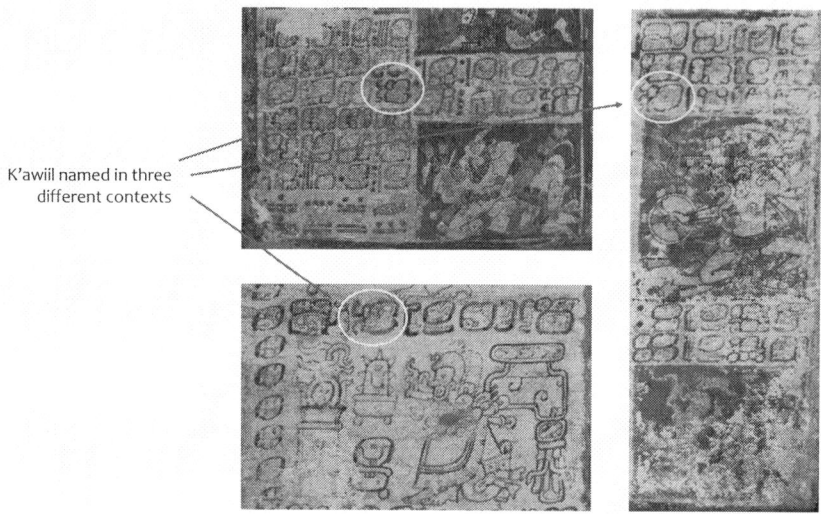

K'awiil named in three
different contexts

FIGURE 6.4 Deities named in the Venus Table and in other almanacs throughout the Dresden Codex. Courtesy SLUB

For the fourth figure—the one in the east—we now find that Tawiskal Uwoojil provided additional information.

It is not intuitive to a modern reader, but for the east the names of the deities listed in the fourth column on each page correspond to the full-figure representations that Tawiskal Uwoojil painted in the top right illustration of the next page (Thompson 1972, 68; Bricker and Bricker 2011, 189–94). On page 46, for example, the fourth column gives the name of a Death God, Kimil, acting in the east with Chak Ek'. Then on page 47 we find a full image of Kimil sitting on a sky-band throne at the top right of the page. This pattern works for every page, with 4-Pawahtun in the east in the text of page 47 and an illustration of 4-Pawahtun at the top of page 48; Xajaw Uj (or the Moon Goddess) in the east with Chak Ek' on page 48 and illustrated on page 49; Jun Ajaw with Chak Ek' on page 49 with Jun Ajaw on page 50; and, bringing it all back around, the Celestial Dragon with Chak Ek' on page 50 and his illustration on page 46.

That there is a shift between page text and image must strike the reader as odd, but the problem goes away with the lower text. In the lower text, the deities have been shifted such that the names and their illustrations show up on the same page. The implication is that the lower text results from some kind of shift in practice relative to the upper text, and in his copying of the table Tawiskal Uwoojil made sure that the images accorded with the shift, not the upper material. We will explore this shift in detail in chapter 10 where we discuss the long-term correction of the table and its use at the late Postclassic city of Mayapán.

Returning to the exploration of the imagery, we notice that Tawiskal Uwoojil depicted Kimil on page 47 and the other deities on these pages in animated positions, sitting on celestial thrones and holding some type of ritual object in one hand. For our purposes, it is this handheld object that turns out to be key.

In 1930 Carlos Villacorta and Jose Antonio Villacorta Calderon published a hand-copied and commented version of the entire Dresden Codex. In Jose Antonio's commentary, he proposed that the handheld object in these Venus pages might be a cup of some kind. Concerning the image of Kimil (or Schellhas's "God A") he wrote:

> Una divinidad se haya sentada al estilo asiático sobre los signos astronómicos de Júpiter y Marte. Lleva en la mano derecho una copa espumeante. (Villacorta and Villacorta [1930] 1990, 95)

He seemed to stretch the Asian connection a bit too far for the image on page 46, though, as he wrote:

> Divinidad sentada al estilo asiático sobre signos astronómicos del Sol, la Luna, Venus y Marte. Adorna su cabeza suntuoso tocado y lleva en la mano una copa, en la que introduce su trompa la cabeza de un elefante. (Villacorta and Villacorta [1930] 1990, 93)

Fortunately, we can now iconographically identify the headdress as directly in accord with the glyph naming the deity. This image represents the Celestial Dragon we saw in the cosmogram at Copán in chapter 4 and not some member of an errant lost tribe of elephants from across the Pacific. The point, though, is that Villacorta saw these handheld items as cups filled with liquids and possibly beverages.

In his own commentary, Thompson took a different approach and suggested that some of these handheld objects might be rattles (1972, 68). Fortunately, we can now go beyond iconographic speculation; we now have the archaeological record to turn to, which resolves the ambiguity. Specifically, we can now turn to the mid-twentieth-century Carnegie-sponsored excavations in Yucatan to find that the handheld objects in the upper images take the same form as unadorned ceramic incense burners identified by yet another Harvard archaeologist, Robert Smith (1971, 15, 47, 48) (figure 6.5).

Robert Eliot Smith and his younger brother, Augustus Ledyard Smith, graduated from Harvard in the 1920s and began their archaeological careers at the central Guatemalan Classic period site of Uaxactun (Willey 1988, 684). In 1951 Robert Smith moved from ceramic analyses at the site where Ricketson excavated the first solar-year-marking E-Group to the massive Carnegie Institution project at the Late Postclassic Yucatecan site of Mayapán. Like Chich'en Itza, Mayapán finds representation both in the historical documents of the early Colonial period and through modern archaeological excavation. As noted above, we will return to this site in chapter 10 as the probable home of Tawiskal Uwoojil. Here it is important to note that the ceramic forms that are ubiquitous at Mayapán find substantive precedence at Chich'en Itza and other sites from the Terminal Classic and Early Postclassic as far away as Central Mexico and Oaxaca (R. Smith 1971, 253).

Among the ceramic artifacts he analyzed, Smith collected a large proportion of *incensario* or "incense burner" or "censer" fragments. At the bottom of some of the complete tall, vase-shaped censers, Smith recovered the residue of incense (R. Smith 1971, 15).

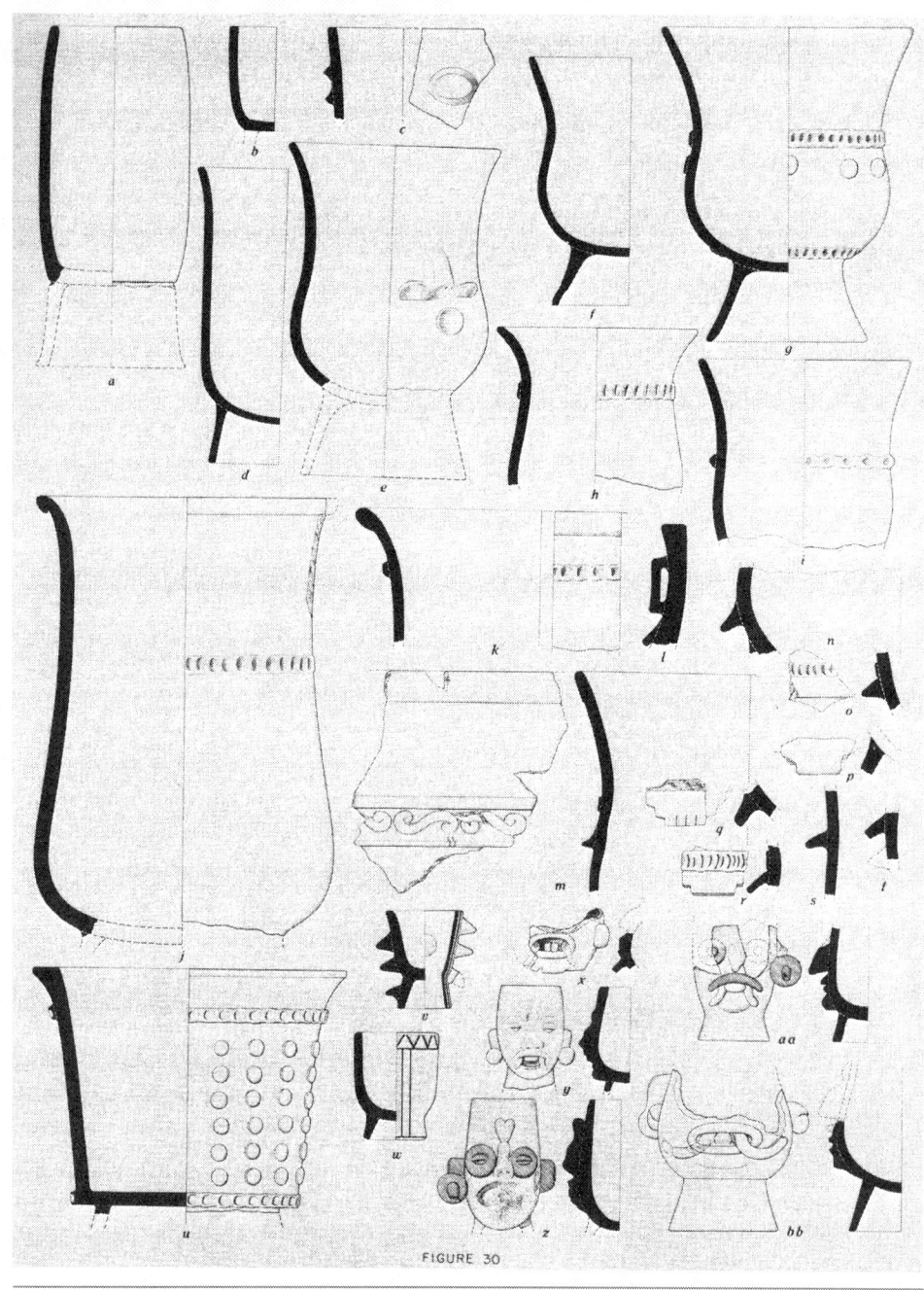

FIGURE 30

FIGURE 6.5 Smith's illustrations of pottery forms found at Mayapán (R. Smith 1971)

As noted in the introduction to this book, incense burning was a ritual activity with great time depth in Mesoamerica. Hieroglyphic statements describe rulers engaged in incense burning as a critical part of *k'al* period-ending rituals during the Classic period (see figure 3.2). For his dedication of the Triad Group structures at Palenque for the 9.13.0.0.0 8 Ajaw 8 Wo period end, for example, Kan B'ahlam included in his list of titles *ch'ahoom*—"person who censes" (Tokovinine 2013, 59). Given the vast number and elaborate construction of incense burners at that city, such would have been a particularly poignant reference (Miller and Martin 2004, 205).

We also encounter the mythological origin of the ritual use of copal incense in the sixteenth-century *Popol Vuh* recovered by Brasseur de Bourbourg. In the mythological portion of the text, the authors tell us that rather than be sacrificed and her heart offered to the Lords of the Underworld, the young mother of the Hero Twins tricked the Lords by sending them a ball of incense in the shape of a heart. Her captors and liberators tell her that they will "deliver this apparent duplicate of your heart before the lords" (D. Tedlock 1985, 116).

> And when they came before the lords, they were all watching closely:
> "Hasn't it turned out well?" said One Death.
> "It has turned out well, your lordships, and this is her heart. It's in the bowl."
> "Very well. So I'll look," said One Death, and when he lifted it up with his fingers, its surface was soaked with gore, its surface glistened red with blood.
> "Good. Stir up the fire, put it over the fire," said One Death.
> After that they dried it over the fire, and the Xibalbans savored the aroma. They all ended up standing here, they leaned over it intently. They found the smoke of the blood to be truly sweet. (D. Tedlock 1985, 116–17)

Thereafter incense as the blood of trees becomes the proper offering to the Xibalban Lords of the Underworld. Xk'ik' makes clear that this is the motivation of the deception:

> "And hereafter, as for One and Seven Death, only blood, only nodules of sap, will be theirs. So be it that these things are presented before them, and not that hearts are burned before them. So be it: use the fruit of a tree." (D. Tedlock 1985, 116)

The suggestion at this point is that Tawiskal Uwoojil painted the deities depicted at the top of the Venus pages each with an incense burner in hand. Accordingly, they would have been part of a ceremonial activity with broad purchase throughout Mesoamerica.

What makes Smith's work on the ceramics at Mayapán particularly useful here is that he found another pottery form associated with the plain vase incense burner. The potters of Mayapán additionally produced numerous vases of this shape that served as bases for extremely elaborate ceramic effigies of deities (figure 6.6). These were quite complex constructions; some were fashioned with built-in tubes so that incense burned in the bowl would generate smoke coming out through the effigy's hands (R. Smith 1971,

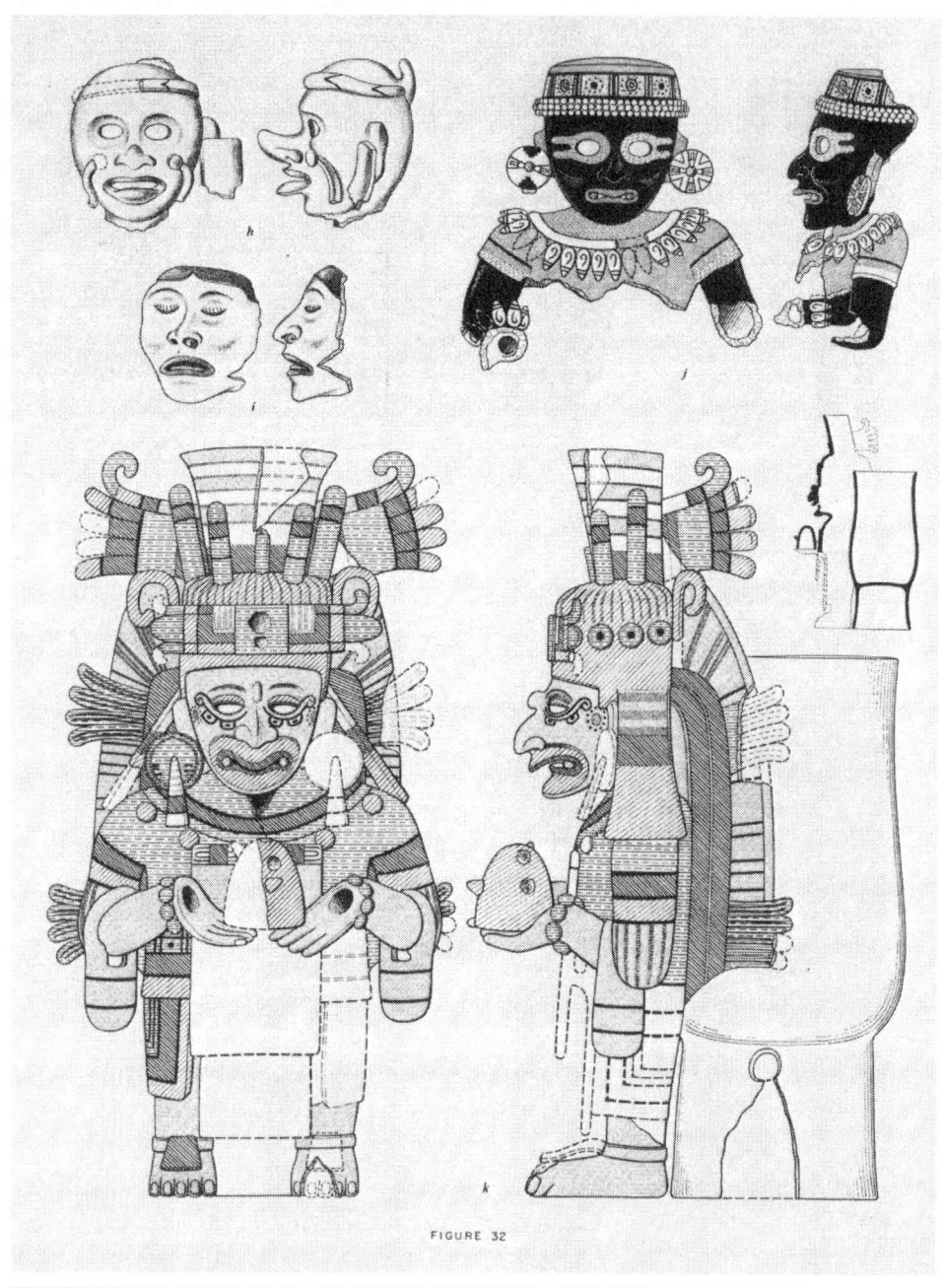

FIGURE 32

FIGURE 6.6 Smith's illustrations of full-figure effigy censers from Mayapán (R. Smith 1971)

100–15). We do not have to go far to find a description of just such ceramic figurines in Landa's manuscript recording Yucatec ceremonial traditions.

In *Relación de las cosas de Yucatan*, Landa describes the making of "idols" for annual ritual activity. In his account of the rites performed for each month of the 365-Day Count, Landa writes that for the month Yax "[artists] renewed the idols of clay and their braziers; for it was the custom that each idol should have its little brazier in which they should burn their incense to it" (Tozzer 1941, 161). In his rich annotation of Landa's manuscript, Alfred Tozzer commented—even before such figurines had been archaeologically excavated—that throughout the text "the idol and its brazier are not clearly differentiated. It is possible, in some cases, that the two are combined" (1941, 161n840). He goes further in a footnote to Landa's references to indigenous "gods":

> In all but two of the dozen or more places where the idols of these deities are mentioned they are called "images," and we learn that the image is "a hollow figure of baked earth." (Tozzer 1941, 137n646)

And he states that

> we read in a *Relación* that copal was burned in the clay idols: "They worshipped some idols made of clay like small jars and pots of sweet basil (with) . . . faces made on the outside of them. They burned in these a resin called copal of a strong odor." (Tozzer 1941, 110n 502)

These descriptions accord extremely well with the incense burners from Mayapán.

In accord with Landa's description and Tozzer's inferences, therefore, we confront the nuance of Tawiskal Uwoojil's illustrations in the Venus pages. In particular, we may infer that the illustrations in the upper images on pages 46–50 actually depict burning ceramic effigy censers. In other words, Tawiskal Uwoojil probably did not intend to depict observable celestial phenomena with these illustrations, as purportedly implied by the sky band; rather, he appears to have painted the scenes in the Venus pages as animated representations of the artifacts that Smith excavated from Chich'en Itza and Mayapán and that Landa described as having witnessed in the sixteenth century. The upper illustrations, then, do not depict strictly abstract, immaterial deities, but physical objects burning incense, witnessed by the public.

Evidence for precisely this convention for representation can be found elsewhere in the Dresden Codex. Ceramic deity effigies involved in space-time ritual activity also show up in the "New Year pages" of the manuscript, which record a set of ritual activities that have a long history of study within Mesoamerican religions. By turning to comparison with other codex records, we push forward the opportunity of seeking to better understand the Venus pages by contextualization relative to other indigenous documents.

THE NEW YEAR PAGES

A comparison of the artwork and glyphic handwriting throughout the Dresden Codex suggests that Tawiskal Uwoojil did not paint only pages 24 and 46–50 of the manuscript while in his study. He also appears to be the scribe who painted pages 25–28 (figure 6.7). These pages have been long recognized in modern scholarship as describing and illustrating activities that intricately link space and time (Thompson 1972, 89); they are not, however, commonly recognized as connected to the Venus pages. In part, this is likely because in the New Year pages there is no record of explicitly astronomical phenomena.

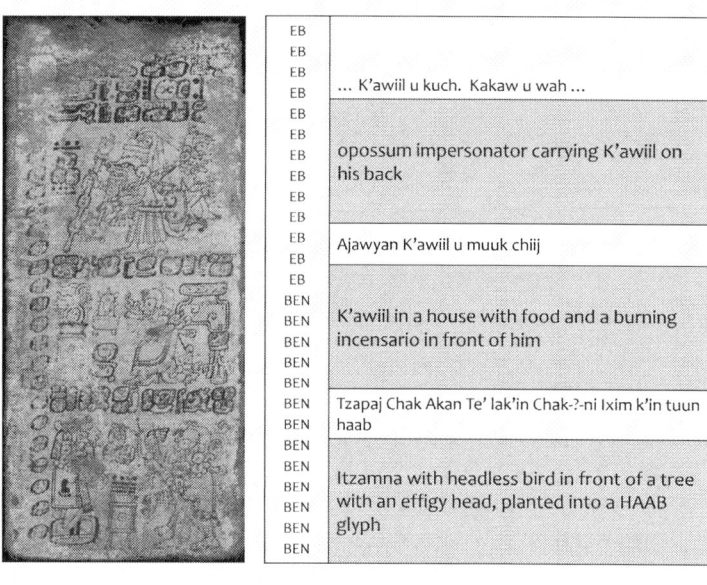

EB EB EB EB EB	... K'awiil u kuch. Kakaw u wah ...
EB EB EB EB EB EB	opossum impersonator carrying K'awiil on his back
EB EB EB	Ajawyan K'awiil u muuk chiij
BEN BEN BEN BEN BEN	K'awiil in a house with food and a burning incensario in front of him
BEN BEN BEN	Tzapaj Chak Akan Te' lak'in Chak-?-ni Ixim k'in tuun haab
BEN BEN BEN BEN BEN BEN	Itzamna with headless bird in front of a tree with an effigy head, planted into a HAAB glyph

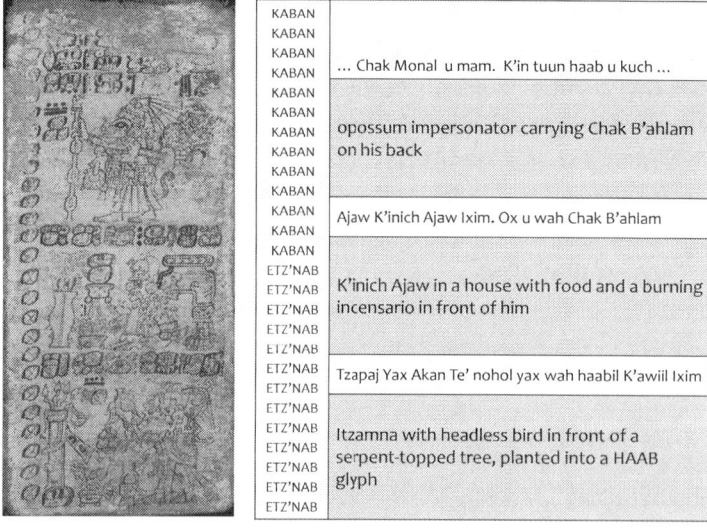

KABAN KABAN KABAN KABAN	... Chak Monal u mam. K'in tuun haab u kuch ...
KABAN KABAN KABAN KABAN KABAN	opossum impersonator carrying Chak B'ahlam on his back
KABAN KABAN KABAN	Ajaw K'inich Ajaw Ixim. Ox u wah Chak B'ahlam
ETZ'NAB ETZ'NAB ETZ'NAB ETZ'NAB ETZ'NAB	K'inich Ajaw in a house with food and a burning incensario in front of him
ETZ'NAB ETZ'NAB ETZ'NAB	Tzapaj Yax Akan Te' nohol yax wah haabil K'awiil Ixim
ETZ'NAB ETZ'NAB ETZ'NAB ETZ'NAB ETZ'NAB	Itzamna with headless bird in front of a serpent-topped tree, planted into a HAAB glyph

FIGURE 6.7 Pages 25–28, the Dresden Codex "New Year pages" (*a, b: above; c, d: facing page*). Courtesy SLUB

Accordingly, we here confront a different kind of ritual continuity between the Classic period record in Copán Temple 11 and the ceremony of the Venus Table. By exploring the connection between the Venus pages and the New Year pages, we find that Tawiskal Uwoojil and his colleagues were drawing from a deep record of ritual activity that demonstrates an underlying sameness to these, the Copán Venus record and numerous calendric ceremonies.

Pages 25–28 are referred to in the modern scholarly literature as the New Year pages or as documenting the Wayeb Rites (Thomas 1882, 56, cited in Thompson 1972, 89). Here again interpretations have been anchored primarily to calendric pattern

C

IK'	
IK'	
IK'	
IK'	... u kuch lemhaab ?-na ... yax ti' wah ...
IK'	
IK'	
IK'	opossum impersonator, standing in a cenote, carrying Ixim on his back
IK'	
IK'	
IK'	
IK'	Ajawyan Itzamnah
IK'	
AK'BAL	
AK'BAL	Itzamnah in a house with food and a burning incensario in front of him
AK'BAL	
AK'BAL	
AK'BAL	Tzapaj Yax Akan Te' chik'in Kimil yax haabil k'in tuun haabil
AK'BAL	
AK'BAL	
AK'BAL	Kimil with headless bird in front of a serpent-topped tree, planted into a HAAB glyph
AK'BAL	
AK'BAL	

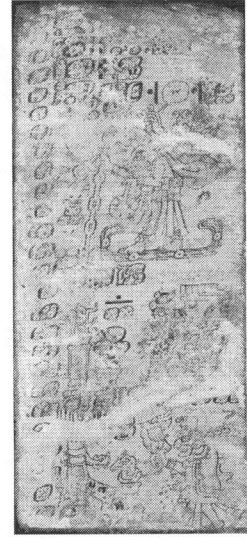

D

MANIK	
MANIK	
MANIK	
MANIK	... Ek Monal u mam Kimil u kuch kimil haabil ...
MANIK	
MANIK	
MANIK	opossum impersonator, standing in a cenote, carrying Kimil on his back
MANIK	
MANIK	
MANIK	
MANIK	Ajaw 4-[Deity] yotoch
MANIK	
LAMAT	
LAMAT	Ak'ab-[Deity] in a house with food and a burning incensario in front of him
LAMAT	
LAMAT	
LAMAT	Tzapaj Yax Akan Te' Ixim yax haabil Itzamnah
LAMAT	
LAMAT	
LAMAT	Itzamnah with headless bird in front of a serpent-topped tree, planted into a HAAB glyph
LAMAT	
LAMAT	

FIGURE 6.7 *(continued)*

  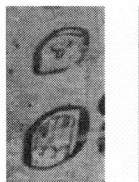 

Page 25	Page 26	Page 27	Page 28
Eb, Ben	Kaban, Etz'nab	Ik', Ak'bal	Manik, Lamat

FIGURE 6.8 Day Sign transitions on pages 25–28. The 365-Day Count selects four of these as "yearbearers": Imix, Ik, **Ak'bal**, K'an, Chikchan, Kimi, Manik, **Lamat**, Muluk, Ok, Chuwen, Eb, **Ben**, Ix, Men, Kib, Kaban, **Etz'nab**, Kawak, Ajaw. Courtesy SLUB

interpretation, which is really all that is necessary to see what was pointed out by Cyrus Thomas in 1882. At the far left of each page, our scribe painted an isolated column of 260-Day Count Day Signs, along with the glyphs for the cosmic regions in the lowest row of text on each page (figure 6.8). The specific Day Signs included and their correspondence to the cosmic regions turn out to be sufficient for interpretation, but there is also much more.

For the dates, on page 25 Tawiskal Uwoojil painted a column of twenty-six Day Signs at the far left of the page. In this column, the first thirteen were all repeats of the Day Sign Eb and the bottom thirteen were all Ben. On page 26, he painted thirteen Kaban Day Signs at the top of the column, followed by thirteen Etz'nab Day Signs. He propagated the pattern through page 27 with thirteen Ik' followed by thirteen Ak'bal Day Signs, and on page 28 he recorded thirteen Manik Day Signs followed by thirteen Lamat Day Signs. The pattern across these pages is straightforward: the Day Signs on each single page are sequential, and those on subsequent pages are evenly separated by three days. As for the cosmic regions, the sequence runs from the east on page 25 to the north on page 26, the west on page 27, and the south on page 28.

For the intent of this pattern, we recall from chapter 1 that for a 360-day period (the haab), all twenty Day Signs cycle through evenly (i.e., 360 is divisible by 20 without remainder). In a repeating sequence of haab, therefore, the same Day Sign will start every 360-day cycle. Accordingly, the attentive reader will also have noticed that during the Classic period the Long Count was anchored to the Day Sign Ajaw—each haab, winikhaab, and pih (tun, katun, and baktun) ended on Ajaw.

For a 365-day period, the "extra" five days means that instead of just one Day Sign, the 365-day period will end on one of four different Day Signs. During the Classic period, these Day Sign "yearbearers" were Ik', Manik, Eb, and Kaban, meaning that the "seating of" Pohp would have landed only on one of these four Day Signs. With 0 Pohp (or 5 Wayeb) landing on the Day Signs Eb, Kaban, Ik', and Manik, this would put the first day of the New Year or 1 Pohp on Ben, Etz'nab, Ak'bal, and Lamat. This is precisely the sequence that Tawiskal Uwoojil painted on pages 25–28 of the Dresden Codex with Eb/**Ben** on page 25, Kaban/**Etz'nab** on page 26, Ik'/**Ak'bal** on page 27, and Muluk/**Lamat** on page 28. Every year the appropriate Day Sign would show up on one of these pages; every four years the sequence would repeat.

Scholars have argued for a "shift" or a calendric revision during the Colonial period, which demonstrates an alignment between these four pages and the narrative description of "the Wayeb Rites" provided by Landa. Taking into account his terminology of a "dominical letter" as the symbol assigned to the first day of the year, and recalling that the first day of the year follows the "unlucky" last five days of the previous year, Landa places these components into ceremonial processional activity:

> It was the custom in all the towns of Yucatan that there should be two heaps of stone, facing each other at the entrance of the town, on all four sides of the town, that is to say, at the East, West, North and South, for the celebration of the festivals of the unlucky days, which they observed in this way every year.
>
> In the year of which the dominical letter was Kan, the omen was Hobnil, and, according to what they said, they both ruled in the region of the South. (Tozzer 1941, 139)

Scholars since before Förstemann have noticed that if we shift one day from Ak'bal to K'an (and so also from Lamat to Muluk, Ben to Ix, and Etz'nab to Kawak), then there is a match between the sequence of dates and the glyphs for the cosmic regions in the Dresden Codex and in Landa's description (Coe 1999, 117). Pages 25–28 of the Dresden Codex highlight the "dominical letter" Day Signs following the "unlucky days," and each is associated with one of the cosmic regions.

It is a very powerful connection that scholars have recovered in recognizing the resonance between the Precontact New Year pages and the Postcontact *relacion* by Landa. If we venture a bit further into the hieroglyphic text and imagery, we find even richer connection.

Tawiskal Uwoojil recorded in hieroglyphic text that each page depicts the planting of a tree. For the east, his text states that "the Great-Akan(?)-Tree is planted in the East," whereas for the other three directions "the First-Akan(?)-Tree is planted" in its appropriate direction. At the bottom of each page, our scribe illustrated a decorated stone pillar correspondingly planted in a **HAAB** glyph. The last three stone pillars, corresponding to the south, west, and north, are identical: iconographic symbols mark them as made of stone; each pillar is covered at the top by a cape and a serpent wrapped around the top. The pillar for the east, on the first page in the sequence, is the only different one—this pillar wears a much longer cape along with a jade necklace and is topped by the head of Chaak, who wears a headdress and earspools. Like the Venus pages, the ceremonial activity of the New Year pages privileges the east.

Landa's description continues:

> In this year then they made an image or hollow figure of the god of clay, which they called Kan u Uayeyab, and they carried it to the heaps of dry stone which they had raised at the southern side. (Tozzer 1941, 139–40)

On pages 25–28, we find that in the middle of each page, Tawiskal Uwoojil painted a deity sitting in a decorated temple, with a brazier burning incense placed in front of him. Foreshadowing the point, we may note that, although they vary in style, the incense burner on page 27 takes the same shape as those in the Venus pages and matches the shape of the vessels archaeologically recovered from Mayapán and Chich'en Itza.

Each of the deities on these pages is named in the hieroglyphic text; Tawiskal Uwoojil referred to each as the *tz'ak-ajaw*, or the "lord" of "ordering." This is the same verb we saw in Yax Pahsaj Chan Yopaat's Copán Temple 11 text, guiding the reader through the proper organization of the narrative. In the New Year pages, the row of glyphs giving this title is positioned in line with the transition between the thirteen Day Signs of 5 Wayeb and the thirteen Day Signs of 1 Pohp.

Landa's description goes on with offerings of food placed in front of these deities:

> The image having been incensed, they cut off the head of a hen and presented or offered it to him. This having been done, they placed the statue upon a standard called *kante*, also placing upon its shoulders an angel, as a sign of water and that this year was to be good; and they painted these angels, and made them frightful to look upon. . . . When they reached the dwelling of the chief, they placed this image opposite to the statue of the god, which they had there, and thus they made to it many offerings of food and drinks, of flesh and fish. . . . They made a heart out of bread and other kind of bread with seeds of gourds, and they offered these to the image of the god Kan u Uayeyab. Thus this statue and the image were kept during these unlucky days and they perfumed them with their incense and with the grains of maize ground with incense. (Tozzer 1941, 141–42)

Tawiskal Uwoojil painted the figure in the bottom set of illustrations on all pages 25–28 carrying a headless bird in his right hand and sprinkling incense from his left. It is worth noting here, however, that Landa describes the activity performed by a priest, whereas the performance appears to be conducted by a deity in Tawiskal Uwoojil's manuscript. We will return to this issue below.

The central features of this ceremonial activity have led scholars to posit a core under-lying cosmological model. Ritual time is intricately connected to ritual space, and the combination is charged as the responsibility of a set of deities—or of mortal human authorities working with deities. In fact, both Michael Coe and Evon Vogt have argued that this was not an isolated pattern, but something ubiquitous across Mesoamerica (Coe 1965; Vogt 1964). The magnificent murals of San Bartolo attest to something akin to the Wayeb Rites, going back to the late Preclassic (Bassie-Sweet 2008, 70; Stuart 2004, 6). Also, before the 819-Day Count was understood as an astronumerological tool, its cycle through the cosmic regions was well recognized (Lounsbury 1978; Thompson 1978, 212–17). We can now read the text itself, for example, on the tablet Kan B'ahlam commissioned for the smallest structure in his Triad Group: 19 K'IN 14 WINIK 1 KIMI 7 YAX WAL-ja-ya K'AWIIL-NAL OCH-K'IN-ni or *19 k'in 14 winik 1 Kimi 7 Yax wa'alaj K'awiilnal ochk'in*—it was "299 days after 1 Kimi 7 Yax the K'awiilnal was stood up in the West." In this count, a different K'awiilnal was set up every 819 days in one of the cosmic regions so that the full cycle actually takes 3,276 days. That this cosmic circuit, too, was connected to physical objects is attested by the elaborate effigy censers excavated from Kan B'ahlam's Triad Group at Palenque.

RITUAL CIRCUITS AND PROCESSIONS

The artistic elaboration of these ceramic vessels, classified as "composite censers," at Palenque occurred in the construction of the base. Potters built these bases as tall cylindrical tubes, which allowed for an inverted cone-shaped bowl to sit securely at the top (Miller and Martin 2004, 229) (figure 6.9). Ritual assistants could then place red-hot charcoal into this bowl, so that priests or rulers could sprinkle copal incense on it to release bright-white smoke and a sweet perfume. The assistant would then place a lid on top to regulate the airflow and so the rate of burning. At Palenque, artists decorated the tall cylindrical bases with large deity heads wearing elaborate headdresses. In most cases, the heads were of the patron deities of Palenque's royal dynasty (Cuevas Garcia 2004, 253).

FIGURE 6.9 Composite effigy censer from Palenque. Photo courtesy LACMA (Schele Drawing Archive #21073)

So many of these composite censers were found in the construction fill of Kan B'ahlam's Triad Group that archaeologists believe they were made and then buried after use in a ritual circuit. The next cycle would require the construction of a new composite censer, just as Landa described for sixteenth-century Yucatan. For the needs of the 819-Day Count alone, circulating around the platform in the center of the Triad Group, a new censer would have been required every three years. If other ritual circuits used these censers as well, then it becomes clear why so many have been found in the archaeological record. In any case, these censers at Palenque make clear that cosmic circuits enclosing time and space were not restricted to the length of the solar year and New Year rituals.

We can move even further afield to find that aspects of the principal characteristics of this ritual activity have been preserved in different forms over time. The Berkeley linguist William Hanks found evidence for abstractions of these cosmic circuits used to provide rhetorical and political legitimacy in Colonial period documents. Hanks, for example, translates a passage from the sixteenth-century Yaxkukul Document 1.

> Southward I go counting out (ordering) boundary stones.
> All (the way) it goes
> until arriving at the foot of Turtle Mount,

(where) there is the boundary stone.
There I leave off (the) Cumkal people,
there too I pick up (am joined by) my father Ixkil
Itzam Pech,
(who is) from Sicipach and (along with) his counsellors.
Pairwise I proceed accompanied by them ##. (Hanks 1987, 674)

The narrator charts out a large geographic square arranged according to the cosmic regions. As the narrator walks in procession along the sides of the square, one or more representatives of that region accompany him. When he changes direction, his accompaniment changes as well.

In the abstract, this bears great resemblance to the other space-time enclosing ceremonies we have seen: there is one entity responsible for setting the overall trajectory, but different partners accompany that entity for each of the cosmic regions. Resonance with the Venus pages of the Dresden Codex seems clear, with the exception that in the Yaxkukul document the narrator is processing along a geographic path, not a temporal one (Aldana 2011d, 58). There is even a strong parallel to the Féjérvary-Mayer frontispiece, in which time performs the enclosing of space, with the period defined by the 260-Day Count.

This is not to suggest that the Féjérvary-Mayer frontispiece actually depicts either the Wayeb Rites or some kind of Venus ceremony. Rather, this suggests that during the Late Classic and Postclassic, ritual activity took on a certain protocol, and it can be seen underlying these various applications to distinct periodizations. Although it requires further treatment that would take us too far afield here to explore in detail, we can also see this basic structure within the period-ending ceremonies that are the driving force behind the erection of stelae during the Classic period (Aldana 2011d). For stelae, or *lakamtuun*, groups of 360-day periods (or haab) are celebrated with k'altuun rituals by the k'uhulajaw. These are performed *with* the patron deities of the dynasty (see, e.g., Palenque), and the k'uhulajaw often is depicted impersonating a deity on the front of the stela. As I have developed elsewhere (Aldana 2011d, 62), this may have been the motivation behind Waxaklajun Ub'aah K'awiil's self-representations throughout the great plaza at Copán. Here too we would have deity impersonation, patron deities, and the space-time enclosing rituals for the k'altuun (cf. Newsome 2001, 214–18).

Although there is a strong resonance between all of these examples and the New Year pages, the latter give us much more information about the accompaniment by deities—and in particular by ceramic effigies. At the top of each page, Tawiskal Uwoojil painted a finely clothed opossum carrying a staff and walking toward the column of Day Signs. Each opossum carries a small animated figure on his back. The text is heavily eroded at the top of the page, but on it Tawiskal Uwoojil provided the name of the deity figure being carried by the opossum, and in each case, he described it with two terms: *u mam* and *u kuch*. The latter term *u kuch* is often translated as "one's burden" or "one's charge" (Coe 1999, 150; Thompson 1972, 91). In hieroglyphic texts, it usually refers to the burden of a political or religious office. This phrase and similar depictions occur throughout the Dresden Codex.

On the other hand, the first term, *mam*, literally means "maternal grandfather," but also carries a more general meaning of "revered ancestor" (Tozzer 1941, 136n632). Tozzer explores this concept in his annotations to Landa's manuscript:

> There is an interesting statement in Pio Perez (1843 ed., 437) connecting Mam with the rites of the Uayeb days, "The Indians feared those days, believing them to be unfortunate, and to carry danger of sudden death, plagues and other misfortunes. For this reason these five days were assigned for the celebration of the feast of the god Mam, 'grandfather.' On the first day they carried him about, and feasted him with great magnificence [sic]; on the second day they diminished the solemnity; on the third they brought him down from the altar and placed him in the middle of the temple; on the fourth they put him at the threshold of the door; and on the fifth, or last day, the ceremony of taking leave (or dismissal) took place, that the new year might commence on the following day, which is the first of the month Pop." (Tozzer 1941, 137n 646)

In these cases, "Mam" is the deity who is "carried about and feasted." This is made possible because—as noted above—indigenous priests treated the ceramic effigies as the deities themselves.

Not to venture too deeply into the connection, but there are contemporary rituals practiced within Mayan communities that resonate with this structure and terminology. In twentieth-century Highland Guatemala, anthropologist Michael Mendelson encountered Maximon and the Catholic saints as statues and bundles in the region around Lake Atitlan. Mendelson's extensive account resonates with the bulk of this chapter, spelling out how the community leaders sponsored a festival to the San Martin bundle.

> In any given year, each *cofradia* is located in the house of its *alcalde* who is responsible to the head of the village for the performance of the saint's ritual in that year. The next year, the *cofradia* paraphernalia moves on to another *alcalde*'s house and a different set of *cofrades*. Public ritual is performed by the *cofrades* as representatives of the *principales* and the Municipality on the saint's day (*fiesta*) with prayers, drinking and feasting in the *cofradia* and processions of the saint's statue or statues to and from the church. (Mendelson 1958, 121)

Mendelson then recounts the ritual dance central to the festivities.

> I now describe [the Dance] as I witnessed it during fiesta San Martin, 1952. . . . The marimba played and the four dancers moved in circles, hopping from foot to foot and swaying from side to side, occasionally whirling round in one spot, the "tigers" emitting long whistles and sharp cries and pawing the backs of the "deer" with the squirrels. Four times the group knelt abruptly, one behind the other, and crossed themselves, thus salut-

ing the four cardinal directions at three-to-four-minute intervals in the dance. . . . All the while, one individual said to be the leading "tiger" and "very wise in the dance" swung an incense burner over and around them. This man, with one assistant, now repeated the dance as the "deer," and, in the courtyard, a real battle was enacted, the "deer" striking with his horns and the "tiger" assistant with teeth and paws. Eventually the "deer" died, climbed onto the "tiger's" back and was carried into the *cofradia*. (Mendelson 1958, 122)

The connections of Mendelson's account to the New Year pages are strong, although Mendelson goes further to provide a fuller account and appreciation of the role of the public. Finally, Maximon and the statues of the Catholic saints are referred to with the term *mam*, which also means "grandfather" in K'iche' Mayan. We will see in the next chapter that there also are political issues that are addressed through this deity. Overall, the ethnographic and ethnohistoric literature seem to portray a consistent relationship between time, space, processions, and ceramic effigies—be they of Precontact deities or modern Catholic saints—and they remind us that a strong engagement of the public, in theater and in procession, was expected.

Returning to the New Year pages in the Dresden Codex, Thompson suggested that the figures on the opossum's back actually represented animations of ceramic figurines; they are the patron deities of the ceremony, but here in naturalistic postures "perched rather precariously on the conventionalized representation of a carrier frame on the bearer's back" (Thompson 1972, 90). If we allow for these figures to represent ceramic effigies in ritual activity, then it is hardly a stretch to see these pages as something of an "instruction manual" for the performance of New Year rites. Tawiskal Uwoojil appears to have recorded human ritual performance within his illustrations.

From this perspective, it is not hard to see the opossum on page 25 itself as a deity impersonation—the hands, feet, and body architecture, after all, are those of humans. Thompson noted that he specifically considered "evidence for identifying these persons in opossum guise as the Bacabs" (1972, 90). Accordingly, the figures in the middle register may well have been animations of the ceramic effigy incense burners excavated from Postclassic archaeological sites in Yucatan. In this interpretation, the opossum is a deity impersonator carrying a ceramic effigy to meet another ceramic effigy housed in its own temple—the text on page 28 above the middle register says *tz'ak ajaw 4-[deity]; yotot* or "4-Deity is the tz'ak ajaw; it is his house." This now sounds even more like Landa's description (cf. Thompson 1972, 90–91).

Meanwhile, in the bottom register of pages 25–28, another priest as deity impersonator scatters incense on the "stood-up" stone tree in the appropriate cosmic direction. Similarly, the deities in the bottom register of each page are performing the activities assigned by Landa to the priests charged with conducting the ceremony, as we see in the above quotation. In accord with Thompson, therefore, the proposal is that in his illustrations Tawiskal Uwoojil depicted the relationships among the deities themselves, while deity-impersonating priests with effigy incense burners performed these rituals for public engagement.

Such a perspective allows us to take two observations from the New Year pages and the other time-space *k'al* rituals back to the Venus pages. First, all of the specific deities represented as effigy censers that Tawiskal Uwoojil included in the New Year pages also show up in the Venus pages (see figure 6.4). Second, he painted the figures in the middle register of the New Year pages, sitting in their temples, with smoldering incense burners in front of them—some of the same types held by the figures in the Venus pages. In both cases, Tawiskal Uwoojil was capturing ritual activity adhering to the same underlying conceptualization, but one was guided by the period of the solar year while the other followed the Venus Round.

In other words, returning now to the right-hand sides of the Venus pages we may place them in a broader ritual context, suggesting that we are seeing material culture in the illustrations and hieroglyphic text. More, if the ceramic effigies in the New Year pages were placed in their "houses" and if the sky bands on the Venus pages refer to architecture, then we might be led to wonder if the illustrated figures at the top of the Venus pages weren't placed in some conspicuous building. We may well suspect that there was some architectural construct that would accommodate the movement of the ceramic effigy censers around a ritual circuit, temporally guided by the subperiods of Venus's cycle and maintained by designated lineages or ch'ibals of the city. In other words, we might expect to find a "Venus building" or something of the sort from either the Terminal Classic or the Postclassic periods.

For those familiar with Mayan architecture, that leads us—and has led others—directly to the "Caracol" at Chich'en Itza (Aveni 2001, 275).

CERAMIC EFFIGIES, THE VENUS PAGES, AND A VENUS BUILDING

When we turn our attention to the construction of the Caracol—now not as tourists joining Landa's walks around the largely abandoned site, described in chapter 1, but as visitors to the site in full bloom—we return to the Terminal Classic period, just before and during the period that has notoriously been described in popular literature as the "collapse" of Mayan civilization (cf. Sharer 2006, 59). While there is no doubt that the region underwent substantial changes during the period—including military strife evidenced at various sites—recent, more systematically accumulated data suggests a complicated and varied process of transition (Sharer 2006, 503). Chich'en Itza, for example, was in the drier area relative to the massive cities of the Peten, but it survived and even thrived through the documented droughts that hit the peninsula. There is also evidence at multiple sites that commoners and perhaps a "middle class" remained for some time after the construction of stone monuments ceased at some cities in the southern lowlands (Sharer 2006, 520; W. Fash 1991). Drought almost certainly played a role, but its impact was neither uniform nor universal (Sharer 2006, 515).

One factor that we have already seen in the transition to the Postclassic is that an internationalism crept into artistic representations. Along with this internationalism, it may be that the economic engine of the Mayan region shifted. Broadly considered, the data that we can rely on at this point reveals a move of trade routes—and so the core of elite culture—from the southern Mayan lowlands to the northern lowlands and coasts of Mexico and Central America. Other cities in the north such as Coba, Dzibilchaltun, and Uxmal, for example, prospered during the Terminal Classic and into the Postclassic period (Sharer 2006, 549). Along the northern coast and all through Belize, population centers remained strong, even while some of their neighbors failed. Politically, the Chontal Maya residing along the southern Gulf Coast begin to play a much more prominent role in regional politics during this time (Sharer 2006, 528). Here it is worth noting that the Puuc-style architecture at Chich'en Itza demonstrates its connection to the northwestern Yucatan Peninsula and away from the south—a point we will follow up on below.

So it may be, for instance, that the extended drought impacted the water levels of the Usumacinta or the rivers of Belize, hampering the transport of trade goods by canoe throughout the region. Even an impact on the smaller rivers of the eastern watershed would have placed the Usumacinta's role as the superhighway of transportation in and out of the Peten in grave peril. Or it may be that political strife among the cities controlling the Usumacinta reached a level that made it more profitable to seek alternative routes. Whatever the case, we do know that by the early Postclassic, trade had shifted to ocean canoe travel around the peninsula. Cities with river access to the Caribbean via the east coast flourished, and Chich'en Itza would become a powerful force in the northern Mayan region (Sharer 2006, 558).

During this time of transition, significant change occurred at Chich'en Itza as well. The masons active before the transition built structures in the immediate vicinity of the Caracol dominated by the Puuc architectural style. Before the Terminal Classic proper (i.e., AD 900), the Temple of K'uk'ulkan described by Landa had not yet been built. There had been a structure in its place, but it was far simpler and more conventional—a pyramidal base with a staircase on one side (the north) and a structure with a single entrance at the top (Sharer 2006, 565). Archaeologists have characterized the whole city as more provincial at the time with ceremonial activity distributed across different local "centers" (Cobos and Winemiller 2001). The Caracol and its neighbors certainly constituted one such "center," attested by hieroglyphic inscriptions associated with them (figure 6.10). The latter inscriptions carry the Long Count dates that reflect a shift in calendric practices—the ones highlighted by Morley and Spinden as providing the link between the Classic period "Long Count" and the "Short Count" adopted in the Postclassic, which in turn preceded the much later "Katun Count."

When we look a little more closely at these texts, though, we detect another change. There appears to be a political shift underway, in this case one that deemphasizes the public role of a paramount authority (Sharer 2006, 569). The inscriptions do describe activities between authority figures and deities, such as structure dedications aligned with the Short Count. But the most prominent human figure in these texts is not named as the k'uhulajaw, and there are no dynastic histories to be found.

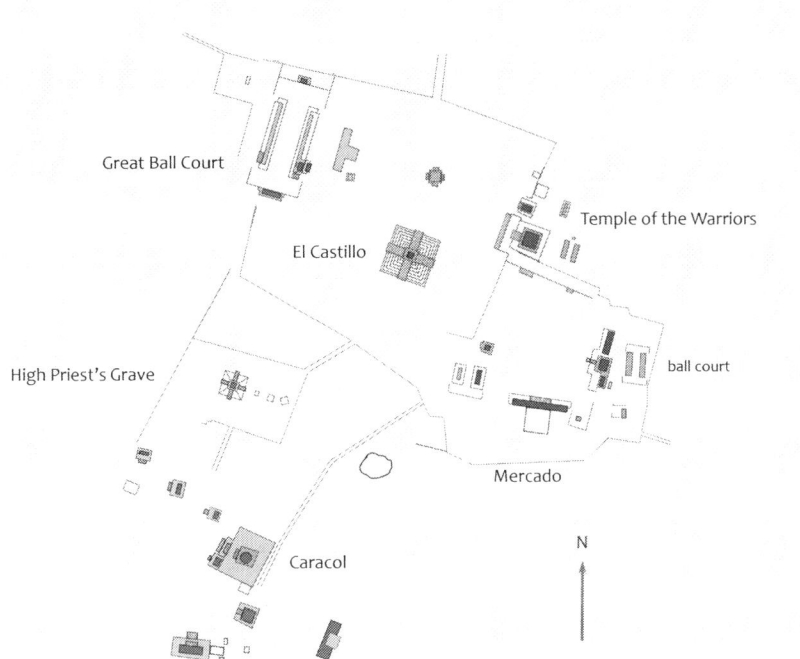

FIGURE 6.10 Caracol "center" at Chich'en Itza in the southwest region of the Late Terminal Classic city

Scholars are divided on the actual political infrastructure at Chich'en Itza, some arguing for rule by an alliance of elite "houses," and others promoting a traditional kingship, though without aggrandizement of the ruler as individual (Martin and Grube 2008, 227; Ringle 2004; Schele and Mathews 1999, 197; Sharer 2006, 580). We have seen that Juan Josef Hoil commented in the Books of Chilam Balam on the rulership of Hunac Ceel, but he also included in the katun prophecies that governance was broken up.

> 4 Ahau was the katun when the four divisions were called <together>. The four divisions of the nation, they were called, when they descended. They became lords when they descended upon Chichen Itzá. The Itzá were they then called. (Roys 1933, 77)

So while there is scholarly dissent about the specific form of political infrastructure, there is agreement that some shift occurred at Chich'en Itza (Lopez Austin and Lopez Lujan 2000, 26; Ringle 2004, 168). Something transpired, transforming the small city from a provincial center carrying heavy Puuc architectural influence to the regionally impressive metropolis with extensive artistic ties to Central Mexico.

It also appears that the population of Chich'en Itza after the shift was itself diverse in composition (Lopez Austin and Lopez Lujan 2000, 26; Ringle 2004, 168; Sharer 2006, 569). The archaeological record from much smaller cities to the south at this time demonstrates significant mobility of the lower socioeconomic classes. As much as

25 percent of the population of these smaller cities were newcomers—families moving into the region from some distance away (Prufer et al. 2017, 65). One can imagine, therefore, that at a city the size of Chich'en Itza the population would have been at least as dynamic and probably drew from cities and villages some distance away, creating an international metropolitan atmosphere. In such case, it seems reasonable that public ceremony would be pulled away from lineage patron deities and more productively moved to honoring deities inspiring broader interest. It is within this context that we consider the role of the structure known as the Caracol. It may be that it served a political structure that no longer privileged a bloodline connected to a lineage patron deity, but now it may have supported a priesthood or alliance that maintained accessible deities to a large and diverse public audience.

The Caracol or Observatory

The interpretation of the Caracol, as we have seen already, has been greatly influenced by Sylvanus Morley himself. Morley's impact on the field of Maya archaeology went far beyond his own albeit formidable published research; he also enjoyed an extensive network of associates from which he was able to recruit researchers for Carnegie once he took on the directorship of the anthropology program in 1914. Coe writes colorfully of the unconventionality of his recruits:

> It is pleasant to look back on the kinds of people whom Morley brought into Carnegie, and the life that they led. Few of them had that union card of modern times, the Ph.D. . . . It is said that the Smith brothers, Bob and Ledyard, were recruited for the Uaxactún dig by Oliver Ricketson at the bar of Harvard's very social Fly Club. Gus Strömsvik, later to direct the Carnegie project at Copán, was a rough-hewn Norwegian sailor who had jumped ship in Progreso, Yucatán, and who began work at Chichén repairing the expedition's trucks. Ed Shook entered his career as a Carnegie draftsman. (1999, 127–28)

For another, Frans Blom had been working for an oil company in Chiapas and Tabasco (Mexico), through which he found the occasional opportunity "to visit ruins" (Byers 1966, 406). From one such visit to Tortuguero, Blom sent a drawing (with notes) of Stela 1 to Morley, who was so impressed that he invited Blom to work on a Carnegie excavation (Byers 1966, 406). Blom joined Oliver Ricketson on the trip to Uaxactun, which resulted in the recognition of the "Group of the Solar Observatory" that we saw in chapter 2.

Ricketson's 1926 work at Uaxactun was not his first in the field nor his first investigation into Mayan astronomy. Having been initially discouraged by his travel across the Yucatan Peninsula with Morley in 1921, Ricketson found new inspiration in 1924, going straight from Uaxactun with Blom to Baking Pot in British Honduras (to conduct his own excavation), and then to work with Morley at Chich'en Itza (Lothrop 1953, 70). Together Ricketson and Morley trained their attention on the structure they called "the Observatory," but which became better known as the Caracol.

As we saw in chapter 2, much of the upper levels of the structure had already crumbled by the time they began work there. Undaunted, Ricketson took compass data on the windows that were still intact, complaining that his efforts to use a theodolite were foiled by the cramped space (1928a, 442). Noting that he was "without even a rudimentary knowledge of astronomy," he sent his compass readings to Louis Bauer, director of the Carnegie Department of Terrestrial Magnetism (see figure 2.1). Bauer's computations corroborated Ricketson's 1924 direct observation of the vernal equinox sunset sighted along the "right inner jamb to left outer jamb" of Window 1 (1928a, 442–43). Likewise, the "right inner jamb to left outer jamb" of Window 3 pointed due south (1928a, 443). This meant that the windows were not centered on the cardinal directions, but slightly skewed. Bauer also informed Ricketson that two sightlines from the windows marked the extreme declinations of the Moon, leading Ricketson to believe that astronomical motivation might be found behind many more Mayan structures (Ricketson 1928a, 444).

Ricketson's interpretation of an astronomical motivation behind the Caracol remained strong over the following decades. His call to alignment measurement took on a new life with the technological innovation of the late twentieth century. Exploring this architectural application of astronomy, we eventually find our way back to the Caracol at Chich'en Itza, but we first consider a scholarly development across the Atlantic Ocean.

Enter Archaeoastronomy

During the 1960s, the astrophysicist Gerald Hawkins took an interest in the great monolithic ruins of Stonehenge in England. The idea of exploring astronomical knowledge encoded within ancient architecture was not new within European scholarship; Stonehenge itself had been considered astronomically relevant since the Middle Ages, and it was periodically reassessed as such into the twentieth century (Fernie 1990, 103). When Hawkins took it up, proposals had already been debated concerning its use as a calendar and its alignment to solar phenomena (Fernie 1990, 104; Hawkins 1965). What Hawkins introduced in the 1960s, through his training as an astrophysicist, was the exhaustive computational power of an IBM computer. Hawkins employed this computational power to follow up systematically on Ricketson's call; he checked every possible architectural alignment at Stonehenge against as many rising and setting positions of the Sun, Moon, planets, and bright stars as possible. The result was a proposal that Stonehenge encoded more than just solar alignments; it also encoded the motion of the Moon (Hawkins 1965; Aveni 2003, 150–51).

Hawkins's finds did not escape critique—archaeologists challenged his statistical analysis as well as his apparent disregard for archaeological context (Fernie 1990, 104). But the work came at an auspicious time, bringing together "two magically appealing subjects at least one of which [the editors of *Scientific American*] invariably tried to include in every issue: archaeology and astronomy. The romance of space and the mystery of the past!" (Gingerich 1980: ix). Part of the appeal to Hawkins's work, then, appears to have been its resonance within Euro-American popular culture(s). Once again, like

[14]C, American technological developments would shape the interpretation of the Meso-american past.

The Space Age was launched with Sputnik on October 4, 1957. The orbiting of the Earth by John Glenn in 1962 and the technological race culminating in Neil Armstrong's steps on the Moon on July 20, 1969, filled the media in various forms. Very quickly, astronomy and "outer space" became ubiquitous in American popular culture. In the thick of it, Walter Wingo of the "Science News-Letter" referred to NASA's establishment of "a public relations program unrivaled in the history of the U.S. Government [intended] to sell the people on the benefits of the space program" (1963, 341). Alton Frye of the Harvard University Center for International Affairs described the "American public" as potentially "saturated into apathy by news media seeking to keep up with the most visibly exciting area of technology" (1966, 103).

The public interest in space, of course, was not restricted to matters of international politics; "by the 1970s, science fiction had become an established part of popular culture" (Consolmagno 1996, 129). Science-fiction titles, mixing fantasy and outer space, made it onto national bestseller lists. This intriguing interplay between astronomy, the Space Race, science fiction, and ancient history was in part captured by Michael Coe's overview of *Native American Astronomy* in 1977:

> The public on both sides of the Atlantic has been led to believe in the existence of voyagers from outer space, in sunken continents, in white "culture gods," and in heaven knows what else, a state of affairs heavily exploited by book publishers and television producers. (1977, ix)

And that the interest made it into academic circles was anecdotally attested by John Eddy in his 1977 review for the *Journal for the History of Astronomy*:

> This volume [*Archaeoastronomy in Pre-Columbian America*] compiles 18 of the 26 papers presented at a joint Mexican-U.S. meeting on pre-Columbian archaeoastronomy held in Mexico City in June 1973.
>
> I am a little surprised at what a popular book it has proven to be. Both of the copies in the University of Colorado libraries seem perennially checked out, I see a number of private copies around, and somebody is always borrowing mine. (1977, 497)

It may well be that this public interest is what helped archaeoastronomy survive the onslaught of what we've already seen of the advancing hieroglyphic decipherment and develop in its own right. Years later, Anthony Aveni implied as much, referring to the public's role in the development of the field in the 1960s and 1970s:

> the flood of trade and popular works on archaeoastronomy, though useful in bringing new ideas to a wider audience, did little to contribute to its professional status. Although archaeoastronomy has shed much of the burden of the sensationalist baggage it once

acquired in the aftermath of the Stonehenge controversy, popular works that advocate an extraordinary and oft-difficult-to-document role for astronomy in shaping human culture still reach the level of trade text publications (e.g., Bauval, 1995; Sullivan, 1996; Ulansey, 1989). Many of these works exhibit both millenarian and deterministic qualities in which seminal cosmic events drive the course of civilization. (2003, 151)

A revitalized interest in astronomy—after an alleged lull—appears to underlie Horst Hartung's introductory remarks in 1975:

Contrary to the reluctance characteristic of the forties, fifties, and sixties, in the seventies scholars of Mesoamerican cultures generally accept the idea that there existed a consideration of astronomical events in pre-Columbian architecture and planning. (1975, 111)

Hartung, it turns out, was to play an important role in the preservation of astronomical interpretation within Mesoamerican cultures into the 1970s and beyond and contributed specifically to the interpretation of the Caracol.

Horst Hartung was born in Germany and immigrated to Mexico at the age of thirty-two to form part of the new faculty of architecture at the University of Guadalajara. Noted for designing some important buildings in Jalisco, Mexico, his greater contribution grew out of an interest in ancient Mesoamerican architecture (Díaz-García 2006). Hartung's first publication on the subject, *Die Zeremonialzentren der Maya* (1971), took an architect's consideration of urban design back to Chich'en Itza, adding to it investigations into the urban centers of Piedras Negras, Yaxchilan, and Uxmal. In his work, Hartung looked for celestial orientations defined by the alignment of architectural features to other monuments—very much in the spirit of the decades-earlier work of Ricketson and Blom's interpretation of the E-Group at Uaxactun, or Morley's work at Copán.

Hartung's work met criticism. Katherine Haramundanis of the Smithsonian Astrophysical Observatory, a specialist in scientific measurement, responded to Hartung's book immediately and without much sympathy. She found his results unconvincing primarily because he worked from maps that Haramundanis considered to be "of insufficient accuracy for drawing conclusions concerning astronomical orientations" (1973, 202). This concern was, in fact, the same one raised more generally by Jonathan Reyman (1973), echoed in his review of the (re-)nascent field (1975, 210).

Haramundanis's critique would not be sustainable for long. In 1969 the Yale archaeologist Michael Coe connected Hartung to a recently minted PhD in astronomy, Anthony Aveni (Aveni, pers. comm., 2013). Aveni's early work, too, had received criticism. Anthropologist John Reyman wrote:

The search for alignments, at times, seems to reflect a haphazard, almost random "groping," and the accompanying explanations have tended to be after-the-fact (see Aveni and Linsley 1972). In short, archaeoastronomers have all too rarely used anything approaching the scientific method. (1975, 208)

With a new partnership, Hartung and Aveni were able to address both critiques.

Four years after they first met, Hartung and Aveni collaborated on two fronts. For one, their task, sponsored by both the American Association for the Advancement of Science and the Consejo Nacional de Ciencia y Tecnologia of Mexico, consisted of the "first organized gathering of archaeoastronomers" to consider Western Hemisphere astronomies (Aveni 1977, xii). They thus explicitly brought the project inspired by Hawkins across the Atlantic for a (re)new(ed) interest in Mesoamerican astronomy. Aveni makes this explicit, writing of his new NSF-sponsored project that "in all cases the guidelines set up by Hawkins (1962) and Reyman (1973) have been followed" (Aveni et al. 1975, 163). Aveni also found his second calling here, editing the conference proceedings and initiating a publishing trajectory from the center of the field, which coalesced in a follow-up meeting two years later at Colgate in 1975, where Aveni would be promoted in that year to full professor of astronomy.

The second collaboration spoke directly to Haramundis and Reyman's concerns. In her review of Hartung's work, Haramundanis had thrown down a clear challenge:

> It is unfortunate that to this date, although an enormous amount of work has been done and a vast literature has grown up around the Maya, there exists no definitive work that can answer the question of whether Maya buildings had astronomical orientations or even if the Maya themselves made astronomical observations. (1973, 202)

During the winters of 1973 and 1974, Hartung and Aveni worked with historian of science Sharon Gibbs—a researcher at Colgate where Aveni was developing a teaching career combining astronomy and anthropology that would land him in *Rolling Stone* magazine as one of the U.S. "Top 10 Professors"—to come back to Ricketson's work and revisit the measurements of the Caracol at Chich'en Itza (1975, 977). Aveni seemed to be responding directly to Haramundanis, describing his research as an "organized study of the possible extent of astronomical orientations throughout ancient Mesoamerica," involving

> direct measurement with a transit instrument of particular alignments at the archaeological sites and their subsequent matching with local astronomical rise-set phenomena utilizing a set of computerized tables. (Aveni 1975, 163)

The challenge had been accepted, setting the Caracol up to become the canonical example of Mayan observational astronomy.

Through their collaborative work at Chich'en Itza, Aveni, Gibbs, and Hartung showed that the orientation of the Caracol was measurably skewed relative to other buildings at the site, and it was this deviation that pointed to alignments with planetary phenomena (Aveni 2001, 274). Going beyond Ricketson's results that revealed an equinox alignment to one window, Aveni, Gibbs, and Hartung also found a much more compelling celestial referent in Venus. This in particular seems to have been behind Haramundanis's closing remarks.

A recent analysis of the Dresden Codex (by J. Eric Thompson) suggests that it contains a Venus table in addition to its astrological texts; and accurate site surveys which can determine if reasonable orientations exist to astronomical objects are only now being made. (Haramundanis 1973, 202)

In the end, Aveni, Gibbs, and Hartung used their "accurate site surveys" to bring the Dresden Codex Venus Table back into the interpretation and to argue that

the provisions for correction of the formal Venus tables in the Dresden Codex suggest that observations of Venus were indeed made. We may suppose that the Caracol windows were placed to aid such observations and specifically to preserve the direction of the most predictable disappearances of Venus before heliacal rise. (Aveni et al. 1975, 983–84)

Aveni, Gibbs, and Hartung had reached all the way back to Förstemann's original insights to combine them with the archaeological investigations of Morley and Ricketson and revivify them with Hawkins's computational methods (cf. Aveni 2001, 274). Mesoamerican "archaeoastronomy" had been born, and Maya Studies again was impacted by twentieth-century physicists—this time establishing the astronomical credentials of the skywatchers working from the Caracol at Chich'en Itza. In the view developed by Aveni, Gibbs, and Hartung, the Caracol was built to facilitate observations of Venus.

In addition to confirming Ricketson's measurements, the team's precise instruments did demonstrate an alignment of a niche in the front stairway of the Caracol toward the maximum northern setting position along the horizon of Venus as Eveningstar (Aveni 2001, 274). This alignment, for example, would have been visible once every five Venus Rounds (eight years), and so it may have coincided with the dates of one column of *k'al* records in the Venus Table. Also there may have been a connection in inspiration to the visibilities of Venus not unlike those of Waxaklajun Ub'aah K'awiil's Temple 22 window visibilities at Copán. Of course they served a different function at Chich'en Itza, now only marking a single event within Venus's eight-year cycle and not capturing a period of twenty days.

On the other hand, given the evidence for Venusian concern, we may return to Ricketson's find that the Chich'en Itza architects built two of the windows to carve out an arc of ninety degrees, aligned with the cardinal directions (Aveni 2001, 276). It is not unreasonable to propose that there were originally four windows on the top level, with the two lost to erosion completing the pattern of alignment to the cosmic regions. A third surviving window faces the intercardinal direction of the southwest, suggesting that there may have been four of these as well. Aveni and Hartung noticed further that the intercardinal window points to a second specific event: Venus's southernmost setting position along the western horizon (Aveni 2001, 276). This all brings us back to our original discussion: if we were looking for a "Venus building" from the Terminal Classic or Postclassic, the Caracol at Chich'en Itza would have to be a prime contender.

PERFORMANCE AT CHICH'EN ITZA

So there is good reason to see an astronomer at Chich'en Itza, say K'uk'ul Ek' Tuyilaj herself, making use of the windows of the Caracol for observations, but we would be remiss to stop there. Just as we are beginning to see the material culture and even public participation within the Venus records in the Dresden Codex, so too can we realize that there may have been aspects of the Caracol that were not restricted to those privileged enough to view the planet from its windows.

The size and prominence of the building themselves suggest a public audience. K'uk'ul Ek' Tuyilaj and her colleagues entering, acting within, and leaving such a public space would have been highly conspicuous if not intentionally so. If we now incorporate the above recognition of effigy incense burners depicted in the Venus pages and ritual circuits, it makes sense to propose that this structure also provides the possibility of an analogue to Landa's description of the Wayeb rites (figure 6.11). In other words, if *k'al* functions as the verbal description of a cosmic space-time enclosing ritual, if the original version of the Venus Table were drawn up at Chich'en Itza, and if effigy incense burners were the referents of the deities named in the Dresden Codex, then ceremonial activity at the Caracol follows.

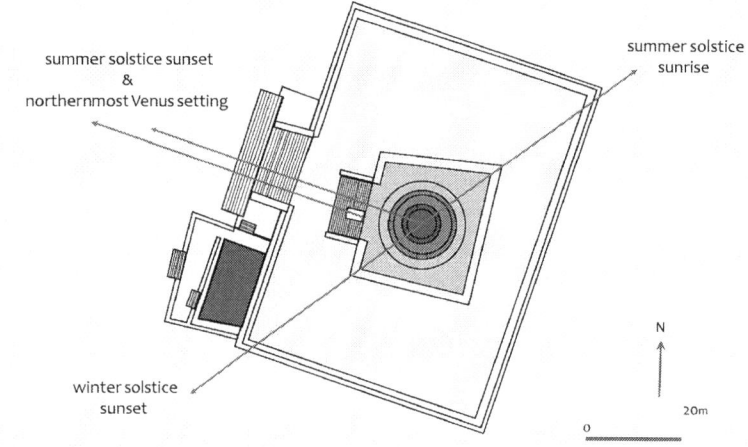

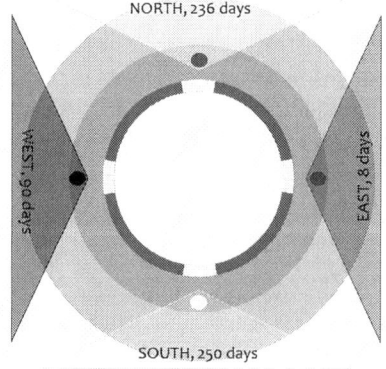

FIGURE 6.11 (*above*) Sight lines built into the architecture of the Tz'iknal (after Aveni 2001, 273–76); (*left*) effigy censer placement around the Tz'iknal for k'al-enclosings of Venus periods

In the case suggested here, the entity that enables the passage of time, or the deity who counts out the time period, is Chak Ek'—Venus. K'uk'ul Ek' Tuyilaj's version of the Venus pages also recorded that Chak Ek' was accompanied by one of twenty deities (two of whom appear to be repeated, so that this is really only a set of eighteen) for any given subperiod of a Venus Round. Therefore, she could have used the Caracol at Chich'en Itza as an architectural theater for the events listed in the Venus Table.

On the first 260-Day Count date, Ajk'uhuun K'uk'ul Ek' Tuyilaj would place a ceramic effigy of the deity listed in her table within the window looking out onto the associated cardinal direction. Venus and that deity would ritually share responsibility for the specified time period, or as the text states: "On [date] Chak Ek' and [deity] enclosed the East."

Parallel to the New Year ceremony transitions, on the date of completion of the enclosing, the effigy incense burner would be moved to the north window to meet the next effigy incense burner. That effigy censer—the one named in the next column of the Venus Table—would take over for the duration of the next Venus subperiod, carrying the cargo for that enclosing. Over the rest of the Venus Round, the windows would have facilitated specific Venus event observations, as suggested by Aveni, but they also would have provided a ceremonial stage for the public celebration of Venus events.

There appears to be evidence in the inscriptions themselves speaking to this role of the Caracol. Several epigraphers working on the challenging inscriptions of Chich'en Itza have argued that the name of the structure shows up in associated texts (Voss 2001; Boot 2005; Bíró and Pérez de Heredia 2016). Mostly associated with a historical figure carrying the title "Tz'ak Ajaw," they argue that during the Terminal Classic the building was known as the Tz'iknal. The -*nal* suffix is unproblematic, identifying the term as referring to a structure of some sort. Alexander Voss then found somewhat oblique corroboration of association with the Caracol by examining the root of the term *tz'ik*. Yucatec dictionaries give *tz'ik* the gloss of "bravo" and "valiente" but also "desgraciado" and "siniestra" (Barrera Vásquez 1995, 883; Voss 2001, 8). Voss argued that this was the result of the original Catholic priest compilers' deprecation of astronomical knowledge as being "pagan" and so worthy of denigration (Voss 2001, 8).

> Es probable que en tiempos precolombinos el lexema ts'ik o ts'its' más bien describiera eventos que se relacionaban a lo oculto y poseían calidad enigmática como augurios, oráculos y profecías. De esto suponemos que el Caracol fue un edificio relacionado al vaticinio. Esta interpretación coincide con las reconstrucciones arqueoastronómicas propuestas para este edificio (véase Aveni 1991: 292–301) hechas previas al desciframiento epigráfico. (Voss 2001, 9)

There is another option for the translation of *tz'ik*, though, that also may be of use. A second gloss for *tz'ik* is "llevar guiando; guiar" (Barrera Vásquez 1995, 884). This translation transparently connects the ritual we have recovered with the name of the building. Venus sets the pace of the movement of ceramic effigies around the windows, and so Venus acts *as a guide* for deities and activities through the structure itself. Such a translation also works on a metaphorical level when we consider the oracular knowledge generated by the Venus Table in the next chapter. From that perspective, Venus guides the entire community through priestly intercession. In both theatrical and metaphorical cases, the "Caracol" is well represented as the Tz'iknal.

When we return to the illustrations in the Venus Table, we find that the final image on page 50 may even have included a playful clue that we are on the "right" interpretive track here. The upper image on page 50 includes two deities instead of only the one named for the time period enclosing the east. The image shows the Maize God holding an incense burner while standing in front of an enthroned Jun Ajaw. Each of these deities is named in the *k'al* passages of the Venus Table: Jun Ajaw encloses the east immediately preceding the enclosing of the north by the Maize God. Tawiskal Uwoojil may have meant this to represent an animation of the type of "changing of the guard" described by Landa. Here the Maize God takes up his duty from the previous responsible deity, Jun Ajaw, who is about to step off his throne—an act that would have been represented by two effigy censers facing each other in front of the east window at the top of the Caracol (see figure 6.3).

In this way, the Caracol as Tz'iknal would have been the setting for the ritual display of deity effigy censers. K'uk'ul Ek' Tuyilaj would still be able to use it for observations, but the building also would have had a prominent public role. Residents of the city had only to look up at the structure to find out where Venus would have been in its cycle. When a burning effigy censer was in the south window, they knew that Venus could not be seen in the morning or evening sky, but that they should expect it soon as Morningstar in the east. When they saw it in the east window, they would know that Venus was Morningstar and might be viewable in the predawn hours. Also, they would have anticipated their impending participation in or witness of large ceremonies and/or processions when the effigy censers were changed. There is even room here for theater beyond ritual procession and the changing of the occupied window. To see this, we move on to the next illustrations on the right-hand sides of pages 46–50.

THE WARRIOR IMAGERY

Whereas most treatments of the Venus Table since Seler begin with the warrior imagery on pages 46–50, within the reconstruction presented here we only come to it now, near the end of our interpretation. As we saw in chapter 1, the lower two illustrations on the right-hand sides of pages 46–50 have been interpreted as representing an act of violence relatively consistently since Seler's work in the early twentieth century. The focus is on a spearing event.

On page 46, on the right-hand side, Tawiskal Uwoojil illustrated a warrior with painted face and elaborate headdress for the central image. The warrior holds a shield in his right hand and a spearthrower in his left. Below this image and below a section of hieroglyphic text, Tawiskal Uwoojil painted a nonhuman being, identifiable by his nose and eye as the deity K'awiil. In this illustration our scribe depicted K'awiil lying on his back with a spear in his abdomen. This pair of illustrations serves as the point of origin for all modern scholarly characterizations of Venus as a warrior or affiliated with war (figure 6.12). On the other hand, we have seen that some scholars have considered the warrior deities to be independent of Venus—not identities that the planet takes on.

Tawiskal Uwoojil painted the "caption" for this event in the hieroglyphic text directly above the illustration of the warrior. The text begins with the now familiar *k'ahlaj lak'in* phrase—"the east is enclosed." We have already seen that this refers to the completion of the enclosing of the east, by which the Venus Round is enclosed as well as just the east as a cosmic region. The next statement follows the pattern of the columns of text on the left-hand side of the page, providing the pairing of a deity name with Chak Ek'. Instead of the deity listed in the fourth column of the left-hand side of the page, though, here we find the hieroglyphic name of the warrior figure in the illustration (figure 6.13). The name on page 46 is logographic and as yet has not been deciphered, so scholars continue to refer to him according to the designation made by Schellhaus—this is "God L." Next, the text goes on to address the victim. The next glyphic statement is: *K'awiil. U jul.*—"K'awiil. It is his spearing."

So the glyphs appear to tell us exactly what Seler identified in the imagery of the Borgia Codex and what we saw in chapter 1. No surprises. Continuing along with the hieroglyphic text, we confront a key term, constituted by the syllabograms *mu* and *ka*. In Yucatecan, *muk* is readily translated as a verb meaning "to cover," or "to bury," which is one interpretation presented by Linda Schele and coauthor Nikolai Grube (1997, 82, 148–56; see also Aveni 2001, 192–93). With a long vowel, though (possibly captured by its disharmonic spelling *mu-ka* [J. Robertson et al. 2004]), *muuk* in Yucatec takes on the meaning of "news—either good or bad" (Barrera Vásquez 1995, 534). The term's

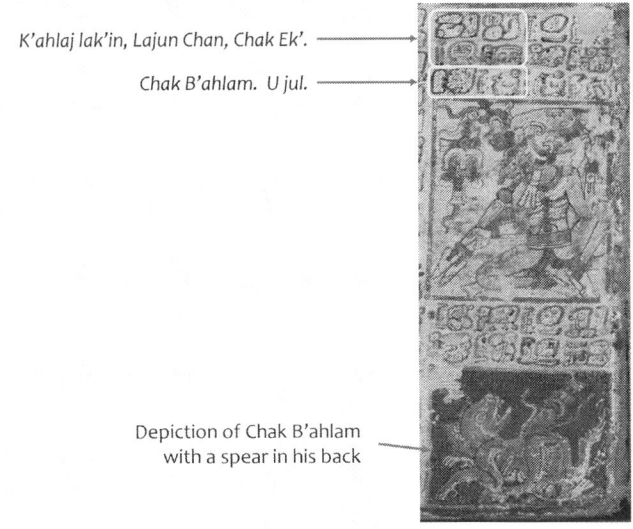

K'ahlaj lak'in, Lajun Chan, Chak Ek'. ⟶

Chak B'ahlam. U jul. ⟶

Depiction of Chak B'ahlam
with a spear in his back ⟶

FIGURE 6.12 Spearing statements in the Venus Table. The text above the image of Lajun Chan as warrior gives the same phrase as in the text on the left-hand side of the page, but with a different deity: "K'ahlaj lak'in Lajun Chan Chak Ek'." It is followed by the statement: "Chak B'ahlam. U jul." Or "Chak B'ahlam. He is speared." Courtesy SLUB

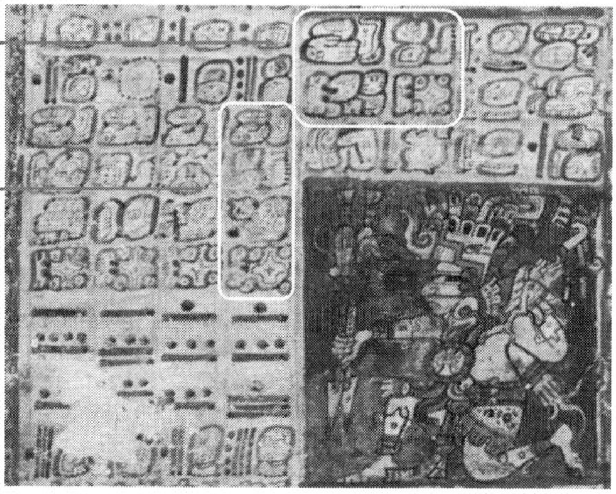

K'ahlaj lak'in, Tawiskal, Chak Ek'.

K'ahlaj lak'in, Jun Ajaw, Chak Ek'.

FIGURE 6.13 *K'ahlaj* statements for the east (*lak'in*) in the Venus Table. Courtesy SLUB

extensive use throughout the codex makes it clear that the "news" translation—often rendered as "omen"—is more productive here. More recently Mathews and Bíró (2008) have corroborated this reading, crediting the "omen" meaning of *muk* to David Stuart and John Robertson and *mu'k* to Alfonso Lacadena and Soren Wichmann. Here, then, *u muuk* becomes "it is the omen of," or "it is the news of," and the referents are the named entity or entities that follow. Thus Seler's analogy is complete. A deity spears a victim at the first morning visibility of Venus, generating an omen.

Following the grammar, though, it becomes clear that it is not Venus as warrior but a warrior deity timed by Venus's periodicity who commits the act of violence, which in turn generates an omen or set of omens. In line with the Féjérvary-Mayer frontispiece, the spearing of a victim by a warrior is one event that results from the final enclosing of all subperiods. The artist behind the Féjérvary-Mayer frontispiece also implies that the warrior deity is not identified as the entity that sets the overall period—in that case the 260-Day Count. Instead, the completion of the period triggers the designated deity to spear his victim. Here we find that the text more closely aligns with Seler's proposal of these warriors as regents of Venus periods and not with Thompson's proposals that they represent Venus itself.

In turn, this spearing event generates an omen, which characterizes the time period. This too, then, constitutes a shift in the overall interpretation of the Venus pages. Now we don't see the first appearance of Venus as mechanically triggering a spearing that produces an omen to be feared. Just as Venus did not mechanically trigger war during the Classic period, now Venus is not the warrior who automatically causes destruction (Aldana 2005). Now we see a ritual connected to the abstracted subperiods of Venus, resulting in a charge to one specific deity. Here again we turn to Landa's words to find a resonant interpretation:

> They believed that if they did not observe these ceremonies, they would be sure to have certain sicknesses, which they have in this year. . . . When once the ceremonies were ended and the evil spirit was chased away . . . they considered the year as a good one. . . . But as it happened frequently that they had calamities, . . . so that when these misfortunes came, they should lay the blame on their ceremonies and on those who performed them. (Tozzer 1941, 142)

On this level, the omens are considered typical of the period defined by Venus and characterized by the deity patron. Landa thus provides insight into non-Western politics and economics. Tawiskal Uwoojil and his colleagues would have seen the omens generated by Venus periods as carrying along with them specific ceremonial responses. Those ceremonies would have brought various sectors of society together to conduct them, thus providing and/or generating political and economic activity.

Reflecting also on Landa's description of the Wayeb rites, we may push the interpretation of the public's participation in Venus event rituals even further. In his description, Landa included the different societal roles that made the ceremonies possible.

> They chose a chief of the town in whose house this festival was celebrated on these days and to celebrate it they made a statue of a god, which they called Bolon Dzacab, which they placed in the house of the *principal*, adorned in a public place where everyone could go to it. (Tozzer 1941, 142)

Landa tells us that ritual activity didn't start with the festival date. Preparation was a social activity with hierarchically ranked members playing specified roles as hosts.

> This having been done, the lords and the priest and the men of the town assembled together and having cleaned and adorned with arches and green the road leading to the place of heaps of stone where the statue was, they went all together to it with great devotion. (Tozzer 1941, 142)

Here too is another peek into ritual activity. The *principales* probably didn't do all of the collection of greens (the boughs for arches) or sweep the roads themselves. This was probably performed by the youth and those of lesser rank within the lineage. Here the commoners also were involved.

> And when they came there the priest incensed it with forty-nine grains of maize ground up with their incense, and they distributed it in the brazier of the idol and perfumed him. They call the ground maize alone *sacah* and that of the lords *chahalte*. (Tozzer 1941, 142)

At the end, therefore, the priests and lords did take over, but only after substantive involvement of the public.

Although Landa was specifically describing the ritual activities of the Wayeb Rites, we can easily see how much of this would have applied to other ceremonies—and even how it resonates with the twentieth-century celebration of the San Martin bundle in the communities of Lake Atitlan in Guatemala. As noted above, during the Terminal Classic period, before construction of the Temple of K'uk'ulkan, the Tz'iknal sat at one of the centers of Chich'en Itza. It is not hard to imagine that it served as a "public place where everyone could go" in order to view K'uk'ul Ek' Tuyilaj install an effigy censer and burn incense in it. Accordingly, there may have been a second procession when a deity completed its term, with the effigy censer carried along on the procession by K'uk'ul Ek' Tuyilaj wearing the regalia that allowed her to impersonate a deity as in the Dresden New Year rituals.

Of course we need to take into account the difference in scale between what Landa described—what either Xiu or Cocom witnessed—and what the citizens of Chich'en Itza would have experienced during the Terminal Classic period. For some measure of the difference, we can turn to the murals at Bonampak, painted to commemorate historical events near the end of the eighth century AD (Sharer 2006, 450). There we witness large-scale public ritual, replete with musicians, performers dancing, and nobles in regalia. The involvement and the expense are considerable, and this is all for a city a third the size of Chich'en Itza.

The streets would be cleaned and adorned on the way to the house of the noble who was charged with keeping the effigy censer until it was called upon to take up its term of residence atop the Tz'iknal. Landa's description continues:

> And thus they carried it with much rejoicing and dancing, to the house of the *principal* where the other statue of Bolon Dzacab was standing. And they brought to the nobles and to the priest on the road from this chief's house, a drink made of 415 grains of parched maize, which they called *picula kakla*, and all drank of it. When they reached the dwelling of the chief, they placed this image opposite to the statue of the god, which they had there, and thus they made to it many offerings of food and drinks, of flesh and fish; and they divided these offerings among the strangers who were present and they gave the priest the leg of a deer. (Tozzer 1941, 142)

It is notable that Landa refers to the "strangers who were present," allowing us to see that the ceremony was intended to bring together a large segment of the citizenry, not just the almehen family charged with keeping the effigy censer. This resonates with the modern processions of Maximon through the streets of Santiago Atitlan as well as the procession in the Mazatec highlands that I witnessed, described in the preface to this book.

Moreover, this kind of activity is well documented in the ceramic record of the Classic period. Scenes painted on tall drinking cups depict performances and even deity impersonations by rulers. Some illustrate ritual events or celebrations explicitly, while others represent preparations for such events. The "Holmul Dancer" vessels show k'uhulajaws performing in elaborate regalia, while images on other ceramic vessels show artists painting the masks used in such activity (figure 6.14).

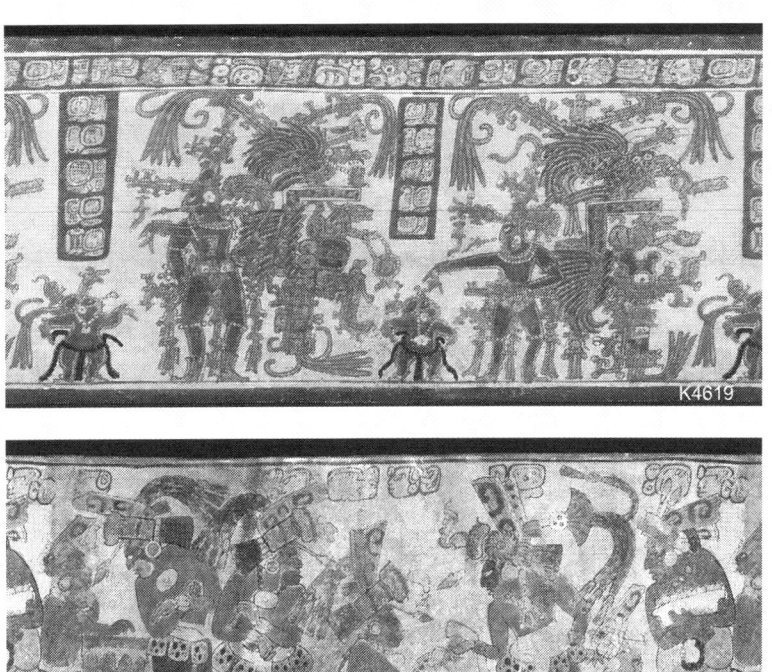

FIGURE 6.14 (*top*) A k'uhulajaw impersonates the Maize Deity in a representation known in the scholarly literature as the "Holmul Dancer" (Kerr #4619); (*bottom*) a group of dancers are accompanied by musicians (Kerr #3009). Courtesy mayavase.com

Finally, we might also follow up on Landa's statement that theater was performed at Chich'en Itza to suspect that it played a role in the activities described in the Dresden Codex Venus Table. Michael Coe reminds us that Landa describes a huge dance including hundreds of warriors as a critical part of a procession. "All sorts of special dances proper for the New Year could be performed on the way. For Muluc years, for instance, a great war dance was staged with as many as 800 warriors participating" (Coe 1965, 102).

The platform of the Tz'iknal would have provided a suitable place for a performance visible to a large segment of the public. Perhaps the lords themselves, or professional actors, dramatized a battle between the central warrior figure and his victim on pages 46–50, culminating with the faked spearing of a costumed warrior—again, not unlike the modern Atitlan ritual battles between "deer" and "tigers." Ajk'uhuun K'uk'ul Ek' Tuyilaj would have been responsible for training the warriors in their performance, culling her instructions from the Venus pages along with the mythology of each individual warrior or victim. They probably were accompanied by dancers representing the different noble houses, each with instructors charged with training their members in their specific roles.

Again, if we follow the analogy through, there is a perfect location on the Tz'iknal structure for a warrior effigy to have been placed, commemorating the theatrical performance. In the stairway leading up to the main building, there is a niche that is well suited to housing a ceramic deity effigy. This is the niche that Aveni noted in his study is aligned with a maximum northern setting of Venus along the horizon. In this way, the warrior deities could have been intimately associated with Venus from the public's perspective, without actually having to be identified with Venus on a technical level.

This exploration of the material culture supporting the rites of the Venus Table highlights the economic investment necessary. For a city the size of Chich'en Itza in the Terminal Classic, this would very likely have been incredibly useful. Twenty different effigy censers may have meant as many as twenty different "principales" at whose houses the effigy incense burners would have resided when not in office within one of the Tz'iknal windows. The different duties associated with keeping the effigy censers and cleaning and decorating the roads could have been distributed throughout the families of each ch'ibal. Moreover, the duties of preparing food and gifts for the festivities accompanying the rituals would have required greater contributions such that we can imagine that these ceremonies served as something of an economic engine for the city.

Twenty different noble houses may be too many for a city of thirty thousand inhabitants, and that probably would have been awkward given that some incensarios would only sit atop the Tz'iknal for short periods of time. But if each noble house was charged with keeping the effigy censers for a complete Venus Round, or four each, then it would take only five ch'ibals. And these could have been distributed among the lesser families within a house by rank. As the city grew, however—and we know that by the late Terminal Classic, the city center expanded northward toward the Great Cenote perhaps to a population of fifty thousand—many more opportunities for political and economic engagement would have been necessary. To this specific issue, we will return in chapter 8. For now, we recognize that a single set of five illustrations in the Venus pages of the Dresden Codex may actually represent a window into the lively ceremonial activity of the city.

REMAINING ANOMALIES

We have at this point moved away from an interpretation of the Venus pages that sees them made up of very subtle and esoteric computations and restricted to the interests of a few astronomer-advisors to a ruler. We can now appreciate that an emphasis on the verb *k'al* brings along with it a ritual life underlying many different time-space ritual activities that may well have had a large public presence. In fact, this would help us make sense of an anomaly within the pages that has received uneven treatment—in the general scholarship, but also here in this book. The subperiods of the Venus Rounds are with one exception rather difficult to make sense of.

As noted by Ernst Förstemann in 1894 and referred to in chapter 1, the subperiods of the Venus Round do not align with the observable events they are purported to capture.

The only period that works well is the period of invisibility between last Eveningstar appearance and first Morningstar appearance. This period is on average eight days, and that is the interval recorded for it on all five pages of the table. The interval given for Morningstar visibility is 236 days, but on average, it should be closer to 263. This shortening means that superior conjunction—the time interval between last morning visibility and first evening visibility—has been increased to 90 days, whereas it should be 50 days. Finally, evening visibility is given in the table as 250 days, when again it should symmetrically reflect the 263 days of morning visibility.

Efforts to address these deviations by modern scholars appeal to other planets and the possibility that the author of the original table intended to capture conjunctions of Venus with other planets or the relationship between Venus and the Moon (Aveni 1992). But we should now recognize that there is a distinctly simpler rationale that has emerged. For one, it is unlikely that most of the public would have been observing Venus on a regular basis, so not many (if any) would have recognized a deviation between the ritual event and the actual visibility of Venus. Second, we have already noted that the precise viewing of Venus would have been limited to those with views from the Tz'iknal or other very tall structures. From the plaza floors, the public would have to take the astronomer-priests—or their relatives who were privileged to accompany these astronomers—at their word. Of course, days after a first morning or first evening visibility event, a commoner could readily corroborate the likelihood that the prediction was accurate. Midway through either its morning or evening visibility, Venus would be easily viewable and recognizable by public viewers on a casual basis. Verification that the priests were doing their jobs was possible by seeing Venus high above the western horizon near sunset and seeing that a ceramic effigy sat in the western window. What we are getting to, then, is that accurate capturing of the subperiod visibilities of Venus may not have been important to the public life of the Venus pages. As a proxy for close observation, commoners would be able simply to have a look at the Tz'iknal to see which window was inhabited by a deity effigy to know where Venus would be in its circuit.

In fact, this helps with another factor already discussed: the variability of Venus's visibility from one Venus Round to the next would have made accuracy of prediction nearly impossible anyway—at least without incorporating geometry into the mathematical/calendric model. What would have been important is long-term accuracy as noted above. In turn, this means that there was a degree of flexibility available to K'uk'ul Ek' Tuyilaj as she broke up Venus Rounds into subperiods. And it appears that she took advantage of it.

A closer look at the Venus Table reveals that there is a pattern to the specific Day Signs listed. When we take them as a set, we find that not all Days Signs show up even though there are twenty available spots. Instead, precisely half show up. This suggests that the subperiods were not strictly intended to capture observable phenomena; rather, they were selected to preserve specific paths through the 260-Day Count. The omens of each Day Sign, therefore, were being taken into account. In the next chapter, we look more closely at these omens and the use of the Venus Table for the noble class as well as the ruler himself or herself.

For now, we may summarize this chapter to start to see more of K'uk'ul Ek' Tuyilaj's daily life. She may well have lived in her parents' house not far from the Puuc architecture around the Tz'iknal. In the late afternoon, she ate tamales with her family before packing up her materials and leaving for work—intending to stay awake most of the night, watching the sky from her perch in the windows and on the ledge of the Tz'iknal. She would keep her own count of the hours of the night, even if it wasn't her official duty, in order to follow the movements of the celestial bodies. After a long night and morning of observations, she would return home and have breakfast with her family, probably an atole, maybe flavored with chocolate and chiles. She would then spend the morning catching up on sleep in her room, back and away from the central patio of the extended family's residence. While the rest of the family went about their daily work, K'uk'ul Ek' Tuyilaj would sleep through the morning to rise in time for the afternoon meal. After this meal, she would return to the city center for meetings with almehen and perhaps the k'uhulajaw himself or herself. Just what those meetings would have considered, we turn to in the next chapter.

CHAPTER 7

Oracular Science

Now concertedly venturing into the reinterpretation of the Venus Table at Chich'en Itza, it is worth recalling that what we have as the Venus pages of the Dresden Codex is Tawiskal Uwoojil's copy of an earlier manuscript. The earlier version, which itself may not have been the first version, was probably written centuries before our scribe took up his brush. This introduces an important complication. As with Hoil's work on *The Book of Chilam Balam of Chumayel*, we should consider, How much of the content of the Venus pages was Tawiskal Uwoojil copying, and how much was he authoring? We know from chapter 4, for example, that the lower portions of the columns of text on the left-hand sides of pages 46–50 contain some kind of incongruity or "shift," pointing to the implementation of a correction to the Venus Table relative to the 365-Day Count (Bricker and Bricker 2007; Lounsbury 1983; 1992a; Thompson 1978). This suggests—and scholars have consistently taken this as evidence—that the upper portion was constructed for an earlier time period, and the lower portion was meant for Tawiskal Uwoojil's own prognostications. As we have seen in chapter 3, this provided the evidence for which Thompson and most scholars since have implemented the corrections on page 24 (Bricker and Bricker 2007, 102).

To increase the resolution of our understanding of how this shift occurred and what it meant, this chapter takes up the project of considering the intellectual life of Tawiskal Uwoojil's predecessor as a member of the royal court or as an advisor to the rulership of Chich'en Itza, working at the Tz'iknal. That is, we take the upper portion of the Venus pages as having been written first and preserved in copies up until Tawiskal Uwoojil's time; then we take the lower section to have constituted Tawiskal Uwoojil's own adaptation of the text. Here, then, we enter a consideration of the multiple meanings of astronomical inquiry within ancient Mayan culture. In turn, this provides a structure to much of the Venus pages that would have already existed long before Tawiskal Uwoojil's work, even if he did "update" it. So we consider the origins of the Venus Table at Chich'en Itza along with the daily contributions of the astronomer to court life and how it all set up the possibility of scientific discovery in the ancient world.

A YOUNG ASTRONOMER

Landa makes no reference to any women priests or scribes in his description of Yucatec society before the arrival of Europeans; that is just as likely his own chauvinism as the possible self-editing of his consultants, Xiu and Cocom. The archaeological record, on the other hand, attests to the highest status of women in Mayan society—for example, with the burial of Na Batz' Ek' within the tallest structure at Caracol in southern Belize (Martin and Grube 2008, 91), or the daughter of B'ahlaj Chan K'awiil at Dos Pilas, who went to Naranjo to serve as paramount at the city (Martin and Grube 2008, 74). While in retrospect the hieroglyphic record gives relatively few examples of women k'uhulajaw, that in no way would have diminished the presence of Na Wak Chanal Ajaw in her own time, represented on stelae in ceremony and as warrior, standing on captives (figure 7.1).

We also know that robust scribal communities attended to the needs of the k'uhulajaw during the Late Classic at each Mayan city (Coe and Kerr 1998; Miller and Martin 2004; Sharer 2006, 120–25), more often than not collaborating in teams on the public monuments that have endured the test of time. As we saw in chapter 4, the stone-working scribes at Copán under Yax Pahsaj Chan Yopaat, for example, had served the dynasty's needs for generations; by the Late Classic, they were creating works for their own homes that rivaled those of their patrons. At other Late Classic cities, the hieroglyphic signatures of some scribes show up to demonstrate that the most notable were allowed to travel from one city to another—perhaps their work was provided as a favor or gift to a peer k'uhulajaw under a system of patronage (Reents-Budet 1998, 74). Within these signatures, we again find that in the painted hieroglyphic record women scribes signed their artwork (figure 7.2).

FIGURE 7.1 Naranjo Stela 24 depicts Na Wak Chanal Ajaw standing on a captive

It is not unreasonable, therefore, to suggest that the author of the original Venus Table was a woman. If

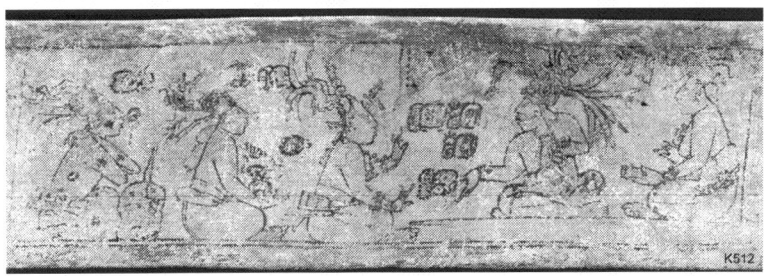

FIGURE 7.2 Female scribe at a royal court (Justin Kerr, Kerr #512). Courtesy mayavase.com

so, she may have followed an educational path at Chich'en Itza akin to that referred to in this book's introduction, in which priesthoods were separated by gender for both males and females and the most talented were recruited into an academic path during their youth. If we identify this author of the first Venus Table as K'uk'ul Ek' Tuyilaj—the evidence for which is covered in the next chapter—then we can picture her at Terminal Classic Chich'en Itza, perhaps from a family of established scribes, living in one of the more affluent residential complexes of the city such as Mak Chanil at Copán; or perhaps she came from the more remote, commoner homes, now making a name for the family as suggested at the end of the last chapter. Either way, she may have visited her family in the suburbs on special occasions, but primarily spent her working hours in the city center, studying with other priests, preparing and celebrating ceremonies, and advising elites and rulers when necessary. To the latter job duties we can now add information specifically from the Dresden Codex.

PROGNOSTICATION

For the primary ceremonial duties with which K'uk'ul Ek' Tuyilaj would have been charged, we return to Landa's words from the last chapter:

> They believed that if they did not observe these ceremonies, they would be sure to have certain sicknesses, which they have in this year. . . . When once the ceremonies were ended and the evil spirit was chased away . . . they considered the year as a good one. . . . But as it happened frequently that they had calamities, . . . so that when these misfortunes came, they should lay the blame on their ceremonies and on those who performed them. (Tozzer 1941, 142)

From this description, it might be easy to impose a modern perspective and dismiss "these ceremonies" as superstitious and thus move on to other considerations. There are two primary points within this statement, however, that provide material for a more interesting interpretation through a richer form of cultural or epistemic translation. In other words, in order to approach more closely how K'uk'ul Ek' Tuyilaj may have lived Landa's description, we first consider the nature of omens in ancient Mesoamerica; second, we explore the broader sociological use of oracles in Mesoamerica and cross-culturally.

Concerning the first, we see in Landa's statement that there is a sense that the omens are characters of the time period in question, and that character is evoked by the deities who preside over the period. And this takes us back to the Dresden Codex.

As we saw in the last chapter, the parallel to Landa's case comes from the middle text on the right-hand sides of pages 46–50 of K'uk'ul Ek' Tuyilaj's manuscript. For the content on page 48, for example, she wrote:

> The East is enclosed by Tawiskal and Chak Ek'. He speared Ixim [the Maize God]. It is the omen of the east of the land, of the caves. It is the omen of [a place].

We have seen that other scholars have interpreted this violence—of spearing the Maize God—to be something that is universally to be feared (Bricker and Bricker 2007, 107; Lounsbury 1978, 789). Harvey and Victoria Bricker write: "The violent theme shown in these pictures suggests that the heliacal rise of Venus as a morning star was an event dreaded by the Maya" (2007, 109). Again, were we to cast the intellectual context as mere superstition, we might be inclined to recognize the "warrior" character of Venus as dangerous or at least threatening to humanity.

On the other hand, it is worth recognizing that the violence that K'uk'ul Ek' Tuyilaj included throughout the table is not necessarily something injurious to humanity. There is clearly a negative outcome for the entity who is speared, but that does not require that the overall omen is a bad one for everyone or for humanity in general. Bricker and Bricker's own statement continued with an opposite attestation—even though they chose to disregard it: "This situation is quite different from what Barbara Tedlock (1999, 43) has described for the people of Momostenango, who regard the heliacal rise of Venus as morning star as a lucky event" (Bricker and Bricker 2007, 109).

Turning to our scribe's text, if the Maize God is speared—as, for example, on page 48—we *might expect* that it would have been interpreted as a failure of crops, which in turn would have been bad for the community of humans dependent on that corn. But it is much less straightforward what she meant by the spearing of "Chak Balam" (Great [or Red] Jaguar) as on page 47. K'uk'ul Ek' Tuyilaj wrote that this spearing would have accompanied (or perhaps generated) *u muuk yohl k'uh, k'uhul wayis ajaw*—"the omen of the k'uh heart of the k'uhul wayis ajaw." But if that is something to be avoided or celebrated, she does not tell us. Likewise, it is not clear why the spearing of the Maize God generates an omen for "the land and the caves of the east," whereas the spearing of an (undeciphered) deity is what generates *u muuk Ixim*—"the omen of the Maize God." Clearly, we would have to know if the undeciphered deity were positive or negative to determine whether the people, the deities, and the Maize God would want him speared or not (figure 7.3). Either way, no omen is a priori characterized as good or bad for the city or its inhabitants.

u muuk statements

FIGURE 7.3 Omen statements in the Venus Table. Courtesy SLUB

In fact it appears that, at her community's level, the opportunity to respond ritually was more important than the ritual violence or whether or not an omen was "good" or "bad." As Landa suggests, the omen prescribed a ritual response, and that response would be intended to mitigate any negative effects of the omen, or enhance any positive ones. K'uk'ul Ek' Tuyilaj's hieroglyphic texts recorded the violence in order to prescribe the necessary ceremonies through which the populace would respond. Framed in this way, omens are generated by the time period and the deities involved, but the ajk'ins and ajk'uhuuns understand this and so conduct the appropriate ceremonial activities to avoid any possible negative effects. The public trusts the priesthood to conduct the proper ceremonies such that the result should be a net positive or null effect. As long as the ceremonies are performed properly, an outcome of normal daily life should result. Looked at this way, omens and their bearers are not to be unquestioningly feared, but are to be understood as necessitating ritual response.

As long as the effects of the deities can be planned for, the community should be protected and allowed to maintain a form of balance internally and externally. The point is that whether or not we as "modern" scholars characterize specific beliefs as superstitious, the societal structure that results allows for rational engagement by the public and community leaders (cf. Latour 1987, 179–213).

And this is what we see in the rest of the Dresden Codex. By recognizing that the Venus Table is rife with divination, we find that it resonates with the rest of the content of the document—it is not just an isolated astronomical text within an otherwise opaque manuscript. Moreover, by reviewing this other material, we see that K'uk'ul Ek' Tuyilaj was probably only able to squeeze in her observations of Venus and the night sky in general within a schedule full of divinatory consultations for official political purposes or perhaps even for personal requests of the almehen. Hers was not a life of leisurely contemplation, but one full of public and private ceremonial engagement.

Taking scale into account, we may consider that even the village ritual in rural Yucatan documented ethnographically by Robert Redfield and Alfonso Villa Rojas can provide substantive insight into K'uk'ul Ek' Tuyilaj's job duties (Redfield and Villa Rojas 1962, 138–43). Redfield and Villa Rojas were able to witness both large and small ceremonies through Sylvanus Morley's recommendation in the 1930s—Chan Kom being not far from Chich'en Itza where Morley was working (1962, x) (figure 7.4). While the large ceremonies they witnessed such as the *cha'chaak* required involvement of the whole

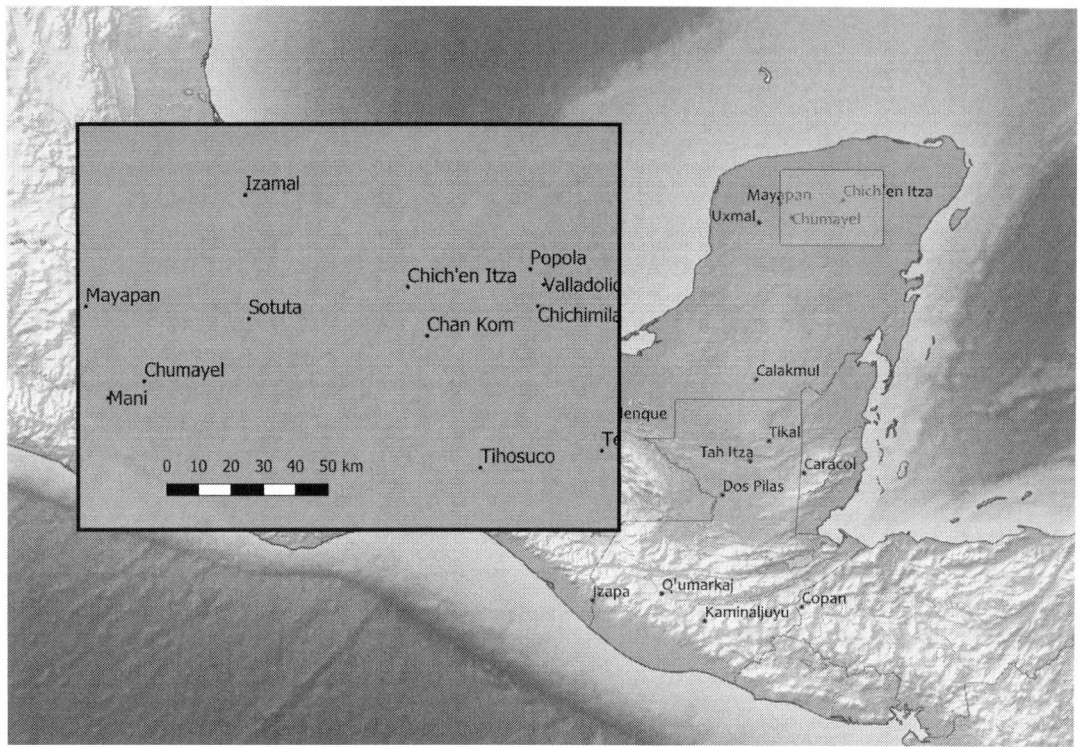

FIGURE 7.4 Chan Kom near Chich'en Itza in northern Yucatan

town's population, under the ritual specialist's direction even the smaller ceremonies such as *u hanli kol* (the dinner of the milpa) for an individual *milpero* (maize farmer) required organization and labor provided by family and friends. That smaller ceremony took only a few hours, but preparations took days, involving far more than those with specific ritual roles.

Central to the u hanli kol event, for example, was a small altar. This altar was built for the event as a table from branches and saplings taken during the clearing of brush and constructed by the *jmeen* with help from the milpero and his friends. On the altar were placed maize breads of different kinds. These too were made by the milpero and his associates under the direct supervision of the jmeen. They were accompanied by ritual drinks such as *balche*—an alcoholic beverage that took several days to prepare. Also critical to the ceremony was the ritual killing of chickens and turkeys. The fowl would be handled by four individuals selected by the jmeen from the group; he named them "Chaaks" in a form of deity impersonation. The four Chaaks assisted the jmeen by holding the fowl while he poured balche down their throats to drown them. After the drinks, breads, and fowl were all cooked and/or prepared, the jmeen offered them to the deities for supernatural consumption along with burning incense. Following an appropriate amount of time, the food and drink were divided among those present for human consumption (Redfield and Villa Rojas 1962, 143). Children, adolescents, and adults not participating in the ritual activity did join for the food and festivities. Overall, with its prepared location, offerings of incense and food, deity impersonation, and reliance on multiple forms of labor coming from across a local community, the basic structure of village ritual in twentieth-century rural Yucatan still accords well with the New Year ceremonies described in the Dresden Codex in chapter 6 of this book.

For much larger-scale processions and feasts to be performed appropriately at Chich'en Itza, a more sizeable segment of the population would have to contribute. During her tenure, K'uk'ul Ek' Tuyilaj would of course have been charged with attending to the omens generated by Chak Ek' interacting with other deities. Her particular duties would have included preparing for and directing processions carrying the appropriate effigy censers for the time period. Receiving effigy censers from the lords of noble houses and rotating these effigy censers around the windows of the Tz'iknal were symbolic actions within logistically complex social relationships and activities. The latter social activities would bring in the public at varying levels politically and economically again, as we saw in the last chapter. Moreover, beyond their participation in the labor of preparation, it is perhaps relatively straightforward to note that the political relationships between ch'ibals and between almehen and macehuales within the populace would have to be attended to in order to ensure their respective participation in these ceremonies. Lineage heads might vie for specific roles in the ceremony, such as deity impersonations, and lineage members might compete among themselves for the roles charged to their ch'ibal (figure 7.5).

Recognizing these layers of social use of religious omens actually brings up the second aspect of Landa's comment, which resonates cross-culturally. In the quotation above, Landa notes that if a favorable result does not follow the ceremonial event, then the failure is attributed to the conduct of the ceremonies or the human authorities charged with them. This is where another agency of the priesthood may have come into play. Ajk'ins would not have had their work strictly bound to observations, computations, and ritual performance—K'uk'ul Ek' Tuyilaj

FIGURE 7.5 Images of craft production for ritual purpose: (*top*) the artist in the upper left works on a mask; (*right*) an ajaw wears a mask in a performance or ceremonial event. Courtesy maya vase.com

and her colleagues also had a political charge. Here we appeal to a resonant practice in ancient China, recovered by the sinologist Hans Bielenstein.

"POLITICAL SCIENCE"

For "Western" cultures, the Scientific Revolution served as that process by which the conduct of European science was removed from intellectual submission to religion. In Medieval Europe, theology maintained by the Catholic Church held the highest position in the generation of knowledge; strictly below theology was philosophy, which included natural philosophy and what we recognize today as "the hard sciences" (Westfall 1989, 218). Astronomy and mathematics were part of the basic education of the elite and the clergy, along with music, all to be studied after mastering logic, rhetoric, and grammar. These all, however, were hierarchically lower than philosophy, which itself was only a prelude to theology.

It was Galileo Galilei in the seventeenth century who captured the notion that scientific investigation should be held to its own standard. His primary concern, as expressed in his (not-so private) "Letter to the Grand Duchess Cristina" of the Medici family, was that scientific investigation might contradict knowledge derived from theology, and that would be detrimental both to the overall doctrines of Catholicism and to natural philosophy (Drake 1957, 186). For his letter, Galileo borrowed a catchphrase from his colleague Cardinal Baronius: "The intention of the Holy Ghost is to teach us how one goes to heaven, not how heaven goes" (Drake 1957, 186). In Galileo's view, knowledge of the physical world followed its own logic, independent of the biblical record.

While this notion of the separation of religion and science often is taken for granted today, the point is that it was the norm in the ancient world for cultures to maintain views similar to that of pre–Scientific Revolution Europe. David King (1993) described the situation in Medieval Islam as "science in the service of the state." Similarly, Hans Bielenstein found considerable political and governmental use in the observation of natural phenomena in ancient China. Of specific concern here, Bielenstein noticed a provocative pattern when looking at historical records: more celestial and terrestrial portents were recorded for difficult political times than for times of peace or prosperity (1950, 1984). Bielenstein suggests that the connection did not arise from an actual difference in the frequency of natural "signs," but from the priestly class's use of them to critique the emperor. In other words, priests could not question the daily decisions of the emperor *directly*; but the emperor was understood to preside over the social *and* natural worlds, so if natural disasters or unusual phenomena occurred in the natural world, they could be interpreted as signs of improper rule (Bielenstein 1950). A resonant mechanism seemed to be in place within Landa's description seen above, where the enactors of ritual response were held accountable if the negative effects of an omen came to pass.

One of the sections in Juan Josef Hoil's *Book of Chilam Balam of Chumayel* itself begins with a record of a ruler's authority and its reliance on prophecy.

Then began his prophecy. Then they began to declare him ruler. Then he was set in the seat of the rulers by them. Then they began to declare him head-chief. He was not the ruler formerly; that was only the office of Ah Mex Cuc. Now the representative of Ah Mex Cuc was declared ruler. The eagle, they say, was his mother. Then, they say, he was sought on his hill. Then they began to take the prophecy of this ruler after it was declared. Then they began to set aloft the house on high for the ruler. Then began the construction of the stairway. Then he was set in the house on high in 13 Ahau, the sixth reign. Then began the hearing of the prophecy, of the news, of the setting up of Ah Mex Cuc, as he was called. Then he carried nearly to Baca the news of Ah Mex Cuc. He was placed there. Then he began to be treated as a lord; then obedience to the name of Ah Mex Cuc began. Then he was obeyed; then he was served there at the mouth of the well. Chichen Itzam was its name because the Itza went there. (Roys 1933, 28)

The term used here attests to the prior metaphoric meaning of *k'in* in Mayan languages, semiotically supporting the reading of "the prophecy . . . the news" as resonant with the term in the Venus pages: *u muuk*. In the middle of the passage, for example, we confront the specific translation choices that Ralph Roys made:

Then began the hearing of the prophecy, of the news, of the setting up of Ah Mex Cuc, as he was called. (Roys 1933, 28)

Caɔuni : v Kuchul v yabil vthan vkin : vua ah mex cuc : v kaba. (page 7r of original facsimile; see Gordon 1913)

What Roys has translated here as "prophecy" and "news" are literally "his words" (*u than*) and "his day/Sun/time" (*u k'in*). Again, these resonate with the *Popol Vuh*, wherein the K'iche' term *ubixic* is used for divination, which Dennis Tedlock translates as "'its saying' or 'its announcement,'" (1985, 2). More to the point, according to *The Book of Chilam Balam*, the ruler was responsible for prophecy, and undoubtedly the ruler relied on his or her ajk'ins and ajk'uhuuns to assist in developing them. Reading signs in nature was also politics in the human realm.

Given this broader review of the political and social relevance of oracular knowledge, we encounter a richer context for astronomical knowledge. Absent a post–Scientific Revolution perspective to guide it, Mayan astronomy was most likely subordinate to or woven within other intellectual pursuits—as we have seen at Copán, where Mak Chanil brought together astronomy, history, art, and architecture. Certainly astronomical observations would have generated patterns that could be interpreted as signs, but the signs themselves held the most weight for the priesthood and its role in society. Just as scribes were not creating earthquakes in China, but they were ascribing meaning to those earthquakes and other natural events, so Mayan priests were charged with determining the meanings of the portents witnessed in the sky. Robert Sharer captured an aspect of this in describing the religious character of rulership:

For each bumper crop of maize or each victory over a rival power demonstrated that the gods looked favorably on both the king and his kingdom. The allegiance to the ruler by the ruled was strengthened, and the morale of the entire kingdom was bolstered. Likewise, failures diminished the power and prestige of the king. Minor failures could be explained by other factors, rather than be taken as signs of supernatural disfavor toward the ruler. Thus, as long as the belief in the king's supernatural connections remained intact, the system would not be threatened. But major failures, such as conquest by a rival power, catastrophic epidemics, or droughts, could shake belief in royal power and place both king and kingdom in jeopardy. (Sharer 2006, 89–90)

An important layer of nuance comes into the picture when we venture further into indigenous records of oracular knowledge. One in particular offers substantial evidence for revision of popular culture representations. Another passage in Hoil's *Book of Chilam Balam*, for example, attests to the responsibility of the ruler and brings back explicitly the role of prophecy.

> Then began their reign; then began their rule. Then they began to be served; then those who were to be thrown (into the cenote) arrived; then they began to throw them into the well that their prophecy might be heard by their rulers. Their prophecy did not come. It was Cauich, Hunac Ceel, Cauich was the name of the man there, who put out his head at the opening of the well on the south side. Then he went to take it. Then he came forth to declare the prophecy. (Roys 1933, 28).

The latter is a fascinating passage on many levels. At first blush, it appears to resonate with popular culture representations of Mayan practices of human sacrifice: men being thrown into a cenote. But the text here provides details that suggest otherwise. The Yucatec of the fuller passage here is as follows:

> Cahopi v tepal lobi : Ca hopi ti yahauli lobi : Ca hopi v tan la baleb : Ca hopi v kuchul u pululteob : Ca hopi v pulicob y ch ch'een : ca vyabac v thanob tumenel yahaulili : mahal v thanob lay Cauich hun hunac Ceele : Lay Cauich v kaba v uinicile : ti cu thical tu hol cheen : ti nohol catun bini chabil catun hoki yalab v than : ca hop u chabal vthan : ca ɔuni v than : ca hopi yalabal ahauil : ca cul hij : ta cuchil ahau vob : tumenob ca hopi ya labal halach vinicil : ma ahau cuchij chen u bel ah mex cuc. (pages 6v–7r of original facimile; see Gordon 1913)

> Then began their reign; then began their rule. Then they began to be served; then those who were to be thrown (into the cenote) arrived; then they began to throw them into the well that their prophecy might be heard by their rulers. Their prophecy did not come. It was Cauich, Hunac Ceel, Cauich was the name of the man there, who put out his head at the opening of the well on the south side. Then he went to take it. Then he came forth to declare the prophecy. Then began the taking of the prophecy. Then began his prophecy.

Then they began to declare him ruler. Then he was set in the seat of the rulers by them. Then they began to declare him head-chief. He was not the ruler formerly; that was only the office of Ah Mex Cuc. (Roys 1933, 28)

Men were "thrown in" according to the passage, so their prophecy could be heard. A close reading recognizes that Hunac Ceel—a ruler whom we will encounter again in the coming chapters—had to put his body in a specific position to hear the prophecy that resulted from people being thrown in to the cenote. What this highlights here is the question of what he expected to hear. We will return to this question below, when we consider the interpretation of divination itself. At this point, the main takeaway is the role of prophecy in establishing political authority.

As to where this oracular knowledge would have been situated sociologically, we return to the overall political organization of Mayan society, which we saw in chapter 1. Recognizing that different priesthoods were dedicated to different relationships between natural phenomena and individual deities, we now confront a sociopolitical scene that would have looked something like a hybrid of modern academia and modern political parties. Each priesthood would have a specialized "field of study" in the signs to be read, but each would also possess political influence relative to the rulership and the almehen in how they distributed responsibilities for different ceremonies. Such a context would mean that the "sciences" would be pursued according to their own internal logics, but in the end the fruit they bore would have been woven through with religious and political interests (cf. Aldana 2007). Such a perspective allows us to see divination sitting alongside astronomical observation—precisely as we have seen recorded in the hieroglyphic text of K'uk'ul Ek' Tuyilaj's Venus pages.

To this point, then, the reader may have noticed that an internal inconsistency appears to have emerged within our treatment of Mesoamerican oracles. On the one hand, we have followed Landa's suggestion that specific omens were the regular and expected outcomes of deities in charge of specified time periods. On the other, we have seen that the unexpected is what allowed for the priesthood to intervene politically within governmental pursuits. If the omens were regular and expected, what room might there be for the unexpected? To address this issue, we now turn to what we know about oracular knowledge in Mesoamerica. Fortunately, the practice is heavily documented in historical and ethnographic records.

Divination

When we turn to descriptions of ritual specialists recorded in colonial times, divination is included within their official duties. A core reference thus far has been Landa's *Relación de las cosas de Yucatan*, in which he writes:

"The sciences which they taught were the computation of the years, months and days, the festivals and ceremonies, the administration of the sacraments, *the fateful days and seasons, their methods of divination and their prophecies,* events and the cures for diseases, and their antiquities and how to read and write with the letters and characters, with which they wrote, and drawings which illustrate the meaning of the writings." (Tozzer 1941, 27–28, emphasis added, quoted in Zender 2004, 94)

Here the reference to "fateful days" and "their methods of divination" along with the specific "prophecies" appears specifically to reflect the content we have already seen throughout the Dresden Codex. Attesting to the persistence of these job duties even when confronted with Spanish colonialism, Andrés de Avendaño wrote more than a century after Landa that the indigenous books he encountered—which were used for songs—also

están pintados por una parte y otra con variedad de figuras, y caracteres . . . que indican no sólo la cuenta de los dichos días, meses y años, sino las edades, y las profecías que sus ídolos y simulacros les anunciaron.

are painted in different parts with a variety of figures and letters . . . that give not only the count of said days, months and years, but also the ages and prophecies which their idols and images announced to them. (1997, 42; [cf. Zender 2004, 85–86])

This description also fits well with the contents of the Dresden Codex.

In his *Study of Classic Maya Priesthood,* epigrapher Marc Zender goes further, referring to divination as the basis of priestly activity across specific titles:

In addition to divination, most priests seem also to have been capable of performing healing rites, and the knowledge of herbs and curative bloodletting seems to have been a staple of recondite priestly knowledge (Chuchiak 2001:136–42; Roys 1943:88, 93–95; Thompson 1970:312). (Zender 2004, 93, emphasis added)

Furthermore, Zender translates the Yucatecan term *ajk'in*—often rendered "daykeeper"—as "'diviner' (glyphic **AJ-K'IN-ni** for *ajk'in*, literally 'he of the forecasting')" (2004, 93), where he emphasizes the more abstract, temporal character of *k'in.* From the cognate title in K'iche', *ajqiij,* we find that indigenous divination has persisted within the more traditional communities of the Mayan region.

Barbara Tedlock's and Dennis Tedlock's ventures into K'iche' lands and cosmologies already have paid substantial dividends in various venues. Here we consider their contribution to an opportunity, hinted at by Aveni (1980, 190), though not yet fully explored. Aveni refers to the work of Sharon Gibbs, which addresses the pattern observed at the end of the last chapter, showing that only certain 260-Day Count Day Signs appear in the Venus Table; they propose that this may reflect a practice in which

the priests were compelled to refer the Venus observations in the table for the purpose of staging the religious ceremonies and enunciating the prognostications that would have attended the Venus appearance and disappearance. . . . The celebration of actual Venus observations on special named days that may not necessarily coincide with the actual event is rather like our custom of officially assigning certain holidays (Memorial Day, Lincoln's birthday, etc.) to the Monday of the week in which they occur. (1980, 190–91)

Although it has not yet been explored by others studying the Venus Table, Gibbs and Aveni suggested a counter to the predominant assumption that we saw growing out of Teeple's work—that accuracy of prediction was paramount within Mayan astronomy. Here they are explicitly noting the possibility that a constraint to specific Day Signs may have been more important for the ritual meaning of a Venus event (figure 7.6). This proposal introduces further options for considering the relationship between astronomy and divination within the above context of generating "signs" for political use. To see it, we turn to the Tedlocks' modern ethnographic work on Mayan divination.

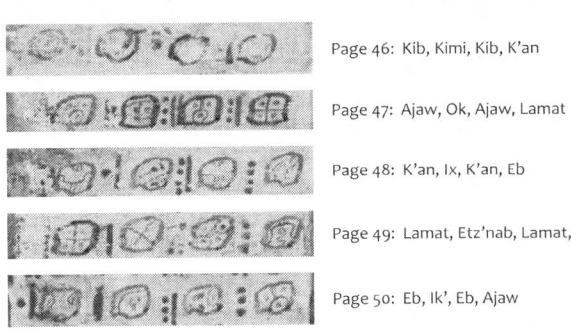

Page 46: Kib, Kimi, Kib, K'an

Page 47: Ajaw, Ok, Ajaw, Lamat

Page 48: K'an, Ix, K'an, Eb

Page 49: Lamat, Etz'nab, Lamat, Kib

Page 50: Eb, Ik', Eb, Ajaw

FIGURE 7.6 Day Signs on each page of the Venus Table. Precisely half of the Day Signs show up on these pages: Imix, **Ik'**, Ak'bal, **K'an**, Chikchan, **Kimi**, Manik, **Lamat**, Muluk, **Ok**, Chuwen, **Eb**, Ben, **Ix**, Men, **Kib**, Kaban, **Etz'nab**, Kawak, **Ajaw**. Courtesy SLUB

The Tedlocks traveled together to Momostenango in the highlands of Guatemala for extended periods from 1975 to 1976, entering the community of ritual specialists by themselves becoming trained as daykeepers (figure 7.7). While Dennis Tedlock focused on working with local ajqiij Andres Xiloj to take up a new translation of the K'iche' version of the *Popol Vuh*, Barbara Tedlock emphasized ethnographic work and the activities performed by K'iche' ritual specialists (B. Tedlock 1992, 47–53). From this experience, Tedlock describes a divination she witnessed in detail in part to challenge previous interpretations (for example, by Charles Wagley, Oliver La Farge, and Benson Saler) that argued for conscious manipulations of the process or even dishonesty by the diviners (B. Tedlock 1992, 170–71).

The overall divination process she describes involves

i.   taking a handful of seeds from a pile without conscious regard to the specific number of seeds;

ii. partitioning these seeds into groups of four, with the remainder forming its own pile (with further conditions restricting the number of seeds in the final pile);

iii. counting through the piles according to a progression through the 260-Day Count;

iv. addressing the days themselves as agents with "will"; and

v.   listening for the blood "to speak" throughout the process. (Aldana 2015, 90)

The process is repeated to ensure that the outcome of the divination can be verified. In abridged form, Barbara Tedlock describes a divination to assess a proposal of marriage:

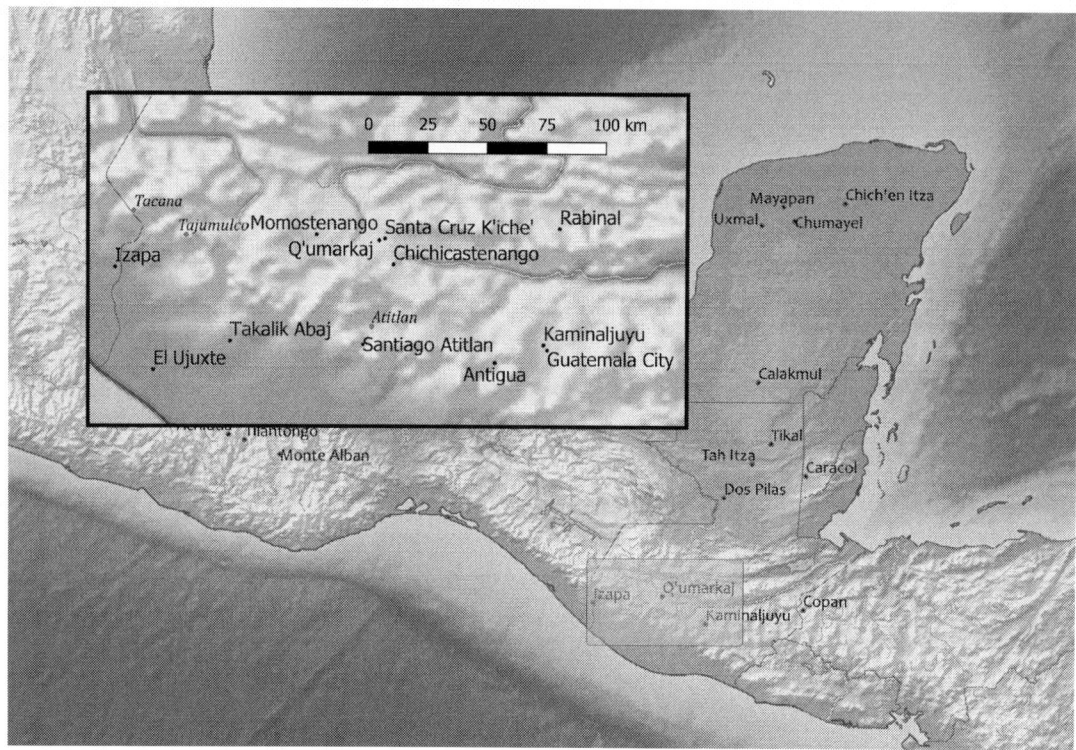

FIGURE 7.7 The K'iche region of the Guatemalan highlands visited by the Tedlocks

When all the seeds of the first arrangement have been set out, the diviner addresses the first group of seeds by the name of the day on which he is divining, for example, . . . "Come here Lord 1 Quej, you are being spoken to" . . . After the first day has been addressed and summoned, the diviner will repeat or allude to the question being asked. In a marriage divination, one might say, . . . "1 Quej, you are being spoken to about the seven good-nesses, seven fatnesses." . . . The counting may or may not be interrupted by the speaking of the blood. The moment it speaks, the diviner stops counting, noting the day name and stating the message. For example, if the blood moves in the right hand on the day 5 Batz', the diviner will say in a low voice . . . "Already he has it grasped, it says" . . . Now there is a moment's pause. If the message does not come again in the same locale or in the paired place . . . then the diviner may ask 5 Batz' . . . "Is it certain that it is you who speaks"? If the blood speaks again the message is confirmed. . . . Now the counting of the piles of seeds will resume. Let us say that in this case, it comes to 4 Ak'abal on the last group . . . This indicates that the marriage would seem to be certain, since one of the mnemonics for Ak'abal is *ak'abil* ("at dawn"), which would be the time of day to begin the asking. (B. Tedlock 1992, 162–68).

Again, our modern tendency to dismiss this type of activity as "superstitious" would stifle its translation as oracle in Mayan cultures to an interesting concept in contemporary "Western" culture. Stanley Tambiah (1990) proposes that practices—which could include this divination account—may be better culturally (or better: epistemically) translated by introducing an additional step into the interpretative process. Tambiah suggests that we build the "conception" of an oracle in modern terms using contemporary scientific tools. This modern/scientific conception can help to translate the "concept" for Mayan cultures or to consider it semiotically without actually "explaining" it from a presumed "correct" cultural position (cf. Taussig 1987).

To follow Tambiah's proposal, we first recognize that K'iche' ritual specialists brought together multiple synthetic and natural phenomena to constitute any given oracle. In the case that Barbara Tedlock witnessed, the 260-Day Count as oracle may be conceptualized as comprising a stochastic component (the grabbing of an a priori indeterminate number of seeds); the element of repeatability (the seeds may be used indefinitely, and in fact the process is repeated to ensure that the reading is one of confidence [B. Tedlock 1992, 163–64]); a language for interpretation that simultaneously constrains the possible outcomes (the 260-Day Count); and an overlying layer of stochasticity in the "speaking blood." When we put these together, we constitute a mechanism or a "conception" of an oracle.

A MODERN CONCEPTION OF A MESOAMERICAN ORACLE
a stochastic initiating event
repeatability
interpretive language
constraining mechanism
additional stochastic aspects

It is important to emphasize, though, that such an interpretation is merely a conception within Western academic thought mapped over from a concept within Mayan thought. The difference, of course, is in the ontology behind each interpretation. The case is pointed to in Barbara Tedlock's divination description: the ritual specialist addresses the 260-Day Count dates directly. The decision, or the will, of the dates is expressed in the speaking of the blood or the patterns made by the seeds. Because "we" as modern scientific observers may not recognize any "entities" behind the oracle, we would be inclined to attribute the "expressions" as chance events. I suggest here that for contemporary "scientific" cultures, the concept of chance or randomness is the result of an imperfect "map" or cultural/epistemic translation from the concept of the oracle within K'iche' culture. What an indigenous conceptualization would characterize as the willful response of an animate oracle, contemporary scholarship attributes to a "random" outcome.

Clyde Kluckhohn also made this point regarding his early twentieth-century Diné (Navaho) consultants:

The conception of "good luck" is hard to translate into the Navaho language. In their scheme of things one is not "lucky" or "unlucky." One has the requisite knowledge (sacred or profane) or one hasn't. Even in what European languages call "games of chance" the Navaho depends upon medicines, rites, and verbal formulas. The same is true with hunting. Getting a deer is never a matter of good fortune; it is a matter of ritual knowledge and one's relations with supernaturals which, again, are controllable. (Kluckhohn 1949, 362, cited in Blackburn 1975, 68)

That this is not simply a contemporary Mayan or Diné conceptualization is corroborated by the historical record of Mesoamerica.

A seventeenth-century divination record comes from Hernando Ruiz de Alarcón's *Treatise on the Heathen Superstitions That Today Live among the Indians Native to This New Spain*. Here too the basic address and the dependence on "chance" are the same as the much more recent K'iche' example given above.

They use this fortune telling in the following manner. From an ear of maize or from among a lot of maize, the fortune-teller selects the most outstanding and beautiful kernels. . . . Having chosen the said kernels, the seer bites off their nibs with his teeth.

. . .

Without our lingering on the number, which is not to the point, and coming to the execution, the seer, after having arranged the kernels on the said cloth, begins his fraud with those that remain in his hand, shaking them in it and tossing them in the air and catching them again many times. And then he begins [an] invocation. . . .

And at the time that he says the invocation he traverses the space that he has created with the stretched-out cloth at full speed with the hand in which he holds the kernels, moving his hand along the edge of the cloth over the maize kernels that he placed on it. *And the invocation is addressed to the maize kernels and to the fingers of his hands, as if attributing divinity to them.* After finishing the words of the incantation, he tosses the maize that he had in his hand into the middle of the cloth, and he judges the fortune according to how the maize kernels fall. The rule that they usually have in judging it is that, if the maize kernels fall face up, the fortune is good—for example, the medicine about which he is consulted will be good, or the lost person or thing that is being sought will show up—and the contrary if the maize kernels fall face down.

. . .

And they are believed like prophets . . . with all the basis for success being in whether the maize kernels fall face up or face down, and also in falling far from or near to the one who throws them. The first is chance, and the second the fortune-teller freely executes, throwing the maize kernels with more force or with less—less if he wants them to remain nearer. (Ruiz de Alarcón 1629, emphasis added)

While this oracle has no association with the 260-Day Count, the dependence on "chance" and on a petition to "inanimate" entities is again fundamental. Furthermore, it follows the structure of a Western conception of an oracle: *stochasticity*—face-up or face-down maize kernels after being tossed; *repeatability*; *language*—here only positive or negative. Perhaps more importantly, it demonstrates that *both* "chance" and the influence of the diviner are important. This is the theme that carries forward through other examples.

Cenote as Oracle

Given this more nuanced development of indigenous divination, we may return to the cenote prophecy that Hoil recorded in *The Book of Chilam Balam of Chumayel* and the authority of the figure known as Hunac Ceel. Cenotes in Yucatan often have water in them and can serve local communities as a water source (Sharer 2006, 53). But the karst environment of the Yucatan also produces vast subterranean spaces that combine dry land and lakes or running water. Accordingly, cenotes can function as water sources, but they can also be portals to the Underworld (Sharer 2006, 726–27). As such, Mayan ideologies represented them as the homes of the souls of ancestors or of deities (Tokovinine 2013, 125). It is most likely the latter aspect that provided an oracular opportunity for Hunac Ceel at Chich'en Itza. We have seen, for example, that in the case of K'iche' divination, the seeds provide the vehicle for the Days to speak; in the corn-tossing divination of Central Mexico, corn itself is the "divinity" addressed. For an oracle originating in a cenote, then, it stands to reason that the entities being addressed were residents of the Underworld—that is, ancestors or Underworld deities. The question then becomes, In the above description, whose voice was Hunac Ceel listening for?

A number of different scenarios arise from an oracle constituted by a person thrown/ leaping into a cenote. At one end of an interpretive spectrum, the process may have been rather mechanical—if the person survived, that produced one omen; if they did not, that produced another. At the other end of the spectrum, that person might be a trained diver aiming to have an experience underwater that would generate a message for communication to the ruler. Perhaps that is why Hunac Ceel had to go to a specific "opening" to listen for a response. Archaeologist William Ringle argues that just such underwater oracles were recorded in the Codex Nuttall. On page 15 of that Mixtec manuscript, for example, we see a woman named 1 Cuauhtli underwater, seated across from a male head and torso emerging from a serpent's mouth. The encounter resonates with the well-known image of "ancestor conjuring" depicted at Yaxchilan on Lintels 15 and 25 (Schele and Freidel 1990, 267, 287). Perhaps, then, the diver was to have an experience within the cenote Underworld and then rise to the surface to communicate that to Hunac Ceel.

In his study of the themes in Hoil's *Book of Chilam Balam*, Roys included

> [a] report by the municipality of Valladolid written in 1579 [that] covers a number of details omitted by Landa. Here we read: "This pyramid (the principal temple of the city) lies between two cenotes of very deep water. One of these is called the Sacrificial Cenote. Chichen Itz was named for an Indian who lived beside the Sacrificial Cenote and who was called Ah Kin Itz. It was the custom of the lords and nobles of all these provinces of Valladolid to fast for sixty days without raising their eyes during this time even to look at their wives or those who brought them their food. And this they did in order, when

they should arrive at the mouth of that cenote, to cast into it at break of day some Indian women of each of these lords. They had told them that they should ask for an abundant year <and> all those things which occurred to them. Thus when these Indian women were thrown in without being bound but flung down as from a cliff, they fell into the water striking it with great force. Precisely at mid-day the one who was to come forth made a great outcry for them to let down a rope to draw her out. When she came up half dead, they built large fires about her perfuming her with copal incense. After she recovered consciousness, she said that down below there were many of her nation, both men and women, and that they received her; but when she raised her head to look at any of them, they struck her with heavy blows on the back of the neck so that she should keep her head bowed. This was all within the water, below which there were believed to be many hollows and pits. And they replied to her whether there would be an abundant year or a bad one according to the questions which the Indian woman asked of them." (Roys 1933, 100–101)

In other words, according to this account, the "Cenote of Sacrifice" was actually an oracle. In the Valladolid case, the woman thrown in retrieved messages from her ancestors (those "of her nation"). The consultation was described as being challenging and physically demanding, which too invokes comparison to representations of conjuring at Yaxchilan. Besides the lintels showing women conjuring their ancestors through serpent portals, Lintel 39 depicts the k'uhulajaw himself, lying on the ground, wrestling with the deity K'awiil. In both cases, consultation appears not to have been a simple and passive engagement, but instead a difficult contest. The text on the Yaxchilan lintel states: *utzakaw k'awiil*—"he conjures k'awiil." The outcome, then, in both cases is an omen, which then would need to be addressed by ritual specialists.

Now we shouldn't exclude the possibility that different practices were maintained for different purposes at the same venue. Perhaps enemy warriors defeated in battle were thrown in to see if they could survive, but specially trained priests would dive in at other times in hopes of engaging residents of the Underworld and coming back to communicate the interaction. Regardless of the specific method, in order to get to the function of oracular knowledge in ancient Mayan society, we have to work against a simplified and mechanistically causal view of religion, which is strong in the scholarship. Robert Sharer, for example, writes:

The Maya cosmos was a living system with deities governing all aspects of the invisible and visible world—all that could be seen in the earth and the sky—and even the underworld beneath the earth. Each individual and social group had its role to play in this system, and the elaborate hierarchy of social roles and classes, surmounted by the elite and rulers, existed to maintain this cosmological order. The ultimate sanction, which reinforced elite authority and allowed kings to control their subjects, was the threat of supernatural retaliation for an offense to the gods. All Maya people, even kings, were subject to these supernatural beings. Any individual, from farmer to king, who deviated from an appointed task or failed in his or her obligations to the gods, would be punished by misfortune, illness, or even death. (Sharer 2006, 92–93)

Sharer's perspective is common, inspired and maintained by Western intellectual postures that dismiss indigenous peoples as technologically backward (witness frequent description as "Stone Age" civilizations) and as religiously overprescribed. While such interpretations facilitate mechanistic investigations into relationships between human communities and their environments, they do not reflect agency, internal discord, or "real-life" interactions within them. Such positions, however, have been continually fed by Europeans justifying conquest by military force and betrayed alliance during the Colonial period and combined economic and military force ever since (L. Smith 2002).

When we set aside our own biases built on these legacies, however, we gain new views on relationships between religion, politics, and science. We find that ancient Mesoamerican "science" as the observation, recording, and interpretation of natural phenomena did occur, but it was put to political and religious purpose. The scientific component itself was focused on the construction of constrained stochastic relationships, which were necessary for the credibility of the practice. This was the mechanism that allowed the outcome of any given divination to be separable from the interests of the diviner. Waxaklajun Ub'aah K'awiil's Venus window, for example, didn't control the movement of the celestial body, but it did constrain the interpretation of its appearance. Scientific efforts were required to construct those constraints. We have also seen how a recasting of oracular knowledge can provide access to a different type of knowledge—one coming from "the gods."

If we now are willing to acknowledge that science was put to the purpose of reading omens as messages from "the gods," it is now upon us to take a closer look at those terms that have been identified within indigenous languages, which scholars have commonly taken as translating to "gods" or "deities." While the latter practice facilitates many interpretations, it also presents complications for interpretation at the philosophical level. That is, we've created a model for how oracles may have worked in the abstract, but we can't expect that K'uk'ul Ek' Tuyilaj saw them in that way. In order to get closer to her own perspective, we take up a short investigation into Classic Mayan theology.

ON DEITY

Alfredo Barrera Vásquez's Yucatec Mayan Dictionary defines the term unambiguously: "**K'U** 1: Dios," and "deidad" according to the seventeenth-century *Motul Dictionary* and the more recent *Diccionario Español-Maya* by Ermilo Solís Alcalá (Barrera Vasquez 1995, 416). The issue that we have to address here is that this translation in language is also already a translation into a European concept; the concept itself is not necessarily present in that form in the original culture/language, and no attempt is made with these definitions to develop a "conception."

Borrowing from written colonial sources as well as archaeological interpretation, archaeologist Michael Smith warns us in our attempts to translate *teotl*, the Nahuatl cognate to Mayan *k'uh*. Specifically, he cautions against simply attributing a human personification to nature or natural phenomena/objects.

> The Nahuatl term *teotl* means "deity" or "sacred power." This is a complex and multifac-
> eted concept that does not fit well with modern preconceptions of ancient polytheistic
> religion. . . . Aztec gods . . . are better viewed as invisible spirits or forces whose roles,
> natures, and forms blended together. (M. Smith 2004, 199)

This "abstracted" personification reflects not only an Aztec or Mesoamerican concept,
but one associated with a wide variety of Precontact indigenous people of the Western
Hemisphere.

Let me be clear that I am not assuming or hypothesizing that all indigenous peoples
of the Western Hemisphere conceptualized divinity or deity in the same way. Rather, I
am suggesting that there are similarities that provide for the opportunity to recognize
common philosophical ground. The gods of Christians, Muslims, and Jews are not the
same, but they are resonant in concept. This kind of resonance is what I'm exploring
here in indigenous religiosity.

In his review of late nineteenth and early twentieth-century Chumash mythology,
for instance, anthropologist Thomas Blackburn notes what he calls "homologies" across
religious concepts of indigenous people throughout North America (1975, 66). In a
sense, Blackburn pushes Smith's abstraction further, referring to "causative agencies"
rather than using the terms "deity," "spirit," or "supernatural."

> Because all causative agencies in the universe are personalized and endowed with such
> human qualities as will, intelligence, and emotionality, and because the beings so con-
> ceptualized are considered to have the potentiality for either positive or negative action,
> unpredictability is an essential aspect of all events and all phenomena. (1975, 69)

While the ubiquity of personalized causative agencies resonates with Sharer's description, the resultant view differs dramatically. Humans are not living in constant fear of
gods and political authorities, but they are surrounded by competing interests, which
provide context for their own.

There are at least three important points in Blackburn's statement that require contemplation here: "causative agency," "positive or negative action," and "unpredictability."
The first point allows us to bring together the above considerations of *k'uh* and *teotl*:
we might think of either of them more abstractly as "causative agencies." This moves
us away from the "bodied-ness" of ancient Roman or Greek pantheons (cf. M. Smith
2004, 199) and allows us to see the same *k'uh* or *teotl* manifested in different forms (cf.
Gillespie and Joyce 1998). The second point speaks directly to the hieroglyphic term in
the Venus pages, *muuk*. As Barbara Tedlock makes clear, omens can be either bad or
good, positive or negative—in fact, they are virtually always both. And finally, because
they are both positive and negative and they possess their own wills, any given outcome
is unpredictable.

I have taken up the question recently (Aldana 2011a, 2014) in order to bring together
the rather abstract philosophical consideration of "causative agencies" in accord with

the mythology recoverable from Mayan cultures. In looking at Mayan "foundational texts," I have suggested that we can identify an indigenous metaphor analogous to the Christian metaphor of "good versus evil." While the latter certainly does not define the entirety of Christian religion, its role as a fundamental cosmological assumption is without question, and its recognition and use in the interpretation of Christian cultures can be extremely productive. The analogous metaphor for Mayan cultures, I suggest, can be found in mythistorical texts such as the *Popol Vuh*.

We saw in chapter 1 that Brasseur de Bourbourg copied and translated a written version of the *Popol Vuh* when he was in Antigua, Guatemala, in the mid-nineteenth century. The manuscript has often been described in popular treatments as the "Mayan Bible" since many of the themes and substories appear to show up in other forms throughout Preclassic, Classic, and Postclassic cultural production (cf. Freidel et al. 1993, 60). But there are complications with any attempt to find a single narrative throughout Mayan mythology parallel to the meticulously codified Christian Bible. Rather, variations on common themes are more likely the "rule" (Stuart 2005, 158). Nevertheless, there is an overall narrative trope to the *Popol Vuh* that shows up in various other Mesoamerican mythological narratives as well.

The *Popol Vuh* tells the story of the Creation of the Sun and Moon and the peopling of the new era that they make possible (D. Tedlock 1985; Christenson 2007). The Creation of the Sun and Moon follows the lives of the Hero Twins—Jun Junahpu and Xbalamk'e—from the failure of their father and uncle, through their trials and eventual death and resurrection. Once the Hero Twins vanquish their foes at the end of their story, they ascend into the sky to become the Sun and Moon. The point here is that, in becoming two major celestial bodies, the Hero Twins have gone through a process of realizing a triumvirate of capabilities. First, in their youth, they are charged with providing nourishment for their siblings and grandmother. Second, they set off on a journey and learn to become warriors, vanquishing creatures who would be harmful to balance in the next era/Sun. And third, they learn to heal—first themselves and then their fathers. Moreover, once they heal themselves, they regularly perform all three capabilities—they wander through Underworld communities destroying and then healing creatures and structures within performances meant to entertain.

The suggestion is that we may go further than abstracting Mesoamerican "gods" as "causative agencies." We may propose that a causative agency is one that simultaneously possesses the three capabilities of damage/poison, repair/heal, and sustenance/nourish. This model can then be abstracted and extended to other "entities" or "causative agencies" such as meteorological phenomena. Chaak is not "the rain god," but the rain itself, which can be destructive, healing, and nourishing. This in turn implies that there are different kinds of Chaaks, which we find attested in the hieroglyphic record. Each has its own "personality." Similarly, the Sun is a quintessential example of *k'uh*, as is the wind (or airflow—allowing it to be associated with "breath" and life as well).

This framework allows us to see Mesoamerican deities/k'uh/teotl in line with Smith's caveat as something other than cross-Atlantic members of a pantheon. Moreover, it provides insight into characterizations of indigenous people and their "spiritual" relationships with "nature." Indeed, we can now see that there is no distinction within such conceptualizations between a "sacred" and a "mundane." All existence becomes both sacred and mundane (cf. Highway 2002).

This intimacy between humanity and nature also is attested in the *Popol Vuh*, in a manner that brings this rather abstract exploration of theology back to the 260-Day Count. Throughout the primordial history in the *Popol Vuh*, there are numerous actors involved—"the gods," or the "causative agencies" in existence even before the world came into being. They play different roles throughout primordial times, but at each of the attempts to create people they provide particular insight. Allen Christenson translates the passage for the creation of the wood people:

> Huracan, along with Sovereign and Quetzal Serpent, then spoke to the Master of Days [Aj Q'ij] and the Mistress of Shaping, they who are seers:
>
> "It shall be found; it shall be discovered how we are to create shaped and framed people who will be our providers and sustainers. May we be called upon, and may we be remembered. For it is with words that we are sustained, O Midwife and Patriarch, our Grandmother and our Grandfather, Xpiyacoc and Xmucane. Thus may it be spoken. May it be sown. May it dawn so that we are called upon and supported, so that we are remembered by framed and shaped people, by effigies and forms of people. Hearken and let it be so."
>
> "Reveal your names, Hunahpu Possum and Hunahpu Coyote, Great She Who Has Borne Children and Great He Who Has Begotten Sons, Sculptor and Wood Worker, Creator of the Green Earth and Creator of the Blue Sky, Incense Maker and Master Artist, Grandmother of Day and Grandmother of Light. Thus shall you all be called by that which we shall frame and shape. Cast grains of maize and tz'ite to divine how what we shall make will come out when we grind and chisel out its mouth and face in wood," so it was said to the Masters of Days. (2007, 80)

In this passage, the authors of the text inform us that the means for creating people should come from divinatory knowledge generated by the "Master of Days," which is Christenson's translation of the title Aj Q'iij. Among themselves, these primordial beings, there are some who hold the same position as the "daykeepers" of K'iche' society. In this passage, they also reveal one of their primary intents with the creation of people: they should be "called upon and supported, so that we are remembered by framed and shaped people." Perhaps most provocative, though, is that there are twenty named primordial beings in the list of names to be remembered.

The authors of the Popol Vuh go further in developing the expectations of people and these *k'uho'ob* in their account of the first "peopling" of the world—in that case with animals. Again in Christenson's translation, the account is as follows:

> Then it was said to the deer and the birds by the Framer and the Shaper, She Who Has Borne Children and He Who Has Begotten Sons:

"Speak! Call! Don't moan or cry out. Speak to one another, each according to your kind, according to your group," they were told—the deer, the birds, the pumas, the jaguars, and the serpents.

"Speak therefore our names. Worship us, for we are your Mother and Father. Say this, therefore: 'Huracan, Youngest Thunderbolt, and Sudden Thunderbolt, Heart of Sky and Heart of Earth, Framer and Shaper, She Who Has Borne Children and He Who Has Begotten Sons,' Speak! Call upon us! Worship us!" they were told. (Christenson 2007, 76)

Key here is that Christenson himself acknowledges in a footnote that "worship" is a bit of a mistranslation. He explains:

> *Q'ijarisaj* (to worship) is derived from the root *q'ij* (day or sun) in the transitive imperative verb form (cause to be). If such a word existed in English, it might be something like "dayify" (to honor their day, perhaps through calendric ceremonies) or "sunify" (to glorify the gods like the glory of the sun). The gods' purpose in carrying out the creation seems to be to provide beings who will be able to speak intelligibly. Only in this way could the gods be worshiped properly—through the articulation of their names with human speech. (2007, 76n83)

More provocatively, "worship" means to "dayify"—the "primordial gods" exhort their creations to turn them into "days" or "characters/aspects of the Sun." It may be, therefore, that the "ritual calendar" of ancient Mesoamerica has twenty Day Signs because each one of them represented "the face of" one of the primordial k'uho'ob. Accordingly, different days would be appropriate for different activities/events, and it would be the charge of the ajqiij/ajk'in to plan accordingly on behalf of human communities.

At the same time, it means that some *interactions* with the "gods" may take the form of reading "signs" in "nature," which can come in different forms. In response to signs read, ritual activity then amounts to generating signs in response, within a greater discourse between humanity and "nature." Ceremonial activity, from this perspective, becomes conversation between humans and the other causative agents that make up a local environment. In this sense, we encounter Blackburn's third point. Oracles are technologies that allow for a nuanced form of communication with different "causative agencies."

"PRIESTS" AND "SHAMANS"

With this reconstruction, I suggest that Mayan astronomy was driven as much by its potential political and theological applications as by the insights it provided into the "nature" of the universe. At some level, this is no provocative proposal; as we saw in this book's introduction, Eric Thompson long bemoaned the characterization of Mayan astronomy as "astrology."

In the last chapter, we saw a level of Venusian astronomy for the public at large. Ceremonial responses at Chich'en Itza were prescribed to engage the public and demonstrate a contract between the polity's leadership and the environment. In turn, ritual activity presented the opportunity for elite participation in politics and economic stimulus. At the philosophical level, I suggest that the specific formulation presented here resonates even more clearly with Precontact concepts of religion as exposed through Marc Zender's treatment of the Classic period ritual specialist title, *ajk'uhuun*, and the history of its decipherment (figure 7.8). Zender's insightful contribution is to go beyond previous interpretations by prioritizing consideration of hieroglyphic representations of the title over time in order to disambiguate its relevant components and then to recognize the linguistic link between the Classic Ch'olti'an term *ajk'uhuun* and the Tzotzil *ch'uun* (2004, 181–82, 189). At end, Zender argues that

> both Tzotzil *ch'u-un* (from Proto-Tzeltalan **ch'uh-un*, itself from pre-Tzeltalan ***k'uh-uun*) and Classic Ch'olti'an (i.e., the hieroglyphic language) *k'uh-uun*, then, are best analyzed as derived transitive verbs with the core meaning "to worship." (Zender 2004, 190)

Whereas his linguistic analysis is above reproach, Zender's final step, which is to translate these terms as "to worship," rests on previous translations and not on discernable job duties. And it is here that I suggest that the problem of cultural/epistemic translation arises again as evident in the examples that Zender used for his analysis. Regarding the root of the term, the noun *k'uh*, Zender notes that "the 17th century *Diccionario de Motul* provides the colonial Yucatec terms <ku> 'god', <kuul> 'to worship', <kuul cizin> 'to idolatrize' and <ah kul ciçin> 'idolator who worships the devil'" (Zender 2004, 194). Thus in getting to the term "worshipper," the translation's origin in the perspective of a seventeenth-century scholar is going unproblematized. Certainly, however, this author—along with other indigenous and European scholars writing for European audiences—would have characterized any type of Precontact priestly activity as the worship of idols or of the devil, for that is precisely what it meant to be "pagan" in their worldview. To accept a definition of *k'uhuun* as "worshipper," then, is to suggest that the concept of worshipper in a European worldview maps directly to the same concept in a Mayan worldview.

FIGURE 7.8 The ajk'uhuun title is held by two of the three figures in the border of Janaab' Pakal's sarcophagus lid at Palenque.

Contrastingly, I argue that Zender's own linguistic analysis of *ajk'uhuun* provides better access to the indigenous concept. He breaks the term down (from its Late Classic glyphic representation) as follows:

AJ-K'UH-HUUN-(na)
Ajk'uh(h)uun
Aj-k'uh-uun
Agentive.prefix-GOD-denominative.transitivizer
"worshipper" (Zender 2004, 192)

The challenge now is to reconsider the translation of *k'uh*/"god" as a derived-transitive verb as "to worship." And the link for me is the "deity" term.

If we now understand *k'uh* or "god" as a causative agency capable of unpredictable positive or negative activity, then we may return to the noble title with an alternative translation possibility. Now the derived transitive form of *k'uh*, "to k'uh," as "to causative agentify" or "to empower," suggests a different translation from "to worship." If a god is a being of power, then "to god" something is to empower it, or to reveal its power. Such an interpretation suggests less that of worshipping and more that of revealing the power of the gods. On the one hand, this seems straightforward. Through deity impersonations and ritual performance, priests lead the representation of deity interactions—they make visible the interactions and powers of the gods. On the other hand, at a deeper level, *ajk'uhuun* might actually be a good translation for the double entendre within the term "diviner": one who makes an entity divine by revealing its intent. By interpreting/divining the will of k'uh, the diviner gives that k'uh power. In this sense, the concept of a "diviner" might be homologous (in Blackburn's sense) to the concept of an *ajk'uhuun* across cultures.

This review takes us back to Zender's suggestion that divination was basic to all priestly duties. His point bolsters our argument by again referring to the sixteenth-century Franciscan Diego de Landa, who wrote that "the duty of the Chilans was to give the replies of the gods to the people, and so much respect was shown to them that they carried them on their shoulders" (Zender 2004, 89; Tozzer 1941, 112). This is precisely the semantic meaning that Zender derives from a linguistic analysis of the term *chilan*, from "chi'-il-a'n (speaker, user of language, prophet)" (2004, 89). Juan Josef Hoil's manuscript preserved the memory of one particularly noteworthy prophet: it is remembered as *The Book of **Chilam** Balam*.

Finally, we can take this exploration one step further to place this all back in a cosmological context. The more abstract "causative agency" translation of *k'uh* suggests that it is not clear that all k'uho'ob (all causative agencies) would be superior to humans. Indeed, humanity as a populace (as a community or collective entity) at least would have to be considered a causative agency—one among numerous others—which in turn provides insight into yet another noble title: *k'uhulajaw*, the term most often translated as the "ruler" of a given polity.

If *ajaw* derives from the Proto-Mixe-Zoquean *aj-aw*, as suggested by Kaufman (2003), meaning "one who shouts," then we confront a different conceptualization of Classic Mayan rulership. Paralleling the translation of the Nahuatl term for ruler, *tlatoani* (*tlato*— "to speak"; *-ani*—agentifier) (Siméon 1977), a *k'uhulajaw* becomes (or gets translated as) a person who shouts with causative agency—among other causative agents. The k'uhu- lajaw in this formulation becomes the representative (as Gayatri Spivak [1988, 72] delin- eates regarding *darstellung* and *vertretung*) of the polity, interacting with other k'uhoob according to context. In other words, a k'uhulajaw can interact with his or her "subjects" in a rational/political way as with other k'uhulajaws. But the k'uhulajaw can also interact with the rain, Venus, or his or her royal ancestry (K'awiil) through other (k'uh) modes of discourse—including divination.

Bringing this all back to astronomy, K'uk'ul Ek' Tuyilaj, the person in charge of what would become the Dresden Codex Venus Table, would have been responsible for inter- preting the interactions between different k'uhoob and Venus as Chak Ek'. Ajk'uhuun K'uk'ul Ek' Tuyilaj and her colleagues would have been charged with the ritual activities that maintained social, political, and economic order at Chich'en Itza by giving Chak Ek' "a voice." When we take this perspective back to the Venus Table, we find that in fact there is room for both unpredictability and prescribed ritual activity.

THE VENUS TABLE AS ORACLE

K'uk'ul Ek' Tuyilaj's Venus Table was broken up into canonical periods such that each row in each page predicted last evening, first morning, (something preceding) last morn- ing, and first evening visibilities (Aveni 2001, 186). For the first row of dates on page 46, for example, Venus's first morning visibility would be expected on 13 K'an. According to warning table models, this date would signal that K'uk'ul Ek' Tuyilaj should have initi- ated her surveys of the morning sky since the recorded event should transpire within a few days. From an oracular perspective, though, two previously noted oddities become meaningful. For one, the restriction to only one-half of the Day Signs suggests intention- ality for oracular purposes. Not any Day Sign could be used for any Venus position—only those that were "even" were included. Second, the abovementioned concerns regarding the "accuracy" of the Dresden Codex Venus Table change from constituting a liability of uncertainty into an asset for divination.

For the Venus Table as oracle in conception, K'uk'ul Ek' Tuyilaj would desire a "bounded variability" in the short term, but extreme accuracy in the long run. She would want the predicted dates—those written in the table—to sit right on the average of the planet's synodic period. In that way, deviations observed would be equally positive and negative (i.e., on average, Chak Ek' would appear late as often as early). This variability, then, could have functioned as the "random" element in the oracle. It is worth noting that this opportunity only works consistently for one event—first morning visibility. The others, we have seen, were probably structured by the Day Sign omens of the 260-Day

Count. Accordingly, it is only first morning visibility that carries with it the hieroglyphic text containing *u muuk* statements and takes up the whole of the right-hand sides of pages 46–50.

The other components map over as well so that, in conception, the Venus Table can be considered an oracle as follows:

i.   the outcome (Venus first morning visibility) is stochastic about an expected date;

ii. "consultation" is repeatable (on 584-day intervals or subparts thereof);

iii. there is a language through which a message can be read (the 260-Day Count); and

iv. an overlying layer of stochasticity is introduced by the weather.

Accordingly, we may now put all of the preceding work together to generate a revised interpretation of the Dresden Codex Venus pages.

In preparation for a first morning visibility, K'uk'ul Ek' Tuyilaj would have reviewed the Venus Round text:

On 3 Kib 4 Yaxk'in, the North is enclosed by Ulum and Chak Ek'. It is 236 days. On 2 Kimi 14 Sak, the West is enclosed by Sinaan and Chak Ek'. It is 326 days. On 5 Kib 19 Tsek, the South is enclosed by Chak Tul and Chak Ek'. It is 576 days. On 13 K'an 7 Xul, the East is enclosed by Kimil and Chak Ek'. It is 584 days. . . .

The East is enclosed by God L and Chak Ek'. K'awiil is speared. . . . It is the omen of the Maize God.

As described at the beginning of this chapter, in the days leading up to the event K'uk'ul Ek' Tuyilaj would have charged an almehen lineage with the construction of the ceramic effigy of Ulum and the preparation of the roads for a procession. This noble lineage would also have been responsible for obtaining food and gifts for the celebration. Artisans would be charged with forming and painting drinking cups—some painted with illustrations of the very rituals they engaged. Additionally, a theatrical performance of the battle between God L and K'awiil would have to be staged, with training, practice, and the construction of costumes initiated well ahead of time.

Finally, alongside all of the above preparations, K'uk'ul Ek' Tuyilaj would have risen before the Sun to climb to the top of the Tz'iknal along with colleagues and possibly select members of the nobility. They would have looked for either corroboration of the first morning visibility of Venus on the predicted date or deviation from it—either early or late. The difference between the 260-Day Count date predicted and that observed would provide the unpredictable "sign" to be read for religious and/or political purposes. This layer of information most likely would not have been shared with the public, but maintained within smaller priestly and noble communities to be put to political purpose.

How would she have known what ritual to prescribe? Now, finally, we realize that as cultural outsiders interested in astronomy, modern scholars have been able to focus on the six pages of the Venus Table without considering the rest of the manuscript. But when we bring in the oracular aspect of the Venus Table, we realize that it fits extremely well with the rest.

"ALMANACS"

Most of the rest of the manuscript that we refer to as the Dresden Codex is made up of "almanacs." These are passes through the 260-Day Count, broken up into subintervals, associated with images and short passages of hieroglyphic text. One of these we saw in chapter 1 as providing the data useful to Förstemann in reconstructing the mechanics of the Mayan number system. The New Year pages of the last chapter constitute another almanac, though that one tied to the solar year. Now we may return to that example as a window into each of the seventy-five almanacs in the codex.

The example we used in chapter 1 actually resonates with what we have reconstructed in the Venus Table. As we saw, the text for the almanac describes the "standing up" of Chaak as a Bakab in each of the four cosmic regions. The first sequence reads:

> On 3 Ix Bakab Chaak is stood up in the East. ?-tamal is his food. After 13 days, on 3 Manik, Bakab Chaak is stood up in the North. Fish is his food. After 13 days, on 3 Ajaw, Bakab Chaak is stood up in the West. Iguana tamal is his food. After 13 days, on 3 Ben, Bakab Chaak is stood up in the South.

This could easily be interpreted as instructions for the ritual movement of Chaak effigy censers around a circuit, with different seats in each of the four regions of the city. As Chaak is placed in each seat, he is offered the appropriate ritually prepared food—perhaps not unlike the modern *ch'achaak* ceremonies practiced in Yucatan.

While the other almanacs throughout the codex function in the same way mathematically, their textual content differs. The pages on the "back side" of the Venus Table are predominantly related to Chaak, but all of the pages leading up to the Venus Table on the "front side" are filled with almanacs relating to different deities. In these pages, we find the sixteen deities that Schellhas designated Gods A through P. In many almanacs they are acting independently, but in others they act in concert (figure 7.9). In addition to a fixed number of deities involved, scholars have noticed that there are also a fixed number of "auguries" associated with them (figure 7.10). Among these auguries, for instance, we find the term that we have already confronted in the Venus pages: *u muuk*. This and four others are gathered provocatively along the back of the Celestial Dragon on pages 4b–5b (figure 7.11). While *u muuk* appears to be essentially neutral, scholars have argued for the positive or negative attributions of the others. For example, an "attributive glyph" often associated with Kimil (the "Death God") itself appears to include a "death" glyph within it (see figure 7.10). Another, often associated with Ixim (the Maize God or God E), has been interpreted as representing an abundance of food and water, and so as a positive attributive glyph (Schele and Grube 1997, 84).

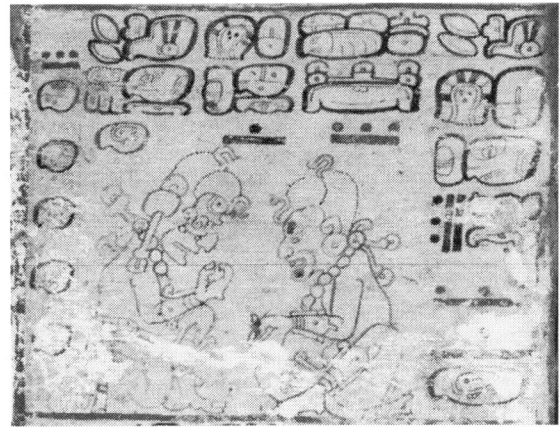

FIGURE 7.9 Deities working together in a Dresden Codex almanac. Courtesy SLUB

augury glyphs

FIGURE 7.10 Auguries in almanac texts. Courtesy SLUB

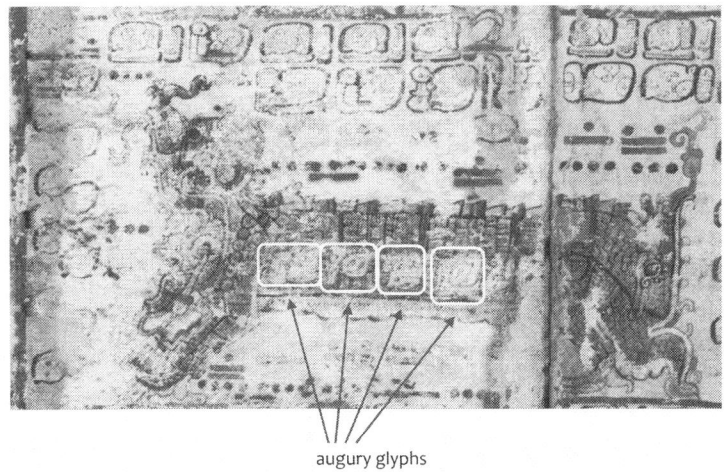

augury glyphs

FIGURE 7.11 The Celestial Dragon with augury glyphs painted on its body on pages 4b–5b. Courtesy SLUB

In the Dresden Codex, Tawiskal Uwoojil and his collaborators copied the various almanacs within groupings, which we can here refer to as chapters of the manuscript, reflecting the deities involved as well as the time intervals each is broken up into (cf. Thompson 1972, 20). The first chapter brings together almanacs that include a diversity of deities, but are most frequently broken up into intervals of fifty-two days. These fifty-two day intervals mean that each section is introduced by a column of five Day Signs (since 5 × 52 = 260). Less common in this section are the almanacs of sixty-five-day intervals, which would accordingly be led by a column of four Day Signs. The text of the first chapter varies significantly from almanac to almanac in its operative verb, but the deities and auguries, as noted above, remain limited to the above sets. A common subject of the verbs within this first chapter, though, is *u-chi-chi* or *u chich*, which means "his/her strength" (Barrera Vásquez 1995) (figure 7.12). It would make sense to see these texts, then, referring to the strength of a given deity's influence over a period of time, and so reflect the impact on the divination in question. Two examples, coming from pages 4 and 5, give *hoch u chich* or "his strength is sparked" and *pehkaj u chich* or "his strength is stirred." Accordingly, the former would

hoch u chich   pekaj u chich

FIGURE 7.12 Almanacs characterizing the strength of a given deity's influence. "Hoch u chich" is the sparking of its strength, whereas "pekaj u chich" is the stirring of its strength. Courtesy SLUB

correspond to a strong effect, while the latter would likely be mild. These almanacs could have been implemented when the influence of the k'uh in question was necessary as part of an oracular response.

A second chapter includes the image of a woman in every illustration. Several proposals have been made for the identity of this figure, from the Moon Goddess to Ix Chel and Kab Ixik and even just women in general (Aveni 2001, 173; Thompson 1972, 47; Schele and Grube 1997). Dominating this section are phrases referring to omens using the term *u muuk*. One fascinating section shows different birds perched on the shoulders of seated women, with each event attributed a different augury (figure 7.13). The following section within this chapter depicts small figures or bundles strapped to the backs of similar seated women while the text above describes the carried item as *u kuch* or "her burden," as we have seen in the New Year pages. Perhaps this chapter contained rituals that were the responsibility of and performed by a priesthood of women.

What we might refer to as chapter 3 of the Dresden Codex is the Venus Table constituting pages 24 and 46–50. As we have seen, this is followed immediately by the Eclipse Table, which we could call chapter 4. Completing this side of the manuscript are slightly more than two pages made up of a table of multiples and a page of illustrated text. Thompson considered the illustrated page to reflect "katun prophecies" largely

oxlajun muwan   k'uk'   mo'

FIGURE 7.13 An almanac of women, with different birds associated with *u muuk* phrases. Courtesy SLUB

based on the fact that there is a winikhaab glyph in the middle of the page (figure 7.14). Perhaps more interesting, though, is that the name Bolon Okte shows up twice in the hieroglyphs on page 60, once in each passage. In the upper passage, the erosion of the page makes it difficult to ascertain the context, but in the lower passage he carries the title "ajaw." Overall, the images are far more narrative in character than the rest of the manuscript.

Continuing on the obverse side, chapter 6 is computationally intense, led off by what commonly are referred to as the Serpent Numbers (Thompson 1972, 80). The text associated with these huge numbers contains several records of the date 4 Ajaw 8 Kumk'u, so it would appear that these were the kinds of computations made use of in astronumerological exercises such as those we saw performed at Palenque—or they may reflect the activities painted on the walls at Xultun (figure 7.15). The following four and a half pages of chapter 7 are devoted to Chaak engaged in various activities.

Chapter 8 of the Dresden Codex seems to combine elements of chapters 6 and 7. Chaak sits atop a Serpent Number to lead off the section, and the prefatory text again refers to 4 Ajaw 8 Kumk'u. The table in this case is dense, without illustrations, but several stacked numbers appear to be dripping with rain (figure 7.16).

The next chapter we have already seen as the New Year pages. In context, we see that they are prefaced by a page dominated by an illustration of the Celestial Dragon accompanied by two deities. A female deity pours precious celestial fluid from a jar, while below her God L carries an atlatl and two darts in a warrior pose reminiscent of the illustrations in the middle of the Venus pages (figure 7.17). Chapter 10 is fourteen pages long and is dominated again by the presence of Chaak. The scribes have returned to the almanac format of the first few chapters, but now the content shifts dramatically from passage to passage.

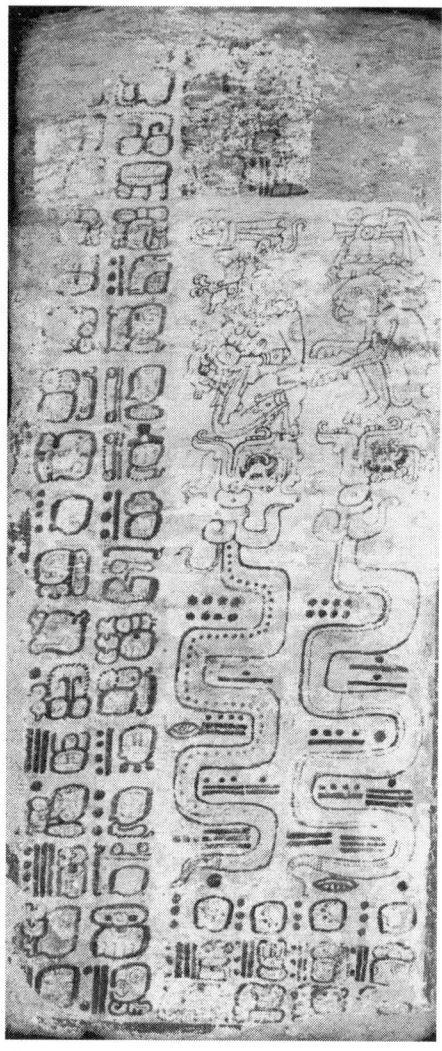

FIGURE 7.14 Page 60 of the Dresden Codex, which is more narrative in content. Courtesy SLUB

FIGURE 7.15 Serpent Numbers of the Dresden Codex. Pairs of Long Count numbers, one in red and the other in black ink, are interspersed in the turns of each serpent's body. Courtesy SLUB

Finally, chapter 11 has been mired in a modern debate over its potential interpretation as a Mars Table, parallel to the Venus Table (Love 1995; Bricker and Bricker 1986b; Thompson 1972, 108). The codex ends with a final chapter reminiscent of the first example we examined: the "standing-up" of Chaak in the four cosmic regions. A key difference here is that the overall interval is not of 260 days, but of 364. The time intervals, then, are of six eights and one seventeen per page, and the illustrations are far more elaborate.

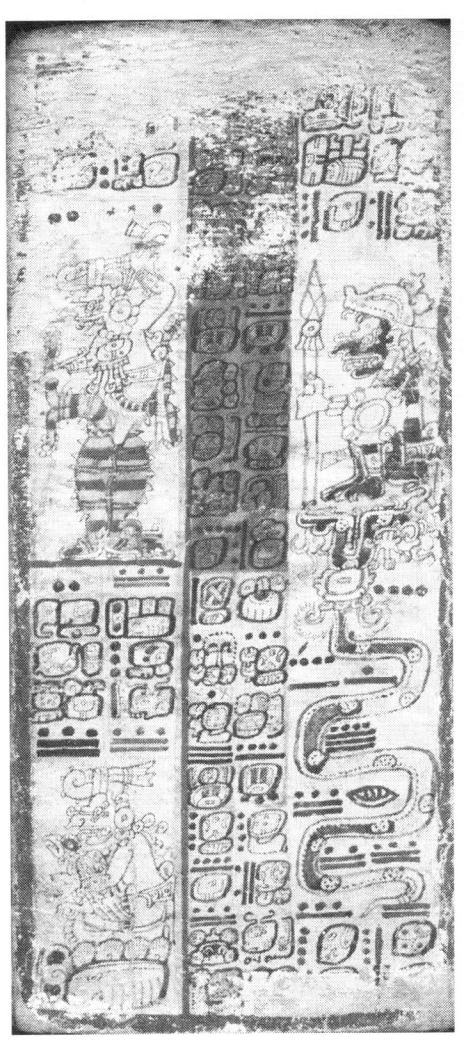

FIGURE 7.16 Chaak Serpent Numbers. Courtesy SLUB

FIGURE 7.17 Dresden Codex page 74 depicting the Celestial Dragon with a female deity and God L. Courtesy SLUB

In sum, all of the content (1) describes a ritual procession, which could be that of a deity effigy, (2) facilitates computations through the 260-Day Count, or (3) describes the omens associated with a given deity's association with a date in the 260-Day Count.

Given the descriptions we saw at the beginning of this chapter for divination practices, we now should be able to see how the Dresden Codex may have been used. Some oracular consultation would have generated an omen. It was the ajk'in's responsibility to determine which k'uh would be properly engaged in a ceremonial response. Based on her assessment, the ajk'in would select a set of almanacs within the appropriate chapter.

A preliminary stochastic process would have generated a date within the 260-Day Count, which probably would reduce the number of relevant almanacs in the set. A second stochastic process would define which almanac and identify the relevant passage within it. Whether the outcome was favorable or not, the ajk'in would give the petitioner instructions as to how to respond ritually. At the level of a large city, this would have been enabled by temples to different deities around the civic-ceremonial center. The person requesting the consultation then would have presented offerings to the appropriate temple according to the ajk'in's prescription.

SUMMARY

Looking back at the Venus Table ceremony prescribed by the passage on page 46, we can now imagine multiple layers of sociopolitical interactions occurring at its performance. A favored almehen house seeking to preserve its status would work to provide as lavish a procession and feast as possible to flatter the k'uhulajaw and please the public. Meanwhile, the priest in charge would have consulted with a few assistants to determine the subtle divined message produced by the stochasticity of actual Venus visibility. This information would have been shared with the k'uhulajaw, setting in motion high-level discussions about the political importance of the divinatory knowledge. These discussions would have involved going back to the pages of the Dresden Codex to determine the specific ritual responses required, using the Venus Table's stochastic result and the specific context.

Given the outcome of the Venus event ceremony and divination, rival houses might have sought to tease information from priests to find potential leads for their own advantage. Any given ceremony would therefore not be strictly a ceremonial performance, but a stage for political theater as well. The upshot is that this framework, at some level, allows us to wrest free Mayan astronomy from characterization as a proto-"Western" science. It now allows for meanings relevant to and within ancient elite Mayan culture itself.

In this case, I have argued that a cultural translation allows us to rethink and appreciate a subtle and rational ritual practice in the Dresden Codex, without dismissing it as "superstitious" or evolutionarily juvenile "proto-science." Likewise, I suggest that even though we may find evidence of oracular activity, it does not require that we consider all Mayan intellectual activity to be of the same character. Oracular knowledge may very well have sat alongside other forms of knowledge. Like revelation in Christianity, or portents in ancient China, oracle-produced knowledge probably fits within a broader political discourse, but it *did not have to dominate it* (see also Bielenstein 1950, 1984). Indeed, *The Book of Chilam Balam* suggests that during times of military strife, generals had almost unchecked authority—prophecy took a back seat. The entire process may be recognized as being just as complex as any more Modern ideology—it simply proceeded from a vastly different set of ontological assumptions. We will return to this observation in chapter 11.

Now, however, we turn our attention to those historical figures who would have engaged these oracular constructs at Chich'en Itza. In both cases, close, daily observation and the recording of observations was necessary, even though that wasn't the sole purpose. Nevertheless, that daily activity is what led K'uk'ul Ek' Tuyilaj to her own ancient Mayan scientific discovery, which we will explore in the next chapter.

Discovering K'uk'ul Ek' Tuyilaj's Discovery at Chich'en Itza

Through the end of the last chapter, we have now moved through the material and intellectual contexts of the Dresden Codex Venus Table. These have provided us with a perspective of a manuscript useful to a high-ranking scribe, such as K'uk'ul Ek' Tuyilaj, working as a counselor to the leadership of Chich'en Itza at its apogee. When we place the Venus Table now into its historical context, we encounter one of its more remarkable features, and we find ourselves returning to the murals of Chich'en Itza, which we saw in the introduction.

By now placing the Venus Table into its historical context, as it was developed at Chich'en Itza and preserved to posterity by copyists, we find that it was made possible by a discovery. This discovery at the intersection of astronomical observation and calendric innovation can be recognized now as fitting well with Terminal Classic Mayan culture. And—as with so many discoveries—it resulted from the coincidence of a happy accident and a keen mind able to perceive the opportunity it presented. To see it, we turn our attention back to page 24 of the Dresden Codex and the contested hypothesis in modern scholarship that the anchor to the Venus Table was not a historical record, but instead it was intended to serve purely numerological purpose. By here allowing for both roles—historical record and numerological purpose—we encounter a discovery related to Venus that sits at a particularly provocative historical position relative to the legendary and mythological figure Quetzalcoatl.

Indeed, the Long Count dates within the full Venus Table not only suggest how it was internally developed; they also place it in the middle of one of the most interesting scenarios of ancient Mesoamerican history. The origins of the Venus Table itself coincide with the "arrival" of Quetzalcoatl to Chich'en Itza. Unpacking this complex tangle of historical, astronomical, and political lines of evidence sets us up to encounter in the next chapter a new interpretation of the Feathered Serpent at Chich'en Itza and Maudslay's century-old desideratum of a more robust connection of the ethnohistoric documentary record to the archaeological record of the site.

A MAYAN VENUS ALMANAC

We can hypothesize at this point, in line with Aveni's suggestion (2001, 191), that the Venus Table recorded in the Dresden Codex had not yet taken its final form by the time of the Caracol's construction. This does not mean, however, that the rituals recorded therein would not yet have been performed. In fact, we can strip down the Venus Table to provide only what K'uk'ul Ek' Tuyilaj would have needed to conduct the *k'al* ceremonies at the Tz'iknal—or even an earlier such structure or platform. When we do, we find that it would have matched the record from Copán that we examined closely in chapter 4, and it would have looked much more like the content of the rest of the codex as reviewed in the last chapter.

Without the 365-Day Count dates, K'uk'ul Ek' Tuyilaj's Venus pages would take the form of a Venus Almanac, giving the 260-Day Count Day Signs, a portion of hieroglyphic text naming the cosmic region enclosed, and illustrations of the deities accompanying Venus (figure 8.1). In these earliest versions, K'uk'ul Ek' Tuyilaj could have painted such an almanac to take up only two-thirds of a page, with the sequence through the 584-day period in the upper register and the warrior spearing its victim along with the omens generated below it. This would give her space to include the Day Signs for each event placed in a column in front of each figure. She could then record the time intervals breaking down the 584-day Venus Round for use to correlate Venus events with ritual activities. She would still need five of these Venus Rounds in order for the Day Signs to repeat.

Normal practice relative to other almanacs would be to follow the sequence listed in the text, but we know that variation of Venus events could be significant. In such case an almanac form would have been beneficial. In an almanac form, that is, she could have adjusted Venus Rounds as necessary in application. K'uk'ul Ek' Tuyilaj could introduce shifts on an ad-hoc basis, moving from one page to another in order to ensure that the proper rituals were aligned with the Day Sign of the impending event. This would mean a less orderly, more oracular progression through the almanac and a greater role of the astronomer's observations on a Venus Round to Venus Round basis, making judgments about which *k'al* sequence to engage when. Such flexibility, we recall, is allowed because the almanac form included no 365-Day Count solar year dates. Accordingly, none of the "correction" work investigated by Teeple would have been necessary.

As mentioned in the last chapter, a Venus almanac would have been appropriate for a smaller city. Four almehen houses, for example, could be charged with the effigy censers for the four subperiods of Venus, and one lineage (the ruling one?) with the warrior effigy censer. Such an approach would have required more attention to the political implications of shifts to calendric sequences and so would have relied more closely on

FIGURE 8.1 Hypothetical reconstruction of the first Venus Round (Day Signs of Kib, Kimi, Kib and K'an) in a Venus almanac as a precursor to the Venus Table of the Dresden Codex

the observation of Venus. In other words, with fewer quantitative constraints, there would be more room for debate over the assignment of ceremonial responsibilities to different lineages. Greater room for debate may have led to greater political instability.

As for the quantitative aspect of this version of a Venus Almanac, it would have looked very much like the content of pages 49 through 54 in the Central Mexican Postclassic manuscript known as the Borgia Codex (figure 8.2). We have already seen the last of the Borgia Codex Venus pages in chapter 1 through Seler's work to develop the connection between Quetzalcoatl and its warrior imagery. Almost a century later, Aveni (1999, S15) too noticed the connection and sought to advance Seler's proposal by introducing a new calendric interpretation to the dates surrounding the images on Borgia pages 53 and 54. He eventually found close connection between the Borgia and Dresden treatments of Venus, such that he claimed to have "shown that . . . pp. 53–54 of the Borgia not only are related to that particular Maya calendar but may have been derived from it" (2001, 196). If we return to his proposal now, with the new content developed here and extended to

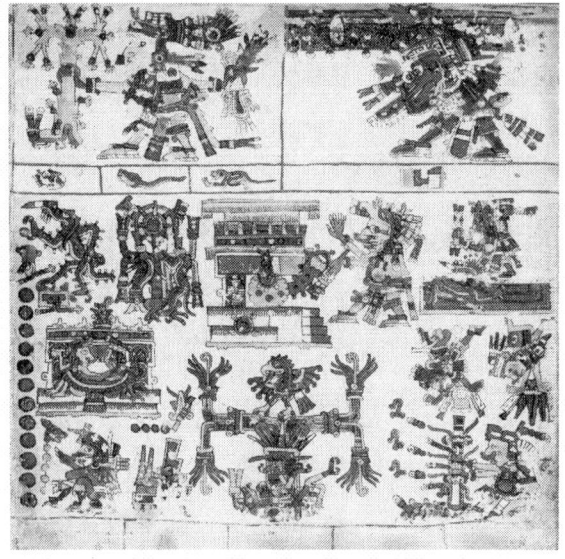

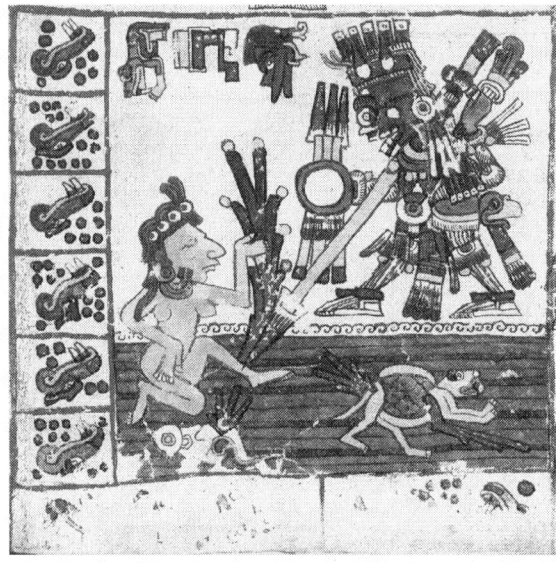

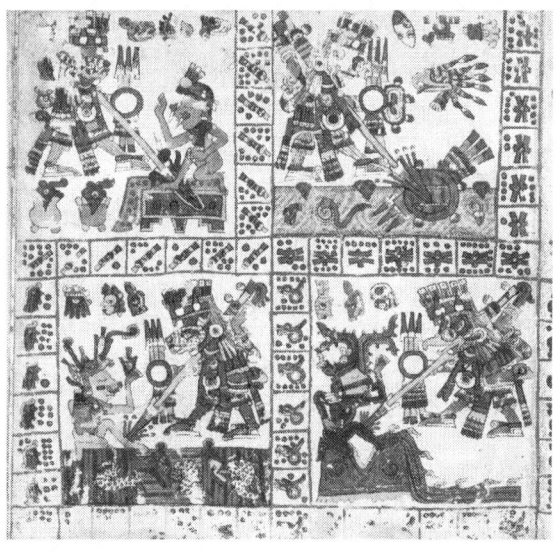

FIGURE 8.2 Venus Almanac in the Borgia Codex: (*top*) Page 51, which follows the same format as pages 49, 50, and 52. Each page corresponds to one of the cosmic directions, identifiable by the bird-tree combination at the bottom center. Page 51 corresponds to the west, with the 9-Wind/Quetzalcoatl deity depicted in the upper band on the right. (*middle*) The lower left quadrant of page 53 depicts one of the five spearings Seler used to find a connection to Venus as Quetzalcoatl. The associated Day Sign is Cipactli. (*bottom*) Page 54 gives the other five spearings, associated with the Day Signs Coatl, Atl, Acatl, and Ollin. LACMA (Codex Borgia); photos courtesy Ancient Americas at LACMA (ancientamericas.org)

include more pages of the Borgia, we can identify an even stronger parallel between the two documents. More to the point, K'uk'ul Ek' Tuyilaj's Venus almanac may have looked a lot more like the version copied into the Borgia Codex, with the recognition that Long Count calendrics eventually pulled the Mayan version in a new direction. The ancestral versions of both almanacs were probably ceremonial siblings, but local technologies and political circumstances led to the divergence captured in their final forms.

Seler and Aveni both focused on just the imagery of Borgia Codex pages 53 and 54, but the interpretation of the Dresden Codex Venus pages developed in this book suggests that we should also consider the preceding four pages of the Borgia manuscript. On those pages, we encounter imagery that should now be familiar: birds atop cosmic-region-associated trees, surrounded by ceremonial activity and/or oracular signs (figure 8.2a). Above these images the Borgia scribe painted well-known Nahuatl deities against celestial backgrounds. Recent work by Felicia Lopez (2016) has found the presence of Nahuatl puns within the oracular content of the pages leading up to the warrior imagery, which would be in line with the divination practices we saw in the last chapter. By now exploring the calendric patterns in Borgia Codex pages 49–52 combined with the illustrations of pages 52 and 53, we will find that we can interpret Borgia pages 49–54 as following a Venus Almanac resonant with a Venus Almanac precursor to the Dresden Codex Venus Table. This, in turn, speaks to the dialogue that appears to have emerged between the leadership of Chich'en Itza and those of Mixtec and Toltec communities to the west.

A CENTRAL MEXICAN VENUS ALMANAC

The scribe who painted the Borgia Codex did not live in the same landscape environment as K'uk'ul Ek' Tuyilaj. Although we don't know his precise location, we know from the style and content of his work that he lived in the highlands of Central Mexico, between northern Oaxaca and the Basin of Mexico. We will see some evidence in the next chapter that the scribe worked at Cholula, but that is still speculative. Either way, this scribe's observational context would have been significantly different from that of astronomers in Yucatan.

Observations of the night sky during the winter, for example, would have been much colder for the Borgia scribe given the regional climate, but more importantly he wouldn't have needed a temple to give him an unobstructed view of the horizon. Plenty of easily accessible hilltops would have provided ideal viewing stations. The same terrain would have created a variegated horizon, greatly facilitating tracking of the solar year by sunrise or sunset. The motions of the stars and planets overhead, of course, would have been the same across regions, and we will see that such motions were handled in similar ways.

There is no debate in the literature that our Borgia scribe painted pages 49, 50, 51, and 52 to represent the cosmic regions. On page 49, he painted a maguey plant as a "world directional tree" with a quetzal dancing at its apex (Lopez 2017, 55). Each of the next three pages he anchored iconographically to bird-tree combinations representing the north, west, and south accordingly. He then completed the set by painting a maize plant on page 53, representing the center around which the four directions cycled. He placed the tree scenes in the lower portion of each page, complemented at the top by larger-scaled images of deities. The content of these pages is summarized in table 8.1.

At the top right of page 49, the scribe painted an image of Tlahuizcalpantecuhtli, the Frost Teotl—the deity that Seler recognized as associated with Venus as the Morning-star. Even more explicit, the scribe included the deity Quetzalcoatl in the upper register on page 51, here in his (Mixtec codex) form as "9 Wind" (Byland 1993, xxvii). Stepping back, then, we find that each of these four pages was assigned to one of the four directions, surrounded by ritual activity and governed by a specific deity—two of whom have been long recognized as associated with Venus. The presence of these deities at least suggest that the scribe intended these four directional pages to be related to the Venus "table" of pages 53 and 54.

The ritual activity depicted on pages 49–52, too, should look familiar at this point. The small, seated figures at the bottom left of each page, for instance, strongly resemble the deity effigies in the Dresden New Year pages. Also, the figure in the center, above the directional tree, is making an offering not unlike the priest figures in the Dresden New Year pages. The layout also is comparable, as the Borgia pages follow the order of east, north, west, and south corresponding to the Dresden Codex Venus Table, with oracular and ritual information assigned to each. Perhaps more important, though, are the Day Signs on each page.

The Borgia scribe painted a sequence of five Day Signs strung out along the bottom of the page associated with the east, page 49. These five Day Signs are Cipactli, Acatl, Coatl, Ollin, and Atl; they correspond directly to the Day Signs encircling the warrior figures on pages 53 and 54. In other words, the Borgia scribe used the Day Signs associated with the east for the Day Signs assigned to the spearings of victims by what Seler called Venus deities. This is precisely the same association made by the Dresden Codex Venus pages, which we have incorporated into the revised presentation of chapters 6 and 7 in this book. On a mathematical level, the fact that there are five Day Signs associated with the east (and with each other direction) also points toward a Venus affiliation since a period of 584 days would correspond to twenty-nine passes through the set of twenty Day Signs with four days left over. A remainder of four selects out five different Day Signs uniquely.

To this point, then, by extending the Borgia Venus pages to include those preceding the scenes of "Venus spearings," we now have Venus deities identified in supervisory positions on pages 49 and 51, the same Day Signs assigned to the east on page 49 and to ritual spearings on pages 52 and 53, and ritual activity associated with the cosmic regions at intermediate Day Signs in the almanac. The correspondence between the two manuscripts is thus quite robust.

In practice, though, the ritual commemoration of Venus Rounds would have differed between the two regions as indicated by a key feature. We have seen that K'uk'ul Ek' Tuyilaj's Venus Rounds in the Dresden Codex were broken up into asymmetric periods that restricted *k'al* events to occur on only a subset of Day Signs. The author of the Borgia

TABLE 8.1 Schematic representation of the Borgia Codex Venus Pages.

p. 54 Rabbit-masked warrior spears an unidentified victim **ACATL**	p. 54 Mictlantecuhtli spears a spear and shield **OLLIN**		p. 49				
			Xipe Totec		TLAHUIZCALPANTECUHTLI		
			EAST (omens)				
			ATL	OLLIN	COATL	ACATL	CIPACTLI
			p. 50				
			Mictlantecuhtli		Xiutecuhtli		
			North (omens)				
			Ehecatl	Itzcuintli	Tecpatl	Miquiztli	Ocelotl
p. 54 Dog-masked warrior spears Cinteotl **ATL**	p. 54 Eagle-masked warrior spears Tezcatlipoca **COATL**	p. 53 *Tlahuizcalpantecuhtli spears a female deity* **CIPACTLI**	p. 51				
			Xochipilli		QUETZALCOATL		
			West (omens)				
			Cuauhtli	Calli	Ozomatli	Quiahuitl	Mazatl
			p. 52				
			Cinteotl		Mictlantecuhtli		
			South (omens)				
			Tochtli	Cozcacuauhtli	Cuetzpalin	Malinalli	Xochitl

Note: The tonalpohualli Day Signs tied to the east on page 49 are the same as those framing the spearing scenes on pages 53–54. This directly parallels the Dresden Codex's k'al events in the east and the spearings illustrated there. Also, Tlahuizcalpantecuhtli is the deity at the top right of the page depicting the east, as Seler recognized at the beginning of the twentieth century. Quetzalcoatl is depicted at the top of page 51.

The sequence of spearings on pages 53 and 54 follow the expected order: a 584-day Venus Round pulls out these Day Signs in order: **Cipactli**, Ehecatl, Calli, Cuetzpalin, **Coatl**, Miquiztli, Mazatl, Tochtli, **Atl**, Itzcuintli, Ozomatli, Malinalli, **Acatl**, Ocelotl, Cuauhtli, Cozcacuauhtli, **Ollin**, Tecpatl, Quiahuitl, Xochitl. On page 49, however, they appear to skip. From Cipactli, the scribe skips Acatl to Coatl and then skips Ollin to Atl. From Atl, the scribe would skip Cipactli to Acatl and then Coatl to Ollin before coming back to Cipactli. While it might appear playful (or a clue that we are missing something), the same pattern is followed by a sequence through the spearings on pages 53 and 54. The Venus Round of 584 days is divisible by 13 with a remainder of 12. That means that the coefficient should slip back by one each Venus Round. To follow this sequence on pages 53 and 54, the scribe starts at the bottom right of the spearing on page 53 with 1 Cipactli, then skips to the middle register of the first spearing on page 54 for 13 Coatl, then skips to the left corner register for 12 Atl. The pattern continues with 11 Acatl in the middle vertical register for the upper left spearing and then 10 Ollin in the top register of the upper right spearing scene. To continue, the scribe now returns to the scene on page 53 and skips to the middle register for 9 Cipactli. Following this pattern, the scribe slowly circulates through all registers on pages 53 and 54 to run through 37,960 days, or 104 years, again just as in the full Dresden Codex Venus Table.

Codex, on the other hand, broke up the full set so that each Day Sign would appear in the almanac only once—that is, five Day Signs per directional page (figure 8.2). The difference is detectable, therefore, but reflects only a variation on the same overall practice—the kind of aesthetic change driven by regional preferences.

That is, the Borgia scribe in the highlands of Central Mexico would have observed a first morning visibility in the east on 1 Cipactli and then turned to page 53 to see that it corresponded to Tlahuizcalpantecuhtli spearing a woman, which in turn led him to page 49 to view the appropriate prognostication and ritual response. At the end of morning visibility, the priest would turn to page 50 for the appropriate rituals and omens for last morning visibility and the north. Pages 51 and 52 would provide first and last evening visibility rituals and omens, so that after a period of 584 days from 1 Cipactli the priest would refer to the next spearing scene, now associated with the Day Sign Coatl. A full run through all Day Signs listed around the spearing scenes would take 104 years—just like the full scope of the Dresden Codex Venus Table.

The reader will have noticed that the conceptualization of Venus periods in both cases resonates with the Féjérvary-Mayer frontispiece and *k'al* space-time enclosing ritual. There were regional variations in implementation such that commemoration of the first morning visibility of Venus in the east would have occurred on the same day in Yucatan as in Central Mexico, but the subperiods would have resulted in different intermediate commemorations. Regardless, Venus in both regions generated similar ritual activity, supervised by an appropriately trained priesthood, very likely in communication with one another as Aveni hypothesized.

THE IMPACT OF THE LONG COUNT

Having thus encountered the precursor to the Venus Table of the Dresden Codex and seen how it may have had broad purchase across Mesoamerica, we may turn to the uniqueness of the version developed at Chich'en Itza. From this perspective, we recognize that the key difference between these almanac versions and what would become the Venus Table in the Dresden Codex is that the latter was intimately connected to the Long Count.

Assuming that she was doing precisely what Hoil would do centuries later, K'uk'ul Ek' Tuyilaj kept track of her observations of Venus using her version of a Venus Almanac, following daily schedules that would not have been set by the seasons of the year. In accord with the planet's morning visibility, K'uk'ul Ek' Tuyilaj would have to rise before the Sun to climb the stairs, leading to the top of the Tz'iknal. At a city the size of Chich'en Itza—even relatively small at the time—she would not have been the only ajk'uhuun serving the community, and so she would not have been charged with keeping track of all celestial bodies. She may have met, for example, the ajk'in there charged with keeping the time of the night and its ti'huun astrological prognostications, and she may have possibly met others. But these early rises would only last for two-thirds of a solar year, after which she could turn her attention to other planets or other matters entirely for two months while Venus was not visible. Then, she would shift to observations of the western horizon at sunset, watching Venus rise higher and higher above the setting Sun. These evening observations would transition to early rises after another 263 or so days, reinitiating the 584-day Venus cycle. K'uk'ul Ek' Tuyilaj would be tracking the planet through the same patterns that Förstemann would again witness in the city of Dresden thousands of miles away and a thousand years later—although hers would follow the Long Count calendar, not the Christian calendar. For us U get a better sense of when in the Long Count she was taking this data, we return to the relationship between the Tz'iknal and the dates in the Dresden Codex.

Aveni's archaeoastronomical work at Chich'en Itza led him to suggest that "the Venus tables provide a means of relating ritual site function to observations of the planet at about the time the Caracol was erected" (2001, 275). A more elaborate discussion he placed in a footnote:

> Compare the approximate archaeological dates of the Caracol (800–1000 A.D.) with the Long Count dates in the Dresden Codex, p. 24. The installation of the Venus table occurred in the tenth century [according to the GMT]. The extant copy of the codex, an updated version, probably dates to two or three centuries later. Furthermore, the extant Dresden Codex table was probably drawn up not far to the east of Chichen Itza. Thus it is possible that the astronomical observations delineated in the Venus table in the Dresden Codex were collected by astronomers perched in the observation chamber of this tower and others like it in northern Yucatan. (Aveni 2001, 361n32)

Aveni's reference to "archaeological dates" appeals to the Tz'iknal's dating to the Terminal Classic via ceramic and architectural styles as well as through radiocarbon dates associated with them. These have been corroborated by more recent studies, taking advantage of next-generation radiocarbon analysis relative to Elizabeth Ralph's work on the Tikal lintels and focused on the Terminal Classic apogee of the city (Hoggarth et al. 2017). But there is even more relevant data here, as noted by Maudslay at the end of the nineteenth century, which has been left out of treatments of the Dresden Codex. Maudslay found that the Tz'iknal is associated with hieroglyphic inscriptions that record historical events related to the dedication of the ritual architecture at Chich'en Itza.

For the inscriptions adorning the Monjas Structure, the Casa Colorada, and the Tz'iknal, all written during what Nikolai Grube and Ruth Krochok (2007, 229) have called the "epigraphic florescence" at Chich'en Itza, K'uk'ul Ek' Tuyilaj's predecessors crafted several narratives. These texts in the structures around the Tz'iknal refer to a figure named K'ahk' U Pakal K'awiil, who is named as the companion of several deities as well as of mortal contemporaries (figure 8.3). In particular, on Stela 1, K'ahk' U Pakal K'awiil is referred to as the protagonist of events occurring on the sixteenth and seventeenth *haab*-endings of *1 Ajaw* as well as

FIGURE 8.3 The name K'ahk' U Pakal K'awiil in the hieroglyphic texts of Chich'en Itza.

the winikhaab end on 10.3.0.0.0. Now that these texts have been deciphered, we realize that we have encountered the links that Maudslay and Morley hoped for: we have a historical figure mentioned both in the hieroglyphic inscriptions of Chich'en Itza and in *The Book of Chilam Balam of Chumayel*.

THE NORTH AND THE SOUTH WITHIN THE MAYAN REGION

In the third Chronicle of Katuns within his manuscript, Hoil recorded that Katun 8 Ajaw "was the katun when Chakanputun perished at the hands of Kak-upacal and Tee Uilu" (Roys 1933, 78). If we follow Morley's practice to assume strict continuity and disinterested historical record keeping, Katun 8 Ajaw would have transpired in the Long Count from 10.5.0.0.1 11 Imix 9 Muwan to 10.6.0.0.0 8 Ajaw 8 Yax. We have seen above that K'ahk' U Pakal K'awiil was commemorating the Tz'iknal before the 10.3.0.0.0 period end, so such a connection would put him in his late fifties or early sixties at the time that Hoil had him defeating the army of Chakanputun. That's somewhat old for him to have been in battle, but not unprecedented for Mayan rulers.

On the other hand, there is another possibility to consider, as suggested by Alfred Tozzer. Responding to Morley's proposal that the Itza Mayans were mechanically motivated by calendric prophecy to repeat the same events over time, Tozzer writes:

> Morley, in discussing this recurrence of a Katun 8 Ahau in Itza history, speaks of a 'chronological coercion' which might be called a compulsion neurosis urging the Itzas to do something when 'the times were ripe for change.' It is far more probable that the events of history were fitted in the katun pattern than that the Mayas felt an urge to make a major change in their life every 256 years. (Tozzer 1941, 21)

In other words, it may have been revision for the purpose of what Grant Jones calls "prophetic history" (1998, 13) that these records were revised. Without the Long Count to constrain them and with a pattern of significant conflict during other Katuns 8 Ajaw, it may be that scribes shifted K'ahk' U Pakal K'awiil's military interventions from a Katun 10 Ajaw, when K'ahk' U Pakal K'awiil was younger, to a Katun 8 Ajaw. While such revisions would make sense of several records that appear to modern readers to be in "error," they reinforce the complications of attempts to find continuity between the Long Count calendar and the Katun Count.

Fortunately, we do not have to rely on simply one reference to this historical leader; there are other sources of corroborative historical context. Linda Schele and Peter Mathews note that the hieroglyphic record at Uxmal one hundred kilometers to the east of Chich'en Itza also includes references to the same K'ahk' U Pakal K'awiil. "The Hieroglyphic Step from the Chanchimez Group at Uxmal refers to the famous K'ak'upakal of Chich'en Itza. . . . Although the context is unclear, the text names the contemporary ruler of Chich'en in an Uxmal inscription" (Schele and Mathews 1999, 259). They continue, noting "that K'ak'upakal's name on Lintel 4 of the Monjas includes *k'ak'nal ahaw*, the same title that appears in Chan-Chak's name at Uxmal. Since this is a rare title, it further strengthens the proposed ties between the two leaders" (Schele and Mathews 1999, 378n6). Evidence from Chich'en Itza, Uxmal, and the *Book of Chilam Balam*, therefore, all corroborate significant leadership of K'ahk' U Pakal K'awiil at Chich'en Itza during the Terminal Classic. And so while the hieroglyphic links between Uxmal and Chich'en Itza have been recovered only recently, their further identification in Colonial period alphabetic texts goes back to the time of Förstemann.

Hubert Bancroft collected numerous historical texts during the nineteenth century, combining them into histories of the regions of the Western Hemisphere. In his work on Mexico and Central America, Bancroft included both colorful local stories about ancient cities as recorded by Landa and others and also some analysis, attempting to place Precontact records in line with Postcontact chronologies. His work on Uxmal, for example, draws from the church historian living in Merida, Diego Lopez de Cogolludo, and his visits to Uxmal near the end of the seventeenth century. In this material, he included alongside historical matters a fable that generated the colloquial names of some of the structures in the civic ceremonial center of the site:

> An old sorceress lived at Kabah, rarely leaving her chimney corner. Her grandson, a dwarf, by making a hole in her water-jar, kept her a long time at the well one day, and by removing the hearth-stone found the treasure she had so carefully guarded, a silver *tunkul* and *zoot*, native instruments. The music produced by the dwarf was heard in all the cities, and the king at Uxmal trembled, for an old prophecy declared that when such music should be heard the monarch must give up his throne to the musician. A peculiar duel was agreed upon between the two, each to have four baskets of *cocoyoles*, or palm-nuts, broken on his head. The Dwarf was victorious and took the dead king's place, having the Casa del Adivino built for his palace, and the Casa de la Vieja for his grandmother. (Bancroft 1883, 632–33)

The "Pyramid of the Magician" is the current vernacular name for the largest structure at Uxmal. Here too, though, those familiar with the narrative of the Hero Twins in the *Popol Vuh* will recognize the technique that the dwarf used to delay the old sorceress as she collected water. The Hero Twins did the same with their mother as they conspired with the rat to find their father's ball game equipment. For the dwarf, the objective is musical instrumentation, but the narrative device is the same. Apparently literary devices—such as astronomical tools—were shared broadly even after the Postclassic period.

Bancroft also recorded Lopez de Cogolludo's first-person accounts of visits to the ruins, which were very in line with Landa's accounts of his visits to Chich'en Itza as we saw in chapter 1. Lopez even included his observation of effigy incense burners within the temples at Uxmal, resonating with the material references in the Dresden Codex Venus pages.

> They have many sumptuous temples in many parts of this Tierra Firme, of which there remain today parts of their edifices, like which are in Vxmal . . . in Chichen Ytza . . . on the top [of one pyramidal edifice] there are situated, separated a short distance, two small

chapels in which are the idols. . . . Some of these have a height of more than one hundred steps, of a little more than half a foot wide, each one. I ascended one time the one of Vxumual, and when I had to descend, I repented because, as the steps are so narrow and so many in number, and as the edifice rises almost straight up, and since the height is not slight going down, one gets dizzy and it is somewhat dangerous. I found there in one of two chapels, offerings of cacao, and marks of copal, which is their incense, burned there but a short time before. (Bancroft 1883, 86–87)

Concerning the connection between Uxmal and the greater Mesoamerican world of the Postclassic, however, Bancroft brings to our attention the structure known now as "the Nunnery."

In Uxmal there is a large patio with many rooms separated in the form of a cloister, where these virgins lived. It is a work worthy of admiration because the exterior of the walls is all of worked stones, where there are brought out figures of armed men in bas-relief, a diversity of animals and birds, and other things, and it has not been made out who were the artificers, nor how they were worked in this land. All of the four fronts of the buildings of that patio (that might be called a plaza) are encircled by a snake worked in the same stone as the walls, the tail terminating under the head, and being in all its circuit four hundred feet [long]. (Bancroft 1883, 84–86)

There is no evidence that the structure functioned as a nunnery, populated by "virgins"; instead, the images carved in stone around the upper façades of the Puuc architecture speak to its municipal purpose. Moreover, the serpent noted by Lopez de Cogolludo will become relevant shortly, as we address the Feathered Serpent in Postclassic Meso-america below.

Regarding the history of the government at Uxmal, occupying that architecture, Ban-croft writes:

All authorities agree on the prosperity attending the reign of the Cocome monarchs [at Chichen Itza] in conjunction with the Tutul Xius at Uxmal. It was perhaps in this period that were built a large proportion of the magnificent structures which as ruins have excited the wonder of the world. (1883, 630)

Here Bancroft references the two primary lineages of the Contact period in Yucatan— the Cocom and the Xiu, reminding us that Landa leaned heavily on Gaspar Antonio Xiu and Juan Nachi Cocom for much of his history. Providing additional detail on the Xius' residence at Uxmal, Bancroft draws from the manuscript that Thompson, Goodman, and Morley would rely on. "Returning to the annals of the Tutul Xius, in 2 Ahau, 981, Ahcuitok Tutul Xiu settled at Uxmal, where his people ruled conjointly with the kings of Chichen and Mayapán for two hundred years, from 981 to 1181" (Bancroft 1883, 629). Bancroft thus endeavored to put the Xiu occupation of Uxmal into a historical period, which corresponded to the early Postclassic.

Other scholars concur with the historical placement of the Xius at Uxmal, but Bancroft's specific proposal for the dating of Katun 2 Ajaw conflicts with the recent work of Schele and Mathews. They suggest an earlier dating by more than two hundred years, claiming that "[a] group of the Tutul Xiw led by Ah-Kuy-Tok'-Tutul-Xiw established themselves at Uxmal and reigned from k'atun 2 Ahaw (9.16.0.0.0, or 751) through k'atun 10 Ahaw (10.5.0.0.0, or 928)" (1998, 259). Schele and Mathews aimed to bolster their claim with reference to archaeological data, stating that "archaeologists now place the beginning of the Puuk polities to the middle of the eighth century" (Schele and Mathews 1999, 259). The reader will note that, here again, we confront the complication arising from the attempt to intercalate Hoil's Katun Count records with Classic period Long Count inscriptions, as well as the further attempt to place these dates within Christian chronologies. Although all agree with Aj Kuy Tok' Tutul Xiu's activity in a Katun 2 Ajaw, precisely when that occurred relative to Long Count or a Christian count is ambiguous.

On the other hand, these various accounts do speak to a key feature of this time in the historical record: it was a time of transition during which the lineage referring to themselves as the Itza left their homeland and arrived at Chich'en Itza, giving the city its name. Hoil's narratives of Itza migrations in *The Book of Chilam Balam* in fact reflect a rich context of shifting demographics and alliances at the time. His records have the Itza traveling as a group from the south around a path encircling the northern peninsula, before settling at the place of their own cenote (Roys 1933, 26–28; Solari 2007). As for corroboration of their foreign origins, the hieroglyphic record at Motul de San Jose across the lake from Tah Itza includes an Emblem Glyph that reads *[i]-tza-'a*, dated to the eighth century AD, suggesting that the ethnic group did hail from the south (Bíró and Pérez de Heredia 2016, 130). With a little more digging, we find that politics in that region speak to the kinds of alliances attempted during this time of transition and a growing rift between north and south.

In the south, at the heart of Classic Mayan geopolitics, an intervention within the local dynasty occurred at Terminal Classic Seibal just as the Long Count switched from 9.19.19.17.19 to 10.0.0.0.0. After military subjugation and a hiatus in monument construction, a new dynasty took over the city (Schele and Mathews 1999, 178). We find record of this shift thanks to the local masons at Seibal, who built up a small platform into a radially symmetric structure—a smaller version and one-hundred-year antecedent of the Temple of K'uk'ulkan at Chich'en Itza. In front of each stairway, artists erected a stela commemorating the new governance of the city. At the east entrance, on Stela 11, scribes recorded that on 9.19.19.17.19 6 Kawak 17 Sip, Aj Bolon Haabtal arrived. His arrival and installation into the k'uhulajawlel, the inscription tells us, were authorized by the ruler of Ucanal (Martin and Grube 2008, 227; Schele and Mathews 1999, 183) (figure 8.4).

The strategic nature of this installation really should not go unnoticed. Seibal lay far to the south of Uxmal, at the origins of the Pasión River, which fed into the Usumacinta. It provided direct access, therefore, to the Gulf Coast and trade with the Mixtec cities of Oaxaca and the Toltecs in the Basin of Mexico (figure 8.5). Ucanal, on the other hand, sat adjacent to the Mopan River, which ran toward the east coast. Not far from Ucanal, Caracol—that behemoth of southern Belize—lay only a few kilometers from the Mopan

FIGURE 8.4 Aj Bolon Haabtal arrives at Seibal, sponsored by Kan Ek' of Ucanal (Seibal Stela 11)

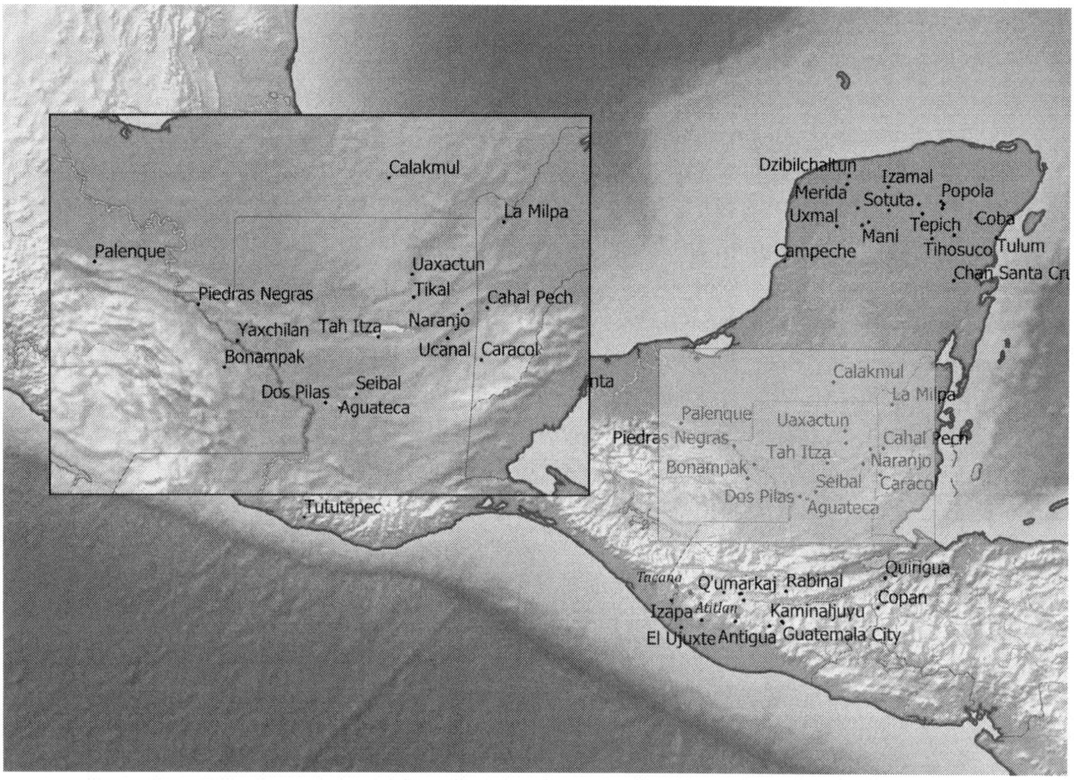

FIGURE 8.5 Seibal, Ucanal, and Tah Itza

and the Macal Rivers, both of which combined and fed into the Belize River, which ran out to the Caribbean. The inscription of Altar 13 at Caracol attests to an alliance between its k'uhulajaw and that of Ucanal during this same time period (Martin and Grube 2008, 98). Overall, then, with the k'uhulajaw of Ucanal gaining a subject ruler at Seibal, he would have been in a position to control trade from the Gulf, across the Peten, to the Caribbean.

It is not only inferred geopolitical strategy that brings Seibal into the Xiu and Cocom histories at Uxmal and Chich'en Itza. The scribes of Seibal Stela 11 included an inscriptional reference to a royal figure named Kan Ek'. Kan Ek' is a name that shows up relatively frequently in the histories of Yucatan, and we will see it again in the next chapter. Its prevalence, however, is not reflective of a single particularly notable individual. Instead, it appears that the rulers of Tah Itza (aka Tayasal, the island city in the center of the Peten) assumed this name on accession to the throne. Given that Tah Itza was a major population center going back to the Early Classic period, it may be even that the Kan Ek' dynastic member who met Hernan Cortes was of the same lineage that witnessed the accession of Aj Bolon Haabtal at Seibal and the same lineage that met with Andrés de Avendaño in the seventeenth century.

More to the point, Tah Itza sat geographically between Ucanal and Seibal, thus suggesting that Kan Ek' of the Terminal Classic and Aj Bolon Haabtal of Seibal sought to reinvigorate the Ucanal alliance that crossed the central Peten. The latter actually claimed as much with the inscription of his Stela 10, recording that his accession was witnessed by the k'uhulajaws of Calakmul, Tikal, and Motul de San Jose (Martin and Grube 2008, 227). The narrative that emerges, therefore, when we turn to the various sources on the Terminal Classic, is one of renewed alliances in the south across the Peten and new connections to Central Mexico in the north via Uxmal. This pattern would eventually put the almehen lineages of Chich'en Itza at a historical crossroads.

CENTRAL MEXICAN CONNECTIONS

Part of the transition occurring during the Terminal Classic period had to do with shifting alliances around southern Mexico and northern Central America. Scholars considering the time period more broadly argue that another of the key features of transition to what we recognize as the Postclassic period was the dissolution of the institution of the *k'uhulajawlel* dynastically related to a set of patron deities (Sharer 2006, 586). Whereas the Classic period hieroglyphic inscriptions emphasized the k'uhulajaw and his or her affiliations with the k'uhulajaws of other cities, in the Postclassic political legitimacy would grow to rely more on the distribution of power across ch'ibals, leading to a much greater role of alliances among *local* nobility (cf. Ringle 2004). It may be that the structure that emerged was more like the confederacies of North America, which would have carried a more democratic aspect. As archaeologist Sarah Jackson has reviewed, the historical record of the Colonial period consistently attributes importance across the Mayan region to "Council Houses," constituted by representatives of the highest-ranking ch'ibals of the city (2011; cf. Alexander 2017). These councils or *cuchteel* would appoint the city's ajaw, not by divine birthright, but via political consensus among cuchteel representatives. Turning to Central Mexico, we confront historical evidence for another alternative to the Classic Mayan k'uhulajawlel.

During the period after the fall of Teotihuacan and before the Mesoamerican Postclassic, a young Mixtec noble in what is now Oaxaca made use of an alternate hegemonic structure to engage his grand ambitions. Similar to the crisis in Spain that led to the crowning of Philip of Anjou, which we saw in chapter 3, the ruler of a prominent Mixtec city, Tilantongo, hadn't married and had produced no heir to his throne (Troike 1974, 420). The situation was not dire, however, as—contrary to the case between Dos Pilas and Tikal—he was on good terms with his half brother. In this case, the younger brother, Na Cuaa ("8 Deer"), had already gained some legitimacy by assuming the rulership of Tututepec, a smaller town on the coastal piedmont of Oaxaca, some one hundred kilometers south of Tilantongo. But Na Cuaa's ambitions were greater; he sought a second form of legitimacy through a nose-piercing ceremony supervised by an independent priesthood. While it may sound like a straightforward-enough procedure, in fact the indigenous codex histories demonstrate that it took some doing.

Iya Na Cuaa was the first son of his father's second wife, so he did not have genealogical primary rights to the throne at home in Tilantongo. He therefore first engaged diplomacy and warfare in his effort to rule Tututepec. Toward his goal, Na Cuaa did obtain the support of his older brother, Ca Qhi, who had taken over from his father. This was a lifelong team effort. Ca Qhi accompanied Na Cuaa into battle and facilitated political meetings for him (Codex Nuttall; Troike 1974, 425).

When Na Cuaa believed himself ready for a nose-piercing ceremony, he had to seek the approval of Qui Huidzu, a local authority at a nearby independent city. He sent two emissaries separately to that local authority, but the latter rejected them both. Na Cuaa was only able to convince Qui Huidzu to grant him the possibility of a nose-piercing ceremony by appeal to a form of oracular knowledge. Qui Huidzu conceded to play the ball game against Na Cuaa, letting the outcome of the competition decide the latter's fate. The Codex Nuttall notes that Ca Qhi played on Na Cuaa's team in this game, and they were victorious; Na Cuaa gained the right to the ceremony and so entrance into a status of *tecuhtli*. Now he was in both ritual and political position to take over the throne at Tilantongo once his brother died, *and* he could maintain his position at Tututepec. He thus gained control over a huge swath of Mixtec lands and noble estates (Byland 1993; Troike 1974, 426).

A critical point here is that it wasn't Iya Qui Huidzu who pierced Na Cuaa's nose (figure 8.6). Codex representations show Qui Huidzu present, but it is his companion who actually conducted the ceremony (Troike 1974, 437–38). Stepping back, we realize that the regional practice, therefore, was one in which local dynasties conceded some of their legitimacy to determination by nonlocal political communities. Qui Huidzu as a member of the priesthood at Achiutla, had decision-making authority over Na Cuaa's ambitions at Tilantongo. This overall structure, we find, resonates clearly with the

FIGURE 8.6 Iya Na Cuaa's nose-piercing ceremony on page 57 of the Codex Nuttall

documented practice at nearby Cholula. An alternate structure of political leadership had emerged to the north and west of the Mayan region.

By the Late Postclassic (AD 1200 to 1500) but with origins in the Terminal Classic and Early Postclassic (AD 900 to 1200), the site in the modern Mexican state of Puebla known as Cholula served as an international center. High-ranking members of noble lineages from Oaxaca through Central America traveled to the city to be inducted religiously into a role of leadership associated with the legendary figure Quetzalcoatl. Thus authorized, they would return home as local rulers—as tecuhtli (Pohl et al. 2012, 26; Ringle 2004). In other words, birthright was a component of accession to a throne, but there were other authorities at remote cities who also played a critical role in determining who could actually rule (Pohl et al. 2012, 26). Annabeth Headrick recognizes that such structures avoided the authoritarian aspect of the Mayan k'uhulajawlel:

> The rulers of the Mixteca Alta established a system of political and religious checks and balances that was strong enough to prevent the emergence of absolute rulers. The existence of multiple oracular sources of legitimacy meant that a claim of divine right could always be challenged (Byland and Pohl 1994, 219). (Headrick 1999, 78)

As hinted at several times, now, the alternate political strategy arising in Mixtec and Toltec lands was intimately connected to the legendary Quetzalcoatl figure—for the Mixtec known as Iya Q Chi (9 Wind). Pages 47 and 48 of the Codex Vindobonensis represent Q Chi's birth, his taking on of all attributes for which the Toltecs were known as artists and scientists, and finally assuming a cosmological place, holding up the sky (Pohl 1994, 8) (figure 8.7). In his Precontact representation, he was a legendary figure and creator of culture.

The same Iya Q Chi figure shows up in Na Cuaa's personal history as well on pages 38, 46, and 65 of the Codex Nuttall. In other words, the tradition recorded into Contact

FIGURE 8.7 Q Chi holds up the sky on page 48 of the Codex Vindobonensis

times found much earlier precedence within codex records. According to most scholarly interpretations, Iya Na Cuaa performed his pilgrimage around the end of the eleventh century. Such timing would place his nose-piercing ceremony near the end of or after Chich'en Itza's period of elite occupation. The origins of the dynasty that Na Cuaa belonged to at Tilantongo, though, go back to AD 940, right at Chich'en Itza's apogee (Jansen and Perez Jimenez 2005, 47). The Feathered Serpent in Mesoamerica and his connection to new political structures emerged in both regions during the transition of the Terminal Classic to the Postclassic.

Turning our attention to this Central Mexico context, then, the Epiclassic sites of the highlands suggest that the iconography of the Feathered Serpent that we associate with Quetzalcoatl emerged there. The trajectory appears to come from a combination of Teotihuacano and Mayan imagery, developed at the geographically intermediate cities of Xochicalco and Cacaxtla, near the modern cities of Cuernavaca and Puebla (figure 8.8). Xochicalco and Cacaxtla were founded and flourished in the aftermath of the internal destruction of Teotihuacan, which corresponded to the onset of the Mayan Late Classic in the late sixth or early seventh century AD. At both cities, feathered serpent imagery emerged that was both intimately associated with rulership and anticipated the feathered serpent imagery used at Chich'en Itza and Tula, as well as by the Aztecs (Sharer 2006, 580) (figure 8.9).

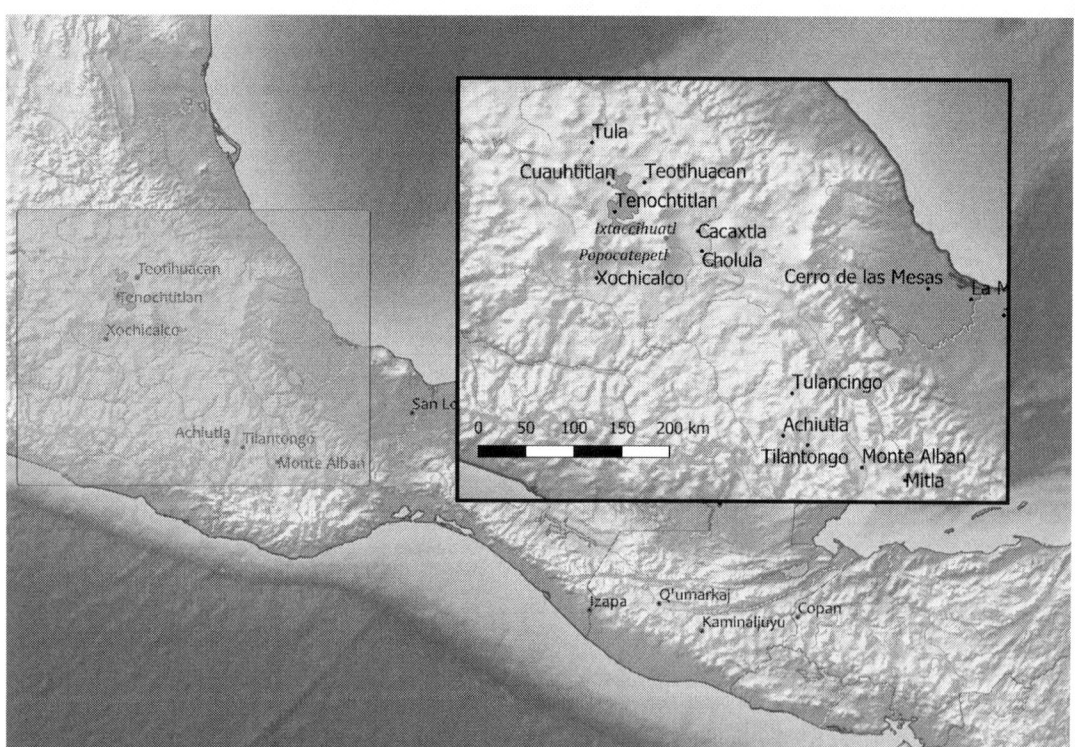

FIGURE 8.8 Cities of the Central Mexican Epiclassic Xochicalco and Cacaxtla, relative to Classic period Teotihuacan and Postclassic Tula

FIGURE 8.9 Temple of the Feathered Serpent at Xochicalco

At Xochicalco, the imagery is straightforward to interpret. Feathered serpents wrap around the Teotihuacan-inspired talud-tablero façade of a structure at the heart of the city (figure 8.9). The positioning of these serpents created open spaces between their bodies' bends and turns, into which artists placed authority figures. To depict the latter, Xochicalco artists used Mayan representational conventions; they look like seated, cross-legged k'uhulajaws with their bodies facing the viewer and their faces in profile. At both Xochicalco and Cacaxtla, it appears that feathered serpents had been pulled away from more abstract contexts at Teotihuacan and now had become more intimately connected to specific leaders (Schele and Mathews 1999, 295; Sharer 2006, 583). While the "historical persona" of Quetzalcoatl has consumed volumes of publications (Carrasco 2001; Lopez-Austin 2015; Nicholson 2001), our concerns here rely primarily on the religiopolitical structure that arose within the Epiclassic period.

It was, for example, during the time period of Xochicalco's florescence that feathered serpent imagery showed up at Uxmal. The central quadrangle structure (what Lopez de Cogolludo called the "Nunnery") was built by the local k'uhulajaw Chan Chaak, and we have seen that it carries feathered serpents on its façade across the length of the North Structure (figure 8.10). What this means is that before the completion of the Tz'iknal at Chich'en Itza and so before the Temple of K'uk'ulkan or even the Great Ball Court there,

FIGURE 8.10 Feathered serpent on the North Structure of the "Nunnery" at Uxmal. The serpent winds around the location masks and the human figures on the structure's façade, not unlike the imagery at Xochicalco. LACMA (Schele Drawing Archive #115037); photo by Linda Schele © David Schele, photo courtesy Ancient Americas at LACMA (ancientamericas.org)

Chan Chaak and the government of Uxmal had adopted an association of feathered serpent imagery in line with the record at Xochicalco and Cacaxtla. What we appear to be seeing, therefore, is the development of a network of rulers legitimized by initiation at a city ritually dedicated to Quetzalcoatl as a precursor to the later Postclassic practice at Cholula.

BACK AT CHICH'EN ITZA

If it resonated with any of these cases, then, the ajaw's role at Chich'en Itza during the Terminal Classic still would have been critical to the city, but possibly not in the individualistic way that supported the k'uhulajaw's aggrandizement during the Classic period. When we return to the Tz'iknal, therefore, we should consider that when K'ahk' U Pakal K'awiil set Stela 1 into the upper level of the platform, he commemorated the completion of the structure (Sharer 2006, 391), which reflected a form of continuity with Classic period practice. But we should recognize also that the absence of his portrait on that monument likely spoke to a shift in the character of his authority.

Given this backdrop, we find that our work thus far on the Venus Table places part of *its* construction during a very provocative period of time—not just for political but also for astronomical reasons. Concerning the temporal implications of these inscriptions, we have seen that the masons of Chich'en Itza completed the Tz'iknal in time for the 10.3.0.0.0 period end (Aldana 2016). Without invoking the day-for-day correlation of the GMT, but recognizing that it is confirmed to be accurate within a few decades (Kennett et al. 2013; Aldana 2015; 2016b), we can narrow down Aveni's window of observations to the ninth century AD. More important for our treatment here, though, is attention to the Long Count dates.

We have already come across Aveni's proposal that the first full table with 365-Day Count dates—the one we here describe as having been written by K'uk'ul Ek' Tuyilaj—was an ephemeris capturing observations in the Terminal Classic. In his interpretation, the corrected tables (through shifted base dates) occurred centuries later. Aveni, therefore, has the masons at Chich'en Itza building the structure and using it by the late ninth century AD, *before* the astronomers installed their first uncorrected table, either 60 years later, following Lounsbury (10.5.6.4.0), or 160 years later, following Bricker and Bricker (10.10.11.12.0). If Aveni's hypothesis held, then the first corrected table would be implemented either 150 (10.9.18.12.0) or 260 (10.15.4.2.0) years after the ritual dedication of the Tz'iknal.

In other words, according to proposals relying on Thompson's interpretation of page 24, an uncorrected Venus Table would have been accurate for the primary occupation of Chich'en Itza. Only after Chich'en Itza fell would the corrections of page 24 have been necessary, as implied by Aveni's reference to "other observation towers in northern Yucatan" (2001, 261). Again, these interpretations are based on the assumption that the anchor to the Venus Table was a numerological contrivance—not a historical record.

If we now hypothesize the anchor of the Venus Table as an historical record as suggested by its continuity with Yax Pahsaj Chan Yopaat's Venus record at Copán and in the scholarly tradition that eventually led to Hoil's hurricane record, then we find new coherence to page 24 and a compelling motivation for its logic. Specifically, if we take 9.9.9.16.0 1 Ajaw 18 K'ayab as a historically recorded first morning visibility of Venus—available to her within the library at Chich'en Itza—then we can compute the rest of its visibilities in the Long Count and so reconstruct what K'uk'ul Ek' Tuyilaj witnessed without addressing any Calendar Correlation whatsoever. In the midst of all the changes going on in Mesoamerica, we also notice that observations from the Tz'iknal at Chich'en Itza would have occurred at a particularly opportune time.

DISCOVERY AT CHICH'EN ITZA

If the Tz'iknal were dedicated by 10.3.0.0.0 1 Ajaw 3 Yaxk'in, then ajk'ins at Chich'en Itza would have had more than twenty-five years of records from the same structure by the time that K'uk'ul Ek' Tuyilaj would make her observations during the winikhaab running from 10.4.0.0.1 13 Imix 4 Wo to 10.5.0.0.0 10 Ajaw 8 Muwan. As she worked through her Venus almanac, scheduling and presiding over the ceremonies we saw in the last chapters, she would have found that a first morning visibility was approaching near 10.4.6.13.0, and she would have realized that this wasn't just any first morning appearance. This would be a first morning appearance occurring on the 260-Day Count date that served as the anchor to the whole table: 1 Ajaw. And that, of course, was a rare occurrence. According to an almanac based on 584-day periods using uncorrected Venus Rounds, the coincidence of a first morning visibility on a date 1 Ajaw should occur only once every 5.5.8.0 days, or 104 years.

This is likely what prompted her to look through the historical records she had access to in her priesthood's library of codices, wherein she found that 9.9.9.16.0 1 Ajaw 18 K'ayab was a documented occurrence of first morning visibility on a date 1 Ajaw. It would have been straightforward calculation at that point to demonstrate that 9.9.9.16.0 1 Ajaw 18 K'ayab + 3 × 5.5.8.0 would get her to 10.5.6.4.0 1 Ajaw 18 K'ayab, which was a date still twenty years in her future. In other words, according to her almanac, she shouldn't be witnessing a first morning visibility on a date 1 Ajaw until 10.5.6.4.0, but here she was approaching one on 10.4.6.13.0, almost a full winikhaab early.

It is very likely that K'uk'ul Ek' Tuyilaj would have been torn at this point. Had other astronomers or interested nobles been apprised, a first morning visibility event on a date 1 Ajaw would have been highly conspicuous. Numerous texts attribute "Jun Ajaw" as the name of Venus, so a first morning appearance on or near this date would warrant special recognition (Aveni 2001, 190). On a political level, whichever noble house was in charge of the ceremony would have acquired additional prestige—and K'uk'ul Ek' Tuyilaj would have been the one to facilitate it for them. If, on the other hand, she chose a day close to 1 Ajaw, but not on it and so foregone special recognition, there certainly would have been many among the nobility and even the commoners who would have noticed it and possibly questioned her or the event itself. At worst, missing the opportunity would have provided an opening for criticism of the rulership of the type we saw in the last chapter.

Then again, if K'uk'ul Ek' Tuyilaj's prediction was wrong and first morning appearance didn't occur on or close enough to 1 Ajaw, then there would be an elaborate celebration of the wrong date. That could be just as bad, if not worse.

It would be fascinating to know how the debate played out or if K'uk'ul Ek' Tuyilaj even brought it to the nobility's attention for deliberation. It may not have been difficult, for example, to shift either the first or last evening appearance commemoration by two days and so move to a less conspicuous date for the first morning appearance. Only K'uk'ul Ek' Tuyilaj, her fellow astronomers, and possibly a few nobles would know if that shift created a significant inaccuracy in observation—and they most likely could be kept quiet. On the other hand, on its technical level what we do know is that the proximity of a first morning appearance of Venus to 10.4.6.13.0 set up the very few observations that would provide all the information necessary for the discovery that would transform her Venus Almanac into the full-blown Venus Table, including its preface.

It is probably unrealistic to assume that K'uk'ul Ek' Tuyilaj would have waited for the actual event before taking up the calendric implications of the calendric coincidence she witnessed. Her excitement would have grown from the moment of last evening visibility of Venus on 10.4.6.12.12 6 Eb 5 Mak. She would have attended to the necessary ceremonial activities charged to the nobles and to perhaps even the descendent of K'ahk' U Pakal K'awiil himself, rotating effigy censers and guiding processions. But her mind would have been on her observation of first morning appearance.

Whether she took a wait-and-see approach or whether at the other extreme she called for an event of great celebration, K'uk'ul Ek' Tuyilaj would have been met with another provocative event seven years later. On 10.4.13.4.0 1 Ajaw 18 Wo, she would have found that last evening visibility had just passed, and another first morning visibility was only five days away.

Stepping back to take this in, this meant that she had records of three first morning appearance Venus events—two that she witnessed and one from historical records. Two of the first morning visibilities on dates 1 Ajaw were separated by three centuries, and the third first morning visibility occurred only five days after a date 1 Ajaw. Put another way, the time interval between the latter two dates would connect two 1 Ajaw dates separated by only five days relative to Venus first morning visibilities. Moreover, the time interval between the latter two events, or between 10.4.6.13.0 and 10.4.13.4.0, was 6.9.0 in duration or 2,340 days. What K'uk'ul Ek' Tuyilaj found here, simply by recording Venus observations during an opportune time period, is precisely what modern scholarship has worked to recover from the Dresden Codex over the last century—the specific mechanism that allowed for the long-term correction of the Venus Table.

BACK TO TEEPLE'S CORRECTIONS

We have already seen in chapter 3 that the table of numbers on page 24 can be used to correct the Venus Table over long periods of time. For the most part, the approach has been to use these time intervals as building blocks in order to develop a resulting single "corrected table" that best matched reconstructed historical observations of the planet during a specified time period (Lounsbury 1983; Bricker and Bricker 2007; Thompson 1972). The key behind all of this has been Teeple's early recognition that the different Correction Intervals are able to do the work based on shifts in multiples of four-day intervals. To facilitate a cultural translation, we will first run through the math of these interpretations using modern terminology; then we will see how it all came together for K'uk'ul Ek' Tuyilaj.

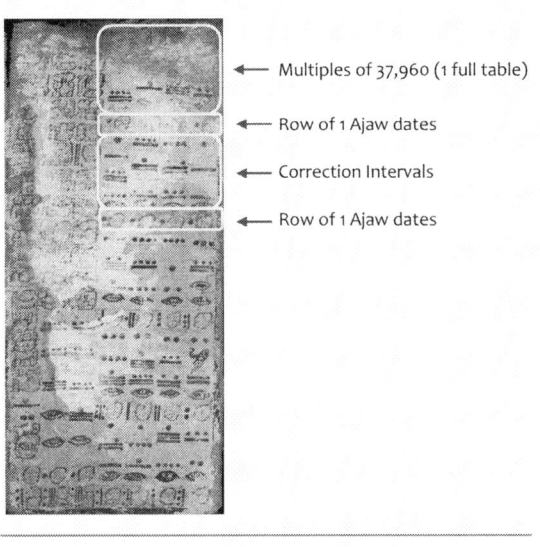

← Multiples of 37,960 (1 full table)

← Row of 1 Ajaw dates

← Correction Intervals

← Row of 1 Ajaw dates

FIGURE 8.11 Correction intervals on page 24. Each interval takes a date 1 Ajaw to another date 1 Ajaw. Courtesy SLUB

For clarity, we use the term "Correction Interval" (CI) to refer to the time intervals listed on page 24 in the second row of the table (figure 8.11). In his continued work on the Dresden Codex, Floyd Lounsbury (1992a, 1992b) deciphered the mathematical construction of these Correction Intervals, recognizing that they were all deliberately built with a common structure. He showed that each Correction Interval takes the form of what we now refer to as a Diophantine Equation (1992b, 208), and he identified the interval of 6.9.0 or 2,340 days as being fundamental to the construction of each. If we then define 2,340 as the Correction Factor (CF) and a Venus Great Cycle (VGC) as equivalent to one run through the entire Venus Table of 5.5.8.0 (or 37,960 days, which is 104 years), then Lounsbury's pattern falls out. With this terminology, Lounsbury found that each Correction Interval in the second row of the right-hand side of page 24 could be described by the algebraic relationship:

$$CI = a \times VGC - b \times CF \ (a,b \in Integers)$$

Specifically, for each term in the row:

$CI_1 = 4.12.8.0 = 1 \times 5.5.8.0 - 2 \times 6.9.0 = 1 \times 37,960 - 2 \times 2,340 = 33,280$
$CI_2 = 9.11.7.0 = 2 \times 5.5.8.0 - 3 \times 6.9.0 = 2 \times 37,960 - 3 \times 2,340 = 68,900$
$CI_3 = 1.5.14.4.0 = 5 \times 5.5.8.0 - 2 \times 6.9.0 = 3 \times 37,960 - 2 \times 2,340 = 185,120$
even
$CI_4 = 1.5.5.0 = 4 \times 5.5.8.0 - 61 \times 6.9.0 = 4 \times 37,960 - 61 \times 2,340 = 9,100$

In each case, the Correction Interval relies on an integer multiple of the Correction Factor. The latter is important because 2,340 is divisible by 260 (9 times), so it will preserve the same date in the 260-Day Count—in these cases it preserves the date 1 Ajaw (figure 8.12). At the same time, the interval 2,340 is close to four Venus Rounds, 4 ×

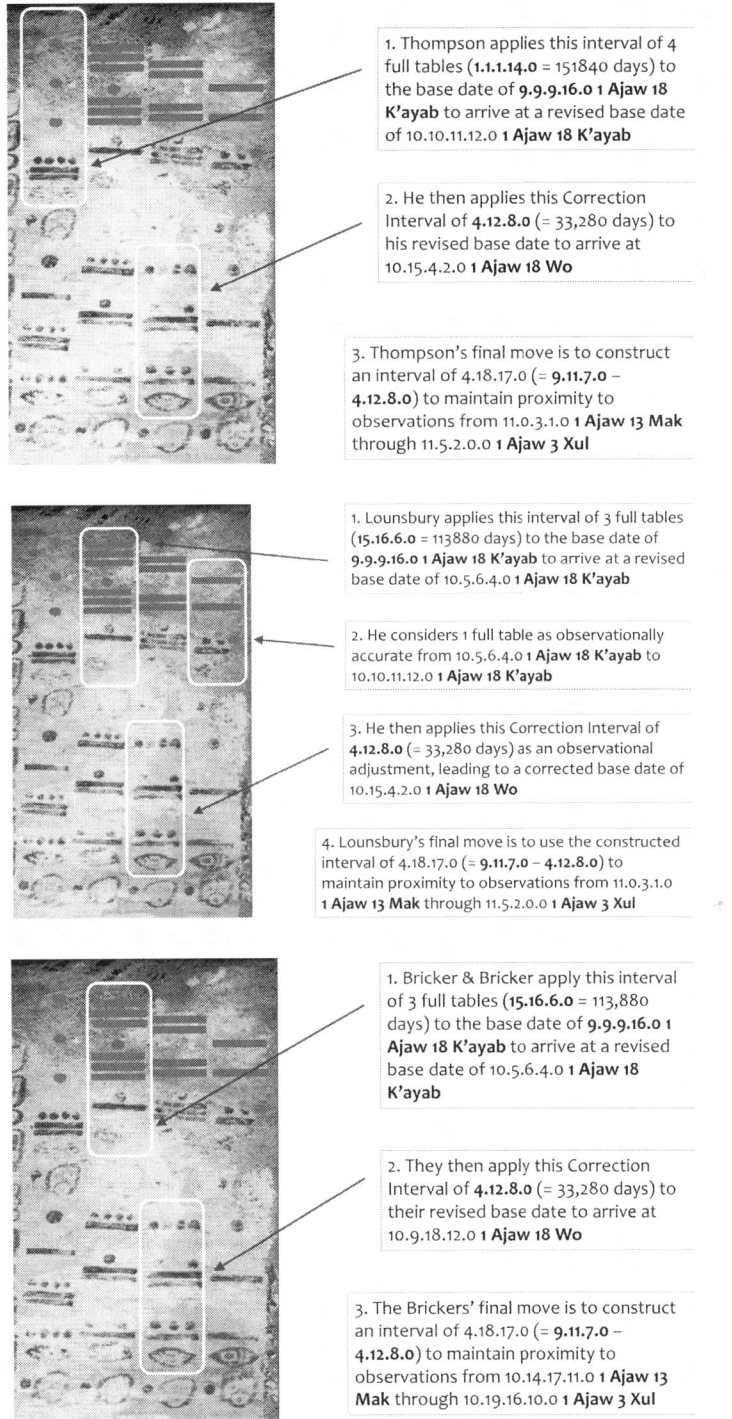

1. Thompson applies this interval of 4 full tables (**1.1.1.14.0** = 151840 days) to the base date of **9.9.9.16.0 1 Ajaw 18 K'ayab** to arrive at a revised base date of 10.10.11.12.0 **1 Ajaw 18 K'ayab**

2. He then applies this Correction Interval of **4.12.8.0** (= 33,280 days) to his revised base date to arrive at 10.15.4.2.0 **1 Ajaw 18 Wo**

3. Thompson's final move is to construct an interval of 4.18.17.0 (= **9.11.7.0** − **4.12.8.0**) to maintain proximity to observations from 11.0.3.1.0 **1 Ajaw 13 Mak** through 11.5.2.0.0 **1 Ajaw 3 Xul**

1. Lounsbury applies this interval of 3 full tables (**15.16.6.0** = 113880 days) to the base date of **9.9.9.16.0 1 Ajaw 18 K'ayab** to arrive at a revised base date of 10.5.6.4.0 **1 Ajaw 18 K'ayab**

2. He considers 1 full table as observationally accurate from 10.5.6.4.0 **1 Ajaw 18 K'ayab** to 10.10.11.12.0 **1 Ajaw 18 K'ayab**

3. He then applies this Correction Interval of **4.12.8.0** (= 33,280 days) as an observational adjustment, leading to a corrected base date of 10.15.4.2.0 **1 Ajaw 18 Wo**

4. Lounsbury's final move is to use the constructed interval of 4.18.17.0 (= **9.11.7.0** − **4.12.8.0**) to maintain proximity to observations from 11.0.3.1.0 **1 Ajaw 13 Mak** through 11.5.2.0.0 **1 Ajaw 3 Xul**

1. Bricker & Bricker apply this interval of 3 full tables (**15.16.6.0** = 113,880 days) to the base date of **9.9.9.16.0 1 Ajaw 18 K'ayab** to arrive at a revised base date of 10.5.6.4.0 **1 Ajaw 18 K'ayab**

2. They then apply this Correction Interval of **4.12.8.0** (= 33,280 days) to their revised base date to arrive at 10.9.18.12.0 **1 Ajaw 18 Wo**

3. The Brickers' final move is to construct an interval of 4.18.17.0 (= **9.11.7.0** − **4.12.8.0**) to maintain proximity to observations from 10.14.17.11.0 **1 Ajaw 13 Mak** through 10.19.16.10.0 **1 Ajaw 3 Xul**

FIGURE 8.12 Procedures proposed for use of the Correction Intervals on page 24: (*top*) Thompson; (*middle*) Lounsbury; (*bottom*) Bricker and Bricker. Courtesy SLUB

584 = 2,336. Therefore, subtracting multiples of 2,340 (or 4 × 584 + 4) from full Venus Great Cycles could be used to compensate for the difference between the synodic period and the idealized Venus Round (Lounsbury 1992b), while preserving the 1 Ajaw date. Lounsbury derived this result by working through hypothetical correction procedures; K'uk'ul Ek' Tuyilaj witnessed the same interval by noticing that Venus's first morning appearance on 10.4.13.4.5 was only five days after a date 1 Ajaw.

In his recovery of the Diophantine structure to these equations, Lounsbury pointed out that the coefficient b of the 2,340 Correction Factor corresponds to the number of four-day corrections built into the Correction Interval (Lounsbury 1992b). The first CI, 4.12.8.0, has b = 2, so it could be used to provide an eight-day correction. It follows that it would be useful for making up an accumulated error of around eight days. We have seen early on, however, that "errors" accrue at a rate of five days per Venus Great Cycle. A single CF of 2,340 days, then, gives an interval (4 days) close to correcting the discrepancy accumulated for one VGC (5.1 days), and an interval of two full Great Cycles (75,920 = 2 × 37,960) would accumulate 10.2 days of error, so a correction of eight days would bring predictions within observable tolerances.

Lounsbury's identification of the Correction Factor and its role within the Correction Intervals, however, only gets us "halfway there." We still are left to confront the nuances of how the Correction Intervals were articulated into a sequence within historical time.

As we have seen, Teeple and Thompson used the Correction Intervals written on page 24 in ways that do not follow directly from the information actually recorded in the preface. They and others have combined the time intervals recorded in ad hoc fashion in order to produce an accuracy of the GMT (Thompson 1972, 63; 1978, 226; Lounsbury 1992a, 185; Bricker and Bricker 2007, 103–4) (figure 8.12). Second, their interpretations require that the anchor of the table deviate from an historical first morning visibility by some sixteen days. So in their interpretations, a table designed to preserve first morning visibilities on dates 1 Ajaw over hundreds of years was anchored to an historical date that did not. In 1983, David Kelley found these points unconvincing:

> It makes no sense to suggest that elaborate tables were constructed with extremely precise correction formulae, accurate to a small fraction of a day, and that the base of these tables was deliberately set inaccurately by 15 to 20 days so that the tables could be used repeatedly without correction until they finally came into step with reality. The alternative—that they were deliberately set at an inaccurate base in the past so that a derivative table (which is not given) would be in step at a contemporary date (which is not mentioned)—seems equally unconvincing, *although the sophistication with which this view has been presented has tended to obscure its basic implausibility.* (1983, 174, emphasis added; see also Aldana 2011d)

Kelley's critique was one of the factors that troubled me when I found myself pulled into a serious consideration of the Dresden Codex Venus Table, which led to my discovery of K'uk'ul Ek' Tuyilaj's discovery.

STARTING OVER

My impression as a newcomer to the field of Maya Studies was that if anything was secure in our reconstruction of ancient Mayan astronomy, it had to be the Venus Table. For my dissertation research in the late 1990s, I took it that way. As a doctoral student in the History of Science Department at Harvard University, I found myself spending as much time in the Anthropology Department in the Peabody Museum, going up and down its cramped, old elevator (built before Morley arrived), or buried among books in the Tozzer Library, as across the street in the Science Center. My advisor in the History of Science Department, Owen Gingerich, was also a member of the Harvard-Smithsonian Center for Astrophysics, so visits to his office—a healthy walk up Garden Street—pulled me even further from Harvard Yard (figure 8.13). But while Gingerich was comfortable advising my astronomical interests, he recommended I seek a coadvisor in anthropology to ensure contextual competence for the cultures I would be investigating.

William Fash welcomed me into the Peabody community of Mayanists and the third-floor lab that seemed always to be bustling with activity: David Stuart inking the illustrations of Piedras Negras stelae and working with Barbara Fash on the hieroglyphic stairway at Copán; and Ian Graham in his corner office, working on his Maudslay monograph or coming out of his darkroom. Working with the Fashes brought me into Maya Studies

FIGURE 8.13 The Peabody Museum and home of the Department of Anthropology at Harvard University. Courtesy Wikipedia

with a firm anchor at Copán; Gingerich's work in ancient astronomy, however, carried the good fortune that he was friends with Anthony Aveni, and that facilitated my move into the history of astronomy.

Like Teeple and Lounsbury and even Morley, my undergraduate training situated me as an outsider to the field. I earned a BS in mechanical engineering from Berkeley and then a master's degree in the same field from UC Irvine before making the switch into the social sciences and humanities. What I came to appreciate most about the History of Science program at Harvard was its insistence on balancing expertise in the technical field studied at the same level as the historical context. One very poignant example came to me when I arrived at Gingerich's office early for a meeting. Gingerich is the world expert on Nicolas Copernicus and his Scientific Revolution–catalyzing manuscript *De Revolutionibus*. At the time, though, he was in the middle of a project with Center for Astronomy postdoctoral fellow Jim Voelkel. Together they were going through Johannes Kepler's work calculating the orbit of Mars, which resulted in his proposal of elliptical planetary orbits and the central concept behind Isaac Newton's theory of gravity. In their study, Gingerich and Voelkel were meticulously working through all of Kepler's mathematical computations—pages and pages of calculations done by hand. That attention to process stuck with me and has informed my work on Mayan astronomy ever since.

For my dissertation, I committed to using Gingerich's approach for all proposed astronomical activity—that is, I would work through the computations myself using Mayan numeration and calculation procedures to ensure that I wasn't simply finding patterns that resulted from the brute-force computational power of modern computers. That was also the approach that served as the inspiration behind my reconsideration of K'uk'ul Ek' Tuyilaj's table of numbers on page 24 of the Dresden Codex. The bulk of this work on page 24 occurred while I was on sabbatical as then a tenured professor at UC Santa Barbara in 2009. The project was accidental, a rabbit hole slipped into while writing up a book chapter on the Maya Calendar Correlation Problem at the request of historian John Steele, a colleague now at Brown University who was a fellow at the Dibner Institute at MIT while I was completing my dissertation. The numerical patterns fell out as I worked between handwritten notes, a whiteboard hung next to the kitchen table to facilitate the kids' homework, an Excel spreadsheet on my laptop, and a pile of beans, sticks, and shells spread out across the table.

It was with this backdrop that I worked through Lounsbury's and Thompson's approaches to matching the long-term correction of the table; both models synchronized at 10.10.11.12.0 1 Ajaw 18 K'ayab, which they assigned to October 25, 1038 AD (see Bricker and Bricker 2007). Lounsbury made his case for this placement by appealing to the Venus Table's "accuracy" in putting the 1 Ajaw 18 K'ayab of 10.5.6.4.0 exactly on a (historically reconstructed) first morning visibility of Venus (mean error of "0"). He claimed that this first morning visibility corresponded to a conjunction of Venus and Mars—in his view, a particularly noteworthy event (Lounsbury 1983, 12).

Harvey Bricker and Victoria Bricker (2007) agreed with most of Lounsbury's work, modifying it slightly by changing the Calendar Correlation by two days and arguing that the ephemeris was more strictly a warning table. While all of these proposals made extremely productive use of the information in the table, it struck me that they did not appeal at all to the logic of the table itself, or the organization penned by its author (see figure 8.12). The primary Correction Interval they all use in fact is not written on page 24 at all, but "implied" by the difference of two Correction Intervals that are recorded prominently (Bricker and Bricker 2007, 102). The one attempt to bring calculations using the implied Correction Interval back to the CIs actually written on page 24 also follows nonintuitive application of the intervals to the base date and has been largely ignored (Closs 1977). While all of these approaches work—that is, they make mathematical sense—they did not appeal to the organization of the page or remain faithful to what was actually written. Finally, it seemed to me that what most constrained these interpretations was their insistence on correcting the sixteen days of "error" that Thompson created with his choice of Calendar Correlation. Again, as we saw in chapter 3, the GMT produced sixteen days of error, so accommodating that is what all of these scholars aimed to fix with different articulations of the intervals on page 24. What I found when going back to the table—without concern of whether or not the GMT was correct—is that there is a very straightforward way to make sense of what is actually written. There also appeared to be a straightforward internal logic that had gone unnoticed.

It turns out that the Correction Intervals explicitly written on page 24 themselves take the Venus Table base date through the anchors recorded:

1 Ajaw 18 K'ayab + 4.12.8.0	1 Ajaw 18 K'ayab + 33,280
1 Ajaw 18 Wo	1 Ajaw 18 Wo
1 Ajaw 18 K'ayab + 9.11.7.0	1 Ajaw 18 K'ayab + 68,900
1 Ajaw 13 Mak	1 Ajaw 13 Mak

These dates, 1 Ajaw 18 Wo and 1 Ajaw 13 Mak, are not just any Calendar Rounds. The two dates generated by these CIs are the two Calendar Round dates written on page 24 and in the table itself (figure 8.14). It seemed to me that we had only to stick to the author's intent for an internally consistent interpretation. In such case, we wouldn't need to reach for an "implied" interval—we could stick to what was written. That didn't address, however, the details of Lounsbury's Diophantine Equations or the issue of Thompson's sixteen days of error. I had to go further with my own mathematical investigation before returning to K'uk'ul Ek' Tuyilaj's work.

The alternate approach to interpreting these Correction Intervals that I hit on opened up a provocative mathematical coherence to the table and generated a more robust interpretation overall. I first translated the elements to include the definition of a new term: the Corrected Venus Interval, which is that time interval used to move from one recorded Calendar Round date to a future Calendar Round date "corrected" for the slippage between the idealized 584-day period and the actual 583.92-day synodic period. We can then take

VGC = Venus Great Cycle = 37,960;

CI_n = nth Correction Interval recorded on page 24; and

CVI_n = nth corrected Venus interval.

Now we have a very straightforward construction of Corrected Venus Intervals based on the information actually written in the upper two rows of page 24 (figure 8.15).

$1VGC + 1CI_1 = CVI_1$

5.5.8.0 + 4.12.8.0 = 9.17.16.0

37,960 + 33,280 = 71,240

This result can be compared to the *uncorrected* interval it is meant to replace, which would be 2 Venus Great Cycles (2VGC):

$2 \times 5.5.8.0 = 10.10.16.0$ $2 * 37,960 = 75,920$

Notice now that the Corrected Venus Interval is listed on page 24 directly below the uncorrected interval it is meant to replace. In other words, the layout of the table itself carries meaning. Moreover, we find that

$CVI_1/2VGC = 71,240/75,920 = 137/146$

This result allows us to compare the first Corrected Venus Interval to the second Corrected Venus Interval:

$1VGC + 1CI_2 = CfVI_2$

5.5.8.0 + 9.11.7.0 = 14.16.15.0

37,960 + 68,900 = 106,860

For the 3 VGCs it is meant to correct

$3 \times 5.5.8.0 = 15.16.6.0$ $3 * 37,960 = 113,880$

So that

$CVI_2/3VGC = 106,860/113,880 = 137/146$

The comparison yields:

$CVI_1/2VGC = CVI_2/3VGC$

It bears emphasis that these all come from what is actually written on page 24, built on Lounsbury's recognition of the Correction Factors.

1VGC – 2 CF to 2 VGC = 2 VGC – 3 CF to 3VGC = 137/146

1 VGC + (1 VGC – 2 CF) = 2 VGC – 2 CF

1 VGC + (2 VGC – 3 CF) = 3 VGC – 3 CF

The result is that each Corrected Venus Interval (CVI_n) relative to its parallel, uncorrected interval produces the same ratio. The same ratio is used according to the same method for correcting projections for the periods of two Great Cycles into the future and that of three Great Cycles.

Finding this relationship pushed me to recognize that there was a strong pattern underlying the intervals recorded on page 24. I wasn't suggesting that K'uk'ul Ek' Tuyilaj actually followed this algebraic representation to develop her model of Venus—*al jabr* itself was still in the process of being moved from a narrative to symbolic representation by Islamic scholars across the Atlantic during the time of this work at Chich'en Itza. Rather, it demonstrated to me that there was a coherence to the information in these tables that had gone unrecognized by Thompson or even Lounsbury.

Moreover, the equivalence of these ratios suggests that they were constructed to be applied independently and not serially. Thompson adopted a serial approach in order to make up for the anchor's "error" and then apply the most efficient means of correction. By following an independent application, the values written themselves become important. It allows for the hypothesis that they were intentionally constructed at least in part around this relationship. Moreover, we do not have to appeal to opaque origins for these intervals. The two Correction Intervals were created with very "Mayan numbers":

$CVI_1 = 71,240 = 137 \times 520 = 137 \times 2 \times 260$

$2VGC = 75,920 = 146 \times 520 = 146 \times 2 \times 260$

$= 73 \times 22 \times 260$

$= 26 \times 2,920$

While

$CVI_2 = 106,860 = 137 \times 780 = 137 \times 3 \times 260$

$3VGC = 113,880 = 146 \times 780 = 146 \times 3 \times 260$

$= 73 \times 2 \times 3 \times 260 = 39 \times 2,920$

The prime number 73 is critical to Mayan calendric manipulations as shown by Lounsbury (1978) since it is a common factor of 365 ($= 5 \times 73$) and 584 ($= 8 \times 73$).

Besides the denominators as whole number multiples of the arithmetic base of all intervals on page 24, these are standard arithmetic relationships readily available to K'uk'ul Ek' Tuyilaj and her Mayan calendric specialist colleagues (Aldana 2007; Lounsbury 1978; Thompson 1972). Nor are the numbers of a scope unattested within the Dresden Codex itself, as they are dwarfed, for instance, by the "serpent numbers" on pages 61–73 of the same manuscript (Thompson 1972, 80–88). Finally, the result is a very respectable 583.934

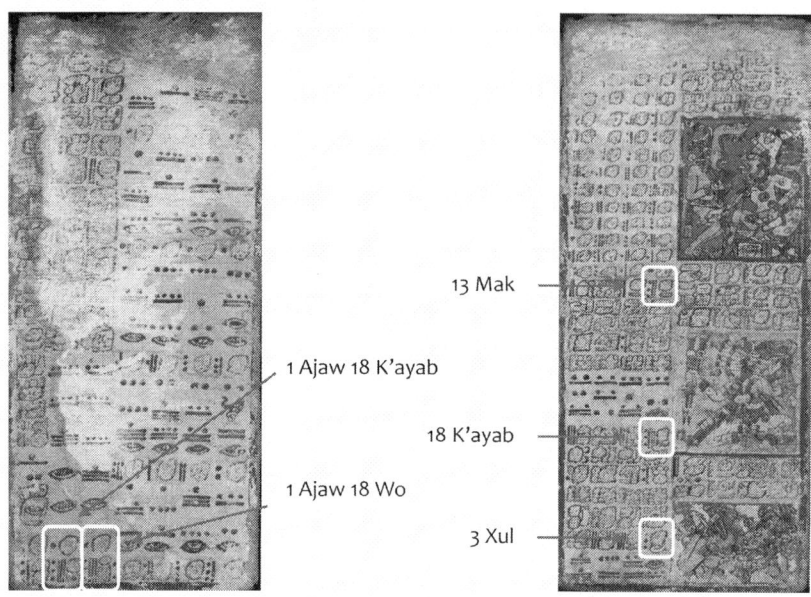

13 Mak

1 Ajaw 18 K'ayab

18 K'ayab

1 Ajaw 18 Wo

3 Xul

FIGURE 8.14 The Calendar Round anchors to the Venus Table. On page 50, 13 Mak, 18 K'ayab, and 3 Xul become anchors with the 260-Day Count immediately above 13 Mak: 1 Ajaw. Courtesy SLUB

(= 71240/122 = 106860/183) average synodic period for Venus, amounting to an error of approximately 20 hours over 104 years—that is, still less than one day of error. This is not as accurate as Thompson's model (583.9203, or 2 hours over 104 years), but it is still very close to the modern value (583.9214), and we find it is productive in other ways.

Bringing this back to what is actually written on page 24, I realized that with this approach there is no skipping around and arbitrarily using inferred Correction Intervals or unscripted applications of one versus another (see figure 8.15). By taking the intended corrections to be applied independently (as opposed to serially), the method is straightforwardly represented in the table of intervals. When we look at the graphical layout of the page, K'uk'ul Ek' Tuyilaj had the 2 CF (8-day) Correction Interval placed underneath the interval for two uncorrected cycles, and she placed the 3 CF (12-day) Correction Interval directly below the interval for three uncorrected cycles. The first correction "fixes" a projection forward of two full tables, while the second correction "fixes" a projection of three full tables. This is a graphical relationship to page 24 that has no parallel in other modern interpretations. When I stepped back from my whiteboard to take in this realization, I felt that I was now recovering the structure of page 24 as Ilaj K'uk'ili Ek' had conceived it. Like Gingerich going through Kepler's computations, I sensed that I too was following the author's logic through the language of mathematics.

It also became clear that K'uk'ul Ek' Tuyilaj had developed a straightforward coherence to nearly all of the right-hand side of page 24. We have seen already that each of the time intervals in the lower three rows takes the base date of 1 Ajaw 18 K'ayab forward

For an advance of **10.10.16.0** or 2 Venus Great Cycles, a correction for slippage is necessary to preserve a **1 Ajaw** base date.

1. Take one Venus Great Cycle of **5.5.8.0** (or 37,960 days) to reach the next **1 Ajaw** date...

2. Add this Correction Interval of **4.12.8.0** to reach the corrected **1 Ajaw** date.

For an advance of **15.16.6.0** or 3 Venus Great Cycles, a correction for slippage is necessary to preserve a **1 Ajaw** base date.

1. Take one Venus Great Cycle of **5.5.8.0** (or 37,960 days) to reach the next **1 Ajaw** date...

2. Add this Correction Interval of **9.11.7.0** to reach the corrected **1 Ajaw** date.

FIGURE 8.15 New procedure for the Correction Intervals on page 24: (*top*) For an advance of 2 Venus Great Cycles (10.10.16.0 or 2 × 5.5.8.0 or 2 × 11,960 days), use 1 Venus Great Cycle and add to it the Correction Interval below 2 Venus Great Cycles. So an anchor on **1 Ajaw 18 K'ayab** + 5.5.8.0 + 4.12.8.0 = **1 Ajaw 18 Wo** as a corrected projection from the base date, taking into account the deviation of Venus's synodic period relative to its canonic period of 584 days. (*bottom*) For an advance of 3 Venus Great Cycles (15.16.6.0 or 3 × 5.5.8.0 or 3 × 11,960 days), use 1 Venus Great Cycle and add to it the Correction Interval below 3 Venus Great Cycles. An anchor on **1 Ajaw 18 K'ayab** + 5.5.8.0 + 9.11.7.0 = **1 Ajaw 13 Mak** as the corrected base date. Courtesy SLUB

in time by multiples of the time interval represented by one row of pages 46–50, or multiples of 2,920 days (five Venus Rounds). These three rows provide multiples one through twelve (figure 8.16). The thirteenth multiple would constitute one full Venus Great Cycle, which is included in the top row—the fifth row from the bottom. Next to one full VGC, K'uk'ul Ek' Tuyilaj included its multiples, or two VGC, three VGC, and four VGC. In between multiples leading up to one VGC and multiples of VGCs, K'uk'ul Ek' Tuyilaj included a row of the corrections necessary to advance by each amount. The fourth row up (or the second row from the top) can be recognized as providing the necessary accommodation for moving beyond one full VGC with the consideration that

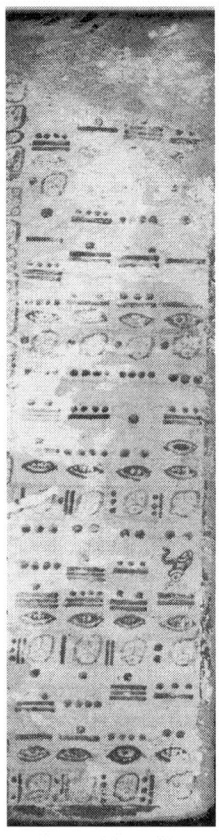

1 1 1 14 0	15 16 6 0	10 10 16 0	5 5 8 0	151,840 4 x 37,960	113,880 3 x 37,960	75,920 2 x 37,960	37,960 13 x 2,920
1 Ajaw	1 Ajaw	1 Ajaw	1 Ajaw	1 Ajaw	1 Ajaw	1 Ajaw	1 Ajaw
1 5 14 0	9 11 7 0	4 12 8 0	1 5 5 0	185,120	68,900	33,280	9,100
1 Ajaw	1 Ajaw	1 Ajaw	1 Ajaw	1 Ajaw	1 Ajaw	1 Ajaw	1 Ajaw
4 17 6 0	4 9 4 0	4 1 2 0	3 13 0 0	35,040 12 x 2,920	32,120 11 x 2,920	29,200 10 x 2,920	26,280 9 x 2,920
6 Ajaw	11 Ajaw	3 Ajaw	8 Ajaw	6 Ajaw	11 Ajaw	3 Ajaw	8 Ajaw
3 4 16 0	2 16 14 0	2 8 12 0	2 0 10 0	23,360 8 x 2,920	20,440 7 x 2,920	17,520 6 x 2,920	14,600 5 x 2,920
13 Ajaw	5 Ajaw	10 Ajaw	2 Ajaw	13 Ajaw	5 Ajaw	10 Ajaw	2 Ajaw
1 12 5 0	1 4 6 0	16 4 0	8 2 0	11,680 4 x 2,920	8,760 3 x 2,920	5,840 2 x 2,920	2,920
7 Ajaw	12 Ajaw	4 Ajaw	8 Ajaw	7 Ajaw	12 Ajaw	4 Ajaw	8 Ajaw

FIGURE 8.16 Full translation of the right-hand side of page 24. The progression of multiples moves from the bottom right to the top left. Notice that the scribe included the Correction Intervals in the second row from the top, between the intervals of less than one complete Venus Great Cycle and the intervals composed of multiples of Venus Great Cycles. Courtesy SLUB

error from observation would have to be accounted for. The base date, plus the element of the right column, first row, plus the correction factor yields the corrected base date, which she should expect to observe.

With this new interpretation of the layout and the method for correction, I then took up a consideration of what the implementation would have produced in Long Count time. As we have seen, K'uk'ul Ek' Tuyilaj would have noticed that a pure projection forward of 3 Venus Great Cycles from the 9.9.9.16.0 historical record to contemporary times did not match observation; the actual first morning appearance would arrive fifteen days before its computed schedule.

9.9.9.16.0 1 Ajaw 18 K'ayab

+ 3 × 5.5.8.0

= 10.5.6.4.0 **1 Ajaw 18 K'ayab**

On the other hand, this uncorrected interval put two 1 Ajaw dates very close to (if not exactly on) first morning visibilities. She would have been able to witness these occurring on 1 Ajaw dates separated by 2,340-day periods.

10.4.6.13.0 **1 Ajaw 13 Mak**
= 9.9.9.16.0 1 Ajaw 18 K'ayab + 183 × (Venus's Synodic Period) + 2.3 days, which is then
= 9.9.9.16.0 1 Ajaw 18 K'ayab + 3 VGC – 3 CF
= 9.9.9.16.0 1 Ajaw 18 K'ayab + 1 VGC + 1 CVI_2

10.4.13.4.0 **1 Ajaw 18 Wo**
= 9.9.9.16.0 1 Ajaw 18 K'ayab + 187 × (Venus's Synodic Period) + 6.7 days, which is then
= 9.9.9.16.0 1 Ajaw 18 K'ayab + 3 VGC – 2 CF
= 9.9.9.16.0 1 Ajaw 18 K'ayab + 2 VGC + 1 CVI_1

From the latter, it follows that the previous 1 Ajaw 18 Wo would have been even more accurate:

9.19.7.14.0 = 10.4.13.4.0 1 Ajaw 18 Wo – 5.5.8.0
9.19.7.14.0 **1 Ajaw 18 Wo**
= 9.9.9.16.0 1 Ajaw 18 K'ayab + 2 VGC – 2 CF
= 9.9.9.16.0 1 Ajaw 18 K'ayab + 1 VGC + 1 CVI1

She therefore ended up with CVI_1 = 1 VGC – 2 CF, and CVI_2 = 2 VGC – 3 CF. And perhaps even more provocatively,

10.4.6.13.0 1 Ajaw 13 Mak
+ 2,340
= 10.4.13.4.0 1 Ajaw 18 Wo
+ 2,340
= 10.5.6.4.0 1 Ajaw 18 K'ayab

The derivation of nearly all of the material on page 24 would have dropped out of the combination of these two observations, seven years apart, with the one base date as a historical record. In other words, with this interpretation, we can leave behind proposals that require jumping around between intervals or the use of "implied intervals" for correction, and we can see the base date as a historical record—as corroborated by the Copán record.

Lounsbury's work demonstrated that the Correction Factor of 2,340 days was fundamental to the Correction Intervals. We have seen here that it would have also been observed directly by K'uk'ul Ek' Tuyilaj. But it even appears that she highlighted this 2,340-day interval on page 24. As mentioned, Lounsbury suggested that the reason behind the 9.9.9.16.0 anchor was that it was numerologically contrived. He considered the Calendar Round on page 24 to have recorded the last 1 Ajaw 18 K'ayab before the end of the previous era and that the interval between it and 9.9.9.16.0 was a "contrived number" made up of several important factors (1978, 787):

$$
\begin{aligned}
9.9.16.0.0 \quad = 1{,}366{,}560 \quad &= 5{,}256 \times \mathbf{260} \\
&= 3{,}744 \times \mathbf{365} \\
&= \mathbf{\textit{2{,}340} \times \textit{584}} \\
&= 584 \times \mathbf{2{,}340} \\
&= 468 \times \mathbf{2{,}920} \\
&= 72 \times \mathbf{18{,}980} \\
&= 36 \times \mathbf{37{,}960}
\end{aligned}
$$

Lounsbury, therefore, proposed that the scribes of the Dresden Codex saw the later 10.10.11.12.0 1 Ajaw 18 K'ayab date as "pivotal." Before this date, Venus Rounds of 584 days were used. After it, corrected intervals were necessary (Lounsbury 1978, 787). But he overlooked the fact that there is a much more straightforward rationale.

As we saw in chapter 2, the dates of many mythological events were contrived by the astronumerologists of Classic period Palenque (Aldana 2004; 2007; 2011d; cf. Powell 1997). I showed that the scribes started with historical dates along with the era base of 13.0.0.0.0, and then they projected backward using integer multiples of astronomical periods (Aldana 2007, 107–8). The same practice can be seen here. The attentive reader will have noticed that a very provocative number resides in Lounsbury's list. When divided by 584—the Venus Round—the interval gives 2,340, exactly the Correlation Factor used in all Correction Intervals. Instead of seeing 9.9.9.16.0 as the enigmatic *endpoint* of numerological contrivance, the revision proposed here follows the practice documented in Palenque's hieroglyphic texts (and Teeple's original hunch to take 9.9.9.16.0 as a historical record of an observed first morning visibility of Venus).

9.9.9.16.0 1 Ajaw 18 K'ayab – 9.9.16.0.0 = –6.2.0 1 Ajaw 18 K'ayab
9.9.9.16.0 1 Ajaw 18 K'ayab – 2,340 × 584 = –6.2.0 1 Ajaw 18 K'ayab

The "Ring Number" on page 24—the one set in mythological times—then becomes the contrived date, generated by using the Correction Factor as a multiplier. Accordingly, K'uk'ul Ek' Tuyilaj was following the tradition of ajk'uhuuns in Kan B'ahlam's court at Palenque, as well as Mak Chanil at Copán. She projected back in time from a historical date, and in this case she thus canonized the key to correcting the table within the Long Count anchor of page 24 (figure 8.17).

Summarizing thus far, the revision that I came upon during my sabbatical of 2009 generated a new interpretation with improved coherence such that

i. it uses only the numbers written in the document for corrections;
ii. Correction Intervals are applied independently to the base date;

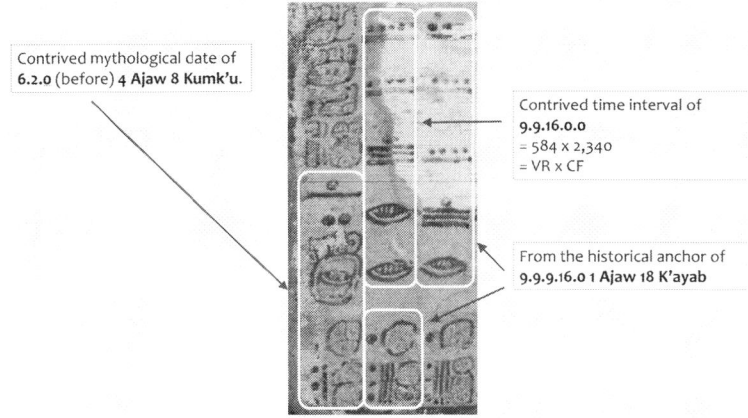

Contrived mythological date of **6.2.0** (before) **4 Ajaw 8 Kumk'u.**

Contrived time interval of
9.9.16.0.0
= 584 x 2,340
= VR x CF

From the historical anchor of
9.9.9.16.0 1 Ajaw 18 K'ayab

FIGURE 8.17 Contrived numbers and the base date of the Venus Table. The convention in the published literature has been to start with the mythological date as set and assume the contrived number projects forward in time. Following the practices of Palenque scribes, we instead may see the historical date of 9.9.9.16.0 1 Ajaw 18 K'ayab as the base, and the mythological date as contrived by using an interval made up of the Venus Round (584) multiplied by the Correction Factor (2,340). Courtesy SLUB

iii. Correction Intervals are built according to the same approximation to the Venus synodic period for projections of 2 or 3 Great Cycles; and

iv. it corroborates and extends Lounsbury's interpretation of the Correction Factor.

It is worth emphasis to note that this all fits into a relatively narrow window of opportunity. Although the Calendar Round dates 1 Ajaw 18 K'ayab, 1 Ajaw 18 Wo, and 1 Ajaw 13 Mak would have recurred every fifty-two years and are always separated by multiples of 2,340 days, this mathematical relationship would not have been worthy of anomalous note during any other time. What caught K'uk'ul Ek' Tuyilaj's attention here was that, at this time, these Calendar Rounds also corresponded to great proximity to first morning visibilities of Venus. And the interval between these dates with its connection to Venus visibility is what gave K'uk'ul Ek' Tuyilaj the Correction Factor—this was her discovery.

VENUS ALMANAC TO VENUS TABLE

Recognizing the discovery recorded on page 24 of the Dresden Codex means that K'uk'ul Ek' Tuyilaj faced a new challenge. While she could now keep track of first morning visibilities of Venus with much greater accuracy over large periods of time, her records would have to change to capture it. And this recognition is what changed the form of the Venus Table from the relatively short Venus Almanac that K'uk'ul Ek' Tuyilaj inherited to the full table that takes up most of page 24 and the upper two-thirds of pages 46–50.

In order to preserve the new accuracy, she would have to include the 365-Day Count portion of the Calendar Round along with the 260-Day Count date. This, it turns out, was not difficult since we have already seen that eight multiples of the 365-Day Count are equivalent to five multiples of the Venus Round. In other words, after five Venus Rounds, the 365-Day Count dates would repeat. On a numerological level, we have seen already that this results from the relationship between Venus and Solar periods—both rely on the prime number 73.

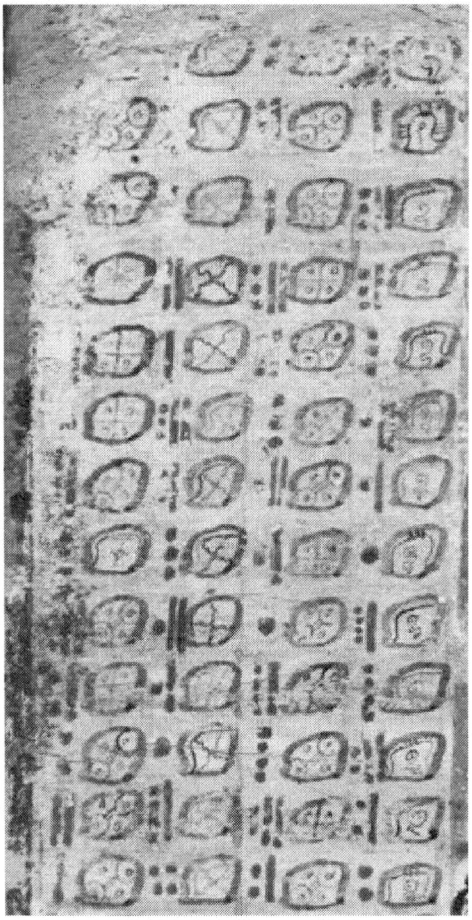

Things did get a little complicated for K'uk'ul Ek' Tuyilaj, however, when she recognized that even though the 365-Day Count dates would repeat, the 260-Day Count dates would not be the same for each run through the five Venus Rounds (figure 8.18). For a proper table accounting for each run, she would need to include the corresponding sequence through 260-Day Count dates. And since there is no common factor to 260 and 365, she had to include every Day Sign with each of its coefficients, 1 through 13. This is why there are thirteen rows of 260-Day Count dates at the top of the table; she could no longer use a concise almanac form. On the other hand, while it made the whole construct more cumbersome, it also allowed for the accuracy in capturing Venus periods that has since made the Dresden Codex famous.

FIGURE 8.18 A Venus Table now needs to include a full set of 13 coefficients for each Day Sign in order to track through a complete Venus Great Cycle. Courtesy SLUB

Whether K'uk'ul Ek' Tuyilaj was older when she made this discovery or still in her twenties, her work must have made some impression. With public ceremonies for Chak Ek' with any sort of prominence at the Tz'iknal, her work advancing the understanding of the planet would have impacts—albeit differential—on different socioeconomic levels of society. It may be, therefore, that the impact resulted in the kind of prestige that Landa described, leading to her public admiration. And with the centrality of Chich'en Itza in regional politics, she may well have accompanied the nobility on their visits to the nearby cities of Ek' Balam, Coba, Dzibilchaltun, or even Cozumel. She may have had some renown, for she had discovered how to keep observations of Venus accurate for hundreds of years into the future.

The upshot is that rather than the standard model's ambiguous placement of the Dresden Codex Venus Table in the Postclassic, the interpretation of page 24 presented in this chapter puts its formulation within twenty-five years of the dedication of the Tz'iknal and possibly within the reign of one of the most important figures in Chich'en Itza's history, K'ahk' Upakal K'awiil. The revision proposed here places the window of opportunity for construction of the Venus Table—through the recognition of the 2,340 CF—at 10.4.0.0.0–10.5.0.0.0, or between twenty and thirty years after it was completed. Twenty-five years of observations from the same building amounts to three full eight-year cycles of Venus, or three rows of the Venus Table on pages 46–50. This would have offered plenty of time for ajk'ins to confirm their basic model of Venus visibilities before encountering the opportunity of 9.9.9.16.0 and the two 1 Ajaw first morning appearances of the 10.5.0.0.0 katun. It appears, then, that Aveni was right in hypothesizing that the Dresden Codex Venus Table was developed using observations made in the Tz'iknal. In addition, we can now see that the Dresden Codex itself very likely captures much of the politics of its time as well.

That is, it is hard to avoid the coincidence of this discovery at Chich'en Itza with another dramatic innovation for which the city is well known. Modern scholars and ancient historians have all attested the religious and political impact of Quetzalcoatl—the Feathered Serpent—throughout Postclassic Mesoamerica, but in particular at Chich'en Itza. In fact, in this book, right from the start, we saw that Seler associated the words of indigenous Mesoamericans characterizing Quetzalcoatl's relationship to Venus with Chich'en Itza and even the Dresden Codex. By considering Quetzalcoatl and the politics of the Terminal Classic and Postclassic, we find that K'uk'ul Ek' Tuyilaj's discovery was likely more than simply coincident with the arrival of K'uk'ulkan at Chich'en Itza. We turn to this investigation in the next chapter.

Stepping back, though, we may also recognize that the interpretations offered by the recalibration work of this book have now pushed us away from a focus on astronomy in a Western conceptualization and certainly away from a strict emphasis on accuracy. By and large, this is an artifact of having emphasized the operation of the table proper and its connection to other documented space-time rituals—a project that now has become realizable with access to a century of archaeology in Mesoamerica and the decipherment of the hieroglyphic record.

CHAPTER 9

Venus, Quetzalcoatl, and the K'uk'ulkan Sodality

Tecun Uman fought valiantly against the allied indigenous and Spanish soldiers, led by Pedro de Alvarado in 1524, as these foreigners entered K'iche' Mayan territory. The nobleman warrior was motivated by the hope that his smaller city would avoid the fate of Tenochtitlan. Privy to word traveling to Qumarkaj much faster than Alvarado's horses, Tecun Uman and his soldiers set up lines of defense against the intruders. Although they were unable to hold off conquest funded by promises of land and laborers from the Spanish Crown, Tecun Uman's efforts developed into legend and are preserved today in allegory, literature, and theatrical dance. The authors of the *Popol Vuh* themselves included reference to him, though without fanfare. They write: "Tecum and Tepepul paid tribute before the faces of the Castilian people. These had been begotten as the thirteenth generation of lords" (Christenson 2007, 296). In the end, with Tecun Uman, it is difficult to discern legend from history.

In that epic manuscript, the K'iche' authors of the *Popol Vuh* included a perhaps underappreciated reference to another great warrior within their historical narrative. Writing among the pine trees of the Mayan highlands, now at the village of Santa Cruz K'iche' in a valley 1 kilometer east from Qumarkaj, these scribes recorded an account that resonates with, without strictly replicating, the Aztec versions of the Quetzalcoatl literature that we saw in the first chapters of this book.

Mormon ethnographer and translator of the *Popol Vuh* Allen Christenson spent years in the highland communities south of Santa Cruz K'iche' around Lake Atitlan. His translation notes that the authors orthographically distinguish two figures who share this name:

The manuscript always spells the name of this lord as Cucumatz, whereas the creator deity with the same, or similar, name is spelled Qucumatz. The variant spellings may be the result of multiple scribes working on the manuscript, each with his/her own creative ways of spelling. Alternatively it may simply be a means of distinguishing between the deity and this lord. (2007, 267n733)

Christenson therefore brings into consideration the fact that within the record of K'iche' mythology and dynastic history, there is a primordial deity, Qucumatz, who was involved in the creation of the world, and a much later nobleman, Cucumatz, who led one of the primary lineages of the K'iche' nation five generations before Tecum, paralleling the case in the Basin of Mexico. Of the latter mortal figure, the K'iche' scribes of Qumarkaj wrote:

> This, then, was the beginning of the increase of the Quiches. Lord Cucumatz founded the grandeur of his descendants. The faces of his grandchildren and his sons were not lost in his heart. He did not do this so that he would be the sole lord. But his nature was enchanted and thus he toppled all the lords of the nations. This he did merely to reveal himself. Yet because of it, he became the head of the nations. The enchanted lord named Cucumatz was of the fourth generation of lords. (Christenson 2007, 276)

The description of their revered ancestor, Cucumatz, bears great resemblance to those we have seen in previous chapters of Quetzalcoatl as a leader and "head of nations."

Because the *Popol Vuh* narrative itself leads up to the time of the scribes who wrote it, however, scholars have been able to trace back historical events to place the reign of the historical Cucumatz near the end of the fourteenth century or early fifteenth century AD. And this places it centuries after the representations of feathered serpents at Cholula in Central Mexico or Chich'en Itza in Yucatan. The narrative itself hints otherwise. According to the scribes:

> Truly [Cucumatz] was an enchanted lord in his essence.
>
> Thus all the lords were frightened before his face. Tales about him were quickly spread abroad and all the lords of the nations heard of the nature of this enchanted lord. (Christenson 2007, 275–76)

If we now, centuries later, had encountered this final statement without its historical context—and so the recognition that it occurred far too late—we might be tempted to propose that the K'iche' at Cumarcah generated the origin of Quetzalcoatl throughout Mesoamerica. But he didn't. Instead, Cucumatz was a member of the genealogy that eventually produced the leaders of the K'iche' nation, to which the authors of the *Popol Vuh* belonged, long after the man/legend Quetzalcoatl of Central Mexico. This review, then, reminds us that while the *Popol Vuh* is grand in scope, treating mythology and history, it is also explicitly *interested* history, providing the credentials of the noble houses of the K'iche'. And that perspective proves useful as we turn our gaze back earlier into the Postclassic.

In this chapter, we consider the timing of the Venus Table's construction relative to changes at Chich'en Itza itself as well as to pan-Mesoamerican appeals to Quetzalcoatl and his counterparts Qucumatz and K'uk'ulkan. We will find that the interplay of ideology and interested history provides an important lens for unpacking the politics of Chich'en Itza. In turn the latter demonstrates that the concept of Quetzalcoatl—its cosmological capacity—was important enough to merit translation, not ossification in a foreign-language name.

FEATHERED SERPENT LEADERSHIP SODALITIES

Associations of great rulers with Quetzalcoatl figures throughout Mesoamerica may be best exemplified by the historical records we have of the city of Cholula in the modern state of Puebla, Mexico (figure 9.1). Mesoamericanists commonly draw from the sixteenth-century Spaniard Gabriel de Rojas, who wrote of the organization as follows:

> [T]hese two [the *aquiach* and the *tlalquiach*] were the ones that governed all the republic and from this religion [of Quetzalcoatl] came the captains named by the *(a)quiach* and *tlalquiach* when they happened to war with their neighbors, just as these two priests were preeminent in confirming in office all the governors and kings of this New Spain. Thus, these kings or chiefs [from outside Cholula], in inheriting a kingdom or principal-

FIGURE 9.1 Cholula, Mixtec highlands, Tilantongo, Tula, Teotihuacan, Xochicalco, Mixteca Alta

ity, would come to this city to pay homage to the idol of this Quetzalcoatl, to whom they offered rich feathers, blankets, gold and precious stones and other things of value, and having made offerings, they were put in a small room dedicated to this purpose in which the two highest priests marked them by piercing the ears, the nostrils, or the lower lip, depending on the kingdom, and they returned to their lands. And because of such things, they were given credit and were respected by the lords. So that they might be respected by these lords, five *indios* from the priesthood also accompanied them home. (Rojas [1581] 1927, 161, quoted and translated in Ringle 2004, 170)

We saw the same structure in the case above of the Mixtec leader Iya Na Cuaa. In Oaxaca, each hilltop fortified village had its own ruler, but Na Cuaa had to convince Qui Huidzu, the representative of another city, that he was worthy of that priesthood's legitimation. Then it wasn't Qui Huidzu but his administrative colleague at Achiutla who actually performed the nose-piercing ceremony (Troike 1974). Not only was the structure the same, but we saw in the last chapter that, in the Codex Nuttall's record

of Na Cuaa's history, the Mixtec "Quetzalcoatl figure" Coo Ndodzo, also known by his Day Sign as personal name Q Chi, acts in the background (figure 9.2).

The point is that in both of these cases a new form of governance is demonstrated in the highlands of Central Mexico. For either case, at Achiutla and Cholula, in order to carry out these rites of investiture the cities of both had a priesthood committed to providing this legitimation service for independent rulers within the region, as well as a local leadership to serve its own city and population. Archae-

FIGURE 9.2 Q Chi, the Mixtec Quetzalcoatl figure

ologist William Ringle (2004) has proposed that the anomalous record at Chich'en Itza reflects the adoption in Yucatan of a similar structure during the Terminal Classic. He argues that the "Gran Nivelacion" or the Great Terrace—that is, the expansion of the civic ceremonial center to the north of the Puuc architecture surrounding the Tz'iknal—represented the city's conversion into a site of investiture into a Feathered Serpent leadership status for the northern peninsular region. In his interpretation, Chich'en Itza was a type of Cholula or Achiutla, serving the leadership of the cities of Yucatan, in an argument heavily informed by the art and architecture of the Great Terrace with special attention to the Ball Court (Ringle 2004, 2009).

We don't know if K'uk'ul Ek' Tuyilaj lived to see the completion of the Great Terrace, but her nieces and nephews certainly would have watched it take shape and find ample use in the late Terminal Classic period. If the area were not access-restricted when not put to ritual purpose, they would be able to wander around the huge plaster platform connecting the buildings of the northern area of Chich'en Itza, centered on the Temple

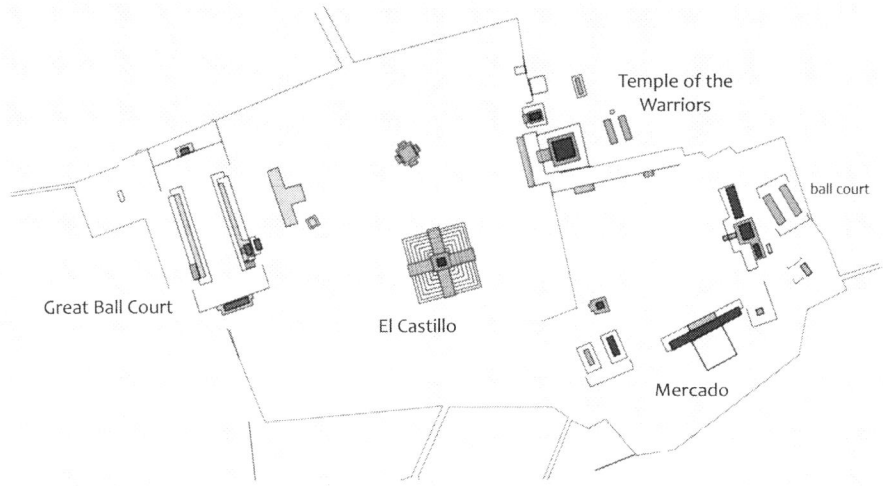

FIGURE 9.3 The Great Terrace at Chich'en Itza. The sacbe running to the north leads to the Great Cenote.

of K'uk'ulkan, as do tourists today (figure 9.3). At the eastern edge of the Great Terrace, they would encounter the Temple of the Warriors and a plaza surrounded by colonnaded palaces, probably filled regularly with priests and politicians going about their daily work. To the immediate north of the Temple of K'uk'ulkan, masons had raised two platforms—those referred to by Landa in his sixteenth-century visit as sites for theater performances—on the path to the sacbe leading to the Great Cenote. To the west of this plaza they would arrive at the Great Ball Court. Although as commoners they most likely would not have been able to see its full scope, the four associated structures within the Ball Court all contributed to a single iconographic mission (Ringle 2004; Schele and Mathews 1999, 254).

Arriving just outside the Ball Court from the Temple of K'uk'ulkan, visitors would find their eyes drawn to the "Lower Temple of the Jaguar," built into the east outside wall at ground level. This structure's principal adornment were the carved columns at the entrance, which framed a jaguar throne (figure 9.4). On the outer walls, artists carved older men carrying rattles and wearing elaborate headdresses, not unlike those worn by Classic period k'uhulajaws. Waterlilies bloom amid long feathers, and fish nibble at the flower petals (cf. Schele and Miller 1986, 46, 115). The carved column on the right depicted four women, each wearing a different skirt. Each, though, is entwined by a serpent emerging from the cleft in the forehead of a personified mountain under her feet. Artists carved four men on the column to the left, also with serpents wrapped around them, which rise from the register below their feet. These figures too have their hands above their heads, appearing to hold up the scene in the upper register of the column. That upper scene, which appears to be replicated on each of the four sides of both columns, is the well-known representation of Ixim ("the Maize God") emerging from a cracked turtle carapace (cf. Kerr Vessel 1892). The message for the public appears to

FIGURE 9.4 Lower Temple of the Jaguar attached to the outside of the Great Ball Court

be one of cosmological status: men and women supporting the sustenance of the city's population (cf. Schele and Mathews 1999, 17). Those dignitaries or other distinguished visitors granted access to enter would find that the interior was completely covered with iconographic carvings; here they would find the mural discussed in chapter 1, and which we will address further below.

Passing into the Ball Court proper with the Lower Temple of the Jaguar on the right, K'uk'ul Ek' Tuyilaj's nieces and nephews would see the South Temple on their left. Masons had elevated this shallow structure off the terrace floor, but the building's character was dominated by the six columns along its entrance. Artists carved each column in upper, middle, and lower registers similar to those in the Lower Temple of the Jaguar. Here, though, the images were of unique human figures, carrying elaborate priestly bags (cf. Zender 2004, 123–24) in one hand, a curved implement in the other, and each wearing a "Toltec" back mirror (Ringle 2004, 206; Schele and Mathews 1999, 245). Schele and Mathews identify them as warriors (1999, 245), but Ringle considers them to be initiates preparing for investiture (2004, 206). Certainly they wear ritual regalia, and that regalia might be meant to carry symbolic meaning, but the specific role these individuals played is ambiguous. On the other hand, what appear to be the names of the individuals have survived erosion on the fifth column, with one such name legible as serpent-star, or possibly Kan Ek' (Schele and Mathews 1999, 245). If this is a member of the lineage we

FIGURE 9.5 South Temple of the Great Ball Court. LACMA (Schele Drawing Archive #82037); photo by Linda Schele © David Schele, photo courtesy Ancient Americas at LACMA (ancient americas.org)

saw in the last chapter, then that might suggest that these were not anonymous figures simply playing roles, but historical figures engaged in a specific historical event. To this possibility we return below.

Turning to face north along the length of the Ball Court, visitors would find their eyes drawn to the Upper Temple of the Jaguar on the right, above the Lower Temple of the Jaguar but on the opposite side of the Ball Court wall. The original architects clearly intended this to serve as a focal point as it bears a highly decorated façade, fronted by two feathered serpent columns (figure 9.6). Access here would have been far more restricted than the Lower Temple of the Jaguar, but those who were able to view inside would have found that artists carved a vast iconographic program on columns as well as interior walls, along with an "atlantean" altar.

Moving past the Upper Temple of the Jaguar, walking along the playing field of the Ball Court, visitors encountered a repeating scene decorating the walls on both sides. Artists carved feathered serpents into bands along the length and around the rings ten-oned into the center of the wall. Of main interest, though, was probably the scene of a ball game in play, with two teams facing each other—a scene repeated six times along each wall. This scene has been interpreted to have depicted repeated ritual activity at Chich'en Itza (Schele and Mathews 1999, 246), but we will see below that it may have referred to an important historical event instead.

Finally, at the northern end visitors reach the North Temple. Like the South Temple, masons raised it on a platform, and artists extensively decorated its interior with murals, carved lintels, and carved columns. But this structure was far smaller.

Overall, the imagery of the Ball Court and that of the rest of the Great Terrace fit together well in style and in content—even though it may be challenging for us today to perceive precisely what its meaning was. The Great Terrace also represented a significant departure from the art and architecture of previous construction at the city, specifically

FIGURE 9.6 Feathered Serpent columns in the doorway of the Upper Temple of the Jaguar. LACMA (Schele Drawing Archive #83002); photo by Linda Schele © David Schele, photo courtesy Ancient Americas at LACMA (ancientamericas.org)

FIGURE 9.7 North Temple of the Great Ball Court. LACMA (Schele Drawing Archive #82026); photo by Linda Schele © David Schele, photo courtesy Ancient Americas at LACMA (ancient americas.org)

that in the vicinity of the Tz'iknal. In the late 1990s, Rafael Cobos and Terance Winemiller examined the sacbes that would have guided visitors and residents throughout the city and argued that the Ball Court and the Great Terrace overall reflect the transformation from a provincial decentered city into a hierarchically based metropolis. "At Chichen Itza the spatial distribution of elaborate versus non-elaborate architecture shows a concentric pattern, whereas the morphology or internal layout of the ancient community with its causeways and architectural groups reveals a dispersed settlement" (Cobos and Winemiller 2001, 288). In other words, Cobos and Winemiller were able to bring along archaeological consensus to view the more provincially directed Puuc-style city of Chich'en Itza as the early version, which corresponded to a more dispersed population. A move was then made concertedly to convert the small city into a new internationally directed metropolis sometime between the time K'ak' U Pakal K'awiil dedicated the Caracol in 10.3.0.0.0 and the abandonment of the city circa AD 1100.

Cobos and Winemiller do not suggest that the older Puuc center itself was abandoned with the growth of the city and the construction of the Great Terrace, but they do suggest that it would now serve a different role. We would do well here to notice that the city of Tula in Central Mexico, which has been traditionally understood to play a role in the life of the historical/legendary Quetzalcoatl, itself underwent a transition between its smaller size from AD 650 to 850 (Tula Chico) and its apogee of influence from AD 850 to 1150 (Tula Grande) (Ringle 2004). Given the tolerances on ^{14}C dates, both transitions would have occurred right around the time that K'uk'ul Ek' Tuyilaj was confirming her new Venus Table, built from the Venus Almanac she inherited, as we saw in the last chapter. It also is worth emphasizing that it is not until the construction of the Great Terrace that we see the explicit imagery at Chich'en Itza of a feathered rattlesnake that becomes associated with Quetzalcoatl/K'uk'ulkan throughout the Postclassic.

CHICH'EN ITZA'S ARCHITECTURAL TRANSFORMATION

While the distinction between the north and the south of the city has been consistently differentiated, overall the extensive mural program within the Ball Court complex has been difficult for modern scholars to build consensus around. Not yet fully in the Postclassic "international style" that Tawiskal Uwoojil would witness, the artists of these scenes mixed traditional Central Mexican with Mayan conventions, making it difficult to determine the authorship and audience for each representation. The mural we saw in this book's introduction within the Lower Temple of the Jaguar, for example, might be viewed as depicting a single procession, with each row comprising a different mix of warriors, priests, and nobles. Or it could be different events separated in time—a sequence of activities performed by the same protagonists, portrayed from one level to another.

Beyond the layout, the iconography of this imagery, too, has proven confounding. The mural artists, for example, depicted a figure wearing Teotihuacan-style eye ornaments associated with a feathered serpent meeting another nobleman (figure 9.8). Was this a

FIGURE 9.8 Detail of the focal point of the mural in the Lower Temple of the Jaguar. The eye-piece-wearing figure on the left is accompanied by a Feathered Serpent that breaks out of the frame. To the right of the central figure facing left is a Xochicalco Day Sign.

foreigner from Central Mexico arriving at Chich'en Itza, bringing with him a Quetzal-coatl ideology? One tantalizing detail might bolster such a hypothesis: there are Day Signs depicted adjacent to the central figures in the Ball Court mural, and the Day Signs use the iconography of Xochicalco and Central Mexico (Schele and Mathews 1999, 255). Given that Day Signs were often used within personal names in the Sierra Madre, such as with Iya Na Cuaa or "Lord 8 Deer," it may well be that artists used this mural to depict the arrival of strangers, not unlike the event at Tikal more than five centuries earlier (Ringle 2004, 214). While that is possible, the rest of the murals in the Ball Court complex help to build a different interpretation.

William Ringle's intensive study of Chich'en Itza's archaeology and art history has led him to propose a coherent alternative interpretation to those prevalent in the late twentieth century. Ringle has worked to resolve the anomalies of Chich'en Itza's art and architecture by drawing extensively from work by Bruce Byland and John Pohl, which in turn placed Mixtec and Zapotec codex histories into architectural and geographic reconstructions. Pohl and Byland (1990) were able to trace, for example, Na Cuaa's journey from Tilantongo in the Mixteca highlands south to Tututepec, where he would take up his first position of authority, and then back to Achiutla where he would earn participation in the nose-piercing ceremony (Ringle 2004) (figure 9.9). Their work on Mixtec dynasties within the Oaxacan highlands demonstrates that the narratives were historical and not mythological—rooted in the architecture of the Sierra Madre landscape.

As in the Oaxacan highlands, Ringle (2004) argues that what occurred at Chich'en Itza was not in isolation. The relationship between Chan Chaak at Uxmal and K'ak U Pakal K'awiil explored by Schele and Mathews also plays a role. Ringle sees a homology between the nunnery depictions of authority at Uxmal and those adopted for the Temple of the Chac Mool at Chich'en Itza (2004, 208). The implication is that the leadership of Uxmal accepted Chan Chaak as a ruler within a feathered serpent sodality—as did Tilan-tongo with Na Cuaa—but that Chich'en Itza's leadership did not want simply to receive a member of a feathered serpent sodality for its ruler. Instead, the nobility of Chich'en Itza aimed for the role of Achiutla or Cholula. In this effort, Ringle argues that the murals in

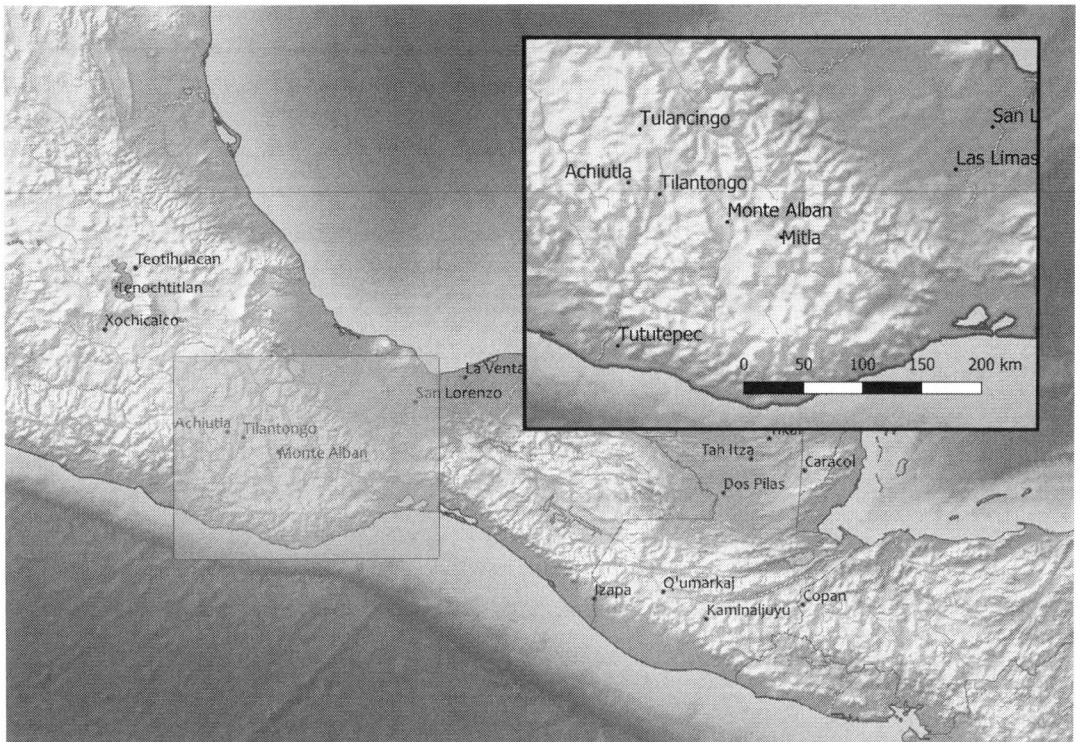

FIGURE 9.9 Locations in Iya Na Cuaa's personal history

the Ball Court structures represent aspects of the investiture of secondary leaders into a warrior sodality associated with the feathered serpent. In sum, Ringle suggests that the Great Terrace was dedicated to the Quetzalcoatl investiture activities of the city while "Old Chich'en was the provincial seat of power" (2004, 210).

Concerning the ambiguity of the mural scenes in the Lower Temple of the Jaguar, then, Ringle views them as a huge procession that the leadership of Chich'en Itza participates in, all centered on an empty jaguar throne (2004, 170). In such case, the eye-piece-wearing figure we saw above is at the center of activity. We note that, in this position, he is shadowed by a feathered serpent rising up, arching overhead, breaking through the artistic format of the mural. And here we confront an important innovation that has only been implicit to this point. In Mixtec and Aztec representations, Quetzalcoatl as a figure bears the feathered serpent symbolism in his name, but not in iconography. The standard representations—as we see in figure 9.10—depict him with a long-snouted mask, with a truncated-cone-shaped hat and a half of a conch shell hanging from a necklace. In contrast, in Yucatan feathered serpent leaders are depicted with an actual feathered serpent wrapped around their body. This may speak to the origins of the role played by this figure—the transparent imagery associated with the name in the Mayan region implying that the conceptual identification was primary. In any case, because the imagery dominates, we will identify this figure as a K'uk'ulkan leader—not necessarily as

FIGURE 9.10 Feathered Serpent–associated figures in (*top*) Aztec representations compared to (*bottom*) Mixtec representations

a single historical figure bearing this name, but perhaps more in line with Kan Ek' at Tah Itza. In this case, K'uk'ulkan figures bore the name to reflect their position in society, more than as a marker of their unique individuality.

Returning to Ringle's view of the Ball Court murals, the K'uk'ulkan figure, along with all others, moves toward the empty throne as a site of investiture. In this and other resonant scenes throughout the Ball Court architecture, Ringle does not see the K'uk'ulkan figures as different individuals. Rather, he claims that "all of these are but a single person at different stages of investiture, culminating in the final scene on the north wall of the [North Temple]" (Ringle 2004, 186). For Ringle, this is not history represented on the walls of the Ball Court; these images depict the idealized template for a designated leader to take on the authority of K'uk'ulkan leadership.

Again, there is strong corroborative evidence for Ringle's interpretation. The Ball Court at Chich'en Itza is disproportionately huge, for instance, suggesting that it was meant to be symbolic, not a functioning venue for the game as traditionally played (Ringle 2004, 170). Also, the Ball Court is entirely appropriate as a site of investiture given that in the *Historia Tolteca-Chichimeca*, among other documents, the ball game/court is associated with the investiture of secondary lords in Postclassic Central Mexico (2004, 189). In line with his view that the murals depicted an idealized event, Ringle also goes back to rescue some of Eduard Seler's work on the Codex Borgia.

In his early twentieth-century work, Seler proposed that Codex Borgia pages 29–46 recorded the mythological birth and challenges faced by the Mixtec version of Quetzalcoatl, Coo Ndodzo. To Seler, they showed cosmological or mythological events, distant from human experience. With Ringle's reading of the murals in the Chich'en Itza Ball Court, he now sees these more as a template for the concept of rulership:

> I believe that the events of the two sections [Terrestrial and Underworld in the Borgia] ultimately trace a theme of death and rebirth with direct application to the body of the ruler, precisely because Quetzalcoatl embodied kingship and the ruler was himself an

avatar of Quetzalcoatl. Thus, on one level, this section of the Codex Borgia is a paradigmatic explanation of the mystery of divine kingship, reconciling human mortality with the persistence of that office through the reincarnation of the spirit of kingship in successive rulers. (Ringle 2004, 172)

He argues that this "suggest[s] they are in some sense embodiments of Quetzalcoatl" (Ringle 2004, 173).

In Ringle's interpretation, then, Borgia page 29 depicts a Coo Ndodzo–sodality initiate's meeting with a mortuary bundle of Tlahuizcalpantecuhtli (or Venus as Morningstar). The imagery of the following pages then tracks the initiate through rites in the Underworld and terrestrial realms before he emerges triumphant as a Coo Ndodzo ruler. Taking the connection still further, Ringle notes that "several Mixtec leaders, including Lords 9 Wind, 12 Wind, and 8 Deer, employ identical face paint" to the figure that emerges from the visual narrative constituting pages 29–46. According to Ringle, then, the Borgia pages were consulted by priests of the Mixtec region for depictions of the ritual activity that was intended to take an initiate's encounter with a mortuary bundle through to his emergence as a member of the same form of authority carried by Iya Na Cuaa and the leaders of other local communities. In this view, the murals of the Chich'en Itza Ball Court architecture depict investiture and serve the same purpose as pages 29–46 of the Borgia Codex for highland Central Mexico.

Bringing their work together, we find that Ringle, Byland, and Pohl have developed a complex but robust interpretation of the Great Terrace at Chich'en Itza. We can now see the expansion of the city from its decentered provincial Puuc character in the south to a vast metropolitan center in the north as an effort intended to incorporate Feathered Serpent authority into the region by creating an architectural center for investiture.

There is important additional information, however, that we have only recognized in chapter 7 of this book. Namely, the pages immediately following what these scholars propose to be depictions of investiture in Codex Borgia pages 29–46 we now have recognized as an extended section concerning Venus. Specifically, the codex depiction of the process for becoming initiated into the Coo Ndodzo sodality—itself carrying heavy iconographic representations of Venus—in turn leads to the Borgia Venus Almanac, which describes rituals affiliated with the visibilities of the planet itself. With this in mind, we may now extend Ringle's, Byland's, and Pohl's interpretation to offer a new reading and use of the Borgia Venus pages, which will help us better understand the archaeological record at Chich'en Itza and the Dresden Codex.

COO NDODZO INVESTITURE

Pages 29–46 of the Borgia Codex have been recognized for some time as constituting an unconventional narrative. The scribe did not arrange the content in strict almanac sequences as did Tawiskal Uwoojil in the Dresden Codex, nor is the material partitioned into smaller "scenes" as in the pages preceding and following it within the Borgia. Rather, full abstract iconographic complexes fill entire pages, several without any apparent human representations at all. A run through much of this material, though, demonstrates how these pages may have served well as a guide for investiture into a Coo Ndodzo (Mixtec "Feathered Serpent") sodality.

As we saw above, Ringle follows Byland and Pohl to argue that on page 29 the Borgia scribe painted the central figure to represent a mortuary bundle wearing the regalia of Tlahuizcalpantecuhtli. This Nahuatl deity is the one Seler used to connect the Frost Teotl to Venus as Morningstar. The Borgia scribe here painted the Tlahuizcalpantecuhtli bundle and the bowl supporting it in the "hocker" pose of giving birth. Death giving rise to new life is thus represented and aligns easily with religious themes present throughout Mesoamerica. Accordingly, around this bundle small, naked anthropomorphic figures with Q Chi heads issue forth from (are born from?) star-marked Coo Ndodzo serpents. From an investiture perspective, this may well represent the birth of the initiates' Coo Ndodzo soul/*nagual*, corresponding to the beginning of the investiture process.

On page 30, the Borgia scribe painted all twenty Day Signs encircling a focus on birth scenes from Q Chi serpents, but now each Q Chi figure holds a priestly bag. Circles highlight every fifth Day Sign, which is pierced with a bloodletter by a priestly figure carrying the same bag as the Q Chi figures (figure 9.11). Here, the representation of Venus in these pages is reinforced. Whereas the first page included an image of Tlahuizcalpantecuhtli,

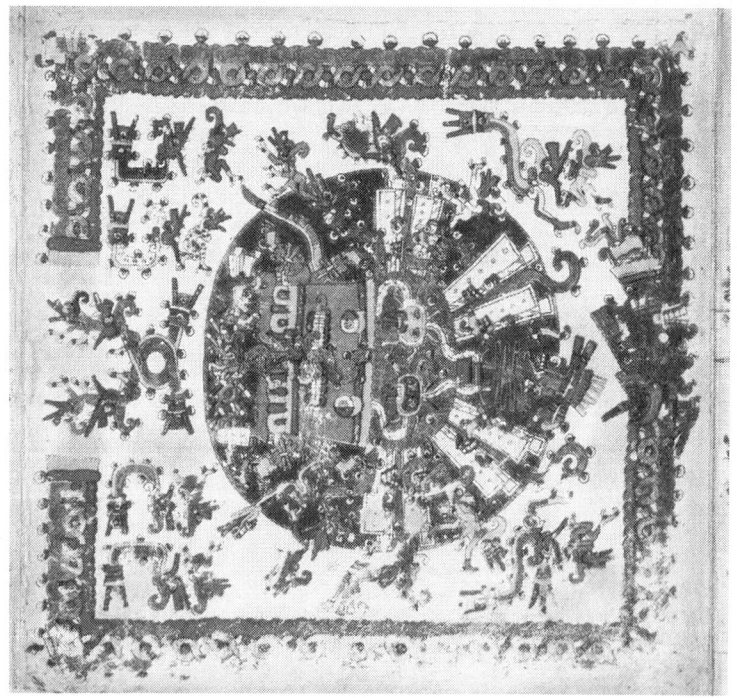

FIGURE 9.11 Borgia Codex page 29. LACMA (Codex Borgia), photo courtesy Ancient Americas at LACMA (ancientamericas.org)

on page 30 the priestly figures carry on their back the directional trees of pages 49–52 from the Borgia Venus Almanac. Moreover, the highlighted Day Signs correspond to the sequence through one Venus Round and so show up on the same pages 49–52 of the Venus Almanac. The implication of the calendric sequence is that priests in charge should have scheduled investitures only on those Venus Rounds that began with first morning appearance on the Day Sign Cipactli.

We encounter here, therefore, a very practical use of the Borgia Venus Almanac in addition to its capacity to keep track of Venus visibilities. Now we can see that a full eight-year (five Venus Round) almanac is necessary in order to plan for ceremonies initiated by Venus Rounds that would begin on Cipactli dates. Of course observations of the planet and omens generated would be useful for other purposes, but this Borgia almanac also would allow for the scheduling of investiture ceremonies well in advance.

As for the specific role of page 30, it is likely that the Borgia scribe painted this image to represent the part of the investiture in which the initiate would take on a new name from the 260-Day Count—such as Na Cuaa (na-cuaa or "8 Deer")—probably ascertained through divination. Likely involving a bloodletting ceremony, the initiate now has a new nagual/soul and a name representing membership in the sodality.

On pages 31 and 32 the scribe depicted three different "rooms," each bearing two significant markers tying them all together (figure 9.12). In each case, they are decorated with star symbols and they carry Day Signs—specifically, twelve of the remaining Day

FIGURE 9.12 Borgia Codex page 30. LACMA (Codex Borgia), photo courtesy Ancient Americas at LACMA (ancientamericas.org)

Signs not circled on page 30. On page 31, the rooms carry the elaborate star symbol that some have associated with Venus (Aveni 2001, 225). Here the initiates wear skeletal masks in place of the Coo Ndodzo masks they had on previous pages. The ritual itself appears to take the form of a ritual bathing—yet another symbol of (re-)birth (figure 9.13).

On page 32, the scene is dominated by *tecpatl* or stone knives. Human figures now hold the skeletal heads of the figures on the previous page, and from the bottom of this page a fully human figure is born out of crossed knives. The entire sequence thus far may represent the death of the previous self and rebirth now as a member of the Coo Ndodzo sodality. Accordingly, the initiate now wears body and face paint similar, though not identical, to the final costume he will wear at the end of page 46. The Borgia scribe painted the initiate now wearing the red mouth paint and a short beard.

On pages 33 and 34, the scene changes (figure 9.14). Here the Borgia scribe no longer paints activities transpiring within some interior space. Now the view is from the outside of temples, much more like the representations of the historical Mixtec codices. Now the focus is on numerous events taking place both within and outside of elaborately decorated temples. These scenes may well relate to the training and education that was

required of the initiates after the first phase of initiation. Here the scribe depicts the initiates' meetings with specific authority figures through formal ceremonies as part of the process of investiture.

On page 34, the scenery becomes far more formal, with now numerous figures wearing elaborate Coo Ndodzo regalia (figure 9.14). A critical aspect of this scene is the ritual playing of the ball game, resonant with Na Cuaa's required process. Page 35, though, appears to return to the interior ritual space, now with the initiates in much more elaborate regalia (figure 9.15). They have now adopted the "stripe eye" face paint to interact with the Coo Ndodzo serpents a second time. By this point, the scribe is giving the sense that the overall process is protracted and may take many days to complete.

Pages 37–46 then depict the various meetings and rituals required of the initiates to become fully vested within the Coo Ndodzo sodality. Here the Borgia scribe gives us an impression of the duration of the process. Page 40, for example, shows a complete run of three passes through the twenty Day Signs, suggesting that either the whole sequence should take sixty days or some portion of the investiture would take sixty days. This may seem excessive, but given that it would only have occurred once every

FIGURE 9.13 Detail of Borgia Codex page 32, showing a figure emerging from a pair of knives. LACMA (Codex Borgia), photo courtesy Ancient Americas at LACMA (ancientamericas.org)

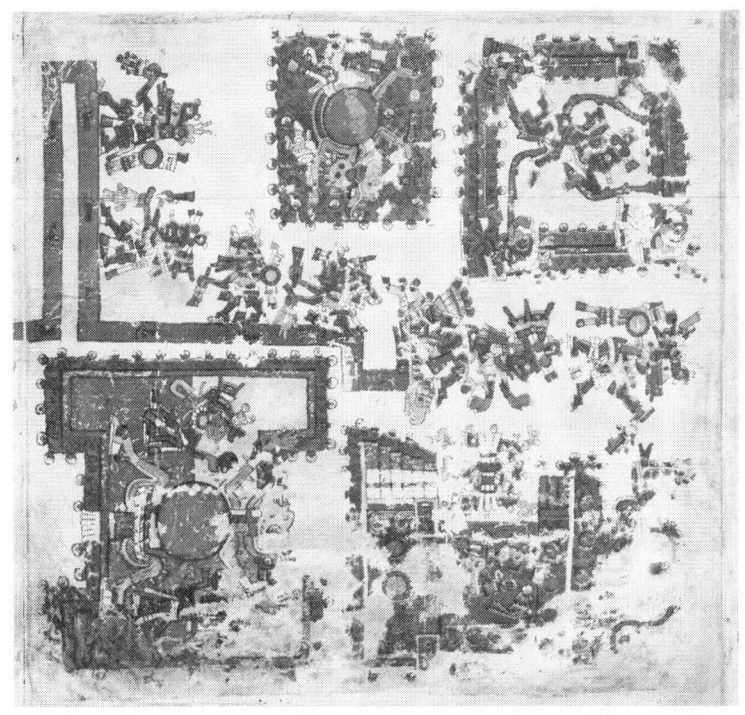

FIGURE 9.14 Borgia Codex page 34. LACMA (Codex Borgia), photo courtesy Ancient Americas at LACMA (ancientamericas.org)

FIGURE 9.15 Borgia Codex page 35. LACMA (Codex Borgia), photo courtesy Americas at LACMA (ancient americas.org)

eight years (for Venus Rounds beginning on a date Cipactli), the scope is reasonable. Moreover, such activity would have generated plenty of economic activity to keep the city bustling for extended periods of time.

In the end, at the left-hand side of page 46, the initiate exits the midsection of an extended body, which is marked with knives and elaborate star symbols. The initiate now is fully dressed in the attire worn by Na Cuaa. The face and body paint as well as the short beard is now the same as that acquired by the Mixtec Quetzalcoatl figure, Q Chi, in the Codex Vindobonensis, just before he dons his long-snouted mask. The final event that the Coo Ndodzo initiate performs before leaving page 46 of the Borgia Codex is a fire-drilling in the presence of two figures. Further attesting that the Borgia scribe was working with pan-Mesoamerican concepts, the figure on the right in this scene is a version of the same warrior figure we find on the frontispiece of the Féjérvary-Mayer Codex and on page 49 of the Dresden Codex.

While the architectural scenes in pages 29–46 are specific to the temples of a given location—Cholula?—we can imagine that similar ceremonies would have taken place in the architecture of the Great Terrace at Chich'en Itza. Only one change in assumptions is necessary to bring these observations together under a consistent narrative.

Ringle, Byland, and Pohl recognize several of the scenes in the Borgia investiture pages as occurring in the Underworld. They therefore make the implicit assumption that they refer to "otherworldly" events, not those engaged by mortal actors. But we have already seen that within indigenous cosmologies, the Underworld is part of local geographies—it can be entered into and returned from while one is still alive. Specifically, caves and other subterranean spaces form places within the Underworld, and they were architected explicitly to enable interactions between the living and the dead as well as myriad other inhabitants of that realm. Ringle himself referred to depictions in the Codex Nuttall of Underworld oracles consulted underwater, possibly in cenotes, as we saw in the last chapter. In such case, we may read the entire set of Borgia pages as taking multiple initiates—not just one idealized king—through a series of rites in the architecture and Underworld of a city dedicated to investiture, to prepare them for leadership roles back at their own homes.

At Chich'en Itza, the movement of initiates would have gone from Underworld spaces such as cenotes and cave-spaces (such as Balankanche) through the Temple of the Warriors and the Great Ball Court, integrating perhaps at the very end a nose piercing at the top of the Temple of K'uk'ulkan. Also, rather than a focus on Coo Ndodzo—as in the Borgia pages—the hero figure would likely have been a Feathered Serpent figure acting with a Cloud Serpent associate.

With such a reading, it would make sense for the Borgia scribe to have painted these pages of initiation immediately preceding the Borgia Venus Almanac. The rites of investiture would have been timed by the observable periods of Venus, in accord with its connection to Coo Ndodzo and Quetzalcoatl. At Chich'en Itza, the terrestrial activities would have been performed in and around the Ball Court structures, while the Underworld activities would have been performed in the caves and cenotes around the city. Instead of Qucumatz, Coo Ndodzo, or Quetzalcoatl, at Chich'en Itza the feathered serpent patron of the sodality was K'uk'ulkan. And of course the observation of Venus would still be charged to the ajk'in working in the Tz'iknal.

HUNAC CEEL REDUX

Ringle's interpretations of the mural program in the Great Ball Court relied on another important shift. Previous scholars considered a specific figure depicted throughout the murals to be in something of an oppositional role to the Feathered Serpent figures in the murals. Ringle works to dispel interpretations that there is internal strife captured in these murals, pitting one "Captain Sun Disk" against a "Captain Feathered Serpent" (2004, 168). Instead, Ringle argues that they are on the "same side" and that the solar representations symbolize the Sun itself. He concludes:

> What is interesting is that the [Upper Temple of the Jaguar] and its murals merge two long-standing strains of Mesoamerican military ideology. One, associated with Venus, has roots extending back to the Early Classic period and to Teotihuacan, if not earlier. The other, sun associated, was later to be most dramatically expressed by the Aztecs. It is thus not a matter of Captain Serpent conquering Captain Sun, or vice versa, but, as with so many other aspects of this site, the creative melding of a wide spectrum of traditions to forge new forms of social and political organization. (Ringle 2009, 42)

With a solar figure and a Feathered Serpent figure working together, and with the additional presence of a Cloud Serpent figure, we now have a more complex representation of authority at Late Terminal Classic Chich'en Itza. Gone is the structure of a single k'uhulajaw appointed by a council of almehen. We now may see a solar-affiliated authority figure connected to a k'uhulajaw lineage residing in the older, southern part of the city, while the Feathered Serpent and Cloud Serpent priesthood presided over the K'uk'ulkan sodality in the north. All worked together, but each had a specific domain of responsibility.

If we combine this infrastructure with the investiture role of the Great Terrace, then we now may view the Ball Court murals themselves as having been intended for an audience of initiates. Accordingly, we might propose a departure from Ringle's interpretation that they were conceptually resonant with the ritual prescription of the Borgia Codex, and instead we see them in line with the narrative of the *Popol Vuh*. In particular, while most readers focus on the mythological portions of the *Popol Vuh*, its authors included a protracted history of how the leadership of Qumarkaj came together. This final part of the *Popol Vuh* discusses internal political strife within lineages as well as battles between the K'iche' armies and those who opposed them.

Indeed, the murals in the Upper Temple of the Jaguar depict grand battles led by Feathered Serpent warriors, which may have represented the required quelling of dissent against the development of a new system of governance. The images of battles or travel across hilly terrain in the Upper Temple of the Jaguar have suggested to some scholars that these occurred in the Puuc region of the Yucatan Peninsula—the most topographically varied region near Chich'en Itza (cf. Schele and Mathews 1999, 234–35). It would make good sense to see these efforts engaged in concert with a K'uk'ulkan sodality already established at Uxmal, perhaps to extend the region of influence.

With this comparison, it makes sense to explore the murals as visual representations of the founding historical narrative of the feathered serpent sodality, intended as part of the experience for initiates. Surely they would have heard some of the history through oral narratives preserved in various forms, but here they would find the "official (interested) history" surrounding them as they passed through the ritual tasks required of their investiture.

If the murals indeed represented founding history, then parts of what they depicted would be replicated at each investiture, but other parts would hearken back to the key events that led to the possibility of their presence there. Within the Lower Temple of the Jaguar, for example, an inductee might look up to see the processions leading up to the first Feathered Serpent's accession to the jaguar throne and recognize that they participated in a replica of that procession. Then they may have looked up in the North Temple to see the first K'uk'ulkan leader have his nose pierced just before he himself lay back on the throne for his own nose-piercing (Ringle 2004, 191). On the opposite wall, they may have seen figures dressed in the regalia of the lords of the Underworld, reminding them of their cave ceremony preparations.

Accordingly, there is an intriguing possibility for a new interpretation if we address a curious anomaly within the scenes in the Great Ball Court walls mentioned above. The only scene in which no Feathered Serpent figure is involved is the decapitation scene repeated multiple times along the interior walls of the Great Ball Court itself (figure 9.16). In these images, two teams stand against each other, with each individual wearing distinct regalia. Between the leaders of the two teams sits a large rubber ball, marked with a giant skull on

FIGURE 9.16 Detail of the Ball Court decapitation scene on wall of the Upper Temple of the Jaguar

its surface. At the head of the line facing right, a warrior stands holding the head of his opponent by his hair. At the front of the opposing line, a warrior kneels without his head, and out of his neck serpents stream out. Scholars often refer to this scene as evidence that the losers of ball games throughout Mayan civilization often were decapitated (Schele and Mathews 1999, 46).

What is interesting about the case here is that the same scene shows up in the mural of the Upper Temple of the Jaguar. Here the full teams are not represented, but the perpetrator of violence stands next to the ball and the victim, again with serpents streaming out of his neck. We may entertain, therefore, the possibility that this was not a repeated ritual event, but instead represented a specific historical event in which an extraordinary and notable instance of violence occurred during a ball game. Moreover, it may be important that there is no figure on either team—on the Ball Court walls or in the Lower Temple of the Jaguar scene—associated with the Feathered Serpent. Throughout the Great Ball Court, feathered serpent imagery frames the scenes, but it does not intrude upon them. Given the greater Mesoamerican context we've reviewed above, it may well be that we are seeing here a violent and extraordinary event occurring before (and so rhetorically instigating) the arrival of K'uk'ulkan.

We don't have to look far to find that the Books of Chilam Balam provide a ready candidate for this interpretation of Chich'en Itza's history. Bringing together Juan Josef Hoil's version (the Chumayel manuscript) and other accounts of strife at Chich'en Itza, we encounter another political controversy, this one involving the historical figure named Hunac Ceel—whom we met in chapter 6 as a ruler consulting the oracle of the Great Cenote at Chich'en Itza. Hoil's *Book of Chilam Balam* tells us that Hunac Ceel was involved in a widely recognized debacle—an episode which was brought to modern scrutiny in the 1920s and 1930s. At the time that John Teeple was working on the time intervals on page 24 of the Dresden Codex, the cultural anthropologist Ralph Roys was investigating Yucatec culture and working on his translation of Hoil's manuscript.

Roys had earned his bachelor of philosophy degree at the University of Michigan, majoring in Medieval French (Thompson 1967, 95). It was his passion for languages, having studied both European and Mesoamerican languages, that led him to Yucatec. His talent brought him to the attention of William Gates, then director of the Middle American Research Institute at Tulane University (Thompson 1967, 96). Roys drew a small salary, working in absentia while living in Vancouver and tending to the family lumber business before joining the Carnegie clan in the Division of Historical Research (Thompson 1967, 96). He eventually connected with Morley, and the two worked together on the Xiu Chronicles, the document from which Eric Thompson pulled some of the fundamental material used in his Calendar Correlation proposal as we saw in chapter 3. Roys's work on Juan Josef Hoil's manuscript, though, is probably what he is best known for, and he acknowledges the help he received from his colleague in the preface. "Dr. Sylvanus G. Morley has spent much time and thought in going over my manuscript and has offered many valuable suggestions as well as searching out and obtaining related material in Mexico and Yucatan" (1933, ii).

When it came to Hunac Ceel specifically, though, Roys was baffled. Hoil had recorded multiple passages making reference to the historical figure, which Roys translated. In "A Song of the Itza," he included:

> However, at the same time there was born in the stone, the black stone of terror, the one named Verbum-tuorum, Ix-coal-tun, Ix-coal-cab, taken by the mistress of the world. Then there was set in its place the thrice seasoned heaven, the seasoned heaven. White

and clean, it lay guarded in the heart of Sustinal Gracia. Thirteen orders of katuns lay prostrate in the stone. Then the ruler, Hunac Ceel, stirred into motion.

The song: Ho! What is so precious as we are? It is the precious jewel <worn on the breast.> Ho! What is the distinction of righteous men? It is my mantle, my loin-cloth. So spoke the god. Then do you mourn for anyone? No one. A tender boy was I at Chichen, when the evil man, the master of the army, came to seize the land. Woe! (Roys 1933, 60)

The song, therefore, appears to give us a bit of the backstory to Hunac Ceel's engagement with the Cenote oracle. It appears that Hunac Ceel arrived as a warrior and that his character was being described. Specifically, the song sets up first the character of a "righteous" man and then contrasts that with the "evil" of Hunac Ceel. We find, then, that he has also provided some foreshadowing.

It is in Hoil's later "chronicles" where value is more transparently ascribed:

8 Ahau was when the Itzá men again abandoned their homes because of the treachery of Hunac Ceel, because of the banquet with the people of Izamal. For thirteen folds of katuns they had dwelt there, when they were driven out by Hunac Ceel because of the giving of the questionnaire of the Itzá.

4 Ahau was when the land of Ich-paa Mayapan was seized by the Itzá men who had been separated from their homes because of the people of Izamal and because of the treachery of Hunac Ceel. (Roys 1933, 75)

In an appendix to his translation of *The Book of Chilam Balam of Chumayel*, Roys summarized his concerns:

The conquest of Chichen Itzá by Hunac Ceel and his allies is without doubt one of the most interesting and puzzling episodes in Maya history. It might well be called the Trojan War of Yucatan, for tradition ascribes as its cause the theft of the wife of a powerful ruler by the chieftain of another great and famous city. As in the case of its classical counterpart, the immediate cause of the trouble may have been the abduction of the wife, but the struggle doubtless originated in political jealousies of long standing and the desire of one city to obtain control of economic resources previously enjoyed by another. (1933, 103)

The complexity of the situation might remind us of the politics we saw in the sibling rivalry between B'ahlaj Chan K'awiil and Nuun Ujol Chaak at Tikal and Dos Pilas as well as the betrayal of Waxaklajun Ub'aah K'awiil at Copán by his colleague at Quirigua. On the other hand, the Hunac Ceel episode appears to have been far more robustly remembered.

Roys's treatment continues, bringing in attestations of the story by the late seventeenth-century priest Andrés de Avendaño as well as the early seventeenth-century evangelists Bartolome' de Fuensalida and Juan de Orbita:

In the second Chumayel chronicle we read: "then came the treachery of Hunac Ceel. Their town (Chichen Itzá) was abandoned, and they went into the heart of the forest to Tan-xuluc-mul, as it is called." This is important, for Avendaño locates this place near Lake Peten, around which the Itzá were living at the time of the Spanish Conquest of Mexico and northern Yucatan. Here they maintained their independence until the close of the Seventeenth Century. They told Father Fuensalida, a Spanish missionary who visited them in 1618, that they had come there from Chichen Itzá. According to their story, in a certain Katun 8 Ahau one of their rulers at Chichen Itzá stole the bride of another more powerful chieftain during the wedding festivities. Fearing the consequences of the act, the offender and his subjects abandoned their city and retired to Lake Peten. The con-nection of this event with the Hunac Ceel episode is further confirmed by the historical fragment in the Book of Chilam Balam of Mani which tells of the same affair. Here it is stated that an unspecified person "sinned against Ah Ulil, the ruler <of Izamal>, against the wife of his fellow-ruler." We conclude that it was the ruler of Izamal whose bride was stolen. (Roys 1933, 104)

So Roys describes a major offense committed by a ruler of Chich'en Itza, causing a sub-stantial disruption to local politics. Perhaps more importantly, the event is extremely well attested: in the local manuscript of Hoil in northern Yucatan, in a copy in the Chilam Balam of Mani probably recorded in the early sixteenth century, and then in the histories pre-served around Lake Peten Itza over a century later. This was an event worth remembering.

None of the stories make any reference to K'uk'ulkan, yet there is a different difficulty that Roys is at pains to explain. Specifically, he is confronted with the problem that the Hunac Ceel episode is attributed in some cases to the Katun 8 Ajaw, which he then wants to place at the end of the occupation of Chich'en Itza. The Katun Count chron-icle leaves Roys with a puzzle he cannot solve. On the other hand, when we recognize that the date attributions are themselves inconsistent within the same document, and when we acknowledge Grant Jones's interpretation of "prophetic history," we lose the constraint that Roys himself adopted. Now we may see the Hunac Ceel episode itself as having occurred earlier in the occupation of Chich'en Itza, in fact most likely before K'uk'ulkan arrived.

If the latter is the case, then Hunac Ceel becomes the symbol of "what went wrong" with individual rulership in the form of the k'uhulajawlel in the Terminal Classic period—individuals were acquiring authority by force without international accountability. The episode in many ways would have both been memorable in itself and served as the per-fect foil for the arrival of a peacemaker—someone like a K'uk'ulkan leader. It is all cir-cumstantial, but this sets up the interpretation that what we are seeing in the Ball Court bench murals is the episode of treachery that the Books of Chilam Balam recorded and that was preserved in the written and oral history of the Itza at Tayasal.

In this interpretation, Hunac Ceel was the ruler of Chich'en Itza—a position he acquired by military force and consolidated by consulting the Great Cenote oracle. His city was hosting a celebration for the wedding of the Izamal ruler, and either within the festivities themselves or perhaps from previous encounters, Hunac Ceel became enamored of the Izamal ruler's bride. Part of the wedding festivities included the playing of the Ball Game by the nobility, and it was at that event that Hunac Ceel took the opportunity for the "treachery" of beheading the ruler of Izamal and stealing his intended bride. As someone who took the throne forcibly, he likely did not have wide support of the nobility, and so they turned against him, leaving Hunac Ceel and his closest supporters to flee "back" to their relatives at Tah Itza.

Such an interpretation fits with the several roles in which we find the story. For one, it would make sense in the foundational narrative within the structures of the Ball Court—a hegemonic charter for the K'uk'ulkan sodality as a countermeasure to the caprices of individual autonomous rulers. Now would-be rulers would have to acquire the legitimacy conferred by the priesthood of the Feathered Serpent before acceding to any throne—just as Na Cuaa would do in the Mixtec highlands decades later. Second, mortal violence within a ball game during a wedding celebration motivated by illicit love could very well be characterized as "treachery" and remembered as such whether or not the details were preserved to posterity. Third, the story would have been sufficiently ostentatious to have been memorialized in several forms: as songs, as oral histories, and as part of an official history, all of which we have here. And finally, the preservation of this narrative would parallel the various Mixtec codex accounts of the murder of the historical figure Ca Qhi in a sweatbath (Pohl 1994, 87; Troike 1974, 297).

ARRIVAL OF THE QUETZAL STAR

Concerning the larger interpretation, then, we appear to have political strife at the beginning of Chich'en Itza's Terminal Classic occupation and then an architectural expansion and extended period of occupation under the influence of a Feathered Serpent presence. These give us general bounds on the role of K'uk'ul Ek' Tuyilaj's work on the Venus Table, but they don't specify further. There is another source of information, though, that ties all of these themes into a small window of opportunity.

Recent epigraphic work by Peter Bíró and Eduardo Pérez de Heredia (2016) suggests that some of the key iconography of the Great Terrace has its earliest appearance at Chich'en Itza in the older, Puuc section of town. Specifically, they argue that the origins of the Terminal Classic shift first show up on a stone-tenoned disk that decorated the façade of the Tz'iknal. This "Caracol Disk," which we saw in the preface of this book as containing a representation of K'uk'ul Ek' Tuyilaj, recorded a meeting that looks like an abbreviated version of those throughout the Ball Court structures (Bíró and Pérez de Heredia 2016).

Bíró and Pérez de Heredia make a bold claim with this monument. They write:

> In a general sense the Caracol Disk relates to a ritual known as *toma de posesión* (taking possession) or *hedz luum* (*asentar la tierra*, setting the land), which is mentioned in the

Chilam Balam and was an archetypical ceremony all over Mesoamerica (Oudijk, 2002; Boot, 2005: 110–123; Nielsen, 2006). This ritual involved the "coming of the torch", the same event that took place at Tikal in the 4th century. This new foundation of Chichén Itzá was marked by the K'atun 8 Ajaw, which will be a particularly significant date in the *Chilam Balam*. (2016, 133)

The interpretation of "taking possession" comes from a combination of their reading of the hieroglyphic text on the Disk along with the imagery on its front (figure 9.17). In the lower image on the Disk, in front of the image of K'uk'ul Ek' Tuyilaj, a central figure carries a torch, and the text begins with the verb "arrives." Perhaps far more provocatively for the current treatment is that the subject who arrives is named K'uk'ul Ek' (Bíró and Pérez de Heredia 2016, 138), or "quetzal-like star." Given the feathered serpent imagery

FIGURE 9.17 Illustration of the scenes on the Caracol Disk after Mark Van Stone (April 13, 1997, in FAMSI Schele Drawing Collection no. 5085) and photograph (see figure P.1). (*top*) K'uk'ul Ek' Tuyilaj wears the Feathered Serpent headdress, with its body wrapped around hers. (*left*) Portion of the hieroglyphic inscription on the disk, which Bíró and Pérez de Heredia read as "the arrival of the quetzal star" (2016, 138)

on the Disk, it is not at all difficult to interpret this monument as a representation of the arrival of the K'uk'ulkan sodality at Chich'en Itza, directly associated with Venus and, specifically, the Tz'iknal structure.

A closer look at the imagery of the Caracol Disk provides important details. The artists carving the imagery broke it up into two scenes. In the upper scene, two figures meet across from an incense burner, each with attendants behind them. The figure on the right wears a quetzal headdress and appears to be the *chahom*, adding incense to the burner. Behind this figure rises a serpent with a rattlesnake tail, and out of its mouth rises a figure identified as Mixcoatl, or "Cloud Serpent," as an ancestor figure. These two perfectly match the Feathered Serpent sodality representatives as reconstructed by Ringle, Byland, and Pohl. The figure on the left in this scene wears a canine headdress, but also a long, bird-snout mask, not unlike Coo Ndodzo masks of the Borgia Codex initiates we saw above. This figure carries a smaller figure, likely a ceramic deity effigy or a mortuary bundle.

A number of these figures are actually named in the text, while others have their titles given. Bíró and Pérez de Heredia read one of the names as

> Aj B'olon K'awil Lem? Taj Tza', meaning "He of the Many Kawiil, the Jeweled Obsidian from Tzaj". This descriptive title lists the toponyms as he was from Tz'iknal (the Caracol Building) from the neighbourhood of Tzaj in Wak[hab'?]nal, which was the name of Chichén Itzá itself (Voss, 2001; Boot, 2005: 345). His title was *aj k'ahk'* ("he of fire") and *k'uhul aj kan* ("divine speaker"; Voss, *ibid.*, 160). (2016, 135)

We find here, again, the indigenous Yucatec name of the Caracol: Tz'iknal. We also find corroboration that the k'uhulajaw title has been dropped, though parts of its concept remain in new titles.

Another key figure identified in the text speaks to regional trade relationships and the geographic tensions reviewed in the last chapter. One of the named individuals is from the Chontal region of the Gulf Coast—the critical area for determining whether routes would go north along the coast or south through the rivers of the Peten. Bíró and Pérez de Heredia write:

> Indeed, the full toponymal title Aj Holtun B'alam Peten Itzamal Tz'ak Ahaw could be translated as "He from the Port of the Jaguar from the island/province/lagoon where there is an abundance of Itzam, from Tz'ak(tam)", where the town, the island and the region are mentioned. This line of reasoning raises the possibility that the lord who arrived to Chichén Itzá came, in last instance, from the Laguna de Términos in Tabasco. (2016, 139)

The Caracol Disk, therefore, explicitly refers to regional associations bearing directly on the issue of trade routes and the pressures to go south, reengaging the Peten, or to go north, via seafaring canoe. It also predates the imagery of the Great Terrace, but carries much of the imagery that becomes central to the Ball Court murals.

Of particular interest, though, is that in the lower scene we encounter the woman wearing a feathered serpent headdress, which carries a full serpent body, wrapping around her lower back (see figure 9.17). This representation of a headdress is uncommon, but it does show up in the codex histories of Iya Na Cuaa's rise to power. On page 29 of the Codex Nuttall, for example, a male figure wears the full serpent headdress in the same fashion as does K'uk'ul Ek' Tuyilaj, depicted on the Caracol Disk. Additionally, if we return to the hypothesis that we have interested history represented on the Disk and in the Ball Court murals, then it makes sense to see parts of them telling the same story.

Returning to the mural in the Lower Temple of the Jaguar, for example, K'uk'ul Ek' Tuyilaj can be seen in the procession of men, where her exposed breasts are unambiguously depicted. In the Temple of the Warriors, she wears the same skirt on one of the hundreds of columns, which depict warriors and priests, young and old (cf. Schele and Freidel 1990, 502n44). Since we now see the Ball Court murals as visual histories, it makes sense to see these as different representations of the same woman—one who played a very significant role in bringing the Feathered Serpent sodality to Chich'en Itza, and one in on the project early enough to have been depicted on the Caracol Disk.

Getting back to the timing of this meeting, we realize that just as critical as the identification of K'uk'ul Ek' in the text is the date itself. On the Caracol Disk, we confront independent corroboration of our interpretation that K'uk'ul Ek' Tuyilaj's observations of Venus at the Tz'iknal were the catalyst for her discovery.

Bíró and Pérez de Heredia write that "the date on the Disk reads 2nd *tun* of K'atun 8 Ajaw, or 10.5.1.0.1–10.5.2.0.0" (Bíró and Pérez de Heredia 2016, 136). If we recall from the last chapter that K'uk'ul Ek' Tuyilaj made her first observation on 10.4.6.13.0 1 Ajaw 13 Mak and then confirmed it on 10.4.13.4.0 1 Ajaw 18 Wo, then, when we put these together, we find that the "arrival of K'uk'ul Ek'" took place around eight years after confirmation of her new Venus Table.

To put a finer point on it, the second winikhaab of Katun 8 Ajaw constitutes the period of time running from 2,801 to 3,160 days after K'uk'ul Ek' Tuyilaj's confirmation date. Within this "second tun," then, is 2,920 days, or exactly one complete set of five Venus Rounds (i.e., one full line of dates through all five pages of the Venus Table). The evidence suggests, then, that the order of K'uk'ulkan arrived when K'uk'ul Ek' as Venus was in the same position as when our astronomer proved to the nobility of Chich'en Itza that her Venus Table worked. The arrival of K'uk'ul Ek' described on the Caracol Disk therefore described an event timed to occur on a first morning appearance of Venus.

It is possible that we have even further corroboration of this interpretation from the full image on the Caracol Disk. Bíró and Pérez de Heredia follow epigrapher Erik Boot to read the overall imagery as a narrative sequence. They interpret the bottom scene as occurring first and at night, hence the use of the torch. They believe that the torch event at night is what would provide Underworld associations, which in turn are in line with founding rituals (Bíró and Pérez de Heredia 2016, 149). The upper scene, then, they see as transpiring during the day, occurring after the torch event. If indeed K'uk'ul Ek' Tuyilaj facilitated the scheduling of this event in line with her Venus observations, then an event beginning in the predawn hours and ending during the late morning would have been perfectly timed to center on the first visibility of Venus, which was the event that guided her construction of the Venus Table.

The evidence comes together, therefore, to see the event on the Caracol Disk as that meeting which brought the K'uk'ulkan sodality to Chich'en Itza in significant part due to the work of K'uk'ul Ek' Tuyilaj. This is why she was included in the imagery, as a critical part of the city's visual history.

The Temple of K'uk'ulkan

K'uk'ul Ek' Tuyilaj's Venus Table makes clear that such an "arrival of K'uk'ulkan" was not simply the one-sided introduction of a Venus-related ideology. The first accommodations were likely visible within the traditional center of the city, at the Tz'iknal itself. It is very likely this introduction of K'uk'ulkan's Feathered Serpent sodality that introduced the Nahuatl warrior deities into the rituals of the Venus Table. Tawiskal, Kaktonal—this Mayanization of Nahuatl names and of Sierra Madre deities was probably negotiated as part of the arrival of K'uk'ul Ek', documented on the Caracol Disk. The ritual theatrical performances of violence explored in chapter 5 now would have introduced the public to a new set of deities and the ideas they brought along with them.

Moreover, K'uk'ul Ek' Tuyilaj's new Venus Table specifically required the inclusion of solar periods in order to create a new balance and a new understanding of the phenomenon. In many ways, the Venus Table itself parallels in a mathematical sense the iconographic balance of Venus and the Sun in the Great Ball Court murals and lintels. The Sun Disk–affiliated figure and the Feathered Serpent figure on the North Temple lintels, for instance, provide visual representations of the balancing that tracks Venus Rounds relative to solar years on Dresden Codex pages 46–50.

Taking this even further, the form of the remodeled Temple of K'uk'ulkan itself may have facilitated the attested period of peace that followed the installation of the K'uk'ulkan sodality. Specifically, the masons of Chich'en Itza built the Temple of K'uk'ulkan by augmenting a smaller structure, which had a much simpler form as noted in chapter 5. The earlier structure possessed a single staircase leading to a structure with one entrance and two internal rooms. The later structure was built to approximately twice the size of its predecessor. Its form was that of a symmetrical pyramid with ninety-one steps for the staircases of each of its four sides (figure 9.18). Architects designed the structure at the top to have four entrances, one at the top of each staircase. Ninety-one steps on

FIGURE 9.18 Temple of K'uk'ulkan showing the serpent balustrades of the north face. The symmetric structure has four staircases, each with ninety-one steps. Including the platform for the structure on top, there are therefore 4 × 91 + 1 = 365 steps. Three hundred and sixty-five is a numerological reference to the Sun, and the serpent is illuminated in accord with the Sun's annual period. The structure provides a visual representation of a balance between the Sun and the Feathered Serpent.

four sides and one step serving as a platform in the middle provides a total of 365—a numerological connection to the solar year, as scholars have long recognized (Aveni 2001, 300; Thompson 1972). So there was certainly an important solar presence within the Temple of K'uk'ulkan.

On the other hand, far more often recognized is that these architects oriented the structure to create a visual display of light and shadow tied to the solar year. On the equinoxes—when day and night were of equal length and the Sun rose and set at its midpoint of extreme motion along the horizons—the setting Sun would cast its rays along the edge of the structure's northwest corner to illuminate the pattern of a diamondback rattlesnake along the northern balustrades (Aveni 2001, 298; Aveni et al. 2004, 124). Although it may be the first time such a shadow display was created with the Sun as the light source, cave explorations in Belize have shown that artists may have developed the technique within the Underworld. Within caves, rock formations dropping from ceilings or rising from the floor were modified to create silhouettes taking recognizable shapes. Torchlight illuminating these formations cast shadows on cave walls, attesting to the "presence" of that being within the subterranean space. The architects at Chich'en Itza did the same thing, only their silhouette was built above ground, and instead of a torch they used the Sun itself.

On the Temple of K'uk'ulkan, the shadow pattern runs the length of the balustrades, terminating in the feathered serpent heads carved of stone and placed there. In other words, on the equinoxes, the feathered serpent could be seen descending the Temple of K'uk'ulkan, headed to the north toward the Great Cenote (Aveni 2001, 299). The visual display annually commemorated the cosmological presence of K'uk'ulkan at Chich'en Itza, connecting the celestial realm with the middle realm and the Underworld. Calendrics, architecture, and mural artwork all pointed to the balance between Venus and the Sun—artistic and architectural representations of the mathematical structure that K'uk'ul Ek' Tuyilaj had achieved. As Ringle suggested, "Captain Sun Disk" and "Captain Feathered Serpent" were part of the same, new governmental structure.

SUMMARY

With their focus on the scenes of the Caracol Disk, Bíró and Pérez de Heredia write:

> We consider that these two "snapshots" of the re-foundation of Chichén Itzá displayed in the Disk also mark a new social contract between locals and foreigners, leading to a complex government system in which power was negotiated through areas of economic, political, religious and military control represented by Aj Tza' (the Caracol area) and the Tz'ak lord, respectively. And then the newcomers and the local elite decided to create a new ceremonial complex, distinctly associated with the Sacred Well: the "Toltec" city. (2016, 149)

Placed in a fuller context relative to the investiture role of the Great Terrace introduced by Ringle and with the evolution of the Venus Table, we now have extended further their interpretation of these "snapshots."

In the interpretation proposed here, K'ak' U Pakal K'awiil and the leadership of Chich'en Itza was facing a challenging political scenario. From the south, there was a renewed alliance attempting to revive (or rescue) cross-Peten trade by canoe. To the west, there was a new hegemony tied to Quetzalcoatl, preferring coastal trade and into which Chan Chaak at Uxmal had already been welcomed. During this time, Chich'en Itza still was anchored to the southern Puuc-style architectural center, surrounding the Tz'iknal and maintaining provincial Venus Almanac ceremonial activity. It was likely during this time that Hunac Ceel engaged in treachery and added a layer of precarity to unchecked individual k'uhulajawlel autonomy.

Within that context, K'uk'ul Ek' Tuyilaj observed a set of patterns that led her to inform the almehen lineages of an impending momentous event for the observation of Venus's first morning visibility on a date 1 Ajaw. She also let them know that this came with a discovery allowing her to compute Venus Rounds with new, unheard-of accuracy, which in turn relied on the incorporation of solar-year tracking into the almanac. K'uk'ul Ek' Tuyilaj's discovery, then, appears to have come at an incredibly opportune moment. The nobility decided to follow it with a new form of hegemony for the city: a balance between the ajawlel—symbolized by the Sun and remaining in "Old Chich'en"—and a new warrior-priest sodality, serving the city of Chich'en Itza, centered on the Great Terrace. This new center also served as the site for cities throughout the region interested in feathered serpent legitimacy and the trade relations that went along with it.

Accordingly, a representative from an appropriate Quetzalcoatl religious center in Central Mexico was invited to establish a new center at Chich'en Itza, and it was timed by the visibility of Venus. Some among the local almehen would have seen K'uk'ul Ek' Tuyilaj's discovery as a sign—divination from the celestial realm that the affiliation with Venus and K'uk'ulkan was warranted. Others likely would have challenged such an interpretation. Certainly those in alliance with the lineage selected to enter the Feathered Serpent leadership role would have supported it, but that would have disenfranchised others. For those in favor of a presence of the Feathered Serpent, nobles might have argued that K'uk'ul Ek' Tuyilaj's discovery was not actually accidental. In line with the discussion in the last chapter, her Venus discovery was the sign that they needed. It was a new calibration of Venus movement and ritual at the same time that Venus-K'uk'ulkan was growing in presence throughout Mesoamerica. On the other hand, there certainly would have been those in opposition to the proposal—almehen houses, for instance, whose influence would diminish with the new structure. Additionally, other cities in the region may have been reluctant to follow the new hegemony, or they may have explicitly allied with relatives or partners pursuing a revived trade across the southern Peten. In effect, the introduction of a new model would have resulted in conflict that well could have required military force to quell. And so it may be these battles—led by the Central Mexican symbolism of the Feathered and Cloud Serpents, in alliance with the ajaw symbolized by the Sun—that were recorded visually in the murals of the Ball Court structures.

Once the strife was settled, the Great Terrace would be built and serve for investiture rituals timed by the Venus Table, moving young initiates through the plazas, into the Ball Court, through its structures to be versed in the sodality's history, and probably through Balankanche and other caves parallel to the prescriptions of the Venus pages in the Codex Borgia. In this way, K'uk'ulkan was brought to Chich'en Itza, not as an individual hero, but as a new order, manifested within the architecture, within societal organization, and guided by the movement of astronomical bodies. K'uk'ulkan may have been memorialized in legend as a god arriving from the west, aggrandized and personified like Tecun Uman, but the architectural and historical records suggest that he was represented by individual leaders, invited to Chich'en Itza by K'uk'ul Ek' Tuyilaj's astronomical calculations.

CHAPTER 10

The Third Correction Interval, Mayapán, and Tawiskal Uwoojil

We have seen in the last two chapters an argument for a first version of the Venus Table, in an almanac form, used by K'uk'ul Ek' Tuyilaj before her discovery of the Correction Factor. A second version was her elaboration of the Venus Table extending across pages 46–50 along with the Correction Intervals that she developed at the Tz'iknal. A third version is the final form that we now have as the "full" Dresden Codex Venus Table, which includes the last two Correction Intervals on page 24 and the bottom section of "shifted text" on the left-hand sides of pages 46–50 (figure 10.1). In this chapter, we return to Tawiskal Uwoojil—the copyist whose version of the Venus Table we have today—and the evidence that his contribution came from the city, politics, and circumstances of the Late Postclassic city of Mayapán.

We saw in the introduction that Tawiskal Uwoojil's manuscript came neither from archaeological excavation nor from documented extraction from an indigenous library or home; this has led most scholars to rely primarily on its internal contents for interpretation. Slight extensions beyond these contents have been made by implication, including, as we have seen, Aveni's proposal that the data for the Venus Table was compiled at the Tz'iknal at Chich'en Itza (2001, 191). But these extensions have not affected how the document itself has been understood. And this has led to a somewhat monolithic approach to the document: in many ways, scholars have approached it as a tool without a history.

The approach developed in the previous chapters of this book, however, allows us to move away from sole focus on the manuscript and how it was put together computationally. In chapter 7, we saw that the middle two Correction Intervals on the right-hand side of page 24 might be best understood within the context of the Terminal Classic at Chich'en Itza to suggest a discovery made there, which enabled both the full 104-year scope of the Venus Table and the computational corrections necessary for its long-term use. Within this material for corrections, we found that while the middle two

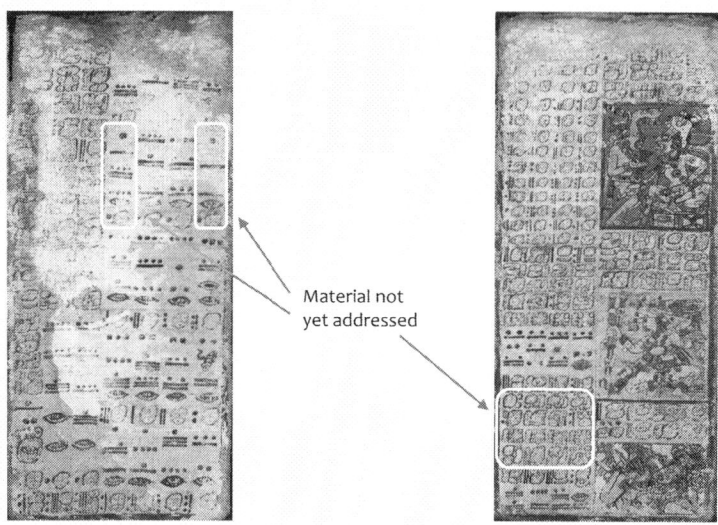

Material not
yet addressed

FIGURE 10.1 "New material" for the third and final version of the Venus Table. Courtesy SLUB

Correction Intervals fit neatly with the other material on page 24, the final two Correction Intervals—placed at the beginning and end of the Correction Interval row—deviate sufficiently to have been recognized as "anomalous" for some time (Bricker and Bricker 2007, 101; Thompson 1972, 63) (figure 10.2).

The first interval in the second row, 1.5.5.0 or 9,100 days (at the far right), is less than half the size of the next smallest Correction Interval and has been considered problematic since it was first treated by Teeple in the 1920s. The last interval in the row, 1.5.14.4.0 or 185,120 days (farthest left), is the largest by almost a factor of three and for the most part has been left out of revision approaches. It may be that because the emphasis in past scholarship has been on seeking the most accurate computational implementations of the information on page 24 and not in preserving what is actually written, this oversight has been ignored (Aldana 2016a).

On the other hand, we have also seen that all scholars treating the manuscript have recognized that there is historical change documented within the Venus pages, reflected in the different base dates guiding progressions through the table (Bricker and Bricker 2007, 102). These shifted base dates mean at least that, at some point, one ajk'in was making accommodations between what had been recorded by another ajk'in working during that other time and what she or he observed. By following the lead of the last chapter, we encounter in this chapter an opportunity to integrate these shifted base dates further into historical interpretation. By having placed the Venus pages in discourse with a specific temporal and architectural setting at Chich'en Itza in the last chapter, we now confront evidence for the final content in the architecture of Mayapán.

This effort leads us to investigate further astronomy and/or cosmology within urban planning, which is a topic that has been engaged for some time (e.g., Ashmore 1989; Aveni 2001; Aveni and Hartung 1986). We have seen even in earlier chapters of this

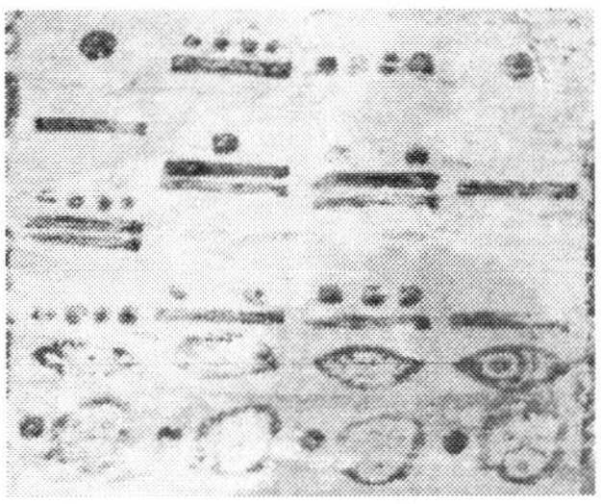

	1.5.14.4.0 (185,120)	9.11.7.0 (68,900)	4.12.8.0 (33,280)	1.5.5.0 (9,100)
	1 Ajaw	1 Ajaw	1 Ajaw	1 Ajaw

FIGURE 10.2 The final two Correction Intervals on page 24 at either end of the row: 1.5.5.0 and 1.5.14.4.0. Courtesy SLUB

book, for instance, how Temple 11 at Copán was meant to "center" the city. The E-Groups identified by Morley and Ricketson provided opportunities for community-wide ritual, combining solar observations and religious procession. In the records of the great Aztec capital of Tenochtitlan, one of the (two) tlatoque named Motecuhzoma confronted something in between the E-Groups of the Preclassic and a Venus alignment at Uxmal. Alfred Maudslay first hypothesized that the Templo Mayor was positioned to allow for an equinox alignment, which Aveni then followed up on with colleagues Horst Hartung and Edward Calnek:

> That the orientation of Templo Mayor may have been associated with the problem of recalibrating the calendar, and furthermore that the starting point was the equinox, is indicated in the following statement made by Motolinia (1971:51):
> *Tlacaxipeualistli.* . . . This festival fell when the sun was in the middle of [the temple of] Huitzilopochtli, which was the equinox, and because this was a little twisted Moctezuma wished it torn down and straightened.
> The reference to structural modifications undertaken by Moctezuma (whether Moctezuma Ilhuicamina [ca. 1440–1467] or Moctezuma Xocoyotzin [1502–1520] is meant is unclear) indicates that at least one astronomical function had been incorporated directly in the temple's architectural design. (Aveni et al. 1988, 290)

Aveni goes on to summarize the notion that additions to the Templo Mayor had to be modified appropriately to adjust for the preservation of this equinox view (2001, 238). This historical example, combined with the archaeological patterns, puts us in a position to better appreciate the case of Mayapán. By following the above leads and putting the Dresden Codex Venus Table into dialogue with the histories of Mayapán, we encounter a new example of ancient urban planning as well as find reason for the remaining content of the Venus pages.

By itself, the size of the far-left Correction Interval in the second row of page 24 of the Dresden Codex at the very least implies a connection to Mayapán. Scholars since the mid-twentieth century have suggested that its size suggests that it was used to correct for a later period, most likely after the fall of Chich'en Itza (Thompson 1972, 15–16; Bricker and

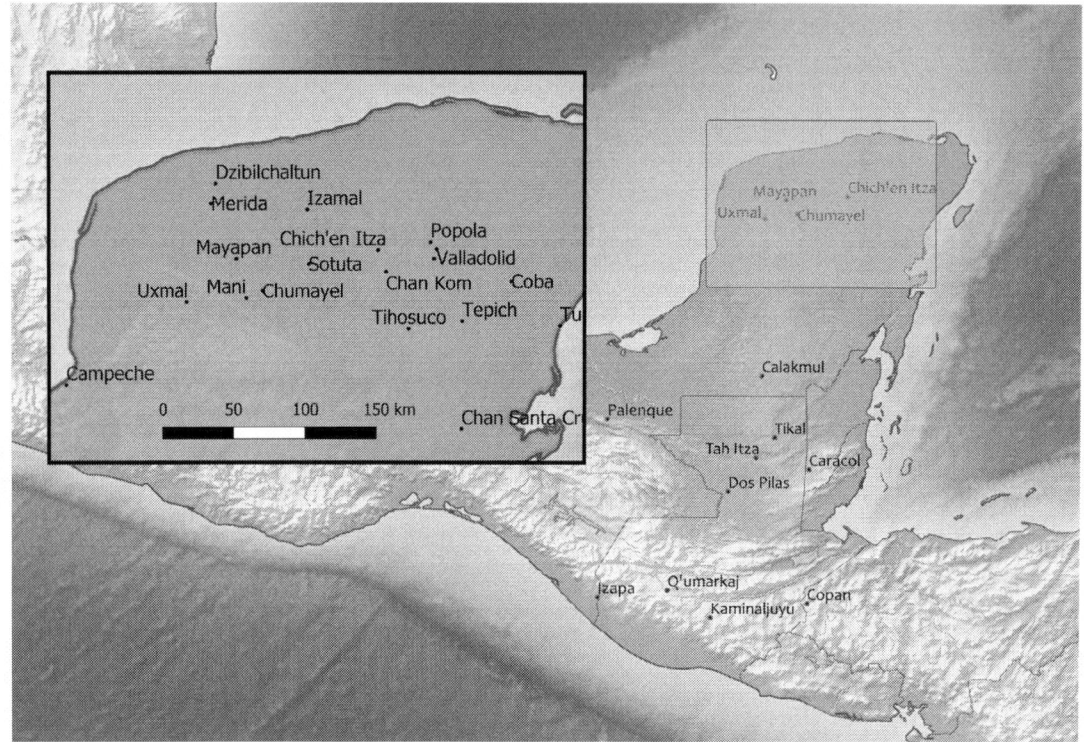

FIGURE 10.3 Potential homes of the Dresden Codex, including Coba, Mayapán, Dzibilchaltun

Bricker 2007, 116). Anthony Aveni, for example, followed Lounsbury to place the final corrected version of the table at around 11.5.1.18.0 1 Ajaw 3 Xul, which in turn he correlated to December 22, 1324 (2001, 191). This temporal push, after the fall of Chich'en Itza, still allows for a number of different Postclassic sites across the northern Yucatan peninsula as the location of the book's use, including, for example, Dzibilchaltun, Ek' Balam, or Coba (figure 10.3). And while temporally these are all possible, we will find that the strongest candidate is Mayapán itself—the last sizable city in the north, and a city attested historically as well as architecturally to have supported an interest in Venus and K'uk'ulkan.

Our investigation now turns to histories of Mayapán preserved in Hoil's *Book of Chilam Balam* as well as in Landa's *Relación* and other sources, which provide a wealth of material for interpretation of the Postclassic city. Additionally, the extensive archaeological excavation by the Carnegie Institution in the 1950s and more recent work by Carlos Peraza Lope, along with his even more recent collaboration with Marilyn Masson, contribute provocative new evidence supporting the writing of the Dresden Codex at that city, as we will see in this chapter. By now examining the relationships between astronomical practice and architectural context in concert with page 24 and the text of the Venus Table, we see that Tawiskal Uwoojil and the Venus Table were very likely instrumental to the founding, the early development, and the ritual life of Mayapán and K'uk'ulkan's role there.

MAYAPÁN'S CONTESTED HISTORY

For most of the twentieth century, the cities of the Mayan Postclassic received a level of scorn from archaeologists. The name of the period itself reflects the notion that all things Mayan achieved apogee during the "Classic" period, and the transition thereafter was one of decline. Tatiana Proskouriakoff herself described the Postclassic as the "dramatic culmination of a long process of cultural decay" (Masson and Peraza Lope 2014, 24).

The documents we have worked with in this book do speak to a level of political strife during the Postclassic, but not one that is out of line with what we have already seen for the Classic period—for example, at Dos Pilas. Also, when we bring together Juan Josef Hoil's book (the Chumayel manuscript) and Diego de Landa's *Relación*, we encounter a "League of Mayapán," which purportedly quelled political unrest in the region for some time, speaking to a continuation of the historical narrative we developed in the last chapter considering the role of K'uk'ulkan in Yucatan. Whatever peace the Feathered Serpent's presence brought to Chich'en Itza during the Terminal Classic, though, appears to have been coming apart by the twelfth century.

One primary outcome of Chich'en Itza's depopulation episode—what Roys refers to as the "conquest"—was that two almehen factions headed to distinct destinations (1933, 103). One group headed south for the center of the Peten to follow Hunac Ceel (now a century later) to reoccupy (or simply return to) Tah Itza. Their descendants would defend the island city centuries later against Spanish and European intrusion until the end of the seventeenth century after a failed attempt at religious conversion by Andrés de Avendaño (Jones 1998). The other group went west to take up residence at Mayapán relatively early in the history of that city. These contexts set the background for the architectural and computational work that Tawiskal Uwoojil would have been charged with early in the history of Mayapán.

Landa describes part of this context, concerning the settling of Mayapán and the fact that it was made up of different lineages. Regarding the Xiu ch'ibal, Landa writes:

> And there, they began to settle and to construct very good buildings in many places; and the people of Mayapan became very good friends with them and were glad to see that they cultivated the land as the natives do; and in this way those of Tutul Xiu subjected themselves to the laws of Mayapan and thus intermarried, and as the lord Xiu of the Tutul Xius was such he came to be very much esteemed by everybody.
>
> These tribes lived so peaceably that they had no quarrels nor did they make use of arms. (Tozzer 1941, 31)

In his translation of Landa's *Relación de las cosas de Yucatán*, Tozzer expands the description by including another *relación* in a footnote:

> In that *guardiania*, near a mission-town called *Telchac*, a very populous city once existed called *Mayapan* in which (as if it were a court) all the *caciques* and lords of the province

of *Maya* resided and there they came with their tribute. Among these were two principal ones, to whom the others acknowledged superiority and vassalage and for whom they had great respect, one was called *Cocom* and the other *Xiu*. (1975, 35n172)

This description attests the changed form of rulership in Postclassic Yucatan, now moved away from an individual paramount genealogically bound to the throne. Now we see the distributed power among different lineages, even though some of them are privileged. In particular, we find that the most important ch'ibals politically at Mayapán were those of the Xiu and the Cocom.

Moreover, this peaceful period is what Tozzer and Roys referred to as the "League of Mayapán." We should, of course, recognize that we are consulting biased sources. As we saw in chapter 1, Landa relied on two local noblemen for most of his information on the region. Gaspar Antonio Xiu was a revered elder in the Xiu family based in Mani, with a regional alliance including the town of Chumayel and the abandoned city of Mayapán. Juan Nachi Cocom was the leader of the Cocom lineage, based in Sotuta, historically connected to Chich'en Itza (see figure 10.3). So the versions of political history that Landa had access to and the material in Hoil's manuscript reflected the perspectives of the descendants of the most powerful families of the region.

That the important families of Chich'en Itza moved to Mayapán appears to be attested further in Hoil's manuscript. In the section titled "A Book of Katun Prophecies," we read:

Katun 4 Ahau is the eleventh katun according to the count. The katun is established at Chichen Itzá. The settlement of the Itzá shall take place <there>. The quetzal shall come, the green bird shall come. Ah Kantenal shall come. Blood-vomit shall come. Kukulcan shall come with them for the second time. <It is> the word of God. The Itzá shall come. (Roys 1933, 92)

There is a suggestion here, therefore, that K'uk'ulkan arrived or was in some way involved with the move to Mayapán. The statement also reflects that this would have been his *second* arrival, implying that the first corresponded to his presence at Chich'en Itza. This is corroborated by Landa.

This Kukulcan established another city after arranging with the native lords of the country that he and they should live there and that all their affairs and business should be brought there; and for this purpose they chose a very good situation, eight leagues further in the interior than Merida is now, and fifteen or sixteen leagues from the sea. They surrounded it with a very broad stone wall, laid dry, of about an eighth of a league leaving in it only two narrow gates. The wall was not very high and in the centre of this enclosure they built their temples, naming the largest, which is like that of Chichen Itza, the name of Kukulcan, and they built another building of a round form, with four doors, entirely different from all the others in that land; as well as a great number of others round about joined together. In this enclosure they built houses for the lords only, dividing all the land among them, giving towns to each one, according to the antiquity of his lineage and his personal value. And Kukulcan gave a name to this city—not his own as the Ah Itzas had done in Chichen Itza, which means well of the Ah Itzas, but he called it Mayapan, because they called the language of the country *Maya*, and the Indians (say) *"Ichpa,"* which means "within the enclosures." This Kukulcan lived with the lords in that city for several years; and leaving them in great peace and friendship, he returned by the same way to Mexico, and on the way he stopped at Champoton, and, in memory of him and of his departure, he erected a fine building in the sea like that of Chichen Itza, a long stone's throw from the shore. And thus Kukulcan left a perpetual remembrance in Yucatan. (Tozzer 1941, 23–26)

The record therefore reflects strife at Chich'en Itza leading to the departure of its nobility, followed by the settling of a new city to the west and the intercession of K'uk'ul-kan. In his *Relación de las cosas de Yucatán*, Landa goes further to discuss a role of K'uk'ulkan at Mayapán that associates him less with a historical figure and more with a deity, resonating with Central Mexican references and the results of the last chapter. Within Landa's treatment of a ritual practiced during the month Xul, Landa wrote:

In the tenth chapter has been related the departure of Kukulcan from Yucatan, after which there were among the Indians some who said that he had gone to heaven with the gods, and on this account they regarded him as a god, and fixed a time for him in which they should celebrate a festival to him as such, and this was celebrated throughout all the land until the destruction of Mayapan. (Tozzer 1941, 157)

There is an important theme here concerning a shift in K'uk'ulkan's presence at Mayapán that we will return to below.

These various accounts in many ways now draw us back to what we already know about the Postclassic. Whatever transitional approach to governance had been developed at Chich'en Itza, while the political landscape of Mesoamerica was in flux, was by the middle Postclassic now in line with the hegemony of Central Mexico. Attempts to resurrect the dominance of trade across the central Peten failed, and the interconnectedness of Mesoamerica now relied on coastal routes around the Yucatan Peninsula (Sharer 2006, 570). Mayapán was built in a location that would allow for continued participation in international trade, but the region no longer held the status that the k'uhulajaws of Tikal and Calakmul once did.

What this also appears to have meant for Mayapán is that its government would now reflect that of the Basin of Mexico—and that meant a greater presence of the Feathered Serpent, now firmly as K'uk'ulkan in Yucatan. At the same time, it does not appear that the K'uk'ulkan sodality formed at Chich'en Itza made the move along with the nobility. The final statement above by Landa gives us a rich opportunity to discover how this shift played out. At least one implication in Landa's statement is that the referred-to "fixing a time" for a figure who "had gone to heaven" could well relate to a specific visibility of the planet Venus. While the historical records reviewed above attest to roles of the Xiu and Cocom ch'ibals along with K'uk'ulkan in the politics of Mayapán, we find in the next section that the architectural record at Mayapán also supports the specific influence of Xiu's and Cocom's respective heritage.

MUTED RESEMBLANCE

Anyone who had visited Chich'en Itza would find the civic-ceremonial center of Mayapán to look familiar. As Landa pointed out, the masons at Mayapán built the principal structure to resemble a miniature version of the Temple of K'uk'ulkan at Chich'en Itza, including feathered serpent balustrades on its north staircase (Aveni et al. 2004, 128) (figure 10.4). Anthony Aveni, along with colleagues Susan Milbrath and Carlos Peraza Lope, went so far as to find that the difference in size is actually measured: whereas the version at Chich'en Itza had 91 steps on each of its four sides, contributing to its 365 total steps, the architects at Mayapán built only 65 steps per side, resulting in a total of 260 (Aveni et al. 2004, 130). Clearly, the architects at both cities were engaging calendric numerology, and both appealed to a common architectural aesthetic; but this was not pure replication, and the numerological link to the solar year had been dropped. On the other hand, visitors might notice that, having come from Chich'en Itza, the Cocom

FIGURE 10.4 The Temple of K'uk'ulkan at Mayapán. The cenote Ch'en Mul lies between it and the circular structure at left.

appear to have had the clout to bring their architecture style with them—and this architecture focused on the feathered serpent symbolism of K'uk'ulkan.

The point that I took away from this comparison during my doctoral work at the end of the twentieth century is that there was definitely a representation of Chich'en Itza's architecture at Mayapán, but it was muted. There was very likely a close affiliation with K'uk'ulkan and Venus, but it had shifted from one city to the next. My interest piqued, I initiated a more thorough comparison of the architecture.

Just east of the Temple of K'uk'ulkan at Mayapán, on a near tangent line across from the opening of the cenote Ch'en Mul, the local architects designed a second structure to resemble the Tz'iknal at Chich'en Itza (figure 10.5). This "replica" was far less the direct imitation of its predecessor; while it preserved the look, the architectural features were modified to suggest an accommodation of its function. They built no internal staircase in the Mayapán structure and no windows at the top. Instead, they put four doorways around the base of the structure, aligned east, north, west, and south. Within the structure, masons built a cylindrical core with four niches, appropriate for effigy censers, in between the doors—that is, in intercardinal positions. So ritual activity still could have incorporated effigy censers and Venus observations, but the difference in visibility suggests that public engagement would have changed.

The orientation of these structures also differed markedly between the two sites. As we saw in chapter 5, the masons of Chich'en Itza constructed the base of the Tz'iknal to contain two alignments to the visibility of Venus at the horizon (Aveni 2001, 273). One

FIGURE 10.5 An imitation of Chich'en Itza's Tziknal at Mayapán

is oriented toward Venus's maximum northerly extreme along the horizon, the other toward its southerly extreme. Neither of these was built into the features of the version at Mayapán. Also for the orientation of the Temple of K'uk'ulkan at Chich'en Itza, we saw in the last chapter the setting sun on the equinox illuminates a feathered serpent along the northwestern balustrade (Aveni 2001, 298; Aveni et al. 2004, 124). There is a similar effect on the version at Mayapán, but it isn't nearly as precise and it occurs near the winter solstice, not on the equinoxes (Aveni 2004, 131). Furthermore, the feathered serpent balustrades of the Temple of K'uk'ulkan at Chich'en Itza faced the Great Cenote. At Mayapán, architects designed the Temple of K'uk'ulkan to sit adjacent to the cenote Ch'en Mul, with the serpent balustrades pointed to the central plaza of the civic-ceremonial center. Of course, the symbolism would have resonated between the two. The feathered serpent could have been used to represent an ability to move between sky, middle world, and Underworld through cenotes and the Temple of K'uk'ulkan.

Taking these observations together with Landa's statement that K'uk'ulkan was the "founder" of Mayapán, I wondered whether or not it might be possible to determine if astronomy had been used explicitly as a tool of statecraft (Aldana 2001, 258; 2003). While the connection to Chich'en Itza and the Cocom was obvious, a second connection—to the Xiu lineage at Uxmal—was only a hint in the back of my mind. A connection did emerge eventually, though, and it depended in large part on Aveni's work of more than a decade earlier.

TRANSPORTING ASTRONOMICAL PRACTICE

After his collaborations with Hartung at Chich'en Itza, during the 1980s Aveni went on to survey possible astronomical orientations of architecture across Yucatan. In this work, he recovered what appeared to be a unique orientation toward Venus at Uxmal, located seventy kilometers (a two- to three-day walk) west of Mayapán in the Puuc hills of Campeche. One of the most impressive of the Puuc architectural style cities in Yucatan, the masons covered the public buildings with the countless *witz* masks we saw at Chich'en Itza in chapter 1.

In his work, Aveni turned his attention to a structure resembling Chan Chaak's municipal palace (the so-called Nunnery we encountered in chapter 8), but built near the edge of the city's civic-ceremonial center. For this structure, known as the House or Palace of the Governor, Aveni noted that the sculptors at Uxmal carved more than 350 star icons into its ornate façade (Aveni 2001, 286). These **EK'** icons were carved into the lower eyelids of the masks, perhaps implying that the masks themselves were viewing the celestial realm and so perhaps providing a clue that astronomical interests contributed to the building's construction (see figure 1.21). As for the House of the Governor itself,

Kowalski's studies of Uxmal's inscriptions reveal that the Palace of the Governor was erected by one Lord Chac [Chan Chaak], ruler of a powerful late eighth-early ninth century dynasty. Most likely the building, a long, low-lying multi-roomed structure conventionally referred to as a "palace" by students of architecture, functioned in part as a royal residence as well as a place for the storage of paraphernalia and the periodic enactment of rituals. Kowalski argues that the powerful imagery of Lord Chac's double-headed jaguar throne in front of the building on axis with the sculpted image of the ruler himself, surrounded by bicephalic serpents and cosmic symbols carved over the central doorway of the outward-looking building, supports the notion that the Palace of the Governor was the chief administrative center not only of Uxmal, but perhaps for much of the surrounding Puuc region as well (1987, p. 236). (Aveni 2001, 283)

Kowalski's interpretation, therefore, accords well with the rising presence of the feathered serpent in Mesoamerican politics during the Terminal Classic that we saw in the last chapters, and it speaks to the prominence of Chan Chaak's reign in that change.

But Aveni's archaeoastronomical interest was spurred by the fact that the urban planners at Uxmal also introduced an anomaly into the placement of this structure—it deviates conspicuously in orientation from the main architectural group (2001, 283). Whereas the other structures of the civic-ceremonial center align to a common northeast alignment, the House of the Governor is skewed (figure 10.6). For his archaeoastronomical interest, therefore, Aveni had identified both iconographic evidence (in the star-eyelids) and an anomalous orientation in the House of the Governor. He was further able to add the fact that, in the plaza to the east of the central doorway of the House of the Governor, a two-headed jaguar throne had been set up in line with the faraway structure (Aveni 2001, 285).

Aveni argued that the motivation for the unique characteristics of the House of the Governor could be found along the eastern horizon. An observer facing directly perpendicular to the steps of the House of the Governor, looking over the jaguar throne, would find themselves viewing the largest temple of the nearest city to Uxmal, which is tall enough to break through the floral canopy (Aveni 2001, 285). Aveni examined what he took to be the intentional alignment of these three constructions and demonstrated that an excellent candidate would have been the rising of Venus at its southern extreme (Aveni 2001, 285). Specifically, the alignment allowed an observer on the steps of the House of the Governor to witness Venus rising out of the main temple of Cehtzuc in line with the jaguar throne as part of its regular pattern of motion once every eight years (Aveni 2001, 285).

Recalling that the Xiu lineage was originally based at Uxmal, its members would have been versed in this Venus commemoration at the House of the Governor. If the Cocom brought to Mayapán the architectural façades of Chich'en Itza, then it made sense to me, as I worked on the last chapter of my doctoral dissertation, to explore the possibility that an alignment was constructed at Mayapán to reflect the traditions of the other major lineage at the city—namely, the Xiu's Venus commemoration at Uxmal (Aldana 2003). Fortunately, the Carnegie Institution had provided the documentation necessary for me to follow up on this hunch.

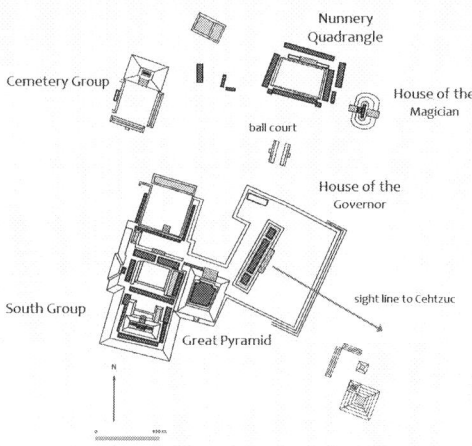

FIGURE 10.6 (*above*) The "House of the Governor" at Uxmal and (*left*) its skewed orientation relative to the other structures of the civic-ceremonial center

Between 1949 and 1951, Morris R. Jones of the Carnegie Institution extensively mapped the architectural remains within the city walls of Mayapán. One very interesting characteristic of this area, readily recognizable in his maps, is the high incidence of cenotes (figure 10.7). We know that cenotes were used as water sources in Yucatan, but we have now also seen in the Hunac Ceel episode that some cenotes served as ceremonial spaces as well. At least in part, this would have been related to the fact that in the karst environment of the Yucatan Peninsula, cenotes were often parts of extensive underground cave networks. They were accordingly, from the Mayan perspective, portals into the Underworld and so laden with ritual significance, as we saw in chapter 1 and in the investiture activities in the last chapter (Brady 1997; Moyes and Brady 2012).

For my doctoral research, I pored over Morris Jones's map of Mayapán and its attendant publications, leading me to hypothesize that there were only two architectural complexes with structures tall enough to pierce the floral canopy and so provide a view for one of the other. The first, of course, was the civic-ceremonial center, from which the top of the Temple of K'uk'ulkan provided views for miles. A smaller complex comprised four

FIGURE 10.7 Site map of Mayapán with locations of cenotes denoted by small circles. The larger circles encompass the city center around the Temple of K'uk'ulkan at left and the structures around Itzmal Ch'en at right. Created by Morris Jones in Smith et al., "Mayapan, Yucatan, Mexico" (Publication 619, 1962), courtesy Carnegie Institution for Science

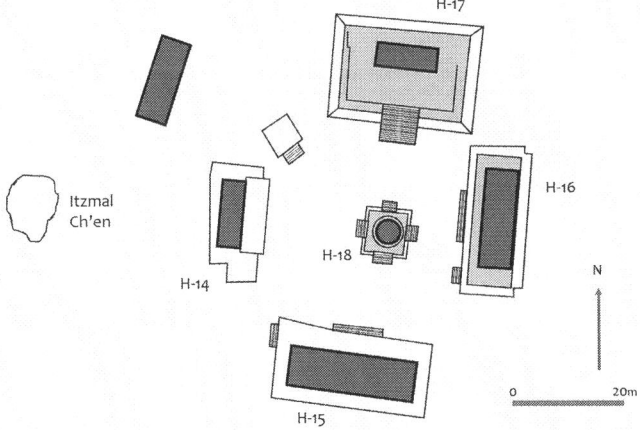

FIGURE 10.8 The Itzmal Ch'en architectural group at Mayapán (after figure 4 in Masson and Peraza Lope 2007)

structures around a plaza, with the northern structure as the tallest. About one-and-a-half kilometers northeast of the civic-ceremonial center, the cenote "Itzmal Ch'en" lay just outside this plaza, adjacent to the western structure (figure 10.8). This meant that the Temple of K'uk'ulkan and Tz'iknal replicas, the two most distinctive structures at Mayapán, were built around the cenote known as Ch'en Mul, and the other tallest structure was built adjacent to another cenote, Itzmal Ch'en (Masson and Peraza Lope 2014, 78, 126). And it is this relationship that at the time I hypothesized might allow for the commemoration of a Venus event in the Uxmal tradition brought by the Xiu.

Calculations based on the map bolstered my hypothesis to suggest that there might have been an intentional Venus alignment at Mayapán (Aldana 2001, 260). Specifically, it appeared that the Temple of K'uk'ulkan and the structure at Itzmal Ch'en were relatively

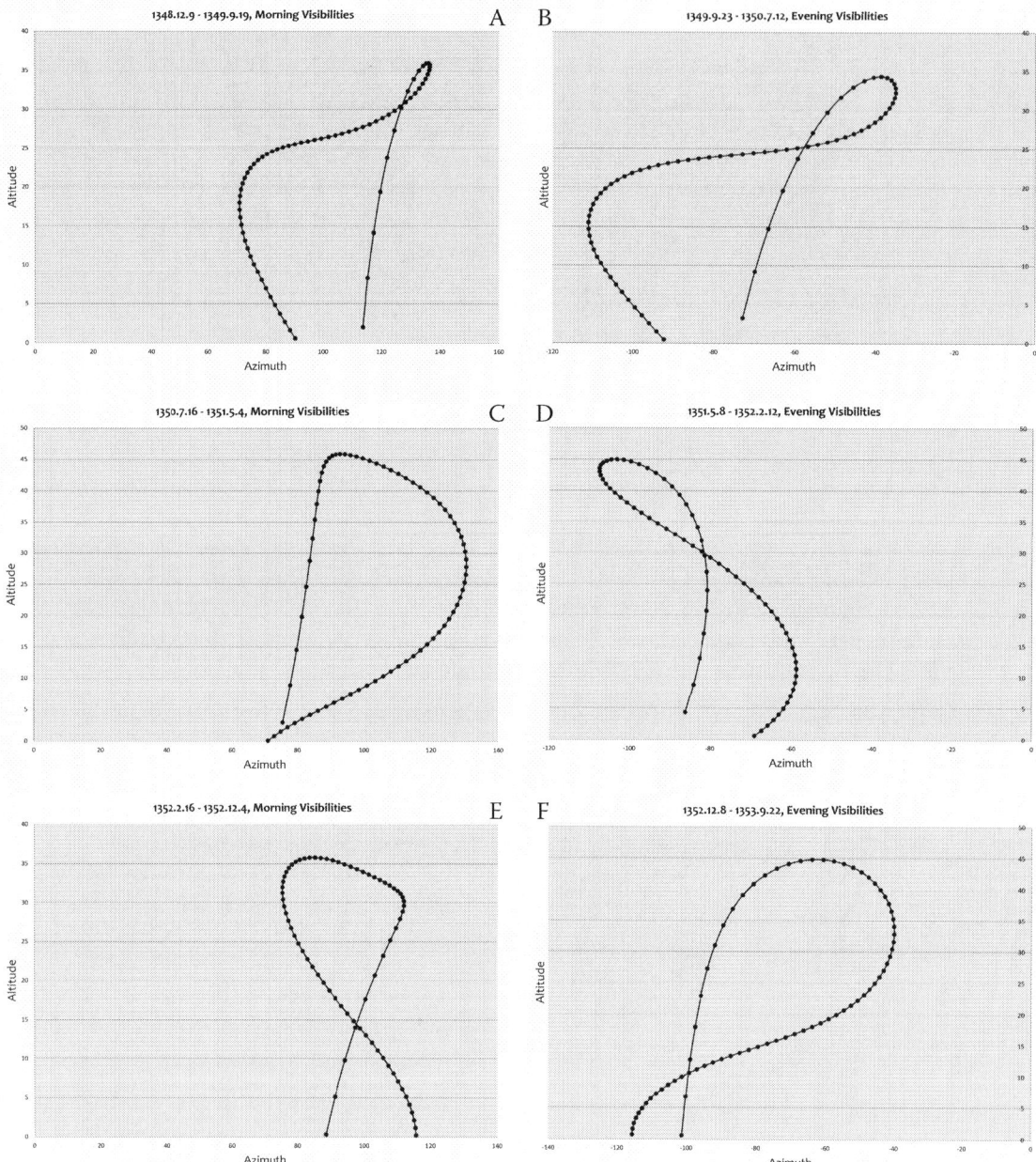

FIGURE 10.9 Venus visibilities at Mayapán

positioned to form an alignment, putting the latter at about twenty degrees north of east at the horizon viewed from the top of the former structure. Reconstructing Aveni's plots for Venus visibilities for the geographic coordinates of Mayapán, I found that "roughly twenty degrees" meant that a Venus event commemoration was possible (Aldana 2001, 261; 2003) (figure 10.9). Aveni's caveats—which we saw in chapter 6—made it clear, however, that computations based on a map would be insufficient to corroborate the

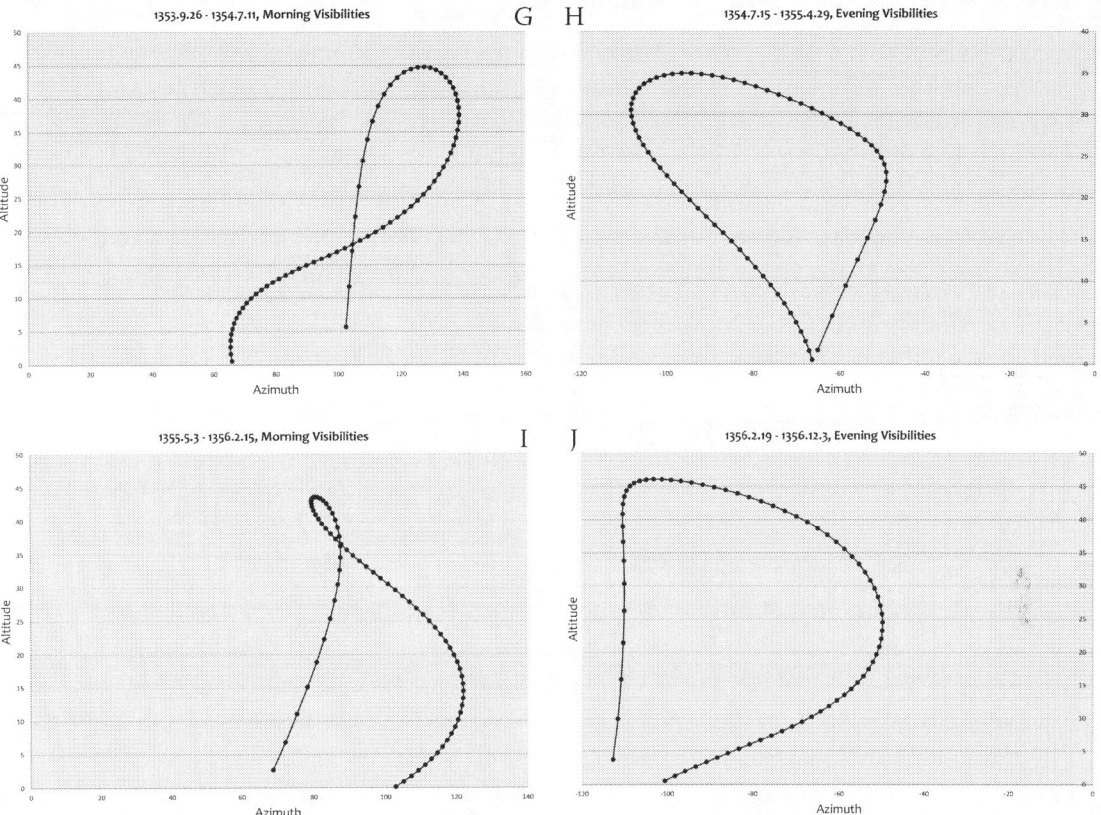

FIGURE 10.9 (*continued*)

hypothesis. What was needed were actual measurements at the site. Fortunately, the David Rockefeller Center for Latin American Studies at Harvard agreed, and I had funding to spend the summer in Yucatan.

Having borrowed a theodolite from a local construction company, I packed up my Toyota pickup in the summer of 1998 and headed south to the U.S. border and along the highways of the Gulf Coast to Yucatan. In Merida, I met with Carlos Peraza Lope and Alfredo Barrera Rubio at the local INAH office, who accepted my request for an appointment in part because I name-dropped that I was working with David Stuart, who held the coveted Peabody lectureship at Harvard, and so from whom I was learning to read Mayan hieroglyphs. Peraza Lope and Barrera Rubio recounted fond memories of Dave and his father, the impressively unique George Stuart, photographer and administrator for *National Geographic* from the 1960s to early 2000s. They granted me permission to work at the site, as long as I took astronomical measurements only.

At that point, my focus was simply on what I could see from the city center, which is where excavation and reconstruction had been focused by that time. Peraza Lope did walk me out to the architectural group at Itzmal Ch'en on the private property of a ranch, but it was completely overgrown, so while we could peer into the cenote and appreciate the size of the tallest building, there was little that could be said about its architecture. Fortunately, we were able to confirm that one could readily see the Temple of K'uk'ulkan from its summit, confirming that these two temples were the only two that were intervisible above the floral canopy. Of course we didn't know it then, but Peraza Lope's work would develop into a fifteen-year project and eventual collaboration with American archaeologist Marilyn Masson that would transform the academic understanding of the city, including an excavation of the Itzmal Ch'en architectural group (Masson and Peraza Lope 2014), which now tremendously pushes this interpretation further, as we will see below.

At the civic-ceremonial center, I used my borrowed theodolite and Aveni's *Skywatchers of Ancient Mexico* as an operational handbook to set up atop the Temple of K'uk'ulkan. Using an astronomical fix, I corroborated the estimates based on Morris's map and more accurately determined that the temple at Itzmal Ch'en was oriented 18.4° north of astronomical east as seen from the Temple of K'uk'ulkan (Aldana 2003, 40). This information allowed me to introduce the computational power of modern computers to reconstruct what would have been visible in the night sky. Turning back time to the Late Postclassic, we find that this orientation would have allowed a viewer at the top of the Temple of K'uk'ulkan to observe Venus rising at the azimuth of Itzmal Ch'en near its first appearance in the morning sky for a few days in a row, once every eight years (see figure 10.9i). Turning the situation around, the numbers suggested that there also were three different Venus Rounds, every eight-year cycle during which near-first-evening visibilities of Venus could be viewed just above the Temple of K'uk'ulkan when viewed from the temple at Itzmal Ch'en (see figure 10.9b, f, h) (Aldana 2003, 42). Hence, on an eight-year cycle, the planet could have been viewed rising out of the structure adjacent to an eastern cenote, and then setting into a structure adjacent to a western cenote—both within the city walls (Aldana 2003, 42). The pattern seemed to me then to have been inspired by the Uxmal orientation—albeit taken one step further.

The initial data I obtained therefore looked promising, but beyond Aveni's concern with the accuracy of maps, arguments attempting to use architectural alignments alone had been questioned consistently since Gerald Hawkins's work on Stonehenge, so of course I needed more. The fact that the architecture paralleled those at Chich'en Itza with Venus affiliations provided useful corroboration, but Peraza Lope's excavations at the civic-ceremonial center also contributed iconographic evidence that corroborated the hypothesis of intent. His work uncovered a low wall that extended orthogonally from the western base of the Temple of K'uk'ulkan. On this wall the muralists of Mayapán painted a series of frescoes, all depicting the same theme: two men carrying banners flanking a central "diving god." The latter figure, I realized, wasn't unique to Mayapán—it showed up throughout Postclassic Yucatan and even in the Dresden Codex itself.

The form of this "diving god" is depicted clearly at the eastern end of the Yucatan Peninsula on lintels of the coastal site of Tulum (Miller 1982). Human or anthropomorphic figures are shown inverted with feet in the air. A more abstract version can be found in the Dresden Codex itself. In Tawiskal Uwoojil's hand, the Eclipse Table ends with a large "diving god" whose head is replaced with the **EK'** (star) hieroglyph. That such a deity figure might depict a falling—or setting—star is thus strongly implied (Thompson 1972, 72).

If we take the diving god as the representation of a falling/setting celestial body, then at the Temple of K'uk'ulkan at Mayapán, we may see the mural as depicting repeating images of figures holding banners on either side of a celestial body descending from the sky. Following the lead that took me to Mayapán in the first place—finding association between K'uk'ulkan and Venus—I seemed to be confronting iconographic corroboration for the architectural alignment. Even better, these images seemed to fit within a broader Mayan artistic practice of using paintings to depict "snapshots" of historical events.

The murals on the exterior walls of the Calakmul Chiik Nahb Structure take the latter practice to an extreme, appearing to depict quotidian scenes from around a local marketplace (Carrasco Vargas and Cordeiro Baqueiro 2012). Much more prevalent are the painted ceramic vessels showing courtroom scenes—either of royal visits or ritual events—such as the ones we saw in chapter 6 (Grube 2006, 247–59; Martin and Grube 2000, 14–15). By analogy, then, the walls of the Temple of K'uk'ulkan suggest that a ritual event involving people holding banners and a setting celestial body occurred at Mayapán and involved the Temple of K'uk'ulkan. This all worked extremely well to support my initial hypothesis. Turning now to Landa's historical records, we find even more attesting a concern with Venus ritual at Mayapán and getting us one step closer to Tawiskal Uwoojil's specific work at the city.

CHIC KABAN

Earlier in this chapter, we saw that Landa noted that after K'uk'ulkan departed and had "gone to heaven," "they fixed a time for him in which they should celebrate a festival to him . . . [at] Mayapan" (Tozzer 1941, 157). Exploring his fuller description, we encounter an event that resonates with the ceremonial patterns implied above.

Landa continued his description with reference to the post-Mayapán celebration.

> After this (city) was destroyed, it was celebrated only in the province of Mani, and the other provinces, in recognition of what they owed to Kukulcan, presented, one one year and another another, to Mani, four, and sometimes five magnificent banners of feather(s), with which they solemnized the feast in the following manner, and not like the previous ones: On the 16th day of Xul, all the priests and lords assembled in Mani, and with them a large multitude from the towns, who came already prepared by their fasts and abstinences. (Tozzer 1941, 157–58)

This passage proves useful on two levels. For one, it fits easily with the recognition that after Mayapán was abandoned, this form of recognition of K'uk'ulkan was maintained at Maní—the post-Mayapán home of the Xiu ch'ibal. Second, Landa's passage recognizes

FIGURE 10.10 Ceramic vessel image of a procession. Courtesy mayavase.com

a distinction between the practice at Mani and the prior one at Mayapán, even though we should also expect some continuity.

More details emerge as Landa's description shifts back to the ritual practice at Mayapán, with the explicit note of the Temple of K'uk'ulkan. This too brings to mind the ritual processions addressed throughout the chapters of this book (figure 10.10).

> On the evening of that day they went forth with a great procession of people, and with a large number of their comedians from the house of the lord, where they were assembled, and they went very quietly to the temple of Kukulcan, which they had previously properly adorned, and having arrived there, and making their prayers, they placed the banners on the top of the temple, and they all spread out their idols below in the courtyard, each for himself, on the leaves of trees, which he had for this purpose, and having kindled new fire, they began to burn their incense in many places and to make offerings of food cooked without salt or pepper and of drinks made of their beans and the seeds of squashes. The lords and those who had fasted remained there without returning to their houses for five days and five nights in prayer, always burning copal and engaged in their offerings, and executing several sacred dances until the first day of Yaxk'in. The comedians went during these five days among the principal houses, playing their pieces and collected the gifts which were given to them, and they carried the whole of them to the temple where, when the five days were ended and past, they divided the gifts among the lords, priests and dancers, and they got together the banners and idols and returned to the house of the lord, and from there each one to his own house (Tozzer 1941, 158).

In this text, we encounter a multiday festival, involving theatrical performances and dances by the political authorities of the day. The numerous ceramic effigies (what Landa refers to as "idols") lining the base of the Temple of K'uk'ulkan and the burning of incense draw parallels to K'uk'ul Ek' Tuyilaj's practices at the Tz'iknal at Chich'en Itza. The fact that there were numerous effigies (and not just one or two commemorating a specific

Venus subperiod) suggests a different approach relative to the practice recovered at Chich'en Itza—as implied by the change in architecture of the version at Mayapán—but one that might still be compatible. Also, Landa doesn't mention them, but there must have been members of the public deeply involved—keepers of the flame for the incensarios, witnesses to the performances and dance, craft-persons behind the regalia.

Finally, the description notes that the banners were taken to the top of the Temple of K'uk'ulkan. Landa concludes:

> They said and considered it as certain that Kukulcan came down from heaven on the last day of these (five days), and received their services, their vigils and offerings. They called this festival Chic Kaban. (Tozzer 1941, 158)

Overall, Landa's description of Chic Kaban, which he must have obtained from Gaspar Antonio Xiu, fits well with the astronomy we have recovered from the architecture of Mayapán and the Dresden Codex Venus Table. Given the alignment of the Temple of K'uk'ulkan relative to the structure at Itzmal Ch'en, the event at Mayapán would have been visible three times every eight years and could have been readily anticipated with the records in the Venus pages.

Following the practice at Uxmal, a viewing station in the east at Itzmal Ch'en would allow for a view of Venus setting into the Temple of K'uk'ulkan. Taking banners to the top of the Temple of K'uk'ulkan would generate an event corresponding to the murals on the wall extending from the base of the Temple of K'uk'ulkan—figures holding banners with a celestial body (Venus) "diving" between them. The introduction of ceramic effigies burning incense may have involved a procession of these "idols" from the niches in the imitation Tz'iknal structure at Mayapán to the Temple of K'uk'ulkan in a way that celebrated a connection resonating with the practice at Chich'en Itza.

ITZMAL CH'EN

In 2003, with this data available, I argued that there was a commemoration of Venus at Mayapán that reflected a political compromise between the Xiu and Cocom lineages. The Temple of K'uk'ulkan and the Tz'iknal architectural imitations at the city center visibly demonstrated a connection to Chich'en Itza and symbolized the primacy of the Cocom lineage; the participation of ceramic effigies at Mayapán may also have reflected the practices at Chich'en Itza. The alignment of the structure adjacent to Itzmal Ch'en relative to the Temple of K'uk'ulkan reflected a ritual practice established at Uxmal and so relied on practices and performances maintained by the Xiu (Aldana 2003, 46). I also suggested the possibility of an oracular character to this compromise. That is, the foundation of the city was built upon an association with the movements of a celestial body in stochastic fashion. This would have produced an oracle akin to that of Waxaklajun Ub'aah K'awiil at Copán in Structure 10L-22. But while the Copán k'uhulajaw's oracle was meant for private consultation, the outcome of Mayapán's oracle was visible for all citizens to see. Namely, if Venus were not visible "coming down from heaven" on the prescribed day to pass into the local Underworld, then the contract between the Mayapán rulers and K'uk'ulkan would have been upset (Aldana 2003, 47). This was the interpretation that I published before Peraza Lope and Masson excavated the remote site of Itzmal Ch'en. What they found—with contributions from Aveni—makes the interpretation stronger.

One of the first contributions coming from Peraza Lope and Masson's work has to do with timing. They ran radiocarbon analyses from their excavations of Itzmal Ch'en and were able to place its construction and occupation between AD 1270–1390 (Masson and Peraza Lope 2014, 63). Second, Masson and Peraza Lope (2014, 64) find that the only evidence of prior occupation at Itzmal Ch'en are Terminal Classic sherds, probably related to use of the cenote as a water source. In other words, there was no great settlement at the site before the construction of Mayapán. They go on to make the critical point that Itzmal Ch'en doesn't appear to be simply the suburban residence of an affluent family—it is anomalous in its neighborhood.

> It is still unclear who constructed and used the Itzmal Ch'en group. This question is difficult to answer since no large elite palaces are found near Itzmal Ch'en. Houses in the neighborhood were engaged in a full range of work activities, like those found in other parts of the city; some were generalized and presumably engaged in agrarian production and custodial activities (H-11) while others produced surplus quantities of fine or mundane craft objects (I-55a and I-57). (Masson and Peraza Lope 2014, 146)

In other words, one might expect that an elaborate architectural group at some distance from the center served as a ritual center for the local neighborhood, but that doesn't seem to fit with the context at Itzmal Ch'en. It seems instead to be an island of elaborate architecture within a neighborhood of normalcy. Moreover, the construction there did not build from prior edifice—the character of the architectural group was designed from the ground up.

Each of the next few sentences in Peraza Lope's and Masson's description speaks directly to the Venus/K'uk'ulkan ceremony proposed in this chapter.

> The Itzmal Ch'en group likely served more than the neighborhood in which it was embedded as the rich symbolic array of materials is well linked to the art and ritual *of the site center*. Outlying focal groups were probably incorporated into site-level celebrations, perhaps on a rotating calendrical basis. They would have been landmarks and may have been of key significance to quadripartite division or other cosmological concepts (chapter 4). Turtle sculptures and censers (among other objects) probably reflect rotating calendrical celebrations that took place at geographic subdivisions of towns, polities, and territories (Masson 2000; P. Rice 2004; P. Rice 2009c:23–25). (Masson and Peraza Lope 2014, 146, emphasis added)

The evidence suggests to them that there was ceremonial purpose to Itzmal Ch'en relative to the whole city, and specifically relative to the city center. They are explicit that the group referenced cosmology and subdivisions of the polity and that it was timed by calendric patterns. All of these are perfectly in line with the rituals we have explored in the Dresden Codex and proposed for the architecture of Chich'en Itza.

Next, they turn to K'uk'ulkan.

> Although the buildings of Itzmal Ch'en exhibit idiosyncratic design elements, their functions and general assemblages mimic the style and content of the site center. An emphasis on Kukulcan inferred from the prevalence of serpent sculptures and a round shrine also links this group to the major founding deity of Mayapán. While the Itzmal Ch'en group may have been a site-level facility, it undoubtedly served as an important defining entity for the surrounding residential zone. It would have been a landmark in the cityscape and perhaps a public square for local and visiting pedestrians. As a public water source, it would have been imbued with both sacred and mundane attributes (Brown 1999, 2005, 2006, 2008). (Masson and Peraza Lope 2014, 146)

That the Itzmal Ch'en Group differed from its local neighborhood architectural groups and that the difference aligned with the civic-ceremonial center is significant. That the specific association concerned serpent sculptures and K'uk'ulkan is perfectly in line with the proposed Venus alignment, reflecting the Xiu commemoration.

The work generated by Peraza and Masson comes together, therefore, to provide physical evidence corroborating what we have assembled from astronomical and historical records: K'uk'ulkan "founded" Mayapán in a way that integrated Venus observations into the urban planning of the city. What this does for us now is demonstrate that there were several motivations for the builders of the city to investigate the visibility of Venus at Mayapán with particular interest in the computational/calendrical characterization of the planet developed at Chich'en Itza. They may not have brought over the governing structure, but they did have an interest in bringing an inspiration related to Chak Ek' and K'uk'ulkan.

Returning to the manuscript that initiated this investigation, we realize, therefore, that Tawiskal Uwoojil would have had a local political/religious/astronomical need, and, with either a copy of K'uk'ul Ek' Tuyilaj's manuscript or the original itself, a strong reference work to take on the urban planning required of such a project. At the same time, some adjustment would have been necessary, which all brings us to the final version of the Venus Table.

TAWISKAL UWOOJIL AT WORK

As Peraza Lope and Masson have determined, construction of monumental architecture at Mayapán probably started in the thirteenth century, and Itzmal Ch'en was in use by the late twelfth century into the thirteenth (Masson and Peraza Lope 2014, 61–71).

In addition, Robert Smith's work on the ceramics of the site, which contributed to the effigy-censer interpretations of chapter 6, also provided further ceramic data of use here. Based on his familiarity with Smith's work, Eric Thompson suggested a date for the writing of the Dresden Codex "of near A.D. 1200–1250" (1972, 15). He based his interpretation on the calendrics of the Venus Table, but also in part on

> illustrations of pottery incense burners. . . . That on page *27b* seems to me definitely early Mayapan period and that with button relief on page *28b*, probably so. In reply to my query, Robert Smith writes that the spiked incense burner on page *26b* is very similar to forms belonging to the Hocaba (early Mayapan) complex. (Thompson 1972, 15)

It is worth noting in retrospect that, even though he doesn't seem to have realized it, the incense burners he selected for this attribution are both similar to the forms in the upper illustrations of the Venus pages.

Ceramics, calendrics, and architecture suggest, therefore, that it was during this time that the almehen of Mayapán were developing the vision of the city and including more than simple imitations of Chich'en Itza's architecture. If the ajk'uhuuns designing the city simply replicated the architecture of Chich'en Itza, that could be interpreted as an over-privileging of the Cocom lineage. In order to bring the traditions of Venus commemoration from both Chich'en Itza and Uxmal and to give them relatively equal standing, there had to be a move to include an aspect of the Venus commemoration at Uxmal. At this level, at least, we can recognize the political founding of the city by K'uk'ulkan—negotiating between two ch'ibals to create a new city amenable to both.

It would not have been difficult for these visioners to consider setting up two inter-visible structures within the Mayapán city limits, in order to construct an alignment similar to the one between the House of the Governor at Uxmal and the tallest structure at Cehtzuc. And that may well have been the problem as it was assigned to Tawiskal Uwoojil during the thirteenth century: given the location of the city's civic-ceremonial center, identify a second location that allowed for Venus to be seen entering into or rising out of a structure in the near vicinity.

The tools Tawiskal Uwoojil had available, we have already seen, had been developed for some time by earlier Mesoamerican astronomers. From a given location, an observer was charged with tracking a celestial body's position along the horizon. This most likely would have resulted in multiple options, but the process would have taken some time: eight years at the very least for Venus.

For Tawiskal Uwoojil's work, temporary scaffolds would have been erected to the east and to the west of the Temple of K'uk'ulkan. These would provide him views of the western and eastern horizons with the Temple of K'uk'ulkan as the backdrop. He would have climbed the new Temple of K'uk'ulkan daily to make observations of Venus along the horizon, similar to K'uk'ul Ek' Tuyilaj's practice, but in his case with an interest in constructing something new—and as I did centuries later, attempting to rediscover something old. To assist in his work, Tawiskal Uwoojil would have directed his apprentices to climb the eastern scaffold before sunrise so that they might position markers of first morning and last morning visibilities on the scaffold itself. They may have used effigy censers as markers or perhaps banners or even paint on a staff fixed to the scaffold. The process would then be repeated for evening visibilities.

After doing this for eight years—or one complete row across the five pages of the Venus Table—Tawiskal Uwoojil would have had at least ten markers in place for each scaffold, marking the beginning and end of morning and evening visibilities. Once they were in place, he would have been able to narrow down the viable candidates. For one, he would follow the trajectory of each alignment on the ground to determine whether or not a suitable cenote lay along it. This would provide a metaphor resonant with Chich'en Itza's feathered serpents, pointing to the Underworld through the Great Cenote. Given the number of cenotes in the area, this probably would have eliminated some trajectories, but certainly not all.

Next, if the option were still available, he would have to consult with political figures to decide whether they preferred Venus to rise out of the Temple of K'uk'ulkan or to set into it. That would determine whether they would seek their secondary site to the east or west. Also, there may already have been families settling into different neighborhoods, so choosing a trajectory would have important political implications, and the jockeying for a favorable location—if not in the area, at least adjacent to procession routes—may have already begun.

At this point, Tawiskal Uwoojil probably consulted with the leaders of the Xiu and Cocom ch'ibals to determine that the secondary site would be placed to the east of the Temple of K'uk'ulkan. The cenote Itzmal Ch'en would have been in contention, but for it to make the final cut more detailed observations would be necessary. The task for Tawiskal Uwoojil would be to focus on those evening visibilities of Venus that took ritual experience into account. For example, as ajk'uhuun, he would not want to choose a Venus Round in which an alignment occurred too close to first evening visibility, because that would make public witness of the event risky. If the alignment were only visible on a specific day within the 584-day cycle, then rain, cloudiness, or any number of random occurrences could negatively impact the event. At the other extreme, if Tawiskal Uwoojil waited too long within a Venus Round, the ritual event could take place after the planet was already easily visible to the public, and the event would not be as impressive. One of the stronger options would be to have an alignment with restricted visibility for some time as a safety buffer, providing lead time to ensure that preparations could be made.

Turning now to modern software, we can reconstruct an approximation to the visibilities of Venus that Tawiskal Uwoojil would have observed. Doing so, we find that there is an evening visibility event in AD 1258 that has Venus skirting the horizon for the first twenty days or so of postconjunction visibility. At 340 days after first morning visibility, it would have been close to dropping into the Temple of K'uk'ulkan viewed from the top of the tallest structure at Itzmal Ch'en. This would have suited Tawiskal Uwoojil's purposes perfectly.

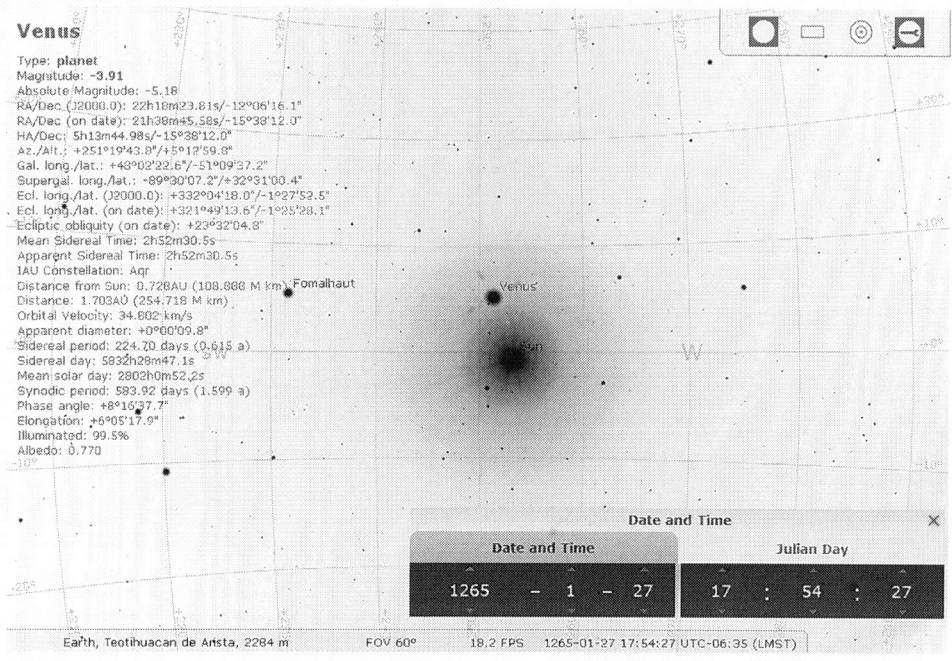

FIGURE 10.11 The Venus trajectory chosen by Tawiskal Uwoojil. The image shows that on January 1, 1265, Venus would have been visible just above the horizon (altitude of just over 5 degrees) and would have set into the Temple of K'uk'ulkan as witnessed from Itzmal Ch'en (azimuth of 251 degrees 19 minutes). Courtesy Stellarium

From his perch at the top of the Temple of K'uk'ulkan, Tawiskal Uwoojil would have witnessed first evening visibility at an altitude of about three degrees on day 320 of a Venus Round, at an azimuth of about eight degrees south of west. For most viewers from the plaza floors or those not looking up and in that specific direction during the last twenty minutes before sundown, Venus would not have been visible—it would have been obscured by the rainforest canopy. Over the following fifteen evenings, Venus would appear slightly further south along the horizon and slightly higher before it initiated its drop into the horizon (figure 10.11). By day 356 of the Venus Round, Venus would appear at nine degrees of altitude and at an azimuth of eighteen degrees south of west. This would have allowed anyone with fairly good eyesight to spot Venus above the horizon and watch it descend over a period of approximately twenty minutes. Even though it would now have been twenty or more days after technical first evening visibility, this Venus event would have been very well suited to serve the needs of a more public celebration of the celestial body. Again, the aim here is the public experience of an event, not the marking of it with technical astronomical precision. This is an important point.

Recall that at Chich'en Itza, the actual observation of Venus over the horizon would have been restricted to relatively few—those with access to the higher structures. In place of direct observation, the movement of Venus at Chich'en Itza was represented for the public by the placement and movement of ceramic effigies around the upper level of the Tz'iknal. The ceramic effigies at Chich'en Itza would have been visible to a far larger segment of the population, and the pageantry would have focused on them as proxies for the visibility of Venus. At Mayapán, according to Landa, the celebration was a commemoration of the "coming down from heaven" of K'uk'ulkan—an event that ostensibly would have relied on the visibility of the celestial body and to a larger group. The architectural group at Itzmal Ch'en would have provided just such a venue, and if a road were cleared in line with the Temple of K'uk'ulkan, the event would have been visible to anyone at the top of the temple, in the plaza itself, or even on the road between the two structures.

Moreover, the event would be conceptually equivalent to the representation of K'uk'ulkan at Chich'en Itza. At Chich'en Itza, the public witnessed K'uk'ulkan descending from the sky annually, when the Sun illuminated the balustrade of the Temple of K'uk'ulkan. Here at Mayapán, K'uk'ulkan as the planet Venus descended from the sky, through the Temple of K'uk'ulkan, into the Underworld via the cenote Ch'en Mul. The representations were different, but the underlying concept was the same.

Operationally, if Tawiskal Uwoojil made this specific Venus Round selection, it would have given him time to confirm that Venus was moving as expected before the ritual event; he would also have time to accommodate a day or two of possible visibility obstruction based on weather conditions. If he waited too long, Venus would have started an upswing in trajectory that would push it higher into the evening sky and make it readily visible to a large segment of the public, so there was some constraint. In the end, the alignment would occur every eight years, and it would be visible approximately 340 days into a Venus Round.

In addition, Tawiskal Uwoojil would have realized that this alignment also allowed for the visibility of Venus rising out of Itzmal Ch'en from the top of the Temple of K'uk'ulkan. Whether by luck or design, this feature added to the overall commemoration of Venus at Mayapán, into and out of the Underworld via two very prominent cenotes.

On the other hand, this new role of observation and the incorporation of evening visibility as well as morning visibility would have required some adjustment to the Venus Table he inherited from K'uk'ul Ek' Tuyilaj.

THE THIRD AND FOURTH CORRECTION INTERVALS

Finally, near the end of this book, we come to Tawiskal Uwoojil's paintbrush on the pages of the Dresden Codex in context. With the cities of Yucatan on the wane, and Toltec alliances in Central Mexico setting the stage for the Aztec Empire, Tawiskal Uwoojil turned his attention to the calendrics that would support the architecture of Mayapán. Of course he would know that it would be impossible to ensure that every observation of Venus setting into the Temple of K'uk'ulkan would fall on a date 1 Ajaw. The first evening visibility event he selected would occur every eight years—every five Venus Rounds—so it would isolate one column in the Venus pages. This would mean that he could preserve a Day Sign for every event, but not the coefficient. This means that at the base level, if nothing else, he could still use K'uk'ul Ek' Tuyilaj's Venus Table to start from, even if he would have to introduce some modifications.

The attentive reader will have noticed at this point that Tawiskal Uwoojil's plan for Venus commemoration at Mayapán itself raises a complication. K'uk'ul Ek' Tuyilaj's Venus Table and the records coming from Central Mexico emphasize the first morning visibility of the celestial body and appearance over the eastern horizon. Western horizon visibilities are certainly captured in the Dresden Codex Venus Table, but the whole table is anchored to first morning visibility, and it is for that event that we have descriptions of omens recorded—not only in the Venus Table itself, but also from Central Mexican sources, as we have seen in previous chapters. With the interpretation developed here, though, it appears that at Mayapán the visibility shifted to observations over the western horizon. The "diving god" motif and the "coming down from heaven" references also suggest the focus on a setting celestial body. It turns out that this becomes an opportunity to address an aspect of the Venus pages that has long been a conundrum for its interpreters. We will see here that whereas in the last chapter the middle two Correction Intervals on page 24 appeared to have been used to adapt a Venus Almanac to observations made at Chich'en Itza, the first and last Correction Intervals are perfectly appropriate for adapting the Chich'en Itza version to the needs of observations at Mayapán.

9,100

One number, recorded in the row of Correction Intervals on page 24, has been considered without exception as an enigma. Eric Thompson wrote: "The number 1.5.5.0 bears no relation at all to the VR; it is 15 VR + 340, whereas every other figure in the table is a multiple of VR without—or in line—with an explicable correction of a few days" (Thompson 1972, 63). Instead, Thompson argued that the number itself was a mistake— probably a copyist's error (1972, 63). He suggested that the 1.5.5.0 (= 9,100) interval recorded in the table should have been 1.6.0.0 (= 9,360), allowing for a still unorthodox, but slightly more useful correction of adding sixteen days. But still this didn't help Thompson's interpretation. Accordingly, most treatments of the Dresden Codex since his *Commentary* either skip this interval entirely (Bricker and Bricker 2007; Paxton 2001) or comment on its oddity and move on (Aveni 2004, 186).

Lounsbury, on the other hand, took the interval at face value and made what appeared to be some progress by realizing that it had the same structure as the other Correction Intervals.

$$1.5.5.0 = 4 \times 5.5.8.0 - 61 \times 6.9.0$$
$$\text{or } 9,100 = 4(37960) - 61(2340)$$

Still, this interval stood out; with a coefficient of 61 for the Correction Factor, this would suggest a correction of 244 days, which would accumulate only after a very long time interval. To Lounsbury, this suggested a turn to the mythological dates on page 24 and an attempt to make the Correction Interval work with the errors suggested by the GMT Calendar Correlation (1992b, 212–14).

Without the constraint of the presumed accuracy of the GMT, my own investigation led me to seek other motivations for the 9,100-day interval. Since we have now left behind the project that would subject the interpretation of the Venus Table to the GMT, and if we take into account the shift in observation at Mayapán, then we gain the opportunity to expand what we think of as a correction. Instead, we now recognize the possibility that this enigmatic Correction Interval would work extremely well as a shift forward. That is, for a cyclical phenomenon, a "correction" (shift backward) of 244 days could also be recognized as a shift forward of 584 − 244 = 340 days. In K'uk'ul Ek' Tuyilaj's version of the table, 340 days after first morning visibility would be very close to canonical first evening visibility. The Venus Table records first evening visibility at 326 days after first morning visibility, so 340 is just 14 days after the table's canonical date. In other words, simply applying the interval 1.5.5.0 to a 1 Ajaw first morning visibility would result in a later 1 Ajaw date very close to first evening visibility. Observationally, it wouldn't be very accurate when applied in isolation, but when combined with a projection forward in time and an appropriate Correction Interval, we find that it may be precisely what we have at Mayapán.

Page 24 records the date 1 Ajaw below the 1.5.5.0 time interval, so we can assume that Tawiskal Uwoojil aimed to anchor the use of this interval to a date 1 Ajaw. We can skip to the chase here, recognizing that the final shift that all interpreters of the Dresden Codex Venus Table have recognized is the one that moves to the Calendar Round anchor of 1 Ajaw 3 Xul. In other words, it would follow precedent for Tawiskal Uwoojil to have observed a first morning visibility at Mayapán on a date 1 Ajaw 3 Xul, and then to have looked to the material he had from K'uk'ul Ek' Tuyilaj's work to make mathematical sense of it and to project forward from it to a first evening visibility.

In fact, the math again is straightforwardly painted on page 24. Combining the fourth Correction Interval with the third Correction Interval, we have

9.9.9.16.0 1 Ajaw 18 K'ayab

+ 1.5.14.4.0 + 4.12.8.0

= 9.9.9.16.0 1 Ajaw 18 K'ayab + CI_3 + CI_4

= 9.9.9.16.0 1 Ajaw 18 K'ayab + CVI_3

= 10.19.16.10.0 1 Ajaw 3 Xul

This Long Count date, 10.19.16.10.0 1 Ajaw 3 Xul, is fifteen winikhaab after the Venus Rounds observed by K'uk'ul Ek' Tuyilaj at Chich'en Itza. That's three hundred years. Three centuries after the Terminal Classic at Chich'en Itza puts us into the thirteenth century at Mayapán, which is precisely when the first construction occurred at the site according to Masson and Peraza's radiocarbon dates.

Having observed a first morning visibility on 10.19.16.10.0 1 Ajaw 3 Xul, then, Tawiskal Uwoojil could have recognized that the interval from 9.9.9.16.0 to one of the recorded data points, 10.4.13.4.0, produced a difference of 15.3.6.0. This interval then added to 10.4.13.4.0 1 Ajaw 18 Wo yielded the first morning visibility that K'uk'ul Ek' Tuyilaj observed at Chich'en Itza and probably was recorded in the earlier manuscript as first morning visibility. In other words, if Tawiskal Uwoojil observed a first morning appearance near 10.19.16.10.0 1 Ajaw 3 Xul, he could have consulted K'uk'ul Ek' Tuyilaj's manuscript to find that 10.4.13.4.0 1 Ajaw 18 Wo was also close to a historically observed first morning visibility. The Correction Interval between the original anchor 9.9.9.16.0 and this historical date would then just have to be doubled to arrive at the 10.19.16.10.0 date that Tawikal Uwoojil directly observed.

For the final two Correction Intervals, then, Tawiskal Uwoojil could have used CVI_3 to first project forward to the thirteenth century, then applied the short Correction Interval of 9,100 days to shift from first morning visibility to first evening visibility. In this way, Tawiskal Uwoojil could use K'uk'ul Ek' Tuyilaj's Venus Table to track both first morning visibilities anchored to 1 Ajaw from the Temple of K'uk'ulkan viewing east, and first evening visibilities of the Venus event setting into the Temple of K'uk'ulkan viewed from Itzmal Ch'en.

Beyond the temporal match and calendric shift, Tawiskal Uwoojil also seems to have edited the narrative in the table to accompany the 3 Xul shifted base date. The rows at the bottom of the Venus pages provide the shift required to go from 13 Mak to the 365-Day Count base date of 3 Xul. The 260-Day Count dates do not have to change, since the 1 Ajaw anchor is preserved regardless of shift (figure 10.12). I have proposed elsewhere that the verb accompanying these lower 3 Xul rows is *tzekya'n* as "to become corrected" (Aldana 2011b, 47–48).

> 15 Kumk'u ts'ekyan Jun Ajaw Chak Ek' lak'in 0 Xul
> 15 Kumk'u [correct-INCHOATIVE] Jun Ajaw Chak Ek' East 0 Xul

What he meant by "correction" is clear when we take into account a ritual process that actually depends on Venus visibilities; now the subperiods matter. It is worth recognizing that this text seems to produce a more intuitive description of the relationship between Chak Ek' visibility and calendar dates. For Tawiskal Uwoojil, "the East is now enclosed for the 236 days," "the North is enclosed for 90 days," and then "the West is enclosed for 250 days," and the match seems to be more aligned with the observable phases of Venus. This would have been useful for Tawiskal Uwoojil. If the practice at Chich'en Itza was more about the Tz'iknal and the incensarios on the structure, then the visibility of Venus in the sky wouldn't be as important. If the shift at Mayapán is to the actual visibility of Venus relative to two structures, then the ritual activity should better align with that.

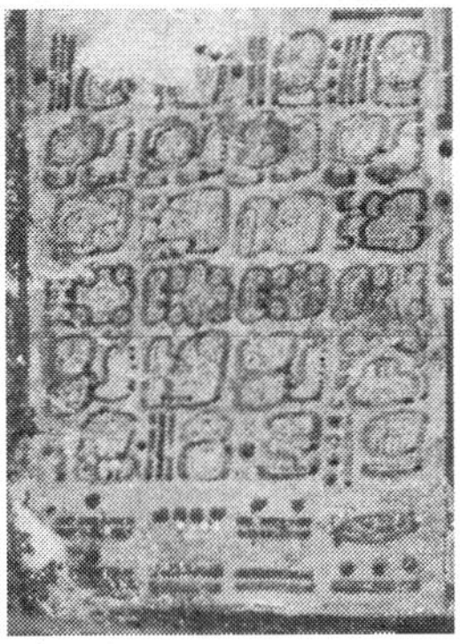

16 Yaxk'in	6 Keh	11 Xul	19 Xul
tsekyan	tsekyan	tsekyan	tsekyan
Uj Xajaw	K'an Te'	Kimil	K'awiil
Chak Ek'	Chak Ek'	Chak Ek'	Chak Ek'
Lak'in	Xaman	Chik'in	Nohol
6 K'ank'in	16 Kumk'u	1 Mak	9 Mak
11	4	12	0
16	10	10	8

FIGURE 10.12 Lower text on page 49 of the Venus pages. Courtesy SLUB

Perhaps related to this shift, the niches for the effigy censers at Mayapán are moved into the "Tz'iknal" building so that their visibility is no longer the purpose; this would very much be in line with a shift to the actual visibility of Venus in the night sky as the focus of ritual activity. And if this is the case, it matches extremely well with the shift in the bottom section of text in the Venus pages.

Understood in this way, there is a historical context and rationale underlying the entire set of Venus pages. Our scribe of the late Postclassic copied the necessary material from the table's first correction, then provided the interval necessary to make it useful for his contemporary times. The revision of page 24 in the previous section showed that the middle two Correction Intervals can be more efficiently used to connect the material actually recorded with the process of correcting the table to match observation. Here the revision provides a rationale for the layout of the Venus Table: the 13 Mak table is provided because it generates the Correction Factor of 2,340; the 3 Xul table is given because it fits the previous data to the copyist's observations and motivations at Mayapán.

Landa wrote that K'uk'ulkan brought peace to Chich'en Itza and then many years later returned to Yucatan to become the "founder" of Mayapán. We now actually have a framework to help us unpack this set of statements. If we follow the hypothesis of the last chapter that one lineage leader became associated with the Feathered Serpent in a manner anticipating the Cholula *tecuhtli* practice, then K'uk'ul Ek' Tuyilaj's intervention could have led to the construction of the Temple of K'uk'ulkan at Chich'en Itza and the move of the city center to the north. Then combining that history with the work that had already been done on Venus observation and ritual at the Tz'iknal, a commemoration of K'uk'ulkan as Venus could have been set up at Chich'en Itza and become integrated into the fabric of the ritual life of the city.

In such case, the "founding" of Mayapán could have been a metaphorical one in which the city was built using the practices established by K'uk'ulkan, invested into his commemoration through the architecture and ceremonies of the city. At Mayapán, K'uk'ulkan was no longer a human member of a political sodality. Here K'uk'ulkan was a way of life.

It may have been that the Cocom chi'bal retained some of the solar authority of the paramount at Chich'en Itza, while the Xiu chi'bal retained an affiliation with a Venusian authority of K'uk'ulkan, but the large-scale separation between the families and their roles was far from distinct. It is now clear that the city's "founding" occurred on the level of narrative, in the practice of ritual activity, in architectural style, in scientific activity, and it all came together through an urban planning project. It now appears that Tawiskal Uwoojil and the Dresden Codex played a critical part in this process.

CHAPTER 11

Science, Astronumerology, and an Abstraction of Capital-ism

A few years after my summer taking alignment measurements at Mayapán, I spent a week in a Zapatista community in the highlands of Chiapas, not far from San Cristobal de las Casas. Then an assistant professor at UC Santa Barbara, I met a colleague at my new academic home who had been working alongside the nonprofit organization "Schools for Chiapas," and that connection resulted in my invitation to visit and assist in an educational project. The modern Tzotzil Mayan teachers of the local autonomous school were interested in incorporating ancient Mayan mathematics into their primary education curriculum. Although it probably appears to the reader as a significant non sequitur, it is part of the latter context that draws my attention at the end of this book.

The intentional autonomy of that Zapatista community carried with it layers of complication. That autonomy was not referring to complete independence or isolation, but instead focusing on the right to maintain their own cultural identities as well as develop sustainable relationships with their local environment. Of course such autonomy came with a cost; communities could no longer rely on the resources of the Mexican federal or state governments in times of crisis or seasonal need. At the same time, they did receive the benefit of independence from that system of governance—independence from the national labor market and constraints that kept them in poverty relative to their peer citizens across the country (Bardacke 1995; Collier 1994).

As an outsider to this community, I was able to find much of the autonomous town relatively familiar from my travels throughout Mexico and northern Central America. Houses and community buildings were made of concrete block or untreated wood; clothing was traditional, but not extremely so. Many of the women dressed in huipiles and woven skirts, while men wore Westernized cotton pants and button-down shirts. Two exceptions, though, were readily noticeable. First were the ski masks or bandanas

FIGURE 11.1 Murals at Oventik (*top*) within a classroom; (*bottom*) with the author facing an image of Emiliano Zapata on the community building

covering faces or at least the nose and mouth. Not everyone wore these all the time within the community, but they were common—increasingly so as one approached the highway that passed adjacent to the town. The second exception comprised the eye-catching murals on the exteriors of so many buildings, as well as on the interiors of the school classrooms and public spaces (figure 11.1). The imagery of these paintings blended ancient Mayan iconography with modern Latin American political symbols. The feel of the community for me was both familiar and inspiringly foreign.

Some of the differences I encountered during that week have had practical impacts on my instructional pedagogy to this day. For instance, Teresa—the liaison between me and the group—informed me that the teachers in my workshop always would work together in small groups. Partially, this was logistical: I did not speak the local Mayan language (Tzotzil), and not all teachers were fluent in Spanish, so working in small groups ensured that information could go back and forth most effectively. More important, however, was the strict policy that Teresa emphasized those small groups held each other to: the class could not move on to new material until every member of every group understood the material that had been presented. It was insufficient for just one member to "get it" so that the group and the whole class could "move on" to other material. If anyone in any group didn't yet understand, the whole class would wait until the members of that group (and/or other nearby group members) could bring along the teacher struggling with the material. The teachers worked as a dedicated community; no one was left behind.

That was a shift for me. I was used to an approach not uncommon in the United States that had us "teach to the B student." This meant that if you, as an instructor, went "too slowly," ensuring that you brought along the students who were struggling, the higher-grade-earning students would be bored. And if you went "too fast," always trying to push the "A students," then more than half the class would likely fail. In Chiapas, we did go as slow as necessary, but the ones who understood the material quickly spent their "free time" helping me help those who needed more time to understand it. And when one person admitted they didn't understand yet, inevitably there was another who was simply too shy to speak up.

The point is that this experience represented a significant difference from my home context. It was a deliberate change from my own experience, but it was not one that was difficult to understand. It constituted a cultural difference that I could comprehend and appreciate and find ways to incorporate into my life back at home. Contrastingly, there were other experiences of difference during my time in Chiapas that I could not understand. Sometimes what I found to be different, and for me unprecedented, was also very hard to describe.

When I was not in the classroom, I would hang out in the store/restaurant at the entrance to the village, chatting with locals as well as international visitors; hunker down somewhere inconspicuous to update my journal; or wander around the public areas of the community. The latter often resulted in my being invited to visit this family or that garden—it allowed me catch a glimpse of the lives of the community members.

And so it was late in my visit, near the end of the week, that I woke and stepped outside the meeting/storage room that was our temporary housing. I stretched and breathed deeply. And it was that act that struck me as different in an unquantifiable way. My first impression—and what I wrote down in my journal later that day—was that "the air tasted different." And while I realized that that captured something of the experience, it didn't really make sense and it didn't capture the feeling. It wasn't "taste" but a different sensation produced by breathing that air. I then tried characterizing it as "the air breathed differently," but that seemed far too mechanical and again didn't get at the experience. Finally, I concluded that there was an experience of my presence or my being in that autonomous space, working in ways fundamentally different from what I had available in my various home- and work-lives in the United States that was tied to my conscious act of breathing and which "felt good." Maybe, I wondered (from an embodied-knowledge perspective?), this was what the idea of autonomy "felt" like.

Outside of that community during that summer and my later returns to autonomous spaces, I have not had that experience. Neither have I been able to forget it. The reason I raise the issue of this experience here, at the end of this book, has to do with the core of what this book is about. At the outset, I aimed to provide a "recalibration" of Mayan astronomy. Such a project in many ways is also reliant on a recalibration of our collective cultural translations of "the Maya"—both living and ancestral. And such a translation should require pushing limits that otherwise just made the different look familiar—perhaps without showing how the familiar can genuinely look foreign.

When I first embarked on my dissertation project investigating the "cultural translation of science," my "project" was largely constrained to one realm of difference: whether ancient Mayans practiced "astronomy" or "astrology." This in turn reflected a relatively simplistic sense of one being "real science" and the other as purely manipulative through emotion, superstition, and patterns in randomness. Those themes emerge throughout this book, but now hold very different meanings. Almost fifteen years later, an event of the first decade of the twenty-first century provided me a rich opportunity to reconsider that binary characterization. Specifically, my perspective shifted discontinuously as I reflected on the role of modern math and statistics within financial markets and the economic crash of 2007—in particular with respect to the kinds of esoteric Mayan intellectual activity we have explored in the preceding chapters.

ASTRONUMEROLOGY AND MODERN ECONOMICS

If we think of Kan B'ahlam's astronumerological calendric innovations at Palenque (in chapter 3) as esoteric, the economic financial markets of the late twentieth century developed in the offices on Wall Street in New York City definitely merit a comparable adjective. A close look at the development of these financial instruments suggests a praiseworthy narrative of innovation driven by opportunity as "the economy" boomed. Such praise ended abruptly, however, when these instruments became transformed into what billionaire investment banker Warren Buffett called "financial weapons of mass destruction" (Shen 2016). We thus now shift our view from noble houses in the great cities of Yucatan and the Peten during the seventh through the thirteenth centuries to the banks and financial service firms of Wall Street—not far from the initial efforts of the Manhattan Project—in the late twentieth and early twenty-first centuries. And whereas Mayan intellectuals were charged at least in part with addressing environmental changes pushed by a prolonged drought, American and British investors were challenged by the need to respond to the environmental catastrophe of an oil spill.

In 1989, Blythe Masters was a financial investor for J. P. Morgan, handling financial instruments known as "swaps" (Lanchester 2009). She was at the time earning a strong reputation for her abilities, which would eventually make her a managing director at only twenty-eight, three years after her critical innovation. In 1989, her role was to assist the firm in responding to a request from the Exxon oil company. For both technological and social failures, an oil tanker that Exxon owned—named the *Valdez*—ran aground off the Alaskan shore and caused an oil spill of 10.8 million gallons. (In Mayan numbers, this would be 3.7.10.0.0.0 gallons, rivaling the Serpent Numbers in the Dresden Codex in scope.) The Alaskan court system charged Exxon $5 billion as a penalty for this crime against the environment—a fee that Exxon could not pay on its own. The oil company needed a loan.

As financial analyst Terri Duhon tells it, J. P. Morgan was reluctant to lend the lump sum to Exxon, as it would tie up their assets under the regulations and financial practices of the time (Wiser 2012). On the other hand, the investment bank had developed a strong client relationship with Exxon and wanted to honor that. Blythe Masters's innovation was to propose a new financial instrument. J. P. Morgan could lend the money to Exxon so they could pay off their debt. Then the investment bank would sell the loan to a third party—the European Bank of Reconstruction and Development, and J. P. Morgan would pay them a fee for doing so. This would take the debt off J. P. Morgan's liability—albeit with all parties well apprised of the terms of the agreement and the risk involved. This exchange they referred to as a credit default swap (Wiser 2012).

But the credit default swaps (CDSs) that Masters invented by themselves were far from sufficient to generate the financial meltdown of 2007. It was their implementation in concert with another financial instrument that unbalanced global economic systems.

Within six years of the invention of CDSs, a large number and wide variety of financial service firms were applying the idea behind swaps to a broader financial instrument known as a Collateralized Debt Obligation (CDO)—a move that evoked surprise in Masters herself. In her history of the financial collapse, Gillian Tett wrote that Masters

> was among those baffled by the C.D.O. boom. "How are the other banks doing it?" she asked. "How are they making so much money?" The answer, Tett says, is that "she [Masters] was so steeped in the ways of J. P. Morgan that it never occurred to her that the other banks might simply ignore all the risk controls J. P. Morgan had adhered to. That they might do so was simply outside her cognitive map." (Tett 2009)

The issue for Masters centered on risk and how it was handled. In the end, this measure of risk would prove to be the core concern.

At the turn of the century, Collateralized Debt Obligations themselves were a recent innovation. And where CDSs came from a need to address environmental destruction, CDOs found their origins in the U.S. Civil Rights Movement. While Eric Thompson was hard at work revising *Mayan Hieroglyphic Writing* for its third edition, the mestizo descendants of the authors of those hieroglyphic texts were being discriminated against in the U.S. housing market.

In 2015, former senator of Minnesota and vice president of the United States Walter Mondale lamented that his work of fifty years before had still not had the impact he hoped for. In 1968, he coauthored the legislation of the Fair Housing Act, which was intended as a fundamental contribution to civil rights in the United States. In 2015, Mondale reflected on the situation in a speech to the Department of Housing and Urban Development:

> When a black family with an income of $157,000 a year is less likely to qualify for a prime loan than a white family with an income of $40,000 a year, the goals of the Fair Housing Act are not fulfilled.
>
> When real estate agents only show integrated schools and suburbs to black and Latino middle-class families, and steer white families away from those same neighborhoods and schools, the goals of the Fair Housing Act are not fulfilled.
>
> When the federal and state governments will pay to build new suburban highways, streets, sewers, schools, and parks, but then allow these communities to exclude affordable housing and non-white citizens, the goals of the Fair Housing Act are not fulfilled. (Badger 2015)

Mondale's frustration stemmed from the obstacles confronted by social-justice efforts in the United States both before the Fair Housing Act and after. The problem in the United States was similar to that engaged by Diego Rivera and Jose Vasconcelos in early twentieth-century Mexico and Tawiskal Uwoojil in thirteenth-century Mayapán: how to effectively bring different communities together via governmental influence.

By the late 1960s, there had been prior failed attempts at Mondale's legislation, but what spurred Congress and President Lyndon B. Johnson to push this into law in 1968 was the calamitous assassination of Martin Luther King Jr. on April 4 of the same year. King had been aiming to maintain his discourse and practice of nonviolent political activism in Tennessee in response to the untenable work conditions of African Americans (Green 2007). On Wednesday, April 3, King delivered his now famous "I Have Been to the Mountaintop" speech that reached hopefully for the position that Mondale would have liked to be able to report on in 2015.

> And then I got to Memphis. And some began to say the threats . . . or talk about the threats that were out. What would happen to me from some of our sick white brothers? Well, I don't know what will happen now. We've got some difficult days ahead. But it doesn't matter with me now. Because I've been to the mountaintop. And I don't mind. Like anybody, I would like to live a long life. Longevity has its place. But I'm not concerned about that now. I just want to do God's will. And He's allowed me to go up to the mountain. And I've looked over. And I've *seen* the promised land. I may not get there with you. But I want you to know tonight, that we, as a people, will get to the promised land!

King returned to his hotel that evening, and the next day he was assassinated.

While King worked within communities and on a national stage, Mondale coauthored the legislation in Congress that was intended to contribute to King's vision by making housing accessible regardless of race, ethnicity, or religion. In the end, their negotiations resulted in a mechanism to motivate material change—it was not enough to appeal to the ethics of bankers, imploring them to provide mortgages to those they might otherwise discriminate against. In the politics and power dynamics of the mid-twentieth-century United States, the trade-off had to speak to the bankers' own interests.

The solution was a mathematical one. The Fair Housing Act created a new institution, the Government National Mortgage Association (GNMA or Ginnie Mae), which would provide mortgages that were otherwise considered "too risky." Financial risk was the rhetorical scapegoat for discrimination, so if that risk could be ameliorated, then so too might the discrimination—this was the origin of mortgage-backed securities. By packaging multiple mortgages together into a new financial instrument, the risk of any one default causing stress on a bank would be mitigated. Banks could continue providing mortgages to individuals and more of them, then sell qualifying ones to Ginnie Mae, which would bundle them and sell bonds to investors. When homeowners made their mortgage payments, the bank would get its due, the investment vehicle would take its cut, and then the bond holders would get their shares. If a homeowner defaulted, that cost would be buffered by the other mortgages still being paid with which it was bundled. Risk could now be minimized and third parties were now sharing the risk that otherwise the bank would have held alone—while making profit.

To a degree, through this scenario, the ethics of fair housing would be supplanted by a proxy of self-interest. Investors could be incentivized with more profit by lending to those they otherwise would not by moving their perceived risk into a financial vehicle that rendered it (at least partially) opaque. Of course what that also allowed for was a mortgage market tool that could be applied to other uses by those with the incentive— and those incentives didn't have to have anything to do with the ethics of fair housing or civil rights. Some might suggest that this is precisely Adam Smith's invisible hand: bankers will do what is best for themselves while also providing new opportunities for others who had been discriminated against. But a word of caution would notice that it didn't require bankers (or their affiliates or ideological associates) to actually deal with the ethics of the situation. They could continue to discriminate or maintain prejudices in all other realms of their lives and relationships. The ideologies supporting prejudice or racism (structural or personal) would go unchallenged and so continue to exist. The point is that this innovated financial instrument was not a simple neutral tool; it carried within it ethics meant to have cultural impact, as Mondale described to his audience at 2015, yet those ethics were replaced with financial incentive. Racism remained, but it would be provided a layer of opacity, buffering its recognition (cf. Taylor 1989).

As a result, the financial instrument of mortgage-backed securities had now been created, even though initially it was constrained for use only by Ginnie Mae. What changed and constituted the next critical step along with credit default swaps to the financial crisis of 2007 was that, in 1978, Lewis Ranieri found a way to innovate with it.

Ranieri was born in Brooklyn, New York, and attended St. John's University as an undergraduate student, but dropped out before completing his degree in 1968. As Martin Luther King Jr. was taking to the streets in Memphis, Ranieri took a job in the mailroom at Salomon Brothers, an investment firm on Wall Street. He was able to work his way out of the mailroom and into investing, by the late 1970s working on the development of a new investment vehicle: the "securitization" of mortgages or Collateralized Mortgage Obligations (CMOs) (McNamee 2004).

The idea of a CMO represented an innovation relative to mortgage-backed securities in that (1) the greater risk of an individual mortgage going into default could be mitigated by the others not going into default at the same time and (2) the CMO now provided payouts as would a more traditional bond, thus creating a new vehicle for an investment market. In retrospect, Ranieri and others claim that what motivated this securitization invention was that they were pioneering a new form of financing that would help baby boomers buy homes and take part in the American Dream (Segal 2017). In this sense, these and other retrospective accounts elide the relationship of CDOs to the Civil Rights Movement, instead appealing to the "race-neutral" American Dream of homeownership. It is just as easy to see, however, that homeownership suggested new markets for profit that Wall Street might be able to capitalize on.

The key here for Ranieri was to privatize what Ginnie Mae had created and sponsored. If the risk of individual mortgages could be mitigated by oversight from a government institution, couldn't another form of that risk management be governed by a market? If it just boiled down to risks and payouts, then theoretically a market itself should also be able to provide a venue with "oversight." Of course this market didn't already exist, so legislation would have to be written to regulate this new market. Not unexpectedly, Ranieri himself was recruited to write the legislation, and that became law in 1982 (Segal 2017). He was blunt in his description of the situation: "Legal investment rules did not exist. This meant we had to write our own rules. I went to the White House and explained what I was trying to do" (Segal 2017). Ranieri then closed his description with this profound understatement: "The security kicked off a lot of innovation. Some of it good" (Segal 2017).

It was then a very short step to go from recognizing that mortgages could be collateralized to recognizing that any debt could be collateralized and provide financial profit. Any type of debt that generated a regular payout could be combined with any others into a new bundled financial instrument that would mitigate the risk of any one individual debt. Yes, mortgages could be bundled into CMOs, but student loans, credit-card debt, and mortgages could all be thrown together into Collateralized Debt Obligations or CDOs. At this point, however, there was one substantial outstanding problem. How to account for the amount of relative risk between the default of one type of loan and another?

Posed this way, we find that what transformed two financial instruments invented to ameliorate ethical lapses (CMOs and CDSs) into Buffet's financial weapons was a mathematical equation.

As Felix Salmon, a blogger on financial markets, wrote it up for *Wired Magazine* in 2009, the Chinese statistician David Li, working in Canada, came up with a way to capture in a formula an otherwise ambiguously defined risk. Before Li's work, financial market analysts were unable to quantify the amount of risk an investor would encounter when assessing the correlation between two events. The problem had always existed; what changed in the late twentieth and early twenty-first centuries is that now there was financial reward accessible to anyone with its solution. Li's approach was to introduce sophisticated statistical mathematics and a critical simplifying assumption. For the formula, Li compared a few different mechanisms and settled on the Gaussian Copula (Li 2000). Li wrote that his article:

> introduces the basic concepts of copula functions into credit studies, and shows how to calibrate the parameter in the normal copula function . . . We have presented several numerical examples to illustrate the use of copula functions in the valuation of credit derivatives, such as credit default swaps and first-to-default contracts. (2000, 53)

Rather than require that analysts investigate the full scope of any given bundle of debt, understand its composition, and then determine its worthiness—as Masters had assumed—according to Li all that was needed was a formula. In addition, beyond the identification of a useful mathematical instrument, Li introduced a proxy. He writes: "In practice, we usually use market spread information to derive the distribution of survival times" (Li 2000, 53). In other words, rather than look to historical information to condition the possible correlation between default events, Li took the prices of credit default swaps as indicators of defaults themselves (Salmon 2009).

There were, of course, others along with Masters who grew concerned at the lack of expertise necessary to use this new financial tool. Terri Duhon described a conversation she had when raising the issue at J. P. Morgan and the lack of data behind their ability to make a decision on a $14 billion international CDO:

> The big problem for us was data. Our quant guy said look: you know, you, the trader. You tell me what you think the price is and I can give you a model that—that when you hit F9—that's the price that pops up (Wiser 2012, min. 30:00).

Neither historical data nor close assessment were necessary. "Quants" could use Li's formula to engineer prices.

The CDO and CDS markets were relatively small when Li published his work in 2000. But Li's use of the Gaussian Copula changed that dramatically. By quantitatively packaging default risk, not only could existing debts be analyzed, but new combinations of debt could be engineered for profit. Certainly CDSs were available for relatively secure checks between large institutions as Masters had intended, but they were also now part of a market open to anyone. Hedge funds in particular began paying premiums on debt they didn't even hold—debt bundled within Collateralized Debt Obligations in the hope that those CDOs would default. For if they did default, the hedge fund would receive the amount of the loan (Tardi 2011). Mortgages could be bundled as Ginnie Mae intended, but Ranieri's securitization approach allowed for mortgages to be bundled with any other debts that generated payouts, and they could be profitably combined using Li's formula. "CDOs were a relatively small segment of the ABS market with only $340 million outstanding issues in 2002 compared to the total CMO market of $4.7 trillion" (Schmidt 2020). Salmon writes:

> The CDS and CDO markets grew together, feeding on each other. At the end of 2001, there was $920 billion in credit default swaps outstanding. By the end of 2007, that number had skyrocketed to more than $62 trillion. The CDO market, which stood at $275 billion in 2000, grew to $4.7 trillion by 2006. (2009)

These numbers are astronomical and challenging to assess—not unlike the Serpent Numbers of the Dresden Codex—but there have been attempts to put them in perspective (figure 11.2). The amount of money invested just in CDSs

was more than the money invested in the U.S. stock market, mortgages, and U.S. Treasuries *combined*. The U.S. stock market held $22 trillion. Mortgages were worth $7.1 trillion and U.S. Treasuries, $4.4 trillion. In fact, it was almost as much as the economic output of the entire world in 2007, which was $65 trillion. (Amadeo 2019)

Here was Buffett's weapon of mass destruction. And while it is intriguing to focus on the numbers, what those numbers represented were huge impacts on the lives of millions of people in the United States and around the world.

In fact, Li's formula didn't accurately capture the risk it was intended to model. It was a numerical fiction. Families moved into houses they couldn't afford and then lost them to bankruptcy—and it all ended with the biggest banks being given money by the government that the banks didn't even want. The point is that a mathematical tool was invented to generate patterns in data, and individuals found ways to use those patterns to justify the modification of their behavior, resulting in large-scale social, cultural, and economic impact.

My interest in bringing this up is not to villainize David Li, or recent secretary of the U.S. Treasury Stephen Mnuchin, or the others who profited from the implementation of a faulty tool. Instead, I was intrigued by the apparent similarity of cases across time and cultures. Both Kan B'ahlam's court scribes at Palenque in the eighth century and twenty-first-century U.S. mortgage lenders had in their possession tools

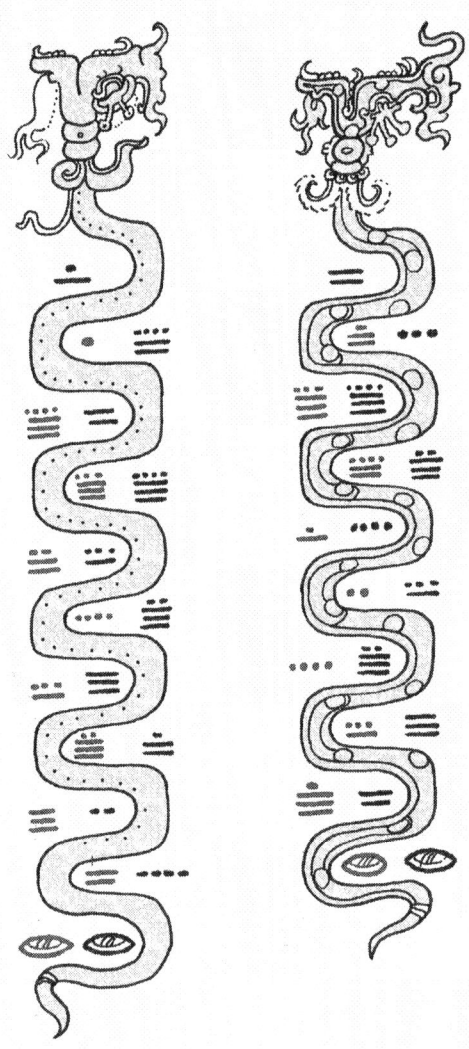

FIGURE 11.2 Serpent Number representations of CDS market-dollar values

that rendered previously intractable mathematical problems accessible. We saw in chapter 2 and again in chapter 5 that those at Palenque were able to link, quantitatively, members of the royal dynasty to mythological ancestors thousands of years in their past. In both cases, the mathematical problem required only a quantitative solution, but it was motivated by and contained within it the values of the sponsoring culture.

Moreover, when we consider the mythological events of interest to Kan B'ahlam's court, we recognize that they did not necessarily have physical correlates to which an accuracy of reconstruction could be measured (Aldana 2007, 118). When Kan B'ahlam was using astronumerology to place the birthdate of one of his patron deities, he may have had no other alternative than to rely simply on what we would consider today to be a pattern in randomness. Similarly, the ancient Ring Number base date of the Dresden Codex Venus Table didn't have to match anything that K'uk'ul Ek' Tuyilaj or her peers could contest with physical evidence. For dates purely within the realm of mythology, few, if any, physical constraints would have allowed for an accuracy check. As long as he and his advisors were convinced that the placement of the event was viable, Kan B'ahlam could take it as such. And importantly, behavioral and material changes would ensue. This mathematical innovation—whether or not it accurately reflected external "reality"—would have cultural and economic effects. So too in the early twenty-first century, "quants" could compute correlations between default risks that were fictitious, but as long as their colleagues were convinced that they were meaningful, they could be accepted as such.

And this is where we return to the problem of "cultural/epistemic translation." That is, we might be inclined—somewhat cynically—to write off Kan B'ahlam's work as "religion" and let the whole matter go as an example of how far modern society has progressed, since we no longer base our government on "astronumerology" or really anything like it. But when we consider its broader context, I am much less convinced that this is a viable stance to take. Specifically, it occurs to me that Kan B'ahlam was not interested in creating history out of mythology for the sake of pure intellectual curiosity. His was not a disinterested "basic science" endeavor. His esoteric scientific project was at the core of what we might characterize today as a political and economic intervention in his local kingdom (and quite possibly neighboring ones as well).

Recovering the birthdates (and other events) of his mythological ancestors generated new incentives to build temples, carve monuments, decorate them, and host feasts or festivals—in other words, these discoveries led Kan B'ahlam to provide the populace with new compelling collective experiences and at the same time, to put a large labor force to work. We also saw in chapter 2 that this project may have been motivated by an accelerating intellectual fashion in the Late Classic period to more ostentatiously integrate astronomical knowledge into public representations. The completion of these structures and their maintenance would have required feasting and festivals, which in turn meant even more labor (craft specialization) and more trade—that is, more "economic stimulus." We in the "modern world" may be inclined to dismiss Kan B'ahlam's "astrology" as primitive religion and based on "patterns in randomness" when considered in isolation, but we cannot dismiss the tangible and quantifiable impacts it had on the city. And it was at this level that it occurred to me that the historian of science Donald MacKenzie pointed to a similar issue when taking a science-studies approach to modern economics.

In *An Engine Not a Camera*, MacKenzie argues that the role of economics as a science has not necessarily been the pursuit of accurate representations of what it calls economic activity. The importance of some economic models is not in their accuracy, but in their ability to generate expectations that lead to changes in behavior, which are intended to have real material outcomes (MacKenzie 2008, 11). Economic modeling (in many cases) has been used to drive economic activity, as an engine, and not objectively represent it from the outside, as a camera would. And this is what intrigued me about the global financial meltdown of 2008. The credit-default-swaps economic modeling of the early twenty-first century seemed just as esoteric as Kan B'ahlam's 819-Day Count—and just as impactful culturally. Mortgage lenders were able to generate numbers that in turn would allow them to link the ability to purchase a home to larger market forces. In *neither* case would it be necessary for the quantification to be objectively accurate. As long as people accepted the link and acted accordingly, the link was validated. And this too brings us all the way back to the first chapter of this book and the interpretation of ancient Mayan astronomy.

RANDOMNESS

While troubleshooting the Calendar Correlation question at the Carnegie Institution in Washington, D.C., in the 1930s, Eric Thompson confronted an awkward situation within the work of his cross-Atlantic colleague Hans Ludendorff. An astrophysicist at the Hamburg Observatory, Ludendorff pooled together a set of dates from Classic period hieroglyphic inscriptions, then checked them against the positions of the visible planets in the night sky (Thompson 1935, 83). The reader will recall that this was the type of study that was common in the wake of Morley's recovery of numerous hieroglyphic inscriptions from the Peten, as we saw in chapter 1. Ludendorff's results showed that a significant number of notable observable astronomical events occurred on the dates generated by the Spinden correlation, and so he took this as evidence that Spinden's correlation was correct and as corroboration of Mayan interests in astronomy (Thompson 1935, 83).

Thompson's review of Ludendorff's work took issue with two aspects. First, he contested the Spinden correlation, as we have already seen. Second, he found that many of Ludendorff's dates actually had been reconstructed incorrectly (Thompson 1935, 83–84). Since the hieroglyphic inscriptions Ludendorff worked from were partially eroded, some parts of the text had to be computationally reconstructed based on the information preserved. Very often, enough information was recoverable to render the reconstructions secure. It was not uncommon at this time, however, that information would be insufficient, or that original readings had to be corrected based on more detailed study. Thompson's analysis first worked to correct those ambiguous or erroneous dates in Ludendorff's sample. He then checked the corrected dates, seeking confirmation of astronomical events within them. Thompson's results showed that there were more "relevant" astronomical patterns among the incorrectly reconstructed dates than among those that were accurately reconstructed (Thompson 1935, 84). The point of his reanalysis, then, was that with the *incorrect* dates, the data suggested that "the Maya" were planning their events (and records of them) according to celestially observable events, but with the *corrected* dates, no such pattern or suggestion results.

In reflecting on Thompson's review, it occurred to me that he had identified an example of "pattern in randomness." Specifically, one of the real challenges facing the interpretation of ancient astronomies—from nonacademic "2012 doomsday prophecies" to the most traditional scholarship on the Dresden Codex Venus Table—has been that presented by "patterns in randomness." As faced by historians of astronomy, the problem resembles Ramsey Theory within the field of combinatorial mathematics. The situation arises from the relationship between the amount of data available and the constraints imposed on it for a given interpretation. For a large-enough data set, with a relatively small number of constraints, verifiable patterns will emerge whether they are intentional/causal or not—these patterns emerge within randomness itself (Aldana 2011c). In terms of the history of astronomy, if we have a large enough set of historical dates, some of them will correspond to astronomical events even though originally they were not so planned.

Ludendorff generated a large data set that was essentially random relative to the patterns he found interesting—we know this because his dates were actually incorrect reconstructions of historical dates. Since the constraints he imposed were sufficiently few (looking for "meaningful celestial events"), patterns were generated simply by randomness, not because Mayan scribes intentionally built them into their hieroglyphic narratives.

The upshot is that it's not that Mayan scribes never included such intervals or records—at the very least, the Lunar Series would contradict such a proposal. The point is only that when we seek to interpret their texts, we need to be aware of false positives. There must be other types of information available to ensure that we aren't just seeing the mathematical correlate of animal figures in clouds. My suggestion at the end of this inquiry is that much of the twentieth century's academic interpretation of Mayan astronomy has been driven by pattern in randomness in ways that have been impactful—just as credit default swaps and astronumerology were impactful. This doesn't mean that "there's nothing there," but it does suggest that the Mayan Apocalypse of 2012 craze may be better understood as a popular culture version of the 2007 financial crisis—one spurred in part by academics themselves—rather than in any way representative of ancient Mayan intellectual traditions.

ABSTRACTING ECONOMICS

The comparison above has been intended to address one form of what we often refer to as "cultural translation." Recognizing that a group of technologists (financial or calendric) could use their innovations to have societal impact allows us to think of cultures (not just individuals) intersectionally. It allows us to recognize "otherness" without totalizing it into "Others." And that does help us to bring nuance and complexity to ancient peoples and their cultures—to appreciate the diversity within ancient Mesoamerican cultures, societies, communities, and perhaps even families.

Even that effort, however, still doesn't address those experiences within the "otherness" that defy parallel. Like the breathing-of-autonomous-air experience at the beginning of this chapter, some experiences may not have analogues that facilitate translation.

Rather than ignore this complication and attempt to reduce communities/societies into universal structures, which then might even further reduce those structures into economic mechanisms interacting with their environments, I suggest a turn back to the work of Pierre Bourdieu for inspiration. As I see it, Bourdieu actually provided a means of addressing the problems of "cultural translation" that render less valuable that effort to universalize and instead create opportunities for accessing intersectional othernesses.

Bourdieu captured part of the problem well in his introduction to "The Forms of Capital." Therein, he cautions against the theoretical or methodological reduction of a given social world "to a discontinuous series of instantaneous mechanical equilibria between agents who are treated as interchangeable particles" (1986, 241). While implementing such an approach might be productive (e.g., for grant-writing purposes), it relies on much of the disingenuousness of economic modeling. For Bourdieu, one way out is to "reintroduce into it the notion of capital and with it, accumulation and all its effects" (1986, 241). The latter point is what allows for further elaboration, which in turn presents the opportunity for intervention here at the end of this book.

Bourdieu takes us back to Karl Marx and even Adam Smith: "Capital is accumulated labor (in its materialized form or its 'incorporated,' embodied form) which when appropriated on a private, i.e. exclusive, basis by agents or groups of agents, enables them to appropriate social energy in the form of reified or living labor" (Bourdieu 1986, 241). Adam Smith reduced societies to systems seeking to satisfy needs and desires through commodities, which in turn allowed him to compare dissimilar commodities through their relative exchange values. The key—at least here—is that this move sets up the next step: all commodities can then be compared to each other through the intermediary of their comparison to human labor. The value of all commodities can be assigned an equivalent value in terms of human labor:

> Labour, therefore, it appears evidently, is the only universal, as well as the only accurate measure of value, or the only standard by which we can compare the values of different commodities at all times and at all places. (A. Smith 2000, 41)

At one level, this seems utterly reasonable and even prudent. How much closer can we come to a human universal than to reduce all costs to equivalent expenditures of human effort?

Next Smith needs some metric for human labor, and it is entirely arbitrary—it can be anything. We could, for instance, measure everything in terms of the number of ergs (units of energy) an average human could output in one day. We could just as easily measure human labor in terms of how many oranges an average person could eat in one day. Instead of either of these, Smith chose to correlate human labor to fiat money—cash (Smith 2000). Because it is arbitrary, on a theoretical level, it is purely a matter of convenience that we use fiat money as the metric. On a practical level, however, the choice is much more important. The abstraction of human labor to money—that is, economic liquid assets, or economic liquidity—means that societies can be reduced to transactions among liquid representations, abstractions of human labor, and commodities. We can talk accordingly about GDP as though it is a meaningful measure of a society. Moreover, accumulating capital amounts to accumulating some proxy for human labor along with the ability to make use of it. We can then consider money the liquid form of capital and so a direct correlate of socioeconomic power (A. Smith 2000, 56–57). Value, according to Smith, is defined in economic terms.

Bourdieu's intervention is not to challenge these views of ultimate value, but to note that there is much more that contributes to an individual's economic capital than simply the labor they have the capacity for combined with the economic assets they have available. Bourdieu recognizes that economic capital also can be impacted by what he classifies as two other forms of "capital": social and cultural (1986, 47). His intervention is to argue that the social should be taken into account when considering one's economic capital. This concept is relatively easy to recognize. Family ties, for example, have the potential to greatly impact the economic capital of an individual (1986, 51). Financial loans or gifts from relatives or members of a social network expand an individual's access to economic capital and liquidity. In the United States, the greatest source of an individual's wealth is inheritance from a parent (Hoffower 2019). The bigger point for Bourdieu, though, is that it is not enough to look at just an individual's explicitly financial assets; we also must take into account the social capital that an individual has to get a better sense of the economic opportunities they have and constraints they may be under.

Bourdieu's second intervention is to recognize that this still does not go far enough. Beyond social capital, there is also cultural capital. Bourdieu recognizes what he terms both embodied cultural capital and objectified cultural capital. The former are those "improvements" one makes to oneself—for example, in education (1986, 49). Certainly training can improve the scope of one's labor output and so increase one's economic capital. Bourdieu points out that there are two factors worthy of recognition. For one, he claims that education cannot be inherited; it must be gained on one's own. Second, the constraining factor is access. The whole family's approach to cultural capital is what provides an individual access or not—and the family one is born into, of course, is completely outside of one's control.

Bourdieu's insight is to note that cultural capital often is hidden. It is not necessarily obvious just by looking at someone that that individual has a college degree, for instance, or that they grew up vacationing with the economic elite. The conversion capability of cultural capital to economic capital may not be visible, but it may be an important factor nonetheless. It's relatively straightforward to recognize that the political aspect of Bourdieu's work here is to argue that, because it is not readily visible, it can easily function as a covert means of restricting access to power—or more subtly a covert privileging of one subculture's access to power.

A relatively recent real-world example illustrates the model. Consider the adoption of online instruction (OI) by institutions of "higher learning" and the role of online degrees. If we recognize only economic capital, then OI could be portrayed easily as a clear means of increasing the value of one's labor and so increasing their economic capital. If the greatest obstacle to higher education is cost, and if an OI degree can be obtained at a substantial reduction in cost relative to attendance at a brick-and-mortar campus, then it becomes possible for those with lower-valued labor to increase it where that wasn't possible before. Given that a bachelor's degree increases the value of one's labor, this increased access should theoretically lead to greater economic capital.

On the other hand, a substantial amount of the cultural capital that comes from the pursuit of higher education results from physical presence within an educational environment. By attending a college campus, students come into social contact with members of different groups and engage in practices that are common within the practices of communities with privilege. Students develop multicultural literacies in order to participate in classroom discussions, new living situations, and professional interactions with peers and supervisors. Very few (if any) of these experiences can be replicated through OI, and so virtually none of the cultural or social capital is accrued.

Bourdieu's intent, then, is to force the consideration of the hidden roles of cultural capital—to make visible the advantages/privileges that some subcultures have over others—unmasking the deception of purported meritocracies. Bourdieu is not saying that individuals outside of the power elite subculture do not have cultural capital or that their cultures do not have capital. Rather, the critical factor is that the capital within these underprivileged subcultures cannot be converted into economic capital (as it is currently defined).

A useful corollary then is that there are many forms of cultural capital, but only some forms—certain forms—can be converted into economic capital. Other forms of capital may be relevant to a given community, but if they cannot be converted into economic capital, they are rendered "valueless" *by modern society at large*. Humans can interact in numerous ways that are meaningful to themselves as individuals or as communities within a subculture, but if those activities cannot be translated into economic capital or an equivalent of human labor, then modern neoliberal society considers them valueless. At least this is the result of economic interpretations relying on final conversions to human labor and/or its monetary proxies.

Such an approach made good sense for Adam Smith as he sought to provide workers with agency over their labor against mercantilist economies, and for Karl Marx as he envisioned the trajectory emplaced by industrialization in the nineteenth century. It also—arguably—makes sense for attempts at both legitimizing and critiquing modern socioeconomic systems, although it critically leaves out considerations of race, gender, and ethnicity. Where the application of this formulation may be called into question, or at least should be called into question, is for noncapitalist, nonmodern societies and civilizations—such as those of Kan B'ahlam, K'uk'ul Ek' Tuyilaj, or Tawiskal Uwoojil. Although it is easy for us today to conceptualize human experiences as convertible from human labor represented by fiat money, it is not obvious that leaders within ancient Mesoamerican cultures/civilizations would have done the same or even would have wanted to. In other words, the problem as I see it is that Adam Smith's and Karl Marx's original formulations of capital-ism and its alternatives have methodologically straitjacketed our conceptualizations of "value"—at least when considering societal level analyses from academic perspectives.

Now, I readily recognize that the Smithian/Marxist approach to economic capital via proxies for human labor certainly has its advantages. It allows, for instance, interpretations of ancient civilizations according to what was done with human labor, and so provides ready interpretations of what types of institutional control would have been necessary to accomplish it. I contend, however, that invoking such models for ancient civilizations that ultimately rely on human labor as the ultimate universal metric leads us always already to replicate versions of the West. Every example will inevitably look like an underdeveloped or even failed attempt at creating the world that Europe and the United States has produced globally (cf. Chakrabarty 1992). Further, through this approach, ideology becomes reduced to a tool of the elite, in the end to manipulate human labor, which again is reinscribed as the ultimate final value. Ideology reifies ancient rulers as proto-capitalists, making exploitative use of religion as the opiate of the masses.

It might be useful for us to portray Kan B'ahlam as such a proto-capitalist, ultimately interested in the accumulation of economic capital, and so for us to focus on his exploits in terms of human labor (and its abstractions). We can, for example, look at the architectural Triad Group at Palenque dedicated to the patron deities of the dynasty as a demonstration of his ability to marshal physical labor and so consider it a measure of wealth and power—indeed, we have recognized above that the impetus provided by astronomy as elite fashion may have provided "economic stimulus." But to restrict it to that scope relies on the perspective that what was important to Kan B'ahlam and to members of his society—of Classic Mayan society—was this placement of value.

Contrastingly, we can see that Bourdieu's formulation of the problem actually makes available an abstraction of economics or of "capital-ism" itself. Currently, human labor and its proxies are useful when economic capital is considered hierarchically superior to both social and cultural capital. But we can also entertain the possibility that it provides simply one variant of the possibilities open to human communities. We also might conceptualize communities that find ultimate value in the liquidity of other forms of capital, such as social experiences or relationships from social capital, or cultural experiences from cultural capital, or even biological-psychological experiences from bio-psychological capital.

If instead we recognize that human economic labor is an important element of economic capital in any society/civilization, but that that society/civilization may have valued it differently relative to social or cultural liquidity, then we may find ourselves in a position to push the investigation of cultural and even epistemic difference. Classic Mayan hieroglyphic inscriptions are replete, for example, with statements of social interactions between the elite, as we saw in the introduction to this book. Ajaws travel to other cities in order to "witness" their colleagues performing period-ending ceremonies. On Stela 17 at Caracol, scribes carved two seated figures regarding each other. Above the scene, the text reads: 5 Ajaw 3 K'anasiy uchok ch'aj yichnal K'awiil Ajaw ch'ok—"On

5 Ajaw 3 K'ayab he scattered incense in the presence of K'awiil Ajaw ch'ok" (Martin and Grube 2008, 99). Some scholars have taken this and similar examples as evidence of the need for political support from a broad base of the nobility. Lesser nobles were included in the k'uhulajaw's ritual events to increase their support (Martin and Grube 2008, 99).

On the other hand, the same phrase is used on Aguateca Stela 1, with perhaps a different meaning. On this monument, artists carved an image of K'awiil Chan K'inich standing, with incense falling from his open right hand. The inscription running the length of the k'uhulajaw's body states: 3 Ajaw 3 Mol tanlamaw uchok ch'ah K'awiil Chan K'inich uchan Ahk Ajaw aj ka baak Mutul k'uhulajaw, yichnal [GI] K'awiil—"On 3 Ajaw 3 Mol it is the half-period end. K'awiil Chan K'inich scattered incense. He is the captor of Ahk Ajaw, he of the two captives, k'uhulajaw of Mutul. He did this in the presence of GI and K'awiil." In other words, the same event is recorded, but in the latter instance patron deities were the ones in whose presence the act was performed. Perhaps, then, it wasn't strictly political legitimation but something of the experience of being in another's presence that was of value in both cases.

We have seen other events—for example, the inscriptions concerning K'ahk' U Pakal K'awiil's activities at Chich'en Itza—occurring *yitah* or "with companions," which could be either mortal or "immortal" (Schele and Mathews 1999, 197). Additionally, many ajaws were bound to others as *yajaw*, including B'ahlaj Chan K'awiil at Dos Pilas in his relationship to Yuhknoom Ch'en of Calakmul (Martin and Grube 2008, 57). Now we could certainly assume that this fits the traditional model, reflecting a preference for economic capital, by assuming that these social connections facilitated access to economic capital. But that would be an assumption on our part. There is as much evidence for it as against it. We instead might suggest that such records illustrate that, in elite Classic Mayan society, value was placed in social relations—not because economic liquidity would result, but because the social relationships themselves were of ultimate value. The experience of witnessing or the experience of companionship with a revered colleague may have been the ultimate value, not the potential they generated for economic enrichment.

The point, then, is that we can hypothesize that different cultures may actually be characterized by different "final conversions" of forms of capital into lived experiences— not just final conversion into economic liquidity. What we refer to as modern "capitalist" societies, accordingly, are those that place ultimate value in final conversions to proxies for human labor, but that is not the only choice available, and it doesn't have to be considered universal.

A key result of this reconsideration, then, is that it allows us to define cultures by how they define their most valuable final conversion(s). A community could be defined, for example, by placing ultimate value on individual ceremonial experience. Cultural, social, and economic forms of capital still would exist in that community, but their relative values would be determined by their relative ability to provide/represent ceremonial experiences. In this type of community/culture, "money" would have value, but only at the level of capital. Money could be used, for instance, to sponsor a ceremony, but that still might not translate into a substantial rise in societal status, as the role played within the sponsored ritual might be of a low rank. The ceremonial experience in this case is the ultimate valued result, and both economic and social capital could be marshaled to provide liquidity within cultural and biological/psychological capital.*

On the one hand, this would seem to provide an opportunity to appreciate better Classic Mayan civilization. If we recognize and begin to excavate these relationships between religion and economic effect, we can better appreciate the complexity of the society they lived in. (Certainly it would be a welcome departure from, for example, depictions of mass hysteria at the sight of an eclipse.) On the other, it provides the possibility of reinvigorating the role of the humanities in archaeological and anthropological investigation.

For a given society, we might shift the ultimate value to a specific religious experience or set of religious experiences. The ultimate value might be a ritually facilitated interaction with one's ancestors—or community-level, state-sponsored ritual experiences. And this recognition turns our attention back to the interest in differentiating astrology from astronomy. It does not test the imagination to propose, for example, that science provides control over the environment and that control is convertible to political control. In fact, we saw in chapter 7 that Robert Sharer says as much. It is then a short step to assume that such political control in turn can be exploited to generate economic control. In this way, astrology becomes a tool for manipulating the population (cf. Aldana 2007, 193–98). On the other hand, if we question the motivation from political control for conversion to economic capital, then we open up different possibilities. In fact, a combination of conversions begins to develop as we consider the specific contexts of Kan B'ahlam, K'uk'ul Ek' Tuyilaj, and Tawiskal Uwoojil.

---

*Diego Durán canonized his confusion in the sixteenth century when he described his interaction with an indigenous man who was lounging in a park.

> Prehendiendo yo á un indio de ciertas cosas y en particular de que había andado arrastrado reco-
> giendo dineros con malas noches y peores días y al cabo de haber allegado tanto dinero y con tanto
> trabajo hace una boda y convida á todo el pueblo y gástalo todo y así riñendole el mal que había
> hecho me respondió Padre no te espantes pues todavía estamos nepantla y como entendiese lo que
> quería decir por aquel bocablo y metáfora que quiere decir estar en medio torné á insistir me dijese
> que medio era aquel en que estaban me dijo que como no estaban aun bien arraigados en la fé que
> no me espantase de manera que aun estaban neutros que ni bien acudían á la una ley ni á la otra
> ó por mejor decir que creían en Dios y que juntamente acudían á sus costumbres antiguas y ritos
> del demonio y esto quiso decir aquel en su abominable escusa de que aun permanecían en medio y
> estaban neutros. (Durán 1880, 268, orthography and punctuation preserved)

Scholars have used this text as definitive of a cultural mestizaje in the form of Nepantla—an in-between space. It strikes me as quite possible, though, that part of that Nepantla had to do with the ultimate values that each had and the inability to comprehend that of the other. This indigenous person used his own labor to generate the resources to patronize a set of religious and social experiences for his family and close community—for him, that may have been enough. He didn't have to strive for greater accumulation of economic wealth. Durán, though, saw a man capable of working hard to generate monetary income, which he then assumed should have been preserved to build economic wealth.

In these now historically developed uses of astronomy/astrology, we might begin by recognizing a hierarchy of ancestors, reaching all the way back to patron and Creator deities themselves. The religious activities required to appropriately honor and respect these ancestors relied on astronomical knowledge. The activities themselves in turn maintained a set of religious activities for the individual members of that lineage (the royal dynasty) as well as for the city population as a whole. In practice, the social capital of each participant generates its own liquidity, which may have been of primary ultimate value. Public recognition of one's lineage might be its own reward. Religious experiences from participating in public ceremonies witnessing Venus "come down from heaven" might provide ultimate value for others. Society in such case is not set up to provide different groups with differential access to economic capital and its liquidity. Instead, different combinations of social, cultural, and economic capital—assemblages of liquidity—become available for each member of society according to their place within it.

We can return here to the notion that we shouldn't endeavor to reduce K'uk'ulkan in the Late Postclassic to an individual political leader. K'uk'ulkan may have been an aesthetic of life for the members of the city of Mayapán. To have witnessed Chic Kaban at Mayapán, then, would have demonstrated an array of values. Lineages of higher and lower rank may have vied for specific interactions with other lineages. Families in different classes would have accessed experiences tailored accordingly within religious ceremonies. And this brings us back to the "title" we've ascribed to both Tawiskal Uwoojil and K'uk'ul Ek' Tuyilaj. As ajk'uhuuns, they were the ones who "divined" the interests of k'uh, or who made manifest k'uh. We may now propose that the way in which they manifested k'uh was to bring the human population into experiences and interactions with their environment. The way of life they intended to bring among their communities was their means of fulfilling their duties. K'uk'ul Ek' Tuyilaj facilitated a new presence of K'uk'ulkan at Chich'en Itza, and Tawiskal Uwoojil adapted that "deity" presence to Mayapán, centuries later.

Certainly their work could be reduced to quantified measures of human labor: hours put into building incense burners, travel necessary to access specific clays, feathers assembled into banners, temples built from stone. The latter is undeniably possible, but it keeps us in a process of always already reading our own culture, our own biases, into the lives of the past—and even into the lives of "Others" today.

To return to the vignette with which I began this chapter, I would like to propose the possibility that what for me was an undefinable experience may have come from the subjective experience of "being" within a community of different ultimate conversions. For the teachers in the Zapatista school, the specific interest was not in my advanced training in engineering mathematics and its potential to increase the economic capital of their students. Instead we may consider the possibility that their interest was in my knowledge of ancient Mayan math and calendrics for its cultural capital. Teachers in these autonomous communities are not valued for their potential to train the next great international economist; they are valued for their knowledge of the past—of *costumbre*—and how to maintain it within new and changing contexts. Mayan math then becomes a form of cultural capital—not because it can be converted into social and then economic capital, but because it can be directly converted into experiences for other community members, which enable connections with their historical identity. Here final liquidity may have been the conversion into a cultural experience.

Such a move might allow for countless reinterpretations of the archaeological and historical records. We can imagine, for example, the debates that may have taken place when K'uk'ul Ek' Tuyilaj explained her discovery to her peers at Chich'en Itza. How different might they look from responses to Galileo Galilei in seventeenth-century Italy? Might there have been a different form of political opposition? Likewise, we might imagine that, from Tawiskal Uwoojil's perspective, the U.S. stock market would be viewed as a type of oracle, capable of and useful for generating economic capital—but perhaps more valuable for the experiences of traders on the trading floor. Tawiskal Uwoojil might be perplexed, though, by the fact that the economic capital generated is not actually used to generate or even facilitate religious experiences for others.

In the end, such intersectional otherness viewed through an abstracted capital-ism allows us to reflect on the various relationships generated by scholarship itself. I cannot but reflect, accordingly, on the fact that to have put this book together, I have had to struggle with some of the same problems that Tawiskal Uwoojil and K'uk'ul Ek' Tuyilaj themselves did—by using many of the same tools and taking up perches to view the same horizons. In turn, I have been driven to work on these problems for and with my colleagues in the academy, but also for those interested in Mayan cultures and astronomies. Ultimately, I am interested too in the experiences this book provides to you as a reader, which undoubtedly must raise critique but also hopefully noneconomic reward.

(EN)CLOSING

For Förstemann, the Dresden Codex Venus Table was a puzzle of foreignness—riddles made up of symbols and obscure logic. There are certainly still mysteries in the writings of Tawiskal Uwoojil, but my hope is that this book might nudge in a different direction where we look for them and what they may mean. At Chich'en Itza and Mayapán, now, we have the opportunity to see how a religious/cultural capital used an astronomical mechanism to generate experiences for commoners and elites. Here cultural capital liquidity was the mechanism for generating the ritual experience. While it may well have been extremely powerful to witness the eight-year "coming down from heaven" of Venus, though, that may not have been enough to overcome the elite political tensions over the management of resources or determinations of authority. Just because a religious system can generate tremendously powerful experiences, that does not mean that they will necessarily be able to overcome societal or political strife. For a noncapitalist community or society (or a community aiming for autonomy within a capitalist society), autochthonous measures of value need first to be determined.

At the end of this project, I still wonder about the life of the author of the Dresden Codex Venus Table, but now I see him copying the words, illustrations, and numbers that K'uk'ul Ek' Tuyilaj first wrote down, both to honor her and to find new insights by moving that knowledge from Chich'en Itza to Mayapán. Rising early to wait for Chak Ek' to appear over the eastern horizon, breathing in the calm morning air, confirming the historical records maintained by his predecessors, and putting them to new use for the residents of Mayapán—I wonder if this experience was what gave him prized reward . . . as it does for me.

REFERENCES

Aldana, Gerardo. 2001. *Oracular Science: Uncertainty in the History of Mayan Astronomy, AD 500–1600*. PhD diss., Harvard University.

Aldana, Gerardo. 2002. "Solar Stelae and a Venus Window: Science and Royal Personality in Late Classic Copan." *Archaeoastronomy* 27 (JHA xxxiii): S29–S50.

Aldana, Gerardo. 2003. "K'uk'ulkan at Mayapan: Venus and Postclassic Maya Statecraft." *Journal for the History of Astronomy* 34:33–51.

Aldana, Gerardo. 2005. "Agency and the 'Star War' Glyph: A Historical Reassessment of Classic Maya Astrology and Warfare." *Ancient Mesoamerica* 16:305–20.

Aldana, Gerardo. 2006. "Lunar Alliances: Shedding Light on Conflicting Classic Maya Theories of Hegemony." In *Viewing the Sky Through Past and Present Cultures*, edited by T. Bostwick and B. Bates, 237–58. Phoenix, Ariz.: Pueblo Grande Museum Anthropological Papers 15.

Aldana, Gerardo. 2007. *The Apotheosis of Janaab' Pakal: Science and Religion at Late Classic Maya Palenque*. Boulder: University Press of Colorado.

Aldana, Gerardo. 2008. "Glyph G and the Yohualteuctin: Recovering the Mesoamerican Practice of Time Keeping and Nightly Astrology." *Archaeoastronomy* XXI (2007–2008): 59–75.

Aldana, Gerardo. 2009. "Essay Review: Aveni Honoured, Skywatching in the Ancient World: New Perspectives in Cultural Astronomy. Studies in Honor of Anthony F. Aveni." *Journal for the History of Astronomy* 40 (1): 109–13.

Aldana, Gerardo. 2011a. "On Deciphering Ancient Mesoamerican Foundational Texts: The challenges of a Non-Logos-Based Creation Narrative." In *Foundational Texts of World Literature*, edited by D. Jullien, 47–68. New York: Peter Lang.

Aldana, Gerardo. 2011b. "The Maya Calendar Correlation Problem." In *Calendars and Years II*, edited by J. Steele, 127–79. Oxford: Oxbow Books, 2011.

Aldana, Gerardo. 2011c. "Behind Astronomical Patterns." *Berfrois*, June 21, 2011. https://www.berfrois.com/2011/06/gerardo-aldana-behind-astronomical-patterns/.

Aldana, Gerardo. 2011d. *Tying Headbands or Venus Appearing: New Translations of /K'al/, the Dresden Codex Venus Pages, and Royal "Binding" Rituals*. B.A.R. International Series. Oxford: Archaeopress.

Aldana, Gerardo. 2015. "An Oracular Hypothesis: The Dresden Codex Venus Table and Cultural Translation of Science." In *Archaeoastronomy and the Maya*, edited by G. Aldana and E. Barnhart, 1–22. Oxford: Oxbow Books.

Aldana, Gerardo. 2016a. "Discovering Discovery: Chich'en Itza, the Dresden Codex Venus Table and 10th Century Mayan Astronomical Innovation." *Journal of Astronomy in Culture* 1 (1): 57–76.

Aldana, Gerardo. 2016b. "¹⁴C and Maya Long Count Dates: Using Bayesian Modelling to Develop Robust Site Chronologies." *Archaeometry* 58 (5): 863–80.

Aldana, Gerardo. 2017. "Classic Mayan Mestizaje." *rEvista: A Multi-media, Multi-genre e-Journal for Social Justice* 5 (2): 71–81.

Alexander, Rani T. 2017. "The Archaeology of Place in Ebtún, Yucatán, Mexico." In *Legacies of Space and Intangible Heritage: Archaeology, Ethnohistory, and the Politics of Cultural Continuity in the Americas*, edited by F. Armstrong-Fumero and J. Hoil Gutierrez, 131–61. Boulder: University Press of Colorado.

Amadeo, Kimberly. 2019. "Credit Default Swaps with Their Pros, Cons, and Examples: How a Boring Insurance Contract Almost Destroyed the Global Economy." *TheBalance.com*, January 11, 2019. https://www.thebalance.com/credit-default-swaps-pros-cons-crises-examples-3305920.

Andrews, E. Wyllys. 1938. "Glyphs Z and Y of the Maya Supplementary Series." *American Antiquity* 4, no. 1 (July): 30–35.

Aoyama, Kazuo. 2008. "Preclassic and Classic Maya Obsidian Exchange, Artistic, and Craft Production, and Weapons in the Aguateca Region and Seibal, Guatemala." *Mexicon* 30, no. 4 (August): 78–86.

Ashmore, Wendy. 1989. "Construction and Cosmology: Politics and Ideology in Lowland Maya Settlement Patterns." In *Word and Image in Maya Culture: Exploration in Language, Writing, and Representation*, edited by W. Hanks and D. Rice, 272–86. Salt Lake City: University of Utah Press.

Aveni, Anthony. 1977. "Concepts of Positional Astronomy Employed in Ancient Mesoamerican Architecture." In *Native American Astronomy*, edited by A. Aveni, 3–19. Austin: University of Texas Press.

Aveni, Anthony F. 1980. *Skywatchers of Ancient Mexico*. Austin: University of Texas Press.

Aveni, Anthony. 1992. "The Moon and the Venus Table: An Example of Commensuration in the Maya Calendar." In *The Sky in Maya Literature*, edited by A. Aveni, 87–101. New York: Oxford University Press.

Aveni, Anthony. 1999. "Astronomy in the Mexican Codex Borgia." *Archaeoastronomy Supplement to the Journal for the History of Astronomy* 24: S1–S20.

Aveni, Anthony. 2001. *Skywatchers*. Austin: University of Texas Press.

Aveni, Anthony. 2003. "Archaeoastronomy in the Ancient Americas." *Journal of Archaeological Research* 11, no. 2 (June): 149–91.

Aveni, Anthony F., Edward E. Calnek, and Horst Hartung. 1988. "Myth, Environment, and the Orientation of the Templo Mayor of Tenochtitlan." *American Antiquity* 53, no. 2 (April): 287–309.

Aveni, Anthony, Sharon Gibbs, and Horst Hartung. 1975. "The Caracol Tower at Chichen Itza: An Ancient Astronomical Observatory?" *Science* 188:977–85.

Aveni, Anthony, and Horst Hartung. 1986. "Maya City Planning and the Calendar." *Transactions of the American Philosophical Society* 76 (7): 1–87.

Aveni, Anthony F., Susan Milbrath, and Carlos Peraza Lope. 2004. "Chichén Itzá's Legacy in the Astronomically Oriented Architecture of Mayapán." *RES: Anthropology and Aesthetics* 45 (Spring): 123–43.

Badger, Emily. 2015. "He Wrote the Law to End Housing Discrimination. Fifty Years Later, He's Still Fighting." *Washington Post*, September 1, 2015. https://www.washingtonpost.com/news /wonk/wp/2015/09/01/he-wrote-the-law-to-end-housing-discrimination-fifty-years-later-hes -still-fighting/.

Bancroft, Hubert Howe. 1883. *History of Mexico*. Volume II. New York: The Bancroft Company.

Bardacke, Frank. 1995. *Shadows of Tender Fury: The Letters and Communiqués of Subcomandante Marcos and the Zapatista Army of National Liberation*. New York: Monthly Review Press.

Barrera Vásquez, Alfredo. 1995. *Diccionario Maya Cordemex*. Merida: Ediciones Cordemex.

Bassie-Sweet, Karen. 2008. *Maya Sacred Geography and the Creator Deities*. Norman: University of Oklahoma Press.

Baudez, Claude-François, and Nicolas Latsanopoulos. 2010. "Political Structure, Military Training, and Ideology at Chichen Itza." *Ancient Mesoamerica* 21, no. 1 (Spring): 1–20.

Berdan, Frances F., and Patricia Anawalt. 1997. *The Essential Codex Mendoza*. Berkeley: University of California Press.

Biagioli, Mario. 1993. *Galileo, Courtier: The Practice of Science in the Culture of Absolutism*. Chicago: University of Chicago Press.

Bielenstein, Hans. 1950. "An Interpretation of the Portents in the Ts'ien-Han Shu." *Bulletin of the Museum of Far Eastern Antiquities* 22:127–43.

Bielenstein, Hans. 1984. "Han Portents and Prognostications." *Museum of Far Eastern Antiquities Bulletin* 56:97–112.

Bierhorst, John. 1998. *History and Mythology of the Aztecs: The Codex Chimalpopoca*. Tucson: University of Arizona Press.

Bíró, Péter, and Eduardo Pérez de Heredia. 2016. "The Caracol Disk of Chichen Itzá (929–932 CE): Some Thoughts on Epigraphy and Iconography." *Estudios de Cultura Maya* 48: 129–62.

Blackburn, Thomas. 1975. *December's Child: A Book of Chumash Oral Narratives*. Berkeley: University of California Press.

Boone, Elizabeth. 2010. *Cycles of Time and Meaning in the Mexican Books of Fate*. Tucson: University of Arizona Press.

Boot, Erik. 2002. "The Dos Pilas-Tikal Wars from the Perspective of Dos Pilas Hieroglyphic Stairway 4." *Mesoweb*, accessed July 19, 2021. www.mesoweb.com/features/boot/DPLHS4.pdf.

Bourdieu, Pierre. 1986. "The Forms of Capital." In *Handbook of Theory and Research for the Sociology of Education*, edited by J. Richardson, 241–58. New York: Greenwood.

Brady, James. 1997. "Settlement Configuration and Cosmology: The Role of Caves at Dos Pilas." *American Anthropologist* New Series 99, no. 3 (September): 602–18.

Brand, Charles. 1945. "Some Fertilizer History Connected with World War I." *Agricultural History* 19 (2): 104–13.

Bricker, Victoria. 2010. "A Comparison of Venus Instruments in the Borgia and Madrid Codices." In *Astronomers, Scribes, and Priests: Intellectual Interchange between the Northern Maya Lowlands and Highland Mexico in the Late Postclassic Period*, edited by G. Vail and C. Hernandez, 309–32. Washington, D.C.: Dumbarton Oaks.

Bricker, Harvey, and Victoria Bricker. 1986a. "Astronomical Implications of an Agricultural Almanac in the Dresden Codex," *Mexicon* 8:29–35.

Bricker, Harvey, and Victoria Bricker. 1986b. "The Mars Table in the Dresden Codex." *Middle American Research Institute Publication* 57:51–80.

Bricker, Harvey, and Victoria Bricker. 2007. "When Was the Dresden Codex Venus Table Efficacious?" In *Skywatching in the Ancient World: New Perspectives in Cultural Astronomy*, edited by C. Ruggles and G. Urton, 95–119. Boulder: University Press of Colorado.

Bricker, Harvey, and Victoria Bricker. 2011. *Astronomy in the Maya Codices*. Philadelphia: American Philosophical Society.

Browman, David. 2011. "Spying by American Archaeologists in World War I (with a Minor Linkage to the Development of the Society for American Archaeology)." *Bulletin of the History of Archaeology* 21 (2): 10–17.

Byers, Douglas, 1966. "Frans Blom, 1893–1963." *American Antiquity* 31.3, no. 1 (January): 406–7.

Byland, Bruce. 1993. "Introduction and Commentary." In *The Codex Borgia: A Full-Color Restoration of the Ancient Mexican Manuscript*, edited by G. Diaz and A. Rodgers, xiii–xxxii. New York: Dover Publications.

Byland, Bruce E., and John M. D. Pohl. 1994. *In the Realm of 8 Deer: The Archaeology of the Mixtec Codices*. Norman: University of Oklahoma Press.

Carrasco, David. 2001. *Quetzalcoatl and the Irony of Empire: Myths and Prophecies in the Aztec Tradition*. Boulder: University Press of Colorado.

Carrasco Vargas, Ramón, and María Cordeiro Baqueiro. 2012. "The Murals of Chiik Nahb Structure Sub1–4, Calakmul, Mexico." In *Maya Archaeology 2*, edited by Charles Golden, Stephen Houston, and Joel Skidmore, 8–59. San Francisco: Precolumbia Mesoweb Press.

Carlsen, Robert S. 1997. *The War for the Heart and Soul of a Highland Maya Town*. Austin: University of Texas Press.

Carlsen, Robert S., and Martin Prechtel. 1991. "The Flowering of the Dead: An Interpretation of Highland Maya Culture." *Man, New Series* 26, no. 1 (March): 23–42.

Carlson, John, and Linda Landis. 1985. "Bands, Bicephalic Dragons, and Other Beasts: The Skyband in Maya Art and Iconography." In *Fourth Palenque Round Table, 1980*, edited by M. G. Robertson and E. Benson, 115–40. San Francisco: Pre-Columbian Art Research Institute.

Caso, Alfonso. 1967. *Los Calendarios Prehispánicos*. México: Universidad Nacional Autónoma de México.

Chafe, Wallace. 2000. "Floyd Glenn Lounsbury, 25 April 1914–14 May 1998." *Proceedings of the American Philosophical Society* 144, no. 2 (June): 225–29.

Chakrabarty, Dipesh. 1992. "Provincializing Europe: Postcoloniality and the Critique of History." *Cultural Studies* 6 (3): 337–57. doi: 10.1080/09502389200490221.

Chambers, Ian. 2010. "The Movement of Great Tellico: The Role of Town and Clan in Cherokee Spatial Understanding." *Native South* 3:89–102.

Chase, Arlen F., and Diane Z. Chase. 2001. "Ancient Maya Causeways and Site Organization at Caracol, Belize." *Ancient Mesoamerica* 12 (2): 273–81.

Christenson, Allen. 2007. *The Popol Vuh: The Sacred Book of the Maya*. Norman: Oklahoma Press.

Clark, John. 2005. "Archaeological Trends and Book of Mormon Origins." *Brigham Young University Studies* 44 (4): 83–104.

Clendinnen, Inga. 1989. *Ambivalent Conquests: Maya and Spaniard in Yucatan, 1517–1570*. Cambridge: Cambridge University Press.

Cline, Howard. 1973. "The Chronology of the Conquest: Synchronologies in Codex Telleriano-Remensis and Sahagun." *Journal de la Societe des Americanistes* 62:9–34.

Closs, Michael. 1977. "The Date-Reaching Mechanism in the Venus Table of the Dresden Codex." In *Native American Astronomy*, edited by A. Aveni, 89–99. Austin: University of Texas Press.

Closs, Michael. 1978. "Venus in the Maya World: Glyphs, Gods and Associated Astronomical Phenomena." In *Tercera Mesa Redonda de Palenque, Vol. IV*, edited by M. G. Robertson, 158–63. Palenque: Pre-Columbian Art Research Center.

Closs, Michael. 1981. "Venus Dates Revisite." *Archaeoastronomy: The Bulletin of the Center for Archaeoastronomy* 4 (4): 38–41.

Closs, Michael. 1994. "A Glyph for Venus as Evening Star." In *Seventh Palenque Round Table, 1989*, edited by M. G. Robertson, 229–36. San Francisco: Pre-Columbian Art Research Institute.

Closs, Michael, Anthony Aveni, and Bruce Crowley. 1984. "The Planet Venus and Temple 22 at Copan." *Indiana* 9:221–45.

Cobos, Rafael, and Terance L. Winemiller. 2001. "The Late and Terminal Classic-Period Causeway Systems of Chichen Itza, Yucatan, Mexico." *Ancient Mesoamerica* 12:283–91.

Codex Telleriano-Remensis. 1901. Loubat Edition. Online resource: https://commons.wikimedia .org/wiki/Category:Codex_Telleriano-Remensis_(Loubat_edition).

Coe, Michael. 1965. "A Model of Ancient Community Structure in the Maya Lowlands." *Southwestern Journal of Anthropology* 21 (2): 97–114.

Coe, Michael D. 1977. "Foreword." In *Native American Astronomy*, edited by Anthony Aveni, ix. Austin: University of Texas Press.

Coe, Michael. 1999. *Breaking the Maya Code.* New York: Thames and Hudson.

Coe, Michael D. 2005. *The Maya.* New York: Thames and Hudson.

Coe, Michael, and Justin Kerr. 1998. *The Art of the Maya Scribe.* New York: Harry N. Abrams.

Coe, William, and Edwin Shook. 1986. "Tikal Report No. 6. The Carved Wooden Lintels of Tikal." In *Tikal Reports, Numbers 1–11*, edited by E. Shook, W. Coe, and R. Carr, 15–112. Philadelphia: University of Pennsylvania Press.

Collier, George. 1994. *Basta! Landa and the Zapatista Rebellion in Chiapas.* Washington D.C.: Institute for Food and Development Policy.

Collins, Anne. 1977. "The *Maestros Cantores* in Yucatan." In *Anthropology and History in Yucatan*, edited by G. D. Jones, 223–47. Austin: University of Texas Press.

Consolmagno, Guy J. 1966. "Astronomy, Science Fiction and Popular Culture: 1277 to 2001 (and Beyond)." *Leonardo* 29 (2): 127–32.

Cornell Alumni News. 1926. "Chemists Honor Cornellian: John E. Teeple '99 to Receive Perkin Medal for Development of American Potash Industry." *Cornell Alumni News* 29, no. 5 (October 28): 58.

Cuevas Garcia, Martha. 2004. "The Cult of Patron and Ancestor Gods in Censers at Palenque." In *Courtly Art of the Ancient Maya*, edited by M. Miller and S. Martin, 253–55. New York: Thames and Hudson.

Desmond, Lawrence. 2001. "Chacmool." In *Oxford Encyclopedia of Mesoamerican Cultures*, edited by David Carrasco, 168–69. New York: Oxford University Press.

Diaz, Gisele, and Alan Rodgers. 1993. *The Codex Borgia: A Full-Color Restoration of the Ancient Mexican Manuscript.* New York: Dover Publications.

Díaz-García, Salvador. 2006. *Monografías de Arquitectos Siglo XX, Horst G. Hartung Franz.* Jalisco: Secretaría de Cultura del Gobierno de Jalisco.

Drake, Stillman. 1957. *Discoveries and Opinions of Galileo.* Garden City, N.Y.: Doubleday.

Dütting, Dieter. 1985. "Lunar Periods and the Quest for Rebirth in the Mayan Hieroglyphic Inscriptions." *Estudios de Cultura Maya* 16:113–47.

Dütting, Dieter, Anthony Aveni, and M. Schramm. 1982. "The 2 Cib 14 Mol Event in the Inscriptions of Palenque, Chiapas, Mexico," *Zeitschrift fur Ethnologie* cvii/2: 233–58.

Dütting, Dieter, and Mathias Schramm. 1985. "On the Astronomical Background of Mayan Historical Events." In *Fifth Palenque Round Table, 1983*, edited by Virginia M. Fields, 261–74. San Francisco: Pre-Columbian Art Research Institute.

Eddy, John, 1977. "Review of Archaeoastronomy in Pre-Columbian America." *American Anthropologist* 79:497–98.

Fahsen, Federico. 2002. "Rescuing the Origins of Dos Pilas Dynasty: A Salvage of Hieroglyphic Stairway #2, Structure L5-49." FAMSI.

Farriss, Nancy. 1984. *Maya Society Under Colonial Rule: The Collective Enterprise of Survival.* Princeton, NJ: Princeton University Press.

Fash, Barbara, William Fash, Sheree Lane, Rudy Larios, Linda Schele, Jeffrey Stomper, and David Stuart. 1992. "Investigations of a Classic Maya Council House at Copan, Honduras." *Journal of Field Archaeology* 19:419–42.

Fash, William. 1991. *Scribes, Warriors and Kings: The City of Copan and the Ancient Maya.* New York: Thames and Hudson.

Fash, William, and Barbara Fash. 1996. "Building a World-View: Visual Communication in Classic Maya Architecture." *RES: Anthropology and Aesthetics* 29 (30): 127–47.

Fergusson, G., and Willard Libby. 1963. "UCLA Radiocarbon Dates II." *Radiocarbon* 5:1–22.

Fernie, J. Donald. 1990. "Marginalia: Stonehenge and the Archaeoastronomers." *American Scientist* 78, no. 2 (March–April): 103–5.

Förstemann, Ernst. 1891. "Explanation of the Maya Manuscript of the Royal Library of Dresden." Translated by N. Thomas. *Bureau of Ethnology* 2022:81–124.

Förstemann, Ernst. 1894. "Page 24 of the Dresden Maya Manuscript." *Zur Entzifferung der Maya-handschriften* 4:431–43. Dresden.

Förstemann, Ernst. 1904. "The Maya Glyphs." Bulletin 28 of the Bureau of American Ethnology 80:393–589, 647–50. In *Mexican and Central American Antiquities, Calendar Systems, and History: Twenty-Four Papers by Eduard Seler, Ernst Wilhelm Förstemann, Paul Schellhas, Charles Pickering Bowditch, Karl Sapper, Erwin Paul Dieseldorff*, 501–13. Washington, D.C.: U.S. Government Printing Office.

Förstemann, Ernst. 1906. "Commentary on the Maya Manuscript in the Royal Public Library of Dresden." In *Peabody Museum of American Archaeology and Ethnology Papers* 4. Translated by S. Wesselhoft and A. M. Parker.

Freidel, David, Linda Schele, and Joy Parker. 1993. *Maya Cosmos: Three Thousand Years on the Shaman's Path.* New York: William Morrow.

Frye, Alton. 1966. "Politics—The First Dimension of Space, a Review of Pride and Power: The Rationale of the Space Program by Vernon Van Dyke; Space and Society by Howard J. Taubenfeld; The Moon-Doggle: Domestic and International Implications of the Space Race by Amitai Etzioni." *Journal of Conflict Resolution* 10, no. 1 (March): 103–12.

Fuls, Andreas. 2008. "Reanalysis of Dating the Classic Maya Culture." *Ameridian Research* 3/3 (9): 132–46.

Gest, Howard. 2004. "Samuel Ruben's Contributions to Research on Photosynthesis and Bacterial Metabolism with Radioactive Carbon." *Photosynthesis Research* 80:77–83.

Gillespie, Susan D., and Rosemary A. Joyce. 1998. "Deity Relationships in Mesoamerican Cosmologies: The Case of the Maya God L." *Ancient Mesoamerica* 9, no. 2 (Fall): 279–96.

Goodman, Joseph. (1905) 1974. "Appendix to A. P. Maudslay." *Archaeology, Vol. VI.* New York: Milpatron.

Gordon, George B. 1913. *The Book of Chilam Balam of Chumayel by Juan José Hoil* (facsimile). Philadelphia: Philadelphia University Museum.

Graham, Ian. 2002. *Alfred Maudslay and the Maya: A Biography.* London: The British Museum Press.

Green, Laurie. 2007. *Battling the Plantation Mentality: Memphis and the Black Freedom Struggle.* Chapel Hill: University of North Carolina Press.

Grube, Nikolai. 2006. *Maya: Divine Kings of the Rain Forest.* New York: Konemann.

Grube, Nikolai, and Ruth J. Krochock. 2007. "Reading Between the Lines: Hieroglyphic Texts from Chichen Itza and Its Neighbors." In *Twin Tollans: Chichén Itzá, Tula and the Epiclassic to Early Postclassic Mesoamerican World*, edited by Jeff Karl Kowalski and Cynthia Kristan-Graham, 205–49. Washington, D.C.: Dumbarton Oaks Research Library and Collection.

Guenter, Stanley Paul. 2003. "The Inscriptions of Dos Pilas Associated with B'ajlaj Chan K'awiil." *Mesoweb*, accessed July 29, 2021. www.mesoweb.com/features/guenter/DosPilas.html.

Hamilton, Earl J. 1938. "Revisions in Economic History: VIII.-The Decline of Spain." *The Economic History Review* 8, no. 2 (May): 168–79.

Hammond, Norman. 1977. "Sir Eric Thompson, 1898–1975." *American Antiquity* 42, no. 2 (April): 180–90.

Hanks, William F. 1987. "Discourse Genres in a Theory of Practice." *American Ethnologist* 14, no. 4 (November): 668–92.

Haramundanis, Katherine. 1973. "Mayan Astroarchaeology: Review of Die Zeremonialzentren der Maya." *Journal for the History of Astronomy* 4:201–2.

Harrison, Peter D. 1999. *The Lords of Tikal: Rulers of an Ancient Maya City*. London: Thames and Hudson.

Hartung, Horst. 1975. "A Scheme of Probable Astronomical Projections in Mesoamerican Architecture." In *Archaeoastronomy in Pre-Columbian America*, edited by Anthony Aveni, 191–204. Austin: University of Texas Press.

Hawkins, Gerald S. 1965. *Stonehenge Decoded*. Garden City, N.Y.: Doubleday.

Headrick, Annabeth. 1999. "The Street of the Dead . . . It Really Was: Mortuary Bundles at Teotihuacan." *Ancient Mesoamerica* 10, no. 1 (Spring): 69–85.

Herschel, John. 1861. *Outlines of Astronomy*. Philadelphia: Blanchard and Lea.

Highway, Tomson. 2002. *Comparing Mythologies*. Ottawa: University of Ottawa Press.

Hoffower, Hillary. 2019. "US Inequality Is Only Getting Worse, and the 'Dynastic Wealth' Bemoaned by Warren Buffett May Be One of the Reasons Why." *Business Insider*, February 23, 2019. https://www.businessinsider.com/is-income-inequality-caused-by-inheritance-generational-wealth-2019-2.

Hoggarth, Julie, Matthew Restall, James W. Wood, and Douglas J. Kennett. 2017. "Drought and Its Demographic Effects in the Maya Lowlands." *Current Anthropology* 58, no. 1 (February): 82–99.

Houston, Stephen. 1993. *Hieroglyphs and History at Dos Pilas: Dynastic Politics of the Classic Maya*. Austin: University of Texas Press.

Hull, Kerry. 2005. "An Abbreviated Dictionary of Ch'orti' Maya." Final Report for the Foundation for the Advancement of Mesoamerican Studies, Inc. (FAMSI), grant no. 03031.

Humboldt, Alexander von, and Helen Maria Williams. 1814. *Researches concerning the Institutions and Monuments of the Ancient Inhabitants of America: With Descriptions and Views of Some of the Most Striking Scenes in the Cordilleras!* Vol. 2. London: Longman, Hurst, Rees, Orme and Brown. Available at J. Murray and H. Colburn, *World Scholar: Latin America and the Caribbean*, accessed August 7, 2018. http://worldscholar.tu.galegroup.com.proxy.library.ucsb.edu:2048/tinyurl/6qYZg5.

Humboldt, Alexander von, and Aimé Bonpland. 1815. *Personal Narrative of Travels to the Equinoctial Regions of the New Continent during the Years 1799–1804 by Alexander de Humboldt and Aimé Bonpland*. Available at M. Carey, *World Scholar: Latin America and the Caribbean*, accessed 7 August 7, 2018. http://worldscholar.tu.galegroup.com.proxy.library.ucsb.edu:2048/tinyurl/6qYr32.

Humboldt, Alexander von, Vera Kutzinski, Ottmar Ette, Ryan Poynter, Giorleny Altamirano Rayo, Tobias Kraft. 2012. *Views of the Cordilleras and Monuments of the Indigenous Peoples of the Americas: A Critical Edition*. Chicago: University of Chicago Press.

Jackson, Sarah. 2011. "Continuity and Change in Early Colonial Maya Community Governance: A Lexical Perspective." *Ethnohistory* 58 (4): 683–726.

Jansen, Maarten, and Gabina Aurora Perez Jimenez. 2005. *Codex Bodley: A Painted Chronicle from the Mixtec Highlands, Mexico*. Bodleian Library, Oxford.

Johnson, Frederick. 1967. "Radiocarbon Dating and Archeology in North America." *Science* 155, no. 3759 (January 13): 165–69.

Johnston, Harold. 2003. *A Bridge Not Attacked: Chemical Warfare Civilian Research during World War II*. London: World Scientific.

Jones, Grant D. 1998. *The Conquest of the Last Maya Kingdom*. Stanford, Calif.: Stanford University Press.

Justeson, John S. 1988. "Non-Maya Calendars of Southern Veracruz-Tabasco and the Antiquity of the Civil and Agricultural Years." *Journal of Mayan Linguistics* 6:1–21.

Justeson, John, and David Tavárez. 2007. "The Correlation between the Colonial Northern Zapotec and Gregorian Calendars." In *Skywatching in the Ancient World: New Perspectives in Cultural Astronomy*, edited by C. Ruggles and G. Urton, 17–82. Boulder: University Press of Colorado.

Kamen, Martin. 1985. *Radiant Science, Dark Politics: A Memoir of the Nuclear Age*. Berkeley: University of California.

Kaufman, Terrence. 2003. *A Preliminary Mayan Etymological Dictionary*. Maya Hieroglyphic Workshops at the University of Texas, Austin.

Kelley, David. 1977. "Maya Astronomical Tables and Inscriptions." In *Native American Astronomy*, edited by A. F. Aveni, 57–74. Austin: University of Texas Press.

Kelley, David H. 1983. "The Maya Calendar Correlation Problem." In *Civilization in the Ancient Americas: Essays in Honor of Gordon R. Willey*, edited by R. M. Leventhal and A. L. Kolata, 157–208. Albuquerque: University of New Mexico Press.

Kennett, D. J., I. Hajdas, B. J. Culleton, S. Belmecheri, S. Martin, H. Neff, J. Awe, H. V. Graham, K. H. Freeman, and L. Newsom. 2013. "Correlating the Ancient Maya and Modern European Calendars with High-Precision AMS 14C Dating." *Scientific Reports* 3:1–5.

King, David. 1993. *Astronomy in the Service of Islam*. Aldershot: Varorium.

Kuethe, Allan, and Lowell Blaisdell. 1991. "French Influence and the Origins of the Bourbon Colonial Reorganization." *The Hispanic American Historical Review* 71, no. 3 (August): 579–607.

Kulp, J. Laurence. 1996. Oral History with J. Laurence Kulp by Ron Doel, Federal Way, Washington, April 11, 1996. American Institute of Physics, accessed June 5, 2021. https://www.aip.org/history-programs/niels-bohr-library/oral-histories/6932-1.

Kulp, J. Laurence, Herbert W. Feely, and Lansing E. Tryon. 1951. "Lamont Natural Radiocarbon Measurements, I." *Science* 114, no. 2970 (November 30).

Lacadena, Alfonso. 2004. "Passive Voice in Classic Mayan Texts: CV-*h*-C-*aj* and -*n*-*aj* Constructions." In *Linguistics of Maya Writing*, edited by S. Wichmann, 165–94. Salt Lake City: University of Utah Press.

La Farge, Oliver. 1947. *Santa Eulalia: The Religion of a Chuchumata'n Indian Town*. Chicago: University of Chicago Press.

Lanchester, John. 2009. "Outsmarted: High Finance vs. Human Nature." *New Yorker*, June 1, 2009. https://www.newyorker.com/magazine/2009/06/01/outsmarted.

Latour, Bruno. 1987. *Science in Action*. Cambridge, MA: Harvard University Press.

Lawrence, Ernest O., and M. Stanley Livingston. 1932. "The Production of High Speed Light Ions without the Use of High Voltages." *Physical Review* 40 (19): 19–35.

Léon y Gama, Antonio. 1832. *Descripción histórica y cronológica de las dos piedras, que conocasión de nuevo empedrado que se está formando en la plaza principal de México, se hallaron en ella el año de 1790. . .* , 2nd ed. Mexico City: Impresora del Ciudadano A. Valdés.

León-Portilla, Miguel. 1988. *Time and Reality in the Thought of the Maya*. Norman: University of Oklahoma Press.

Li, David X. 2000. "On Default Correlation: A Copula Function Approach." *Journal of Fixed Income* 9 (4): 43–54.

Libby, Willard F. 1967. "History of Radiocarbon Dating." Geochemistry and Cosmochemistry of Radiocarbon (Sessions I and II). Radioactive Dating and Methods of Low-Level Counting: Proceedings of a Symposium Organized by the International Atomic Energy Agency in Co-Operation with the Joint Commission on Applied Radioactivity (ICSU) and Held in Monaco, March 2–10, 1967. Vienna: International Atomic Energy Agency.

Life. 1941. "Science and the Future: Search for Knowledge Enters an Era of Great Conclusions." *Life*, October 20.

Long, Robert. 1935. "Appendix III: Remarks on the Correlation Question." *Contributions to American Archaeology* 14:97–100.

Looper, Matthew. 2003. *Lightning Warrior: Maya Art and Kingship at Quirigua.* Austin: University of Texas Press.

Lopez Austin, Alfredo. 2015. *The Myth of Quetzalcoatl: Religion, Rulership, and History in the Nahua World.* Boulder: University Press of Colorado.

López Austin, Alfredo, and Leonardo López Luján. 2000. "The Myth and Reality of Zuyuá: The Feathered Serpent and Mesoamerican Transformations from the Classic to the Postclassic." In *Mesoamerica's Classic Heritage*, edited by D. Carrasco et al., 21–84. Boulder: University Press of Colorado.

Lopez, Felicia. 2017. "Case Study for the Development of a Visual Grammar: Mayahuel and Maguey as Teotl in the Directional Tree Pages of the Codex Borgia." *rEvista: A Multi-media, Multi-genre e-Journal for Social Justice* 5 (2): 41–69.

Lounsbury, Floyd. 1973. "On the Derivation and Reading of the 'Ben-Ich' Prefix." In *Mesoamerican Writing Systems*, edited by E. Benson, 99–143. Washington D.C.: Dumbarton Oaks Research Library and Collections.

Lounsbury, Floyd. 1978. "Maya Numeration, Computation, and Calendrical Astronomy." In *A Dictionary of Scientific Biography*, vol. 15, edited by C. C. Gillispie, 759–818. New York: Charles Scribner's Sons.

Lounsbury, Floyd. 1982. "Astronomical Knowledge and Its Uses at Bonampak, México." In *Archaeoastronomy in the New World*, edited by A. F. Aveni, 143–69. Cambridge: Cambridge University Press.

Lounsbury, Floyd. 1983. "The Base of the Venus Table of the Dresden Codex and Its Significance for the Calendar Correlation." In *Calendars in Mesoamerica and Peru: Native Computations of Time*, 1–26. Oxford: British Archaeological Reports International Series S174.

Lounsbury, Floyd. 1992a. "Derivation of the Mayan-to-Julian Calendar Correlation from the Dresden Codex Venus Chronology." In *The Sky in Mayan Literature*, edited by A. F. Aveni, 184–206. New York: Oxford University Press.

Lounsbury, Floyd. 1992b. "A Solution for the Number 1.5.5.0 of the Mayan Venus Table." In *The Sky in Mayan Literature*, edited by A. F. Aveni, 207–15. New York: Oxford University Press.

Love, Bruce. 1995. "A Dresden Codex Mars Table?" *Latin American Antiquity* 6, no. 4 (December): 350–61.

MacKenzie, Donald. 2008. *An Engine, Not a Camera: How Financial Models Shape Markets.* Cambridge: MIT Press.

Martin, Simon. 1996. "Tikal's 'Star War' against Naranjo." In *Eighth Palenque Round Table*, edited by M. J. Macri and J. McHargue, 223–36. San Francisco: Pre-Columbian Art Research Institute.

Martin, Simon, and Nikolai Grube. 2000. *Chronicle of Maya Kings and Queens: Deciphering the Dynasties of the Ancient Maya.* New York: Thames and Hudson.

Martin, Simon, and Nikolai Grube. 2008. *Chronicle of Maya Kings and Queens: Deciphering the Dynasties of the Ancient Maya.* New York: Thames and Hudson.

Masson, Marilyn A., and Carlos Peraza Lope. 2007. "Kukulkan/Quetzalcoatl, Death God, and Creation Mythology of Burial Shaft Temples at Mayapán." *Mexicon* 29, no. 3 (June): 77–85.

Masson, Marilyn, and Carlos Peraza Lope. 2014. *Kukulcan's Realm: Urban Life at Ancient Mayapán*. Boulder: University Press of Colorado.

Mathews, Peter, and Peter Bíro. 2008. *Maya Hieroglyph Dictionary*. Retreived from http://research.famsi.org/mdp/mdp_index.php.

Maudslay, Alfred P. 1974. *Archæology*. Facsimile edition with introduction by Dr. Francis Robicsek. New York: Milpatron Publishing Corp.

McAnany, Patricia. 1995. *Living with the Ancestors: Kinship and Kingship in Ancient Maya Society*. Austin: University of Texas Press.

McNamee, Mike. 2004. "Lewis S. Ranieri: Your Mortgage Was His Bond." *Businessweek*, November 29, 2004.

McVicker, Donald. 1989. "Prejudice and Context: The Anthropological Archaeologist as Historian." In *Tracing Archaeology's Past: The Historiography of Archaeology*, edited by Andrew Christenson, 113–26. Carbondale: Southern Illinois University.

Means, Philip. 1917. "History of the Spanish Conquest of Yucatan and of the Itzas." In *Papers of the Peabody Museum of American Archaeology and Ethnology* 7. Cambridge, Mass.: Harvard University.

Mendelson, E. Michael. 1958. "A Guatemalan Sacred Bundle." *Man* 58:121–26.

Milbrath, Susan. 1995. "Eclipse Imagery in Mexican Sculptures of Central Mexico." *Vistas in Astronomy* 39:479–502.

Miller, Arthur G. 1982. *On the Edge of the Sea: Mural Painting at Tancah-Tulum, Quintana Roo, Mexico*. Washington, D.C.: Dumbarton Oaks.

Miller, Mary, and Simon Martin. 2004. *Courtly Art of the Ancient Maya*. New York: Thames and Hudson.

Morley, Sylvanus. 1910. "The Correlation of Maya and Christian Chronology." *American Journal of Archaeology* 14, no. 2 (April–June): 193–204.

Morley, Sylvanus Griswold. 1919. "Joseph Thompson Goodman." *American Anthropologist* New Series 21, no. 4 (October–December): 441–45.

Morley, Sylvanus. 1920. *The Inscriptions of Copan*. Publication 219. Washington, D.C.: Carnegie Institution of Washington.

Morley, Sylvanus. 1975. *An Introduction to the Study of the Maya Hieroglyphs*, with introduction by J. Eric S. Thompson. New York: Dover Publications.

Moyes, Holley. 2012. "Introduction." In *Sacred Darkness: A Global Perspective on the Ritual Use of Caves*, edited by H. Moyes, 1–11. Boulder: University Press of Colorado.

Moyes, Holley, and James E. Brady. 2012. "Caves as Sacred Space in Mesoamerica." In *Sacred Darkness: A Global Perspective on the Ritual Use of Caves*, edited by H. Moyes, 151–70. Boulder: University Press of Colorado.

Nahm, Werner. 1994. "Maya Warfare and the Venus Year." *Mexicon* 16:6–10.

Newsome, Elizabeth. 2001. *Trees of Paradise and Pillars of the World: The Serial Stela Cycle of "18-Rabbit-God K," King of Copan*. Austin: University of Texas Press.

Nicholson, H. B. 2001. *Topiltzin Quetzalcoatl: The Once and Future Lord of the Toltecs*. Boulder: University Press of Colorado.

Nobel Media. 2014. "Ernest Lawrence—Biographical." Nobelprize.org, accessed July 7, 2017. http://www.nobelprize.org/nobel_prizes/physics/laureates/1939/lawrence-bio.html.

Bruhns, Karen Olsen, and Karen E. Stothert. 1999. *Women in Ancient America*. Norman: University of Oklahoma Press.

Paxton, Meredith. 2001. *The Cosmos of the Yucatec Maya: Cycles and Steps from the Madrid Codex*. Albuquerque: University of New Mexico Press.

Pingree, David. 1974. "Mayan Astronomy: A Commentary on the Dresden Codex: A Maya Hieroglyphic Book." *Journal for the History of Astronomy* 5:137–38.

Pohl, John M. D. 1994. *The Politics of Symbolism in the Mixtec Codices*. Nashville: Vanderbilt University Publications in Anthropology.

Pohl, John M. D., and Bruce E. Byland. 1990. "Mixtec Landscape Perception and Archaeological Settlement Patterns." *Ancient Mesoamerica* 1, no. 1 (Spring 1990): 113–31.

Pohl, John, Virginia Fields, and Victoria Lyall. 2012. "Children of the Plumed Serpent: The legacy of Quetzalcoatl in Ancient Mexico." In *Children of the Plumed Serpent: The Legacy of Quetzalcoatl in Ancient Mexico*, edited by V. Fields, J. Pohl, and V. Lyall, 15–48. London: Scala.

Powell, Christopher. 1997. *A New View on Maya Astronomy*. Master's thesis, University of Texas, Austin.

Proskouriakoff, Tatiana. 1960. "Historical Implications of a Pattern of Dates at Piedras Negras, Guatemala." *American Antiquity* 25, no. 4 (April): 454–75.

Proskouriakoff, Tatiana. 1993. *Maya History*. Austin: University of Texas Press.

Prufer, Keith, Amy E. Thompson, Clayton R. Meredith, Brendan J. Culleton, Jillian M. Jordan, Claire E. Ebert, Bruce Winterhalder and Douglas J. Kennett. 2017. "The Classic Period Maya Transition from an Ideal Free to an Ideal Despotic Settlement System at the Polity of Uxbenká." *Journal of Anthropological Archaeology* 45:53–68.

Ralph, Elizabeth K. 1965. "Review of Radiocarbon Dates from Tikal and the Maya Calendar Correlation Problem." *American Antiquity* 30 (4): 421–27.

Redfield, Robert, and Alfonso Villa Rojas. 1962. *Chan Kom: A Maya Village*. Chicago: University of Chicago Press.

Reed, Nelson. 2001. *The Caste War of Yucata'n*. Stanford, Calif.: Stanford University Press.

Reents-Budet, Dorie. 1998. "Elite Maya Pottery and Artisans as Social Indicators." *Archeological Papers of the American Anthropological Association* 8 (1): 71–89.

Restall, Matthew. 1999. *Maya Conquistador*. Boston: Beacon Press.

Restall, Mathew, and John Chuchiak. 2002. "Special Commentary: A Reevaluation of the Authenticity of Fray Diego de Landa's *Relación de las cosas de Yucatán*." *Ethnohistory* 49 (3): 652–69.

Reyman, Jonathan. 1973. Comment on "Archaeoastronomy and Ethnoastronomy So Far" by Elizabeth Chelsey Baity. *Current Anthropology* 14, no. 4 (October): 435–36.

Reyman, Jonathan. 1975. "The Nature and Nurture of Archaeoastronomical Studies." In *Archaeoastronomy in Pre-Columbian America*, edited by A. Aveni, 205–15. Austin: University of Texas Press.

Rice, Prudence. 2007. *Maya Calendar Origins: Monuments, Mythistory, and the Materialization of Time*. Austin: University of Texas Press.

Ricketson, Oliver. 1928a. "Notes on Two Maya Astronomic Observatories." *American Anthropologist* New Series 30 (3): 434–44.

Ricketson, Oliver. 1928b. "Astronomical Observatories in the Maya Area." *Geographical Review* 18 (2): 215–25.

Ringle, William M. 1990. "Who Was Who in Ninth-Century Chichen Itza." *Ancient Mesoamerica* 1 (2): 233–43.

Ringle, William M. 2004. "On the Political Organization of Chichen Itza." *Ancient Mesoamerica* 15, no. 2 (Fall): 167–218.

Ringle, William. 2009. "The Art of War: Imagery of the Upper Temple of the Jaguars, Chichen Itza." *Ancient Mesoamerica* 20, no. 1 (Spring): 15–44.

Rivet, Paul. 1923. "Eduard Seler." *Journal de la Société des Américanistes* New Series 15:280–87.

Robertson, John, Stephen Houston, and David Stuart. 2004. "Tense and Aspect in Maya Hieroglyphic Script." In *Linguistics of Maya Writing*, edited by S. Wichmann, 259–89. Salt Lake City: University of Utah Press.

Robertson, Merle G. 1979. "An Iconographic Approach to the Identity of the Figures on the Piers of the Temple of Inscriptions, Palenque." In *Tercera Mesa Redonda de Palenque, Volume IV*, edited by M. G. Robertson and D. C. Jeffers, 129–38. Palenque: Pre-Columbian Art Research.

Robertson, Merle Greene. 1983. *The Sculpture of Palenque, Volume 1: The Temple of Inscriptions.* Princeton, N.J.: Princeton University Press.

Roys, Ralph. 1933. *The Book of Chilam Balam of Chumayel.* Washington, DC: Carnegie Institution.

Rugeley, Terry. 1995. "The Maya Elites of Nineteenth-Century Yucatán." *Ethnohistory* 42, no. 3 (Summer): 477–93.

Ruiz de Alarcón, Hernando. 1629. *Treatise on the Heathen Superstitions That Today Live among the Indians Native to This New Spain.*

Salmon, Felix. 2009. "Recipe for Disaster: 'The Formula That Killed Wall Street.'" *Wired*, February 23, 2009. https://www.wired.com/2009/02/wp-quant/.

Sandstrom, Alan. 1991. *Corn Is Our Blood: Culture and Ethnic Identity in a Contemporary Aztec Indian Village.* Norman: University of Oklahoma Press.

Satterthwaite, L., and Elizabeth Ralph. 1960. "New Radiocarbon Dates and the Maya Correlation Problem." *American Antiquity* 26:165–84.

Saturno, William A., David Stuart, Anthony F. Aveni, and Franco Rossi. 2012. "Ancient Maya Astronomical Tables from Xultun, Guatemala." *Science* 336, no. 6082 (May 11): 714–17.

Schele, Linda. 1999. *Notebook for the XXIIIrd Maya Hieroglyphic Forum at Texas.* Austin: University of Texas.

Schele, Linda, and David Freidel. 1990. *A Forest of Kings: The Untold Story of the Ancient Maya.* New York: William Morrow.

Schele, Linda, and Nikolai Grube. 1997. *Dresden Codex Workbook.*

Schele, Linda, Nikolai Grube, and Federico Fahsen. 1992. "The Lunar Series in Classic Maya Inscriptions: New Observations and Interpretations." *Texas Notes on Precolumbian Art, Writing, and Culture* 29.

Schele, Linda, and Peter Mathews. 1999. *The Code of Kings: The Language of Seven Sacred Maya Temples and Tombs.* New York: Touchstone.

Schele, Linda, and Jeffrey Miller. 1983. "The Mirror, the Rabbit, and the Bundle: 'Accession' Expressions from the Classic Maya Inscriptions." *Studies in Pre-Columbian Art and Archaeology* 25:1–99.

Schele, Linda, and Mary Miller. 1986. *Blood of Kings: Dynasty and Ritual in Maya Art.* New York: George Braziller in association with the Kimball Art Museum.

Schmidt, Michael. 2020. "CMO vs. CDO: Same Outside, Different Inside." *Investopedia.com*, April 30, 2020. https://www.investopedia.com/articles/investing/111213/cmo-vs-cdo-same -outside-different-inside.asp.

Scientific American. 1917. "Potash—A Recapitulation." *Scientific American* 117, no. 16 (October 20): 282.

Segal, Julie. 2017. "War Stories over Board Games: The Man Who Invented Modern Finance." Interview by Institutional Investor Films. https://www.institutionalinvestor.com/article/b15 xp4kclymdnh/the-inventor-of-the-mortgage-backed-security.

Seler, Eduard. 1904. "Venus Period in the Picture Writings of the Borgian Codex Group." In *Mexican and Central American Antiquities, Calendar Systems, and History*, edited by C. Bowditch, 355–91. Bulletin 28. Washington, D.C.: Smithsonian Institution.

Sharer, Robert. 2006. *The Ancient Maya.* Stanford, Calif.: Stanford University Press.

Shen, Lucinda. 2016. "Warren Buffett Just Unloaded $195 Million Worth of These 'Weapons of Mass Destruction.'" *Fortune*, August 8, 2016. https://fortune.com/2016/08/08/mass-destruction -buffett-derivatives/.

Siméon, Rémi. 1977. *Diccionario de la lengua nahuatl o mexicana.* Mexico City: Siglo XXI.

Smith, Adam. 2000. *The Wealth of Nations.* New York: Modern Library.

Smith, Linda Tuhiwai. 2002. *Decolonizing Methodologies: Research and Indigenous Peoples.* London: Zed Books.

Smith, Michael. 2004. *The Aztecs*. New York: Thames and Hudson.

Smith, Michael, and Frances Berdan. 2010. *The Postclassic Mesoamerican World*. Salt Lake City: University of Utah Press.

Smith, Robert. 1971. *The Pottery of Mayapan: Including Studies of Ceramic Material from Uxmal, Kabah, and Chichen Itza. Papers of the Peabody Museum of American Archaeology and Ethnology* 66. Cambridge, Mass.: Harvard University.

Solari, Amara L. 2007. *Maya Spatial Biographies in Communal Memory and Cosmic Time: The Franciscan Evangelical Campaign of Itzmal, Yukatan*. PhD thesis, University of California, Santa Barbara.

Sorensen, Lee, ed. 2018. S.v. "Spinden, Joe." *Dictionary of Art Historians*. Retrieved August 22. http://www.arthistorians.info/spindenh.

Spinden, Herbert. 1924. "The Reduction of Maya Dates." *Papers of the Peabody Museum of Archaeology and Ethnology* 6, 4.

Spivak, Gayatri. 1988. "Can the Subaltern Speak?" In *Marxism and the Interpretation of Culture*, edited by Cary Nelson and Lawrence Grossberg, 271–313. Basingstoke: Macmillan.

Stephens, John Lloyd, and Karl Ackerman. 1993. *Incidents of Travel in Central America, Chiapas, and Yucatan*. New Edition, edited by K. Ackerman. Washington, D.C.: Smithsonian Institution Press.

Stone, Andrea, and Marc Zender. 2011. *Reading Maya Art: A Hieroglyphic Guide to Ancient Maya Painting and Sculpture*. London: Thames and Hudson.

Stuart, David. 1995. *A Study of Maya Inscriptions*. PhD diss., Vanderbilt University.

Stuart, David. 1996. "Kings of Stone: A Consideration of Stelae in Ancient Maya Ritual and Representation." *RES* 29 (30): 148–71.

Stuart, David. 2000. "The Arrival of Strangers: Teotihuacan and Tollan in Classic Maya History." In *Mesoamerica's Classic Heritage: from Teotihuacan to the Aztecs*, edited by D. Carrasco, C. Jones, and S. Sessions, 465–513. Boulder: University Press of Colorado.

Stuart, David. 2004. "New Year Records in Classic Maya Inscriptions." *PARI Journal* 5, no. 2 (Fall): 1–6.

Stuart, David. 2005. *The Inscriptions of Temple XIX at Palenque*. San Francisco: Pre-Columbian Art Research Institute.

Stuart, David. 2006. *Sourcebook for the 30th Maya Meetings*. Austin: The Mesoamerica Center, Department of Art and Art History, University of Texas.

Stuart, David. 2007. *Sourcebook for the 31st Maya Meetings*. Austin: The Mesoamerica Center, Department of Art and Art History, University of Texas.

Stuart, David. 2011. *The Order of Days: Unlocking the Secrets of the Ancient Maya*. New York: Three Rivers Press.

Stuckenrath, R., and Elizabeth Ralph. 1965. "University of Pennsylvania Radiocarbon Dates VIII." *Radiocarbon* 7:187–99.

Tambiah, Stanley J. 1990. *Magic, Science and Religion and the Scope of Rationality*. Cambridge: Cambridge University Press.

Tardi, Carla. 2011. "Collateralized Debt Obligation (CDO)." https://www.investopedia.com/terms/c/cdo.asp.

Tate, Carolyn E. 1992. *Yaxchilan: The Design of a Maya Ceremonial City*. Austin: University of Texas Press.

Taussig, Michael. 1987. *Shamanism, Colonialism, and the Wild Man: A Study in Terror and Healing*. Chicago: University of Chicago Press.

Taylor, Verta. 1989. "Social Movement Continuity: The Women's Movement in Abeyance." *American Sociological Review* 54, no. 5 (October): 761–75.

Tedlock, Barbara. 1992. *Time and the Highland Maya*. Albuquerque: University of New Mexico Press.

Tedlock, Dennis, trans. 1985. *Popol Vuh: The Mayan Book of the Dawn of Life*. New York: Simon and Schuster.

Teeple, John. 1925a. "Maya Inscriptions: Further Notes on the Supplementary Series." *American Anthropologist* 27 (1): 544–49.

Teeple, John. 1925b. "Maya Inscriptions: Glyph C, D, and E of the Supplementary Series." *American Anthropologist* 27 (1): 108–15.

Teeple, John. 1926. "Maya Inscriptions: The Venus Calendar and Another Correlation." *American Anthropologist* 28 (2): 402–8.

Teeple, John. 1928. "Maya Inscriptions VI: The Lunar Calendar and Its Relation to Maya History." *American Anthropologist* New Series 30 (3): 391–407.

Teeple, John. 1931. "Maya Astronomy." In *Contributions to American Archaeology* 1 (2): 29–115. Publication 403. Washington, D.C.: Carnegie Institution.

Tett, Gillian. 2009. *Fool's Gold: The Inside Story of J.P. Morgan and How Wall St. Greed Corrupted Its Bold Dream and Created a Financial Catastrophe*. New York: Free Press.

Thomas, Cyrus. 1888. "Aids to the Study of the Maya Codices." In *Sixth Annual Report of the Bureau of Ethnology to the Secretary of the Smithsonian Institution 1884–85*, by J. W. Powell, 259–376. Washington, D.C.: Government Printing Office.

Thompson, J. Eric S. 1927. "A Correlation of the Mayan and European Calendars." *Field Museum of Natural History, Anthropological Series*, Publication 241, vol. 17, no. 1. Chicago: Field Museum of Natural History.

Thompson, J. Eric S. 1935. "Maya Chronology: The Correlation Question." *Contributions to American Archaeology* 14. Washington D.C.: Carnegie Institution.

Thompson, J. Eric S. 1963. *Maya Archaeologist*. Norman: University of Oklahoma Press.

Thompson, J. Eric S. 1967. "Ralph Loveland Roys, 1879–1965." *American Antiquity* 32, no. 1 (January): 95–99.

Thompson, J. Eric S. 1972. *A Commentary on the Dresden Codex*. Philadelphia: American Philosophical Society.

Thompson, J. Eric S. 1975. "Introduction to the Dover Edition." In *An Introduction to the Study of the Maya Hieroglyphs*, by S. Morley. New York: Dover Publications.

Thompson, J. Eric S. 1978. *Maya Hieroglyphic Writing: An Introduction*. Sixth printing. Norman: University of Oklahoma Press.

Tokovinine, Alexandre. 2013. *Place and Identity in Classic Maya Narratives*. Dumbarton Oaks Pre-Columbian Art and Archaeology Studies Series 37. Cambridge, Mass.: Harvard University Press.

Tozzer, Alfred. 1907. "Ernst Förstemann." *American Anthropologist* 9:153–59.

Tozzer, Alfred. 1941. *Landa's Relación de las Cosas de Yucatan: A Translation. Papers of the Peabody Museum of American Archaeology and Ethnology* 18. Cambridge, Mass.: Harvard University.

Trigger, Bruce. 1989. *A History of Archaeological Thought*. Cambridge: Cambridge University Press.

Troike, Nancy P. 1974. *The Codex Colombino-Becker*. PhD thesis, University of London.

Tsu, Wen Shion. 1934. "The Observations of Halley's Comet in Chinese History." *Popular Astronomy* 42:191–201.

Vail, Gabrielle, and Christine Hernandez, eds. 2010. *Astronomers, Scribes, and Priests: Intellectual Interchange between the Northern Maya Lowlands and Highland Mexico in the Late Postclassic Period*. Washington, D.C.: Dumbarton Oaks.

Van Stone, Mark. 2010. *2012: Science and Prophecy of the Ancient Maya*. San Diego: Tlacaelel Press.

Velasquez, Erik. 2006. "The Maya Flood Myth and the Decapitation of the Cosmic Caiman." *PARI Journal* 7 (1): 1–10.

Villacorta, Carlos A., and J. Antonio Villacorta C. (1930) 1990. *The Dresden Codex: Drawings of the Pages and Commentary in Spanish.* Laguna Hills, Calif.: Aegean Park Press.

Villa Rojas, Alfonso. 1945. *The Maya of East Central Quintana Roo.* Carnegie Institution of Washington Publication 559. Washington, D.C.: Carnegie Institution of Washington.

Vogt, Evon. 1964. "Ancient Maya and Contemporary Tzotzil Cosmology: A Comment on Some Methodological Problems." *American Antiquity* 30 (2): 192–95. doi:10.2307/278851.

Vogt, Evon. 1969. *Zinacantan: A Maya Community in the Highlands of Chiapas.* Cambridge, Mass.: Harvard University Press.

von Schwerin, Jennifer. 2011. "The Sacred Mountain in Social Context. Symbolism and History in Maya Architecture: Temple 22 at Copan, Honduras." *Ancient Mesoamerica* 22 (2): 271–300.

Voss, Alexander. 2001. "La Identidad de los Itzá de Chichén Itzá," 1–13. Merida: Facultad de Ciencias Antropológicas, UADY.

Wald, Robert F. 2004. "The Languages of the Dresden Codex: Legacy of the Classic Maya." In *Linguistics of Maya Writing,* edited by S. Wichmann, 27–60. Salt Lake City: University of Utah Press.

Westfall, Richard. 1989. *Essays on the Trial of Galileo.* Rome: Vatican Observatory.

Willey, Gordon R. 1988. "Obituary: Augustus Ledyard Smith 1901–1985." *American Antiquity* 53, no. 4 (October): 683–85.

Willson, Robert. 1924. "Astronomical Notes on the Maya Codices." In *Peabody Museum of American Archaeology and Ethnology Papers* 6, 3.

Wingo, Walter. 1963. "The Scramble into Space." In *Science News-Letter* 84, no. 22 (November): 341–43.

Wisdom, Charles. 1950. *Chorti Dictionary.* Transcribed and transliterated by Brian Stross. Austin: University of Texas.

Wiser, Callie, producer. 2012. "Terri Duhon." *The Financial Crisis: The Frontline Interviews.* https://www.pbs.org/wgbh/pages/frontline/oral-history/financial-crisis/terri-duhon/.

Zender, Marc. 2004. *A Study on Classic Maya Priesthood.* PhD diss., University of Calgary.

Zender, Marc. 2005. "The Raccoon Glyph in Classic Maya Writing." *PARI Journal* 5, no. 4 (Spring): 6–16.

Zender, Marc, and Joel Skidmore. 2012. "Unearthing the Heavens: Classic Maya Murals and Astronomical Tables at Xultun, Guatemala." *Mesoweb,* accessed July 19, 2021. http://www.mesoweb.com/reports/Xultun.pdf.

INDEX

ABOUT THE AUTHOR

Gerardo Aldana y Villalobos is the dean of the College of Creative Studies and a professor of archaeology and the history of science in the Department of Chicana/o Studies at University of California, Santa Barbara. He is the author of *The Apotheosis of Janaab Pakal: Science, History and Religion at Classic Maya Palenque*, and he specializes in Mayan hieroglyphic history, Mesoamerican art, indigeneity, experimental archaeology, and science and technology studies.

Zeitfracht Medien GmbH
Ferdinand-Jühlke-Straße 7
99095 Erfurt, Deutschland
produktsicherheit@kolibri360.de